THE STUDY OF HUMAN EVOLUTION

1 2 3 4 5 6 7 8 9 0   DODO   7 8 3 2 1 0 9

This book was set in Helvetica by Monotype Composition Company, Inc. The editors were Eric M. Munson and Susan Gamer; the designer was Albert M. Cetta; the production supervisor was Dennis J. Conroy. The drawings were done by J & R Services, Inc.
R. R. Donnelley & Sons Company was printer and binder.

Library of Congress Cataloging in Publication Data

Eckhardt, Robert B
    The study of human evolution.

    Includes bibliographies and index.
    1. Somatology. 2. Human evolution. I. Title.
GN60.E27   573   78-10748
ISBN 0-07-018902-1

THE
STUDY
OF HUMAN
EVOLUTION

Robert B. Eckhardt

Department of Anthropology
Pennsylvania State University

McGraw-Hill Book Company

New York   St. Louis   San Francisco   Auckland   Bogotá   Düsseldorf
Johannesburg   London   Madrid   Mexico   Montreal   New Delhi   Panama
Paris   São Paulo   Singapore   Sydney   Tokyo   Toronto

Contents

List of Supplementary Sections

ix

Preface

Writing a preface to a book is rather like introducing a speaker who is a close friend to a large and somewhat unfamiliar audience. The task must be done briefly, so that all involved can get on to the main event, yet a few extra words of background can often smooth the way for what is to follow. The structure and objectives of this book are presented more fully in Chapter 1 but several comments are in order here.

This book is my attempt to set forth one strategy for studying human evolution. A key feature in this approach is to have the reader participate in the study actively, not passively. But how? On first encounter any field of knowledge seems to consist of a bewildering thicket of facts and theories through which it is necessary to progress in order to understand the subject. Under such circumstances an author can take readers figuratively by the hand and lead them along a pathway through the maze—the path most familiar to the author. This is an approach commonly taken in textbooks, and it has its merits; when done well it can give readers an organized overview of the predominant viewpoint current in the field. There is another approach, however—in which the author serves not as a guide but as a counselor who helps readers develop a set of skills that will equip them to explore a field with a greater degree of independence. I have taken the latter approach here in hopes of pointing readers toward a stimulating intellectual adventure.

This is not to say that I have tried to make the study of human evolution more complex than it must be. In fact, if any objection is raised to the course that I've followed here, it might be that I have occasionally suggested relatively simple explanations where many authorities feel that only complex theories will suffice. However, it is my experience that students should begin with a broad, simple overview—which can grow in complexity as new knowledge is acquired—rather than confronting a situation that is presented as too complex to be understood and must be taken on faith alone. My approach extends to students no greater luxury than that to which professional scientists are accustomed: after all, scientists also construct simplified models to help themselves understand the real world.

A second key feature of this book is that the order in which topics are covered reflects my conviction that our direct and detailed knowledge of the present holds the best key to reconstructing the past. Even though new discoveries are taking place on all fronts, some of our conclusions about human evolution seem more secure than others. For example, the mechanism of inheritance was discovered over a century ago and is now known to be common to all higher organisms, not unique to humans; the forces of evolution that bring about changes in populations of living humans and other creatures have been studied intensively for nearly half a century. In contrast, the fossil evidence for human evolution has been augmented by several revolutionary discoveries in just the last two decades, and the pace at which new human fossils are brought to light is so rapid that it is difficult for even professionals to keep informed of all of the important finds and the changes in interpretation that they sometimes require.

One example of the wealth of new evidence to be dealt with in the study of human evolution is given in the announcement (*New York Times,* January 19, 1979) as this book goes to press of the reconstruction of a primitive human that lived in Africa 3 to 4 million years ago (see also pages 476–479 of this text). Its apelike skull, with small brain and protruding face, contrasts with its ability to stand and walk erect. This find provides us with an excellent illustration of mosaic evolution (see page 318), the idea that different parts of the body can change at different rates. Evolutionary theory can help us to understand these and other changes that are documented in the fossil record.

It should be pointed out that in such a rapidly developing field, any attempt at synthesis entails some risk of error. Thus inferences made here should be taken not as truths to be accepted uncritically, but as working hypotheses that must be tested against the new information that is being accumulated.

ACKNOWLEDGMENTS

While I accept responsibility for any shortcomings of this book, credit for some of its strengths must be shared. I owe a great deal to my former teachers and colleagues at the University of Michigan, particularly

Frank B. Livingstone, who provided me with some of the central components of my scientific training. Many of the ideas and approaches used in this book originated in Ernst Goldschmidt's rigorous and deservedly popular course in introductory physical anthropology at Michigan. I thank him for letting me develop them further in this book. Present colleagues at Penn State have also contributed to the shaping of my perspectives on human evolution. To Paul Baker, for example, I owe my deepening realization of the complexity of human adaptability and the use of experimental approaches to test hypotheses; and my associate Bennett Dyke helped to lend greater rigor to several of the chapters on genetics.

Those who commented on the manuscript make my circle of debt considerably wider. I gratefully acknowledge the careful reading and criticism of Howard S. Barden, Alice M. Brues, Karen F. Davis, Alan G. Fix, Janet O. Frost, Francis E. Johnston, Frank B. Livingstone, David G. Smith, Daris Swindler, and Kenneth M. Weiss.

Although I have tried to use as much original material as possible, no book of this sort could possibly be produced without the help of many others who worked before; I have tried to acknowledge as fully as possible in the text those authors and publishers who gave their permission to use tables, illustrations, and quotations from previous works and continuing research. Ann Hildebrand handled the very substantial task of securing these permissions, and also executed most of the original photographs used here, William Murphy prepared the remainder, and answered queries during the phase of production that took place during my recent period of field research in Peru. Andrew Kramer and Marguerits Sullivan verified and organized the references cited here.

For their guidance in developing the text itself, I thank Jeannine Ciliotta and my editors Eric Munson and Susan Gamer of McGraw-Hill. All have made substantial efforts to ensure that this book will be useful to its intended audience. I enjoyed throughout the advice and good judgment of my wife Carey, a welcome companion during so many quiet hours of research and writing.

Robert B. Eckhardt

1

Introduction

Chapter 1

Human Evolution: The Origin of Our Unity and Diversity

This book is about human evolution, which is a brief phase of a vast, unfolding sequence of events that began long before humans like us ever evolved and which seems likely to continue long after we are gone. *Evolution,* in the most general sense, is the process of long-term directional change that goes on in the natural world. Evolutionary changes operate at several different levels. *Cosmic evolution* is the term given to the awesome changes that shape the universe. Our knowledge of the course of cosmic evolution goes back to about 15 billion years ago, the approximate date of what physicists call the *big bang*—an explosion that involved all the matter and energy in our present universe (Sagan, 1977). The big bang was either a cataclysm of such size that it wiped out all traces of any earlier cosmic events or the unexplainable start of everything out of an unknown void. In either case this event marks the beginning of the universe as we know it.

An early product of cosmic evolution was our Milky Way. This is a galaxy or swarm of over 100 billion stars, many of them like the one that the earth orbits about. Within our galaxy, as elsewhere, cosmic evolution still goes on, giving birth to new stars while old ones continue to change. Our sun will shine at its present level for another 8 billion years before it flares, burns the earth to a cinder, and then contracts to a cooler white dwarf star (Schklovskii and Sagan, 1966). Hydrogen, the simplest of all elements, is the major component of stars. Within each stellar core, where temperatures reach 10 million °C (about 18 million °F) or more, thermonuclear reactions fuse together hydrogen nuclei to produce atoms of helium, carbon, and oxygen and other larger atoms. In the process of cosmic evolution even the elemental atomic building blocks have progressed in a stepwise manner from simple to more complex.

Three to six billion years ago the earth condensed from a vast diffuse cloud of matter. This planet was lifeless at first, but in some settings (perhaps the oceans, though no one knows for sure) *chemical evolution* combined simple substances including water, ammonia, and methane into more complex combinations. These larger molecules include the organic compounds, which are composed of carbon, hydrogen, and oxygen, plus lesser amounts of nitrogen and other elements. All living plants and animals that we know of, however simple or complex, are made up of these organic building blocks, so that any change from one form of life to another is an example of *organic evolution.* By convention a small change that occurs over the course of a few generations (such as the increased resistance of houseflies to DDT) is classed as *microevolution.* The term *macroevolution* refers to larger-scale transformations that take place over thousands or millions of years. There are many well-known cases of macroevolutionary change. Dogs evolved from wolves, and wolves came in turn from animals that bore some resemblance to small flesh eaters like modern weasels and minks. As is true for all other animals, our line of descent can be traced back through hundreds of thousands of generations, though, unlike dogs, we are the descendants of ancestors who were different from any living animals now known to science. Like all these other humans, past and present, you are a product of evolution. This book is about your ancestors and about you.

You and I and all other living people are truly exceptional products of evolution. By this I do not mean merely that we belong to a species (*Homo sapiens*) that is like no other; *all* species are different to some degree. In fact, a *species* is defined as a group of plants or animals characterized by shared biological distinctions and maintained by reproductive isolation. Our earth harbors about one-third to one-half million species of plants as well as nearly a million species of animals. Each one of these diverges from all the others in one or more significant *specializations* or features that have evolved to suit a particular way of life. The human foot, with its straight, nondiverging big toe used in our characteristic striding gait, is one such unique feature. Other specializations can be seen within many groups. Take, for instance, the rodents (rats, mice, and their relatives). Among these, the muskrat, widely distributed over North America, is an excellent swimmer. This scaly-tailed vegetarian is capable of paddling underwater for as much as 18 m (20 yd), as well as cruising along on the surface for longer distances, though on land its webbed feet make it a clumsy walker. The agouti, a South American rodent, has long hind legs on which it can sprint swiftly over short distances like a rabbit. However, it cannot keep up the rapid pace for very long, nor can it climb trees. In contrast, the gray squirrel, distributed over the eastern half of the United States, is an extremely agile jumper and climber. High among the branches of trees it can outmaneuver most other animals easily, even if on the ground it can be run down by many predators.

WHAT IT MEANS TO BE HUMAN

Compared with such specialists as these, humans appear to be ungainly animals at best. The reason for our disadvantage can be explained in structural terms. Most other animals prop up their bodies on four limbs, with one positioned at each corner like the legs of a table. This arrangement gives four points of support and four sources of power for propulsion. We humans, however, support trunk and head above the two hind legs alone, which rise like parallel columns from our platform-like feet. If you think the balancing act involved in standing or walking is easy, remember that a human infant has to spend several months mastering the job. In many of the separate Olympic events designed for humans, animals could beat our finest athletes (see Table 1-1). However, no other animal can match us in sheer versatility. Our edge here comes from the large, flexibly programable brain inside the skull poised atop our spinal column and from the hands that were freed for a greater range of movements once they were no longer needed for locomotion. The Olympic decathlon, which measures performance in 10 different track and field events, could be won only by a human. This victory might not be as certain if the contests that emphasize dexterity in object manipulation were eliminated, since our opposable thumb gives us a grip that is both stronger and more precise than that possible for most animals; still, few other species could match overall human performance even in the events (such as swimming, diving, and gymnastics) that don't require the use of special mechanical equipment. Many of us can swim several hundred yards, run for a few miles, and then climb up a rope or pole. Yet far more significant than our diverse physical abilities is another distinction: humans are the only species that can long retain any detailed knowledge of events like these. Through human speech and the written records that preserve information from centuries past, our species has built up a store of wisdom that cannot even be dreamed of by the other animals with which we share the earth.

You may already have arrived on your own at the conclusion that you are special, even if you are modest enough to keep this belief to yourself. In fact, however, you have in all likelihood underestimated the extent of your individuality. You are unique, different from every one of your

TABLE 1-1   A comparison of the athletic abilities of humans and some other animals

Olympic decathlon: events included*	World record performance in individual event	Year record was set	World record holders		Comparisons with potential animal competitors	
			Name	Country represented	Species	Performance
100-m (109.4-yd) dash	9.96 sec	1968	Jim Hines	U.S.	Cheetah	3.20 sec
Long jump	8.9 m (29 ft 2½ in)	1968	Bob Beamon	U.S.	Kangaroo	12.8 m (42 ft)
Shotput	22 m (72 ft 2¼ in)	1976	Alexander Baryshnikov	U.S.S.R.	Chimpanzee	Inaccurate short-distance throwing only
High jump	2.33 m (7 ft 7¾ in)	1977	Vladimir Yaschenko	U.S.S.R.	Kangaroo	32 m (10 ft 6 in)
400-m (437.45-yd) hurdles	43.89 sec	1968	Lee Evans	U.S.	Cheetah	12.78 sec
100-m (109.36-yd) hurdles	13.2 sec	1972	Rod Milburn	U.S.	None	—
Discus throw	70.86 m (232 ft 6 in)	1976	Mac Wilkins	U.S.	Chimpanzee	Inaccurate distance throwing only
Pole vault	5.69 m (18 ft 8¼ in)	1976	Dave Roberts	U.S.	None	—
Javelin throw	70.8600 m (310 ft 4 in)	1976	Miklas Nemeth	Hungary	Chimpanzee	Inaccurate short-distance throwing only
1500-m (1640.4-yd) run	3 min 32.2 sec	1974	Filbert Bayi	Tanzania	Cheetah	47.94 sec

*The decathlon is a combined track and field event that has been included in the Olympics since 1912. It is held on two successive days, with the first five events listed above on the first day and the last five on the second day. The animal capabilities listed on the right are of course not from specific Olympic contests, but have been calculated from available figures; e.g., cheetahs have been clocked at an average speed of 112.65 km per hour (70 mi per hour) over a distance of 109.36 m (100 yd).

Sources: (1) *World Almanac.* 1978. New York: Newspaper Enterprise Association. (2) Norris McWhirter and Ross McWhirter. 1975. *Guinness Book of World Records.* New York: Sterling.

contemporaries as well as from all your ancestors. Development of your physical appearance and typical patterns of behavior began before you were born and will continue until you die. Many external environmental factors and internal biological processes have acted together to shape these changing characteristics, some of which differ even between identical twins. Two examples, speech and diet, illustrate the pervasive nature of these dual influences.

Diversity of Human Language

Consider human speech. To a speaker of English, the word *papa* is a familiar term for *father.* However, Chileans use *papa* to mean *mother,* and to speakers of Quechua (a South American Indian language spoken by the Incas and their descendants), *papa* means *potato* (as it does in Spanish, which borrowed the word from the Quechua at the time of the European conquest of the New World). The same twice-repeated syllable *pa* stands for very different ideas in each of the three languages; clearly, there is no correspondence between sound and meaning here that is universal for all humans. To be understood in a given language, each person must learn the conventional meanings of the sounds that are used.

The particular language you happened to learn first was the one used by those around you. This was probably some dialect of American English. If you had been born and raised in some other country, you might have mastered Korean, Urdu, Swahili, or German with about as much ease as you acquired English. But all the thousand or so human languages share common features. They all use a small set of sounds, arranged in different ways, to make up the roughly 100,000 words in the vocabulary of an average adult. Moreover, all languages have rules (the grammar of the language) that govern how this large but limited set of words can be patterned into infinite varieties of complex thoughts.

For all humans, language learning begins in infancy. Cries of babies everywhere are similar enough that, for example, British parents can interpret the needs of Italian infants as well as those of their own (Aitchson, 1978). Babies of different lands also seem to babble out at first much the same set of sputters, gurgles, gulps, clicks, and grunts. Like *papa,* other nursery words such as *mama, dada, tata,* and even *baby* itself are worldwide elements of baby talk. Gradually, however, an infant's babbling comes to resemble the sounds it hears in the speech of adults around it. Infants may imitate the sounds they hear, but adults also definitely encourage babies to make the "right" sounds. A baby who repeats the syllable *pa* twice in succession will likely be rewarded by the appearance of a smiling face hovering over his or her crib. Studies of infant behavior have shown that this smile is one of the strongest incentives a human infant can be given. Under such influences, you produced your first words at about a year of age and progressed to a speaking vocabulary of about 1500 words (and an understanding of roughly two to three times as many) by about age 4. Along the way you also put into use nearly a thousand rules of grammar. Some of these rules are very simple. For example, words are divided into separate categories. At first the split is between basic words such as *mine* versus a second group including all less frequently used words. Later, the categories are multiplied, and articles such as *a* and *the* are grouped together and set apart from adjectives, pronouns, and so on. Languages as different as Spanish, Russian, and Chinese have such grammatical road maps (McNeill, 1966). The universality of language, as well as the astonishing rapidity with which both words and rules are learned, argues strongly that humans are biologically programed for language learning—but not for learning a particular language.

Diversity of Human Diet

The diets humans live on are about as diverse as the languages they speak. One conventional bit of German folk wisdom on the subject (*Mann ist was er esst*) translates as "You are what you eat." Few generalizations could be further from the truth. To many Americans, a good dinner might include roast beef, mashed potatoes with gravy, a salad, and perhaps ice cream for dessert. The Chinese counterpart of this meal would consist of a bowl of rice topped with a spoonful of bean curd and seasoned with soy sauce, perhaps with a side dish of cabbage stir-fried in vegetable oil (Chang, 1978). Among the Yanomamö Indians of Venezuela, a major meal served to guests often centers on a thick soup of boiled ripe plantain roots, followed by a basket of boiled green plantains and chunks of smoked armadillo or monkey meat (Chagnon, 1977). The foods differ, and so do the people who eat them. But although the Chinese differ from Americans in some details of physical appearance and although the Yanomamö differ from both, the contrasts in flesh and blood and in hair and bone are surely less than those of diet, and the differences that do exist are not all caused by different diets. Americans do not become bovine from eating beef, nor do Venezuelan Indians become simian as a result of eating monkey meat. In all these peoples, the same biological processes convert different raw materials into the chemical building blocks of very similar (though not identical) human bodies. You are not human because of the foods you eat, since people who eat very different foods all grow up to be human. You are human because inherited growth processes common to all members of our species determine how the raw materials are converted into parts of your body. But you *are* the unique human into which you have developed because of the interaction between heredity and environment. Had you grown up on the diet of the Chinese, you would probably not be as tall or as heavy as you are now. Nevertheless, if you are of American or European ancestry, you would not have developed dark, straight hair or Asian facial features as a result of having eaten more rice or less meat than you did.

Anthropologists are interested in studying the causes of the human diversity that results from the interplay of heredity and environment, as well as in studying the evolution of those human characteristics common to all people.

ANTHROPOLOGY: THE SYNTHETIC SCIENCE

The anthropologist's concern with the determinants of biological and behavioral characteristics is shared by scientists in many other fields. Some of the natural sciences, such as biology or its more specialized branches—genetics, anatomy, and physiology—supply much important information about the characteristics humans share with other animals. Other fields of study, including the social sciences of psychology, sociology, political science, and economics, also inform us about areas of existence that are particularly human. However, most social sciences deal chiefly with just one segment of the world's peoples: the members of societies that are urban, industrial, and literate. Among the disciplines that concern themselves with human characteristics and affairs, anthropology stands alone in two major ways. First, no other science studies such a diversity of human groups or as varied types of information about these groups. Second, anthropology has a coherent viewpoint, a set of perspectives that it uses to organize the great variety of data that anthropologists collect.

The Anthropological Perspective

Anthropology is the study of the biology, society, and culture of the diverse peoples now living all over the world and of the biology, society, and culture of all their human ancestors. *Biology* is the study of all life, from people to the simplest viruses and bacteria. All living things are highly organized systems that exchange matter and energy with the environment and reproduce themselves under the guidance of a hereditary system that maintains stability from one generation to the next while permitting some evolutionary change through time. For anthropologists, a *society* is a group of interbreeding organisms that can last longer than the life span of any of its individual members, whose behaviors are structured by the different roles they fill. Other definitions are possible, but central to all of them is the idea that members of any society share a set of behavioral traits that make up an important part of their *culture*. What is culture? There seem to be almost as many definitions of this term as there are anthropologists. Whole books (Kroeber and Kluckhohn, 1952) have been devoted to a comparison of hundreds of different versions of this concept.

The central idea of culture was formulated scientifically over a century ago by the British anthropologist E. B. Tylor (1871). Tylor characterized culture as "that complex whole which includes knowledge, belief, arts, morals, law, custom, and any other capabilities and habits acquired by man as a member of society." Anthropologists, then, use the term *culture* in a special sense. To many people in our society, a cultured person is one who is interested in the fine arts, such as music, painting, sculpture, theater, and ballet. Someone who does not appreciate these things would, by contrast, be labeled *uncultured.* From an anthropological point of view, however, to be human is, by definition, to be cultured. Humans can develop only within a culture; they cannot survive without the system of beliefs and inventions that can be learned only from other humans. The dependence of all humans on an elaborate culture is a distinguishing feature of our species and is the unifying concept of the anthropological perspective.

The evolution of a culture-bearing species such as our own is the outcome of a stepwise progression from biology to society and culture. Social organization is possible only if certain biological characteristics such as the capacity for behavior are present. Plants, which lack both a nervous system to communicate information rapidly and muscles for rapid responses, do not have any behavior to speak of. Similarly, although many animals learn aspects of their behavior from other group members, no other kind of animal transmits the extent or variety of learned behavior patterns that humans do.

The biological and social prerequisites for culture exist in all peoples of the world. Because of this, the capacity to maintain, modify, and transmit culture is a universal attribute of all human groups. Just as no individual human can grow up outside a culture, no human group exists without culture, and most cultures can be traced back over hundreds or thousands of years. The frequently demonstrated ability of groups and individuals to shift from one cultural tradition to another further underscores the point that what is significantly human is the ability to learn how to participate in a culture. The Vietnamese are only the latest foreign group to enter actively into the complexities of life in the United States; they are part of a process which began several hundred years ago with colonists from Europe and which has been continued by immigrants from all over the world ever since.

Each culture supplies its participants with an interlocking set of values, beliefs, language, traditions, and technologies. The elements in this package enable group members to deal with the challenges of life in the region where they live. A small change in one component of the system can lead over the short run to compensatory changes in others.

If a large-scale change takes place in one realm, however, this can lead to major consequences throughout the system. For example, in 1908 Henry Ford mass-produced the first low-priced automobiles. As the automobile became common, the transportation system changed, as might have been anticipated. However, few of Ford's contemporaries foresaw the changes in population distribution, mating patterns, food preferences, and the age distribution of deaths due to accidents, all of which can be linked to the increasing influence of cars on the American life-style.

Culture changes also take place over longer periods of time. From our present perspective, anthropologists can look back over millions of years and see some of these changes, such as the great elaboration of material culture from crudely chipped stone tools to modern machines. The development of more and more complex technological devices has been paralleled by substantial changes in the biological characteristics of members of the human species. For example, over approximately the last 3 million years, the human brain has about tripled in size. Although the exact causes of this increase in brain size are unknown, it seems that the solution to this problem, as well as the solutions to other problems in human evolution, is more likely to be found through the joint study of biology and culture than through concentration on either area alone.

Subfields of Anthropology

Anthropology is subdivided into four subfields at the present time in the United States: cultural anthropology, linguistics, archeology, and physical anthropology.

Cultural Anthropology.   Cultural anthropology is the study of the culture and way of life of living human groups. Traditionally, cultural anthropologists have concentrated on the so-called "primitive peoples," those who were non-Western and nonliterate and who, as a general rule, lived in small groups with little political or economic power. The description of each group's customs, beliefs, and behaviors constituted an *ethnography,* which was really a systematic catalog of the culture. The comparison of these ethnographies serves as the basis for a body of scientific generalizations that is called *ethnology.*

Cultural anthropology contributes importantly to the study of human evolution by making us aware of the ways of life of peoples who depend on an economic basis very different from the one familiar to us. Industrialized societies have existed for only the last few hundred years. Urban life, and the agricultural base of cultivated plants and domesticated animals on which it rests, goes back no more than about 10,000 years. These developments have taken place against a background of human evolution that spans at least several million years. In marginal areas of the world today, there still exist small groups of *hunter-gatherers,* peoples such as the Australian Aborigines, South African Bushmen, and North American Eskimos. These groups, which lack agriculture, have always been dependent on gathering uncultivated plants and hunting wild animals for their food. By studying them, anthropologists can learn about the conditions faced by our species for almost all its evolutionary history.

Linguistics.   The detailed study of just one aspect of human behavior—language—is called *linguistics.* Most anthropologists have some training in linguistics so that they will be equipped to do fieldwork among peoples whose language is not well enough known to be learned from textbooks. Linguistics has also become a theoretical science in its own right, with a body of generalizations that have come from the comparative study of many languages. For example, to the surprise of some Europeans and Americans who are *ethnocentric* (believing in the intrinsic superiority of their own culture), primitive peoples do not have "simple" languages. All known languages have vocabularies of roughly comparable size, though one may be more elaborated in a given area than another. Eskimos have traditionally had more different words for types of snow than speakers of English have, and they have fewer terms for metal tools. However, with the increased popularity of skiing in the United States, the vocabulary for snow conditions in our language has grown, as is no doubt also true of terms for tools among Eskimos employed in building the Alaskan pipeline. This open-ended aspect of all human languages is one of the important differences between language and the natural communication systems of all other animals. And the ability to produce an unlimited number of thoughts that can be understood by other speakers of the same language is one of the important bases for human culture.

Archeology.   *Archeology* is the study of the material culture (tools, weapons, ornaments, shelters, pottery, food refuse, burial places) of the past. Anthropologists study these physical objects for the clues they provide in reconstructing the behavior of earlier humans. The size of a dwelling, for example, can give some indication of the number of people in an average household; the number of dwellings clustered together can suggest the size of the social group. Food remains can be analyzed in an effort to reconstruct the type of diet on which the members of a group subsisted and the type of food-gathering techniques they used. Stone tools are of crucial importance in mapping the location of earlier populations, since they are more resistant to decay than human bones and are also more abundant; each early human had only one skeleton but may have produced many times his or her weight in chipped stone. Because cultural evidence can tell us so much about the distribution of humans in space and time, the discipline of *prehistoric archeology,* which is focused on human life that existed long before there were written historical records, is difficult to separate from *paleoanthropology,* which is the study of the earliest origins of humans and is usually considered part of physical anthropology.

Physical Anthropology.   The study of the biological characteristics of past and present humans, and of their nearest relatives among the nonhuman primates (monkeys, apes, and a variety of more primitive-looking prosimians), is known as *physical anthropology.* This biologically oriented subdivision of anthropology can be described in terms of how physical anthropologists spend their time and in terms of certain central organizing principles.

The first approach would reveal that some physical anthropologists measure and describe the remains of humans, either the skulls and bones or the mineralized fragments of these known as *fossils.* Other physical anthropologists analyze blood chemicals or peer into cells with powerful microscopes to learn the functions of tiny living particles. Still others monitor the growth of children or the aging process. Among the ranks of this discipline are also those who measure living humans at rest and at work to design clothing and work areas that will provide the greatest comfort and safety. A few of the more adventuresome spend long periods in extreme environments, breathing the thin air at the tops of mountains, baking in deserts, or freezing at the poles while they attempt to discover how people native to these areas survive and reproduce under conditions that document the extreme versatility of our species.

Physical anthropologists who study the ways in which people survive in extreme environments are interested in the interplay of human biology and culture in making possible adaptation, a series of adjustments that permit survival and reproduction. Since humans have adapted to a great variety of environments through time and have become modified in the process, a focus on adaptation could provide a comprehensive approach to understanding the biological and cultural characteristics of our species. The study of evolution provides another comprehensive avenue to the understanding of human unity and diversity. The idea of evolution has been described as one of the most significant intellectual discoveries ever made. It is also one of the most useful ways to link our own species with all living things to show that we are a part of, and not apart from, the world around us.

WHY STUDY HUMAN EVOLUTION?

It is perhaps easiest to appreciate the impact that the idea of evolution has had on contemporary thought by contrasting it with some features that characterized the western world view only 150 or 200 years ago. Then the world was thought to be relatively young, though of paramount importance, having been created only some 4000 years before to occupy the center of the universe. In this world, natural laws were part-time regulations that could be suspended periodically. At any moment, all life could be driven from the face of the earth by some supernaturally caused catastrophe. The very soil bore silent witness to this possibility, since it contained the bones of animals that had perished in the biblical Flood. These bones were sometimes accompanied by curiously shaped stones made of flint, which people believed to be the remains of weapons used in the battle that Lucifer waged against the heavens. Other people, who perhaps considered themselves more enlightened, held that these stones were "petrefactions" formed by thunderbolts passing through the air. The biological realm was a world of fixed types, races, and species. Each of these unique forms had been created separately as part of a conscious overall design. The implication was that since these separations had been intended by a supreme being, mere humans should not attempt to change them. Above all, the world of the past was one in which ultimate control over (and hence responsibility for) events of any magnitude—birth, illness, death, victory, defeat, pestilence, deliverance, success, failure—was believed to lie beyond human power.

In contrast, current scientific opinion holds that the world in which we find ourselves is one of great geologic age (somewhere between 3.2 and 5.7 billion years) and probably of little consequence to any but those of us who dwell on it. Our earth is not the center of the universe; indeed, it is not even the center of our solar system. For that matter, the solar system is not even located centrally in our own galaxy. It has been estimated that there are 10^{20} (a hundred thousand million billion) stars, 20 percent of which are essentially identical to our sun in size, luminosity, and composition and are available for the maintenance of life on whatever planets may surround them. The natural laws that operate in this universe are imperfectly and incompletely known, and some may remain so. Yet these principles are believed to be knowable from the observation of objects and events and to be constant in their action. Our growing knowledge of these laws has placed us in a position that is not (except by delusion) one of complete control, but rather one of responsibility. In accepting the idea of evolution, we have entered a world characterized by orderly change. We should see ourselves both as the products of this change and as the possessors of an awesome capacity to further influence its course. And our influence will be most productive if our knowledge is based on facts and sound theories rather than on superstitions or beliefs.

Facts and Inferences

Scientific issues are decided not on the basis of opinion polls but by examination of the evidence that bears on a given issue. In the case of evolution, knowing the precise meaning of the term is essential.

Evolution refers to changes that occur generation by generation in the genetically controlled characteristics of organisms. A *fact* is a proposition about the world, often a statement that something has definitely occurred, known through observation. As we will see later in this book (Chapter 6), humans have observed inherited changes occurring in some animal populations. In such cases, evolution is a fact; it is as much a matter of observable reality as an object falling toward the earth when released.

Human evolution is less an observable fact than a secure inference. This is true because special problems exist when humans are objects of genetic study. These problems (as well as several unique advantages not discussed here) were set forth in detail by Neel and Schull (1954). The ethics of scientists make it unthinkable for them to arrange human matings just to see how biological characteristics are inherited. The same considerations make it impossible to change experimentally or to standardize the environment simply to see how this would affect the expression of traits. And, quite obviously, any scientist who studies the genetic characteristics of humans can examine over a lifetime of work only a few generations, instead of the thousands that can be studied in some shorter-lived animals or plants. But although it is difficult to observe the evolutionary process in our own species directly, there is little doubt that it does occur. This inference is based in part on the knowledge that humans have the same basic mechanism of heredity as other organisms. It is reasonable, therefore, to assume that evolutionary change could take place in humans in much the same way that it does in other animals.

Another important body of evidence for human evolution is provided by an extensive array of fossil remains. From these, scientists can trace a plausible sequence of gradual changes through time. In geological deposits from the more recent past, within the last few tens of thousands of years, human skeletons are quite abundant and are very similar to our own. Further back in time fossils in this sequence become not only less similar to us but also less abundant. Our evolutionary record does not present an unbroken, generation-by-generation sequence; it is not, in a strict sense, continuous. Because of the discontinuities and imperfections, individual researchers or whole scientific schools may differ, sometimes sharply, in their interpretations of the precise course that human evolution has taken. These disagreements, in turn, are sometimes used by skeptics to cast doubt on the entire concept of evolution or its application to our own origins. How reasonable are these doubts? Some perspective on this point is provided by the realization that imperfections in the record cause similar problems and disagreements in other disciplines as well; see "The Butler Saved the Murdered Man: A Literary Analogy to the Problems of Reconstructing Human Evolution," on pages 16–17. Scholars who study the literature and history of the Middle Ages in Europe are dependent on handwritten

THE BUTLER SAVED THE MURDERED MAN: A LITERARY ANALOGY TO THE PROBLEMS OF RECONSTRUCTING HUMAN EVOLUTION

It is a curious fact that the life and death of King Arthur, the highly romantic and largely mythical medieval British king, first appears in Latin sources. One of the first writers to tell Arthur's story in our own language was Thomas Malory, who wrote his *Morte d'Arthur* (*The Death of Arthur*) around A.D. 1470.

Malory's own written draft of the work has never been found. His story was preserved and made widely available, however, by the work of William Caxton, the first English printer. In Caxton's text, published in 1485, the following sentence appears:

Caxton's text

Sir Lucas saw kynge Angwys-schaunce that late hadde slain Maris de la Roche.

Translation

Sir Lucas saw King Angwysschaunce, who recently had slain Maris de la Roche.

The sentence was reproduced in this form in all subsequent modern editions of Malory's work until the edition of Eugene Vinaver, who noticed a slight problem. The knight Sir Maris, said here to have been slain, reappears later in the text, and yet no explanation is given for his resurrection.

Recently a manuscript copy of Malory's tale was found in the library of Winchester College. In this version, which has been dated to about 1475, the sentence reads a bit differently:

Winchester Manuscript

Sir Lucas saw kynge Angwys-schaunce that nyght had slayne Maris de la Roche.

Translation

Sir Lucas saw King Angwysschaunce that night had slain Maris de la Roche.

Here the sentence is rougher grammatically. Its puzzling message is the same, but its precise wording suggested to Vinaver just how the story of Sir Maris came to be nonsensical. It is likely that at this point Malory's original manuscript differed from the Winchester Manuscript by just one letter (underlined below):

Malory's original

Sir Lucas saw kynge Angwys-schaunce that nyghe had slayne Maris de la Roche

Translation

Sir Lucas saw King Angwysschaunce, who had almost slain Maris de la Roche.

If so, the scribe who wrote the Winchester Manuscript must have mistaken the final e for a t and thus made *nyghe* ("almost") into *nyght* ("night"), killing off Sir Maris, at least literally, in the process.

This logical hypothesis could be conclusively proved or disproved only if Malory's original copy of *Morte d'Arthur* were found. However, no matter how desirable that would be, the discovery of this unique book after about 500 years must be reckoned as extremely unlikely. Its parallel in the study of human evolution might be the discovery of a complete and perfectly preserved human skeleton from 3 million years ago. In both cases some long-standing problems could be resolved directly, with no need for theorizing.

Even if neither of the above discoveries should ever be made, other approaches are possible. One of these is the comparative method, by which characteristics of missing ancestral forms can be reconstructed from analysis of the corresponding traits in a series of descendants, surviving relatives, or even earlier ancestors. In the case of the Arthurian legend, the French sources of Malory's tale can still be examined. From these it is clear that although Sir Maris was severely attacked, he escaped death; he was, in contrast to modern stereotypes, saved by the butler (a character named Lucan the Butler). In modern biology, comparative anatomy—long a reliable source of evidence on evolutionary relationships—has recently been enlivened and conceptually sharpened by the work of Schaeffer, Hecht, Eldredge, and others (see Schaeffer, Hecht, and Eldredge, 1972). Their contributions should make comparative anatomy more precise and useful.

In addition, theoretical studies play an important role in most fields of scholarship, whether humanistic or scientific. The discovery that the Sir Maris who lived to fight another day was dead only apparently, because of the slip of a scribe's pen, reflects Vinaver's scholarly detective work in reconstructing the mechanism by which scribes actually copied manuscripts. As a result, he was able to devise a theory to categorize and explain various types of errors. The above-mentioned shift from an e to a t at the end of a word proved to be an example of one of the more common types. In the study of human evolution, theory plays a similar role. Even the best of theories cannot tell us exactly what happened during some period in the past from which little evidence is preserved. But when even the evidence we have seems confusing, theories from several areas of modern population biology sometimes enable anthropologists to make a reasonable reconstruction of the past. While absolute certainty about evolution may not be possible, at least the extent of uncertainty can be reduced.

Sources: (1) B. Schaeffer, M. K. Hecht, and N. Eldredge. 1972. "Phylogeny and Paleontology," in T. Dobzhansky, M. K. Hecht, and W. C. Steere (eds.), *Evolutionary Biology*. Vol. 6. New York: Appleton-Century-Crofts. Pp. 31–46. (2) E. Vinaver. 1939. "Principles of Textual Emendation," *Studies in French Language and Mediaeval Literature*, Publication no. CCLXVII of the University of Manchester, England. Pp. 351–369.

manuscripts for much of their information about the period. Many of these texts contain sentences that are unclear to modern readers, sometimes because the texts were changed as they were copied and recopied. A careless or poorly trained scribe might misspell a word. The next scribe to copy the work might take the misspelled word for an entirely different one and put other words in the sentence to make it read more smoothly. In the process, meanings were sometimes altered quite radically from the original. It is unlikely that a sentence changed in this manner would make perfect sense in the context of the rest of the passage. As a result, two modern scholars reading the third manuscript might reasonably come to different conclusions about its meaning. Despite their differences, both would be shocked if someone with little or no training in the field announced that their disagreements *proved* that no previous manuscripts had ever existed or that no earlier manuscript, however similar, had had any influence on the text whose meaning was disputed. By the same reasoning, scientists studying human evolution may disagree on particular points. Indeed, biological anthropology or any other field would be in a very bad state indeed if scientists all held to the same dogmas based on small amounts of evidence. Such disagreements about the details of human evolution do not, however, provide a sound or sensible reason for rejecting the entire evolutionary record or abandoning its use in attempting to understand more about the human species.

Hypotheses, Theories, and Laws

If evolution is an observed fact, then why bother with theory? After all, isn't a theory just a tentative, temporary substitute for the certainty that will come after enough facts have been gathered? It is easiest to answer this question by looking over the fence again, this time into another field of science.

An apocryphal story in the history of physics holds that the seventeenth-century natural philosopher Isaac Newton saw an apple fall at his feet one autumn day and formulated a theory to explain why it had done this. However, it is unlikely that Newton would have been remembered as anything other than a neighborhood curiosity had his explanation applied to only that apple, to apples in general, or even to all the falling fruit in England. The usefulness of a scientific theory is measured by its generality, as well as by its adequacy and parsimony (see "Adequacy and Parsimony in Scientific Theory"). The more cases to which a theory can be applied without adding special provisions and qualifications, the greater its value. To explain why objects fall, Newton proposed the existence of an immaterial power that draws things such as apples toward the earth. The same force, he demonstrated, also sets the orbit of the moon around the earth and determines the courses of passing comets. Later Newton expressed the working of this power as his universal law of gravitation: Every particle of matter in the universe attracts every other particle with a force that is proportional to the square of the distance between them (Gamow and Cleveland, 1960, p. 86). This force must therefore govern the paths of planets we cannot even see and the orbits of electrons we can only imagine, as well as apples falling from trees.

Observations made by many different scientists on the strength of the attraction between objects over a wide range of sizes have in fact matched very well with Newton's predictions. This repeated agreement of observations with prediction is the reason Newton's generalization is referred to as a *law*. When used correctly, this term defines a scientific theory that has become established by repeated successful testing.

Until a *theory* attains the status of a scientific law, it often exists as just one explanation among many for the same set of data. A theory can begin with a hunch, and the hunch can then become a formal *hypothesis* to be tested over and over before finally either being rejected or reaching the status of an invariant *principle* or *law*. (A more complete discussion of these terms and how they are used in anthropology and other sciences can be found in Kaplan, 1964, or Kaplan and Manners, 1972.)

Scientific Models

The statement that evolution is accepted by scientists as an explanation for the past and present diversity of life does not mean that the past can be reconstructed in full detail. Detailed reconstruction of the sequence of evolution is no more possible than complete comprehension of the world at present. A physicist cannot begin to describe the movement of each atom in a table, and a geneticist cannot tabulate the complete breeding behavior of each fly in a bottle, any more than an artist can paint every blade of grass in a landscape. The success of artists such as Picasso lies in their ability to abstract the essential elements from the bustling confusion of the real world. The same is true for scientists: mere description, no matter how detailed, is bad science as well as bad art. Part of the scientific procedure consists in abstracting essential elements and combining them into a *model* similar to the original system but simpler in structure. There are various kinds of models, but all share several common features. A model should be easier to work with than the original system it replaces. The dot-and-line example in "Adequacy and Parsimony in Scientific Theory," pages 20–21, is a much-simplified model for scientific data and theory. A model must also correspond to reality in the major respects that are likely to influence experimental results. When dietary requirements are tested, for example, monkeys can usually be substituted for humans, and frequently mice can be used instead of either. All three are mammals with similar

ADEQUACY AND PARSIMONY IN SCIENTIFIC THEORY

A scientific theory is an explanation designed to make sense of a set of observations. At many points, in science as well as in everyday life, however, several alternative explanations may fit the data. At most one—and quite possibly none of the ones thought of—can be correct. To gain acceptance as an explanation, a scientific theory must have two main attributes: adequacy and parsimony. A theory is adequate if it relates all the data relevant to a given problem, and it is parsimonious if it does so in the simplest way possible. These criteria, incidentally, are neither limited to the province of science nor particularly new. Parsimony, for example, was understood to be a fundamental principle of logic long ago. It is sometimes referred to as *Occam's razor,* after the fourteenth-century scholastic philosopher William of Occam, who set it forth formally: If several alternative, equally adequate explanations (or hypotheses) are possible, the simplest one—the one with the fewest parts—is to be considered the best. Like a razor, it cuts away the nonessential.

A simple spatial puzzle illustrates adequacy and parsimony nicely. Imagine that the 10 dots shown opposite represent schematically a set of data (fossils buried in the earth or measurements of stature in different human populations, for example) that requires an explanation.

Each partial explanation of the relationships between the data can be represented by a straight line. The theory that best accounts for the data, then, will connect all the dots with the fewest straight lines (in our example, lines drawn continuously without lifting the pencil from the page). Several alternative theories are also shown opposite.

Theory 1 is clearly inadequate, as it leaves one dot unconnected. Theory 2, with six lines is adequate, but it is not the most parsimonious. Theory 3, which uses five lines to connect all the dots, is more parsimonious and therefore better. But is it the best theory that can be fit to this set of data? Clearly not, because a still more parsimonious theory, requiring only four lines or statements, can be used: theory 4.

These illustrations do not prove that theory 4 is the most adequate

digestive systems, and mice are easier and cheaper to work with. Cows, in contrast, have a more complex four-chambered stomach that enables them to derive nourishment from rather unlikely substances (in Norway cows are fed largely on wood pulp and urea). Information that a given diet allowed cows to thrive would probably be of little significance to a human nutritionist.

Models themselves evolve. As it is tested against additional data, a model that was initially very simple and highly abstract commonly becomes more complex and concrete—it converges more and more on

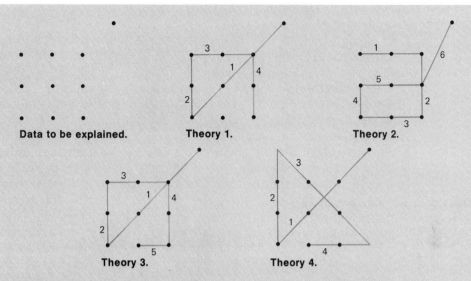

Data to be explained.   Theory 1.   Theory 2.

Theory 3.   Theory 4.

and parsimonious one possible; they prove only that it is the best one developed so far (you are invited to develop a better one).

Theory 4 makes it evident that theories 1 through 3 all operate under an unnecessary, implicit restriction: the assumption that no data points exist outside those already known. To solve many scientific problems it is necessary to go beyond the information at hand. Like so much excess baggage, an unnecessary assumption—for example, that we already have all the important data—often impedes real scientific progress until it is dropped.

You will notice that theory 4 predicts the existence of two more dots of data. A scientific theory is especially useful when it shows researchers where to look for new data which, if found, will help confirm the theory and may also be valuable in other ways. This third characteristic of theory 4 might be called its *predictive ability*. However, this theory's value in doing the immediate job at hand—relating the 10 dots of data to each other—is shown by its adequacy (all dots are connected) and its parsimony (the fewest separate straight lines are used).

reality. But as a model becomes more like the world, it also necessarily becomes harder to work with. The only accurate representation of the real world, after all, is a complete description of it in its full complexity— and the impossibility of dealing with this was the reason for building a model in the first place.

Science is not a search for pure truth; often, it involves the acceptance of what we definitely know to be false. But the models that scientists build are useful because they make it possible to order existing data and to predict the characteristics of discoveries yet to be made.

STRUCTURE AND OBJECTIVES OF THIS BOOK

This book is divided into three parts: Part One, Introduction; Part Two, The Process of Evolution; and Part Three, The Interpretation of Human Evolution.

Part One (Chapters 1 and 2) describes the intellectual climate in which the study of human evolution developed, thus providing insight into the ways in which some of the problems and controversies of the past have shaped thought and work in the present.

Part Two (Chapters 3 through 6) focuses on the process of evolution. Included here is a detailed review of the *genetic basis of evolution:* a discussion of the nature of genes, how genes act to control the development of many of our biological characteristics, and how genes are transmitted from generation to generation in families. Next is a presentation of the *genetic mechanisms of evolution:* the ways in which genes can be changed in frequency over time. Scientists disagree about the relative importance of the mechanisms that can cause evolutionary changes. One of the major debates concerns the relative roles played by influences that are *deterministic* (having results that follow invariably) rather than *random* (having results that are predictable only within a general range). Another concerns the extent to which given human characteristics are shaped directly by environmental factors rather than indirectly through the environment's effect on the survival or reproduction of people with different sets of genes. Such questions can best be resolved by studies of living people, and these form the basis for the principles developed in Part Two. Of course there are limitations to the study of living people, but it offers us more opportunity to test inferences about how evolutionary mechanisms operate than the study of the past. It is chiefly through processes that we see going on at present that we understand evolution through ages distant from our own.

In Part Three (Chapters 7 through 16), some of these debates about the relative importance of various evolutionary forces are projected into the past. But the time scale of human evolution is vast, spanning millions of years. Over long periods, environments changed markedly, populations expanded greatly, and technology proliferated. Against this background, there is no difficulty in explaining any of the smaller-scale events in human evolution, such as the expansion in the size of the human brain or the development of articulate speech. The real challenge is to choose from among the many explanations that are compatible with basic evolutionary theory the one that approximates what actually happened. The aim of this book is to help equip you to make such choices.

The major purpose of the text is to introduce to its readers the kinds of evidence, methods of investigation, and lines of reasoning that are used in the reconstruction of past human evolution and the analysis of past and present human variation. The primary emphasis, therefore, is on explaining *how to study* human evolution, rather than on giving an account of *what happened.* Of course, it would be almost impossible to produce a book that gave examples and methods of study but no conclusions. Inferences about the possible course of past human evolution will be made, but the reasoning behind these ideas will be made explicit, along with their degree of uncertainty. The emphasis is not on what we know, but on how we know it.

It is also impossible for any one book to develop in its readers the skills of scientific inquiry, even as they are used within one field; the job is simply too vast. This is particularly true for the study of human evolution. Although it constitutes only a part of traditional physical anthropology, this subject draws its data and theories from many other disciplines—genetics, ecology, demography, ethology, and geology, to name but a few. But at least a partial defense of the problem-oriented approach followed here can be made. It is just about as difficult to tell the complete story of human evolution between the covers of one book—even without making explicit how that story is known—as it is to explain how inferences are made about human evolution and how still more can be learned. Why, then, choose one approach over the other?

For those readers who will concentrate on the study of human evolution or on some other branch of anthropology, the reason is fairly obvious. They need to learn how to gather and evaluate scientific data, since this effort will absorb their energy and attention for many years. But most readers of this book will not become professional anthropologists. Why should nonspecialists bother to sharpen the skills that will enable them to go from a body of evidence and a set of theories about human evolution to reasoned conclusions about their own place in the world? Perhaps the best answer is another question: How can any of us afford *not* to be able to judge the validity of claims made about human nature? We might be able to take or leave studies that explain the characteristics and habits of ants, elephants, or turtles, depending on our interests. But each year sees the publication of many new conclusions about human origins, human biology, and human behavior. On encountering these claims, will you be able to tell the science from the snake oil? Did humans originate 15 million years ago or only 5 million years ago? Do primitive humans—sasquatch and abominable snowmen—still roam sparsely inhabited regions of Tibet, the northwestern United States, and Canada? Does the IQ have a substantial inherited component, and, if so, does this mean that special education programs are of little value in improving the performance of children in school? Did humans live without violence until they encountered the stresses caused by the effects of widespread agriculture and industrialization—crowded living conditions and rapid social change? Are humans born without any behavioral tendencies, which would mean that all our actions are shaped by the societies in which we live?

You may choose not to bother forming judgments about issues such as these, but that disinterested position will not prevent your life from being affected by the decisions that will be made by others. The decisions will determine how tax dollars are spent on research and education, what subjects and viewpoints school texts will include, and even how laws are framed. A sound knowledge of the distinctive features of humans, and of the origins of these characteristics, should help in shaping a future we all must share. This book is designed to help free you from having to accept the conclusions of other people because you cannot evaluate the evidence and reasoning yourself. It is written with the aim of helping you learn to think critically so that you will come to understand how we know what we think we know.

SUMMARY

1. Evolution, or cumulative directional change through time, operates at a number of different levels including the cosmic, chemical, and organic.

2. The human species, *Homo sapiens,* has arisen through the same processes of organic change that have shaped the other forms of life on earth.

3. Human evolution has emphasized physical and behavioral flexibility rather than narrow specialization; for example, humans are biologically programed for learning language, but not for learning one particular language.

4. Anthropology is the scientific study of the biology, society, and culture of the peoples of the world and of their human ancestors.

5. Anthropology is divided into four subfields: cultural anthropology (with subdivisions of ethnography and ethnology), linguistics, archeology, and physical anthropology.

6. The study of physical anthropology can be approached from the standpoints of several central organizing principles, chiefly adaptation and evolution.

7. Acceptance of the idea of evolution has brought about a revolution in our view of the world and the place of humans in it.

8. Evolution is both fact and theory; differences in gene frequency that are basic to organic evolution have been measured between successive generations, and there are satisfactory scientific explanations for these observed changes.

9. When evolution cannot be observed firsthand, as in the fossil record for human evolution, facts and inferences must be combined into a model which has a satisfactory logical structure arrived at by the use of accepted methods of scientific inquiry.

SUGGESTIONS FOR ADDITIONAL READING

Simpson, George Gaylord. 1964. *This View of Life.* New York: Harcourt, Brace and World. (The view of life referred to in the title is that of an evolutionist, and the author is one of the leading evolutionary biologists of this century. Four major topics are covered: approaches to evolution (historical background), evolution among the sciences, the problem of purpose, and evolution in the universe. Simpson's views are thought through carefully and written with extreme clarity. The section containing notes and references lists a number of other classic works by this author and others.)

REFERENCES

Aitchson, J. 1978. *The Articulate Mammal: An Introduction to Psycholinguistics.* New York: Universe Books.

Chagnon, N. A. 1977. *Yanomamö: The Fierce People.* New York: Holt, Rinehart and Winston.

Chang, K. C. 1978. *Food in Chinese Culture: Anthropological and Historical Perspectives.* New Haven, Conn.: Yale University Press.

Gamow, G., and J. M. Cleveland. 1960. *Physics: Foundations and Frontiers.* Englewood Cliffs, N.J.: Prentice-Hall.

Kaplan, A. 1964. *The Conduct of Inquiry.* San Francisco: Chandler.

Kaplan, D. A., and R. A. Manners. 1972. *Culture Theory.* Englewood Cliffs, N.J.: Prentice-Hall.

Kroeber, A. L., and C. Kluckhohn. 1952. "Culture: A Critical Review of Concepts and Definitions," *Papers of the Peabody Museum of American Archeology and Ethnology,* **47**:1, Cambridge, Mass.

McNeill, D. 1966. "Developmental Psycholinguistics," in *The Genesis of Language.* Cambridge, Mass.: M.I.T. Pp. 15–84.

Neel, J. V., and W. J. Schull. 1954. *Human Heredity.* Chicago: University of Chicago Press.

Sagan, C. 1977. *The Dragons of Eden.* New York: Random House.

Schklovskii, I. S., and C. Sagan. 1966. *Intelligent Life in the Universe.* New York: Dell.

Tylor, E. B. 1871. *Primitive Culture.* London: Murray.

Chapter 2

History of
Evolutionary Theory

Evolution is a revolutionary idea. As in most revolutions, faint stirrings and false starts preceded the decisive overthrow of the previous order of things. This chapter explores the origin of the idea of evolution and traces the concept through about 25 centuries of human thought. The history of the idea is important because the work of each new generation of thinkers tends to be determined by the cultural heritage from the past. New thinkers may build on previous foundations or tear them down, but they never start in a void. Looking backward from our present vantage we can see the progress from the earliest attempts to find order in a confusing world, through battles about the possible causes (supernatural or natural) of this order, to more precise and testable beliefs about how heredity and evolution work to make both stability and change possible in the biological realm.

Biological evolution is still a disturbing concept to some people, but the rapid pace at which changes now occur has helped condition most of us to the idea that modifications take place in the plants and animals that share the earth with us as well as in modes of production and styles of living. This viewpoint marks a sharp break with earlier periods of history, when the pace of technological innovation and cultural change was slower. Many ideas, however, were known long before they were generally accepted and understood, and so our study of the development of evolutionary theory begins with the speculations of ancient thinkers.

EARLY SPECULATIONS ABOUT EVOLUTION

Some notions about evolution appear early in recorded history. Anaximander, who lived in Greece between approximately 611 and 546 B.C., taught that living creatures arose from the primordial mud that covered the earth. Plants and lower animals had appeared first. Humans, he thought, had evolved from a fishlike ancestor that emerged from the water, cast off its skin, and assumed our form. Schemes such as this were of value because they represented attempts to explain the world in *naturalistic* terms—that is, in terms of processes we see going on around us and without the requirement of the intervention of beings with superhuman powers. Indeed, most of the early Greek philosophers are remembered not because their conclusions were correct but because their conceptions were freer of the magical and mystical elements that had dominated the learning of the Babylonians and the Egyptians. This philosophical shift away from mysticism eliminated a whole category of unlikely explanations for natural events, making some ideas proposed by the early Greeks seem much like our own. But closer examination reveals that their explanations were usually lucky guesses among a range of alternatives narrowed by rational deductions rather than due to sound inferences from accurate observations.

TAXONOMY: A SYSTEMATIC VIEW OF THE WORLD

Aristotle

Aristotle (384–322 B.C.) made use of an extensive library to acquaint himself with the thoughts of his predecessors. Unlike previous thinkers, though, he also made numerous personal observations, and it was chiefly on these that he based his generalizations. Aristotle realized that the study of animals could be greatly simplified by arranging them according to their similarities and differences. His system was based on four different types of information: way of life, actions, habits, and appearance. Each separate kind of animal was called an *eidos,* which corresponds in modern biology to the *species* (see Chapter 4 for a complete discussion of this term). All combinations of several similar *eidos* were referred to by the single term *genos.*

Ray

Much later, under the influence of students such as John Ray (ca. 1627–1705) of England, this work of systematic grouping began to develop into a separate branch of the natural sciences called *taxonomy* (the study of classification). Ray progressed beyond Aristotle in recognizing that if various plant and animal species resembled one another to differing degrees, they could be grouped into a hierarchy of taxonomic categories. But Ray did not establish any general categories applicable to *both* plants and animals. His system was also clumsy since each species was identified by a phrase rather than a label, and many of his categories were based on superficial resemblances. For example, he grouped whales with fish and bats with birds. Aristotle had earlier correctly realized that both whales and bats are *mammals,* warm-blooded animals that give milk to their young and have a coat of hair.

Linnaeus

Far more important than Ray, and easily the best known of the early taxonomists, is Linnaeus (1707–1778). His first name, Carl, is sometimes Latinized to Carolus, and he is also known as Carl von Linné. Linnaeus greatly streamlined the system of taxonomy by reestablishing a system of *binary nomenclature.* Instead of being referred to by a description, each animal was given a unique two-part name (*binomial*). The first indicated its membership in a given genus; the second set it apart as an entity different from other species in the same genus. Linnaeus extended this principle even further than Aristotle had by arranging genera hierarchically into other categories on the basis of their similarities and differences. His system is still used today. Despite his lasting contribution to the classification of organisms, however, Linnaeus was not himself an evolutionist. His purpose was to discover how many species had been sent forth in pairs originally, according to the word of God. Nevertheless, Linnaeus' system for arranging living creatures according to different degrees of resemblance seems to have stimulated others to ask why these patterns of similarity and difference exist.

THE CASE FOR EVOLUTION

Buffon

Georges Louis Leclerc de Buffon (1707–1788), a French biologist, was among the contemporaries of Linnaeus who were skeptical of the *creationist* position (the idea that each separate species had been made directly by a supernatural being). Buffon's view of the origin of life was closer to the idea of *spontaneous generation* (the belief that the first simple forms of life came from nonliving chemical systems which gradually became more and more complex).

Buffon also thought that present forms of life merge almost imperceptibly into one another. Although humans are very different from fish, the gap between these two levels is roughly bridged by a series of other creatures, including amphibians, reptiles, and some of the mammals.

Here Buffon actually rested his case on a point that had been documented by the taxonomists themselves: not all species are equally similar to, or different from, one another. What of the gaps that do exist? Here, Buffon's belief in the existence of a virtually continuous series of living forms led him to question the nature of the task in which taxonomists were engaged. He felt that the systems of classification reflected discontinuities that really did not exist in nature. To him, taxonomic systems seemed to impose artificial boundaries on a natural continuum.

In taking on the taxonomists, Buffon was also challenging the Almighty, or at least his powerful earthly interpreters who were on the theological faculty at the Sorbonne. In his writings he only hinted that apes and humans might share a common ancestor, but it is clear that he believed privately that the species of Linnaeus and other taxonomists were not static units, with characteristics fixed for all time according to a divine plan. Instead, Buffon believed he could see evidence of biological change in the world about him. "The pig," Buffon wrote, "has evidently useless parts, or rather, parts of which it cannot make any use, toes all the bones of which are perfectly formed, and which, nevertheless, are of no service to it." Structures that were present as rudiments in some animals suggested the existence of ancestors in which these parts had been larger and more useful, and they allowed for the possibility of descendants in which the structures would be lost. This reasoning implied the possibility of evolutionary change through time. To explain why these changes occurred, Buffon looked to the forces at work in nature, including influences of climate that could be linked to geological processes such as those which brought about changes in the proportion of land and sea on the earth's surface: "In order to understand what had taken place in the past, or what will happen in the future, we have but to observe what is going on in the present."

Though his biological writings had successfully skirted the scriptures, Buffon's geological opinions clashed too openly with dogma. He was forced to publish a recantation acknowledging the infallible authority of the Church, but his private correspondence shows that his views remained unchanged. His beliefs, however, seem to have had little impact on the growth of evolutionary thought. Perhaps Buffon's most enduring positive contribution was the counterpoise he provided to the work of Linnaeus and other taxonomists. By stressing the dynamic and theoretical aspects of natural science, he kept it from becoming a purely mechanical endeavor.

Lamarck

Ideas like Buffon's were much more fully developed in the work of another Frenchman, Jean de Monet (1774–1829), who is usually known by the name derived from his title, Chevalier de Lamarck. Lamarck made a bold frontal attack on the static array of species described by the taxonomists of his day. He emphasized the changing aspect of the natural world and saw some living forms as transitions between others. Lamarck insisted that all these connections had to be explained in terms of an evolution from simpler to more complex forms. His theory was that life on earth began with only the simplest animals and plants, such as algae and worms. From these, more complex forms evolved. Like Buffon, he reasoned that humans were descended from apelike mammals. In place of the image traditional since Aristotle, that of life as a ladder, with the various forms arranged as successive rungs, Lamarck substituted a tree. Living forms were at the tips of branches that could be traced back ultimately to a single, far-removed trunk.

Lamarck thus outlined what proved to be a more appropriate arrangement of types of organisms, and he correctly insisted that evolution was the idea that best explained this arrangement. He also attempted to devise a mechanism by which evolution could occur, but the explanation he provided has found no experimental support. His imaginative and courageous conception of the basic problem, however, was right.

Lamarck's beliefs about how evolution operated are summarized in four propositions:

1. Forces inherent in life tend to increase the size of each living body and all its parts up to a limit that is set by its own needs.

2. New wants in animals give rise to new movements that produce bodily parts.

3. The size of each part is in proportion to the degree to which it is used.

4. Changes that occur during the parents' lives are transmitted to the offspring.

According to these principles, the characteristics of any individual are determined by its own habits and ways of life as well as those of its ancestors. This fits in with observations that can be made in everyday life. A person who exercises regularly will develop larger and firmer muscles than one who leads a more sedentary existence. By analogy, Lamarck accounted for the long neck of the giraffe by continual stretching of that structure to reach the leaves of trees high above the ground. Similarly, he attributed the upright stance of humans to the constant efforts of our apelike ancestors to stand erect.

To explain evolutionary changes, Lamarck speculated that the characteristics developed during an individual's lifetime were passed to the next generation. This would be possible if bits or particles from each organ and structure in the parent's body were transmitted to its offspring. Belief in such a mechanism—called *pangenesis* (from *pangens*, the term for the particles)—was widespread among biologists in the nineteenth century. The changes caused by the addition of new pangens would be very slight in any one generation. Over extremely long spans of time, however, the accumulated changes could give rise to major new structures and organs.

Cuvier

The scientist responsible for discrediting Lamarck's theories was his contemporary and countryman, Georges Cuvier (1769–1832). Cuvier did not attack the idea that acquired characteristics could be inherited, but instead denied that the changes thus produced could be gradual. Unlike Lamarck, Cuvier believed that the earth's geological deposits encompassed only a fairly limited span of time. If his assumption was correct, there would not have been enough time for gradual changes to produce the great diversity of living species known. His deductions were logical, but the conclusion could be no more valid than the premise on which it rested: that the earth was young. Support for this idea existed in the form of a widely known set of calculations made by Ussher, an Irish archbishop, in 1636. Using information contained in the Bible, Ussher laboriously constructed a chronological sequence which established that the world had been created in 4004 B.C. Further work of a similar type by Dr. Lightfoot, who was the vice-chancellor of Cambridge University and a respected religious scholar, fixed the beginning of the process at 9 A.M. on Tuesday, October 23, of that year. Then, as now, it was possible for scholars to be quite precise without necessarily being at all accurate, just as painters can do a neat job while using the wrong color.

By attacking the mechanism proposed by Lamarck, Cuvier was able to discredit the whole idea of evolution. This was ironic because today some of the best evidence for evolution comes from a science Cuvier founded: *paleontology,* the study of ancient life. Cuvier was not the first to study the remains of earlier creatures, but he did manage, for a time, to make the study of fossils respectable theologically as well as scientifically. This was no small accomplishment. Before Cuvier's time, fossils, when they were recognized at all, were commonly the objects of much superstition and conjecture.

Just what are fossils? According to Wendt (1976), the word *fossil* (from Latin *fossa,* "trench") was first used by a German geologist, Georg Agricola (1494–1555). But fossilized remains had been known long before Agricola's time. Some of the ancient Greeks, including Aristotle, recognized fossilized shells as the remains of animals that had once been alive. After Aristotle's time and during the Middle Ages, such naturalistic explanations were unusual, although there was much speculation about how fossils were formed. Some people felt that the soil possessed a "formative quality," a "plastic virtue," or a "lapidific (literally, "stone-forming") juice." Heavenly bodies were believed to exert some influence, though exactly how they did this was not specified. Alongside these ideas about how fossils were made were others to explain why they existed. Albertus Magnus (1192–1280) said that fossils were games of nature; still others held them to be attempts by the Almighty to mock human curiosity. There was evidently no feeling that any experimental demonstration of these ideas was required. It was sufficient, and necessary, that any particular explanation be in accord with authoritative interpretations of the scriptures. The accommodation was not at all difficult, for the Bible presented a convenient explanatory framework into which fossil evidence could be fit. Numbers of animals beyond counting must have perished in the Flood. Unfamiliar bones that turned up were said to have belonged to those unfortunate creatures. Since nearly all discoverers had more imagination than systematic knowledge, fossils were made to fit to known fables. As in most ages, that which was sought was found: giants, dragons, unicorns, and other purely mythical beings were reconstructed.

There were also some dissidents. Two of them, Leonardo da Vinci (1452–1519) and Girolamo Fracastoro (1483–1553), were Italians; a third, Niels Stensen (1638–1686), was a Dane who lived for a time in Italy and held church offices under the Latinized name Nicolaus Steno. Perhaps in response to their common exposure to the enlightened intellectual climate of Renaissance Italy, the three scientists shared two other characteristics. First, all had considerable knowledge of anatomy. Fracastoro and Steno were physicians, and Leonardo was a genius who dissected humans and other animals in order to better represent their forms in art. Second, all three men chanced on fossils in the course of other work and recognized them as the remains of ordinary organisms that had lived and died in earlier times. It is possible that their conclusions were ignored because they were simply so contrary to prevailing opinion.

Cuvier, however, got much attention. He and his pupils combined awesome orthodoxy in religion with the most detailed and painstaking inquiries into anatomical structure and physiological functioning in recent animals. From his studies of the animals of his own time, Cuvier came to believe that each was a whole in which all the parts were interdependent. If one part changes, the other structures must adjust in response. This *principle of correlation,* as it came to be called, gave Cuvier a set of guidelines that could be used to tell which bones in a jumbled mass of fossils belonged together. It could also be used to reconstruct missing parts. (Reconstruction was a necessary exercise, since fossil skeletons are usually incomplete.)

Cuvier's time saw the beginning of systematic, large-scale excavation for fossils, with much of the work being done under his direction. It was soon found that geological deposits were not of uniform composition. Instead, they could be separated, by color, texture, or chemical composition of the rocks, into separate layers, or *strata.* Each stratum, moreover, was found to contain a set of fossils, different from those in strata above or below. All the animals in one stratum make up a *fauna.* All the plants make up a *flora,* and fauna and flora together make up the *biota;* but we are chiefly concerned here with the fossil remains of animals. To an evolutionist such as Lamarck, who believed in gradual change through time, sharp differences between faunas were accidental breaks in a fossil record that had originally been continuous. As Nordenskiöld (1927) has emphasized, however, Cuvier considered himself a strict empiricist, not a theorist. To him an observed faunal difference from one layer to the next was real; it was a fact. Moreover, these facts accorded with —indeed, they supported—the widespread belief in the biblical deluge.

Though later ages have recognized Cuvier as the architect of the *catastrophe theory,* it seems clear that he considered himself to have merely set forth a few principles that followed immediately from his observations. He believed that cataclysmic events periodically destroyed all life in a given region. After each catastrophe, the devastated region was repopulated from some refuge which had escaped destruction. There was no need to search for intermediates between any two successive faunas; none could have survived the catastrophe. And since the whole cycle of destruction and replacement had occurred over only several thousand years, there would not have been time for gradual changes anyway. All species, living and fossil, could therefore be included in the same taxonomic scheme; none was ancestral to, or descended from, any other. Given complete annihilation of all life in a region by each catastrophe, any apparent continuity from one layer to another was misleading. The occasional fossil bone that looked like that of a living horse or lion must have belonged to a different animal. Fossilized human remains, of course, could not exist. Suggestions that any had been found were met with the explanation that modern burials must have been mixed up with fossils from earlier strata by mistake.

The demise of the catastrophe theory was due in no small part to the evidence accumulated by Cuvier himself, his students, and his collaborators. Cuvier had begun with a single deluge to explain why one set of animals was replaced by another. Like an alcoholic's one drink, however, the first catastrophe led to another, and another, and another. By the middle of the nineteenth century, tens of thousands of fossil species had been discovered. More and more deluges were needed to accommodate these, and with a fixed time scale, catastrophes would have had to strike with unbelievable frequency.

The widely accepted catastrophe theory was eventually proved wrong by several British geologists who had the courage to question the theological dogma, accepted by Cuvier and his followers, that the earth and its life were very young.

GIFTS OF GEOLOGY

In retrospect, it is difficult to see why there was such reluctance in the 1800s to believe in the existence of earlier humans. Evidence that the world was more than some 6000 years old was already being obtained by geologists, notably James Hutton (1726–1797) and his better-known successor, Charles Lyell (1797–1875). Parting from the tradition of deriving geological theories from speculative flights of the imagination, each of these men examined the ground beneath his feet. Hutton insisted that geology ought to have nothing to do with ideas about the creation, but should first describe the earth's strata and only then move to thoughts about their origin. The observations that followed showed that formation of geological deposits was not a process completed in the past. Deposition still continues as natural forces (wind, water, temperature changes, and so on) break down existing rocks and redeposit the debris in new layers. From this it was but a small step to the realization that the deposits which have survived from the past may have been formed by processes no different in kind and perhaps not greatly different in rate from those we can see at work today. This was more formally stated in Lyell's *Principles of Geology* (1830) as the idea of *uniformitarianism.* Uniformitarianism is the belief that the forces that shaped the earth and life on it in the past were the same as those natural processes we can observe going on today. In particular, uniformitarianism assumes that no supernatural intervention is required as an explanation for evidence preserved in the fossil record.

Lyell realized that this uniformitarian principle could be used to attempt a reconstruction of the age of the earth. If deposits of a given thickness were formed in a known interval of time, then by assuming an approximately constant rate of deposit, one could arrive at the length of time that would have been needed to accumulate all known deposits. Similarly, if one knew the percentage of extinct species in recent deposits of known age, it would be possible to calculate by proportion the greater age of strata containing a higher frequency of extinct species. Lyell used this second approach and found that the duration of life on earth had to be measured in hundreds of million of years, rather than just thousands.

EXPLAINING EVOLUTIONARY CHANGE: DARWIN

Lyell had demonstrated that the earth had existed long enough for the gradual action of natural processes to bring about enormous changes in the physical world. But the revival of the concept of evolution required a plausible theory of how natural processes could bring about transformations in organisms as well as in their environment. Such a theory was presented by Charles Darwin (1809–1882).

As a young man, after having failed in attempts to educate himself for several professions, notably medicine and the clergy, Darwin got a chance to follow up his boyhood interest in natural history. A friend of the family recommended him for the post of naturalist on the H.M.S. *Beagle,* which sailed on a 5-year voyage to survey the coast of South America. Darwin was thus able to make extensive firsthand observations of a great variety of geological formations, plants, and living and fossil animals. After returning to England, he published *The Voyage of the Beagle,* an account of his findings. The book was a success, and its author continued his scientific studies. He had read Lyell's *Principles of Geology* and was interested in the question of whether species are fixed or subject to change. In 1837 Darwin began to keep a notebook on this subject. The next year part of the answer came when he read Thomas Malthus's *Essay on Population.* Malthus gloomily contrasted the very great fertility of humans with the limited food resources of our planet. Darwin realized that the same problem also existed for other species and that the solution to this problem would give the key to understanding evolution.

In his book *The Origin of Species* (published for the first time in 1859), Darwin outlined his idea that evolutionary change occurs by means of *natural selection.* Darwin's theory of natural selection was that in every generation the numerous members of each species compete for survival and that the ones with the most advantageous hereditary characteristics survive and perpetuate their kind. Several observations are basic to this theory:

1. In every species of animals or plants, each pair of parents typically produces a large number of offspring.

2. Yet the total number of adults in each new generation remains about the same as that in the last.

3. No two organisms are alike; there is variation among individuals.

From the first two observations Darwin reasoned that there must be competition among the offspring of each generation for survival. The third point suggests how this competition could lead to evolutionary change. Certain variations might give their possessors an advantage over other individuals, allowing them to survive longer and reproduce more offspring. If the variations could be inherited, the more numerous offspring would have the advantageous characteristics of the parents.

Darwin was not the only person to come to this conclusion. The idea of evolution, even the more particular idea of evolution by natural selection, was evidently a product of the intellectual climate of the mid-nineteenth century. At least one other scientist, Alfred Russell Wallace (1823–1913), holds a legitimate claim to virtually simultaneous discovery of the idea. But although Darwin and Wallace made similar observations, Darwin's conclusions were formulated after two decades of continuous, painstaking compilation of evidence from the most diverse sources; Wallace made his discovery in a week while delirious from malaria in the Malayan jungle. Wallace deserves credit for conceiving the theory, but it was Darwin's mass of supporting evidence that served as the rock against which the criticism dashed and then ebbed away.

Darwin was not right on all points. His conception of the material basis of heredity, for example, was incorrect. Darwin believed that the traits of the parents, carried in particles (*gemmules*) from each cell, were blended together in the offspring, as black paint from one bucket and white from another bucket would make gray paint when mixed. The offspring in turn were thought to transmit their blended characteristics to their children. However, such a process would constantly work to eliminate the variations that are basic to the theory of natural selection. To maintain a suitable level of variation, Darwin received Lamarck's idea that acquired characteristics could be inherited, but evidently he accepted this explanation only for want of a better one.

EXPLAINING HEREDITY

Mendel

Quite unknown to Darwin, the solution to the problem of inheritance had already been worked out by Gregor Mendel (1822–1884), an Augustine monk who lived in Brno, Austria (a region now part of Czechoslovakia), and was interested in the laws of inheritance. Mendel worked in the monastery garden for 8 years, breeding common garden peas. From the results of these experiments, he was able to infer that some physical units are transmitted from parents to offspring and that those units contributed by one parent are not changed by contact with units from the other parent. These units later came to be called *genes*. Mendel's success was based on two things:

1. He chose his experimental material carefully. Seeds of many different pure varieties of peas were available. These differed clearly in various easily observable characteristics such as seed color and plant height, and each different variety bred true. Thus, in crosses between different varieties it was easy to detect the influence of the parents.

2. He kept careful records of the types of progeny that resulted from each cross and of the numbers of each type.

His first paper, proposing the theory that the similarity of offspring to their parents is due to the regular transmission of certain particles from one generation to another, appeared in 1866. Yet its significance went virtually unnoticed for over three decades. Perhaps some biologists thought that the regularities Mendel observed occurred only in pea plants. Others may have believed that these hypothetical unseen particles did not exist at all. Whatever the specific reasons were, there can be little doubt that Mendel was ahead of his scientific contemporaries in the design of his experiments as well as in the analysis and comprehension of their results. During the period of eclipse, Mendel's paper was read and cited, but as just another study of plant hybridization. It is therefore all the more dramatic that the significance of his work was rediscovered in the same year, 1900, by at least three other men: Hugo De Vries of Holland, Carl Correns of Germany, and Erich von Tschermak of Austria. Following the publication of their work, Mendel's principles were confirmed and extended in studies of other plants and animals all over the world.

CELL THEORY

In the period between the publication of Mendel's paper and its rediscovery, advances were being made in other areas of biology, particularly microscopic anatomy. Discoveries in this field have proved to be of major significance in the advancement of evolutionary theory. The most fundamental of these was the recognition of the cell as the basic structural unit of life. This discovery came only after the microscope had been developed to a sophisticated level.

The technological base of microscope making extends in a direct sequence at least as far back as the sixteenth century. According to Nordenskiöld (1927), two Dutch eyeglass makers named Janssen, father and son, made the first multiple-lens magnifying instruments. Their microscopes, consisting of a tube with a plate for the specimen at one end and the lens at the other, were very crude. The whole apparatus was held like a telescope, with the lenses near the eye and the plate aimed toward a light source. Magnification of about 10 times was the limit of these devices. But even such primitive magnifiers opened a window to a world previously unseen. The most energetic early explorer of this realm was the Italian Marcello Malpighi (1628–1694), who is generally credited as being the founder of microscopic anatomy. Malpighi studied a great variety of animal and plant specimens. He discovered that in plants the visible parts (bark, wood, buds, leaves, flowers, fruits, and so on) were all composed of microscopic units which he called *utriculi* and which we know as *cells*. We know in retrospect that plants were more suitable than animals for these earlier structural studies because plant cells, being bounded by thick walls made of cellulose, are generally larger and easier to see.

By the seventeenth century, improvements had raised the level of magnification to 270 times. With these more powerful microscopes, the Dutch scientist Antony von Leeuwenhoek was able to see blood cells of humans, frogs, and other animals. He also saw the reproductive cells—eggs and sperm—of many different animals and established that even the tiniest single-celled animals reproduce themselves rather than arising new by spontaneous generation from nonliving matter.

Microscopic anatomy remained on this plateau for about another century, until 1827, when the last major improvement was made in the structure of the light microscope. This was the development of a lens free from chromatic aberration, an intrinsic defect in the manufacture of optical devices. Even the best of the early lenses acted like prisms, breaking white light into a rainbow that made it difficult to see the true color of any specimen. In 1827 Giovanni Battista Amici demonstrated the first microscope with an achromatic lense. This improvement opened the way for a rapid succession of important discoveries. By the late 1830s, Matthias Schleiden and Theodore Schwann had independently made the observations that came to be known as the *cell theory:* (1) all life is composed of cells and the products of cells, and (2) all cells come from preexisting cells. The members of each new generation are formed from the reproductive cells (eggs and sperm) contributed by their parents.

Those who used increasingly powerful microscopes and a variety of chemical stains to study these units discovered a number of more minute parts within them. Among the most noticeable of these were the *chromosomes,* bodies in the nucleus of the cell that could be stained with various dyes. The number of chromosomes stayed the same generation after generation. In addition, the sperm cell of the male consisted almost entirely of this nucleus full of chromosomes. These regularities, among others, led to the suspicion that the chromosomes played some role in heredity.

Chromosome Theory of Inheritance

The precise nature of the role of chromosomes in heredity was formulated independently in 1902 by W. S. Sutton, then a graduate student at Columbia University, and the well-known German cytologist Theodor Boveri. Both pointed out that the transmission of Mendel's particles matched the regular behavior of chromosomes during cell division. From this, they inferred that genes must be located on the chromosomes. The *chromosome theory of inheritance,* as Sutton and Boveri's generalization came to be known, explains how different genes are transmitted to individuals in a regular manner: genes are transmitted from parents to offspring in the ratios observed by Mendel because genes are located on chromosomes and the chromosomes are passed on in a regular manner during cell division.

POPULATION GENETICS

The acceptance of Darwin's theory of evolution, coupled with the rediscovery of Mendel's laws, caused a number of geneticists and mathematicians to become interested in the evolutionary fates of various genes in populations. How did these different genes come into being in the first place? What caused some to increase in frequency, others to decrease in frequency, and many to show no apparent change in frequency at all?

Natural selection, Darwin's contribution, provided an important part of the answer to these questions, but not all of it. Selection could change only the frequency of whatever genes already existed. There also had to be a mechanism by which new variations could be generated. In 1901, De Vries, one of the rediscoverers of Mendel's laws, proposed that a new gene could arise from an old one by a sudden, discontinuous change for which he coined the term *mutation.* A few years later, in 1909, this prediction was borne out when in the laboratory of Thomas Hunt Morgan, a leading American geneticist, a single male with white eyes appeared in a population of the usual red-eyed strain of *Drosophila* flies. Subsequently, other mutants of various types were seen to arise, though very rarely and apparently spontaneously. In the 1920s H. J. Muller showed that the frequency of mutations could be increased greatly above the spontaneous level by exposing organisms to ionizing radiation or certain chemicals. The mutations themselves were thought to be different chemical forms of preexisting genes.

During the 1930s and 1940s the study of mutation, selection, and other factors that could change the frequency of genes gave rise to a whole new field of biology now referred to as *population genetics.* The combination of the previously mentioned chromosome theory of inheritance and population genetics formed the core of the *genetic theory of evolution*.

MOLECULAR GENETICS

Even more recently, chiefly in the last quarter of a century, a whole new series of discoveries has given us quite detailed knowledge about the fundamental physical and chemical structure of genes themselves and about how genes operate to control hereditary characteristics.

How Genes Act

The secret of how genes act was discovered before their own chemical nature was known. Mendel's laws were rediscovered, and the connection between genes and the biochemical characteristics of humans was grasped almost immediately. In 1902 the British physician Archibald Garrod studied patients with a disease called *alkaptonuria.* The urine of alkaptonuric individuals darkens on exposure to air. By studying the pedigrees of these patients, Garrod was able to show that alkaptonuria is inherited, and he originated the term *inborn errors of metabolism* to describe this and other biochemical abnormalities that he found to be controlled by genes. In later studies Garrod was able to describe the precise mode of inheritance of alkaptonuria. (Alkaptonuria is a Mendelian recessive; as explained in Chapter 3, in order to be affected, an individual must receive one gene for the abnormality from each parent.) Garrod's papers on the subject were masterpieces of scientific method but they stood in isolation. For about 30 years their significance was almost as generally overlooked as that of Mendel's first paper had been. Then, nearly the same principles were rediscovered by geneticists working on *Drosophila* flies and, later, on the living bread mold *Neurospora.*

Two researchers, George W. Beadle and Edward L. Tatum, realized that the extreme simplicity of the bread mold made it an ideal organism for research on genetically controlled chemical reactions. Many mutant *Neurospora* strains were found. Each mutant required for its growth some chemical nutrient that the normal plant did not need. Each mutant had a genetically controlled defect—an inborn error of metabolism—that blocked a single step in some crucial biochemical reaction. Beadle and Tatum inferred, correctly, that each gene has one primary function: to govern the production of an *enzyme* (or biological catalyst that controls a chemical reaction in the cell). The additional nutritional requirement arose because a defective enzyme prevented the mutant from producing some substance it needed in order to grow. By this work Beadle and Tatum established that the primary function of each gene is to control the production of an enzyme. This was the core of their *one-gene–one-enzyme* theory, which in a very slightly modified form served as the foundation for molecular genetics.

What Genes Are

The earliest theories about the chemical composition of genes followed from the early cytological work establishing that chromosomes are made up of two major components: deoxyribonucleic acid (DNA) and proteins (these compounds are described and illustrated in Chapter 3). Genes must be composed of either of these substances, but which? Until about 30 years ago the answer was simple: proteins. The reasoning behind this answer was very logical. Most organisms, including humans, have many thousands of genes, all of which differ in structure. Proteins are very complex chemicals, while DNA is very simple. Only proteins, therefore, were thought to be diverse enough to provide the needed structural variety.

Gradually this deductive logic came to be refuted by some disturbing facts that began to accumulate in the 1940s. First, A. E. Mirsky found that, in each species studied, the amount of DNA in a cell bears a constant relationship to the number of chromosomes that are present, while the amount of protein varies widely. Then, O. T. Avery, C. M. MacLeod, and M. McCarty noticed that heredity characteristics could be transmitted from one strain of bacteria to another by using only the donor's DNA—but not with its proteins alone. This was supported by the work of A. D. Hershey, who showed that when a virus takes control of the cell of another organism, only its DNA core enters the host; the protein coat of the virus remains outside as a hollow shell.

In 1952 a brief paper by an American, James D. Watson, and his British partners, Francis Crick and Maurice Wilkins, described the structure of DNA and showed as well how this structure could account for the known property of genes. This discovery was one of the most important in all biology. Its implications for the study of human evolution are explored further in Chapter 3.

OTHER ORGANISMS AS MODELS FOR THE STUDY OF HUMANS

By now the genetic material of a great variety of organisms, from bacteria to humans, has been studied. In all of them the genes are composed of nucleic acids. Likewise, the body of generalizations about the inheritance and action of genes, which makes up the genetic theory of evolution, applies to all organisms which have been studied so far. Because of the existence of this powerful general theory, there is no need for any separate theory of human evolution.

SUMMARY

1. Speculations about evolution are as old as recorded history; the chief advances in recent centuries have come through the subjecting of the numerous alternatives to controlled observation and experiment.

2. Systematic study of the characteristics of animals and plants was given a solid foundation by the work of Linnaeus and other taxonomists; their arrangement of organisms by degree of resemblance seems to have stimulated naturalists such as Buffon and Lamarck to provide an evolutionary explanation for the similarities and differences.

3. The idea of evolution was opposed vigorously on scientific grounds by many scholars, among whom Cuvier, the founder of paleontology,

was especially prominent; Cuvier's belief that the earth was only a few thousand years old led him to reject the idea of gradual evolutionary change as an explanation for the past and present diversity of life.

4. An enormously greater age for the earth was demonstrated by the geologists Hutton and Lyell; Darwin realized that over long periods of time gradual changes in organisms could accomplish major transformations, and he proposed a mechanism—natural selection—to explain how these changes could occur.

5. Gregor Mendel demonstrated that the biological characteristics transmitted from parents to their offspring must be due to the existence of material particles, now called *genes;* cytologists later showed that these particles are part of the chromosomes that can be seen as they are transmitted from one generation to the next.

6. The study of population genetics has built upon Darwin's theory of evolution by natural selection by describing several other factors that can bring about evolutionary changes.

7. Molecular genetics has provided detailed information on the chemical structure of genes; this new knowledge makes it possible to understand how genes control the expression of the biological characteristics of organisms.

8. The modern theory of evolution, built up over the course of several centuries by many scientists studying the most diverse organisms, provides the scientific basis for understanding our own characteristics as well as those of our ancestors.

SUGGESTIONS FOR ADDITIONAL READING

Berry, William B. N. 1968. *Growth of a Prehistoric Time Scale.* San Francisco: Freeman. (This book gives a sketchy but useful account of the slow scientific battle that eventually proved that the earth has a great antiquity; it also provides a decent introduction to some of the terminology used in modern geology.)

Irvine, William. 1955. *Apes, Angels, and Victorians.* New York: McGraw-Hill. (This book remains one of the liveliest accounts of the life and times of Charles Darwin, his predecessors, and his associates; especially worthwhile is the section on Thomas Henry Huxley, chief defender of Darwin's theory.)

Provine, William B. 1971. *The Origins of Theoretical Population Genetics.* Chicago: University of Chicago Press. (This brief volume elaborates an important area of background that was only touched on in the preceding chapter, and it gives a particularly good discussion of the conflict between two schools of thought: those who believed that evolution proceeded by the gradual accumulation of small variations and those who thought that changes occurred by discontinuous leaps. This book makes a potentially complex area accessible and interesting.)

Stent, Gunther S. 1972. "Prematurity and Uniqueness in Scientific Discovery," *Scientific American,* **227**(6):84–93. (This brief article gives an extremely useful perspective on the problem of why some important scientific discoveries—such as those of Mendel and Garrod—are overlooked for long periods.)

Stern, Curt, and Eva R. Sherwood. 1966. *The Origins of Genetics.* San Francisco: Freeman. (This volume presents a brief discussion of the significance of Mendel's work, followed by a series of translations of the original scientific papers by Mendel, some of his contemporaries, and his rediscoverers. Of special interest are two papers, one by Ronald A. Fisher and the other by Sewall Wright—both among the founders of population genetics—on the question of whether Mendel's results are legitimate or too close to theoretical predictions to be believed.)

REFERENCES

Buffon, Georges Louis Leclerc de. 1749. *Histoire naturelle.* Paris: de l'Imprimerie Royale.

Clodd, Edward. 1907. *Pioneers of Evolution.* New York: Cassell.

Darwin, Charles. 1896. *Journal of Researches into the Natural History and Geology of the Countries Visited during the Voyage of H.M.S. "Beagle."* New York: Appleton.

Irvine, William. 1955. *Apes, Angels, and Victorians.* New York: McGraw-Hill.

Lyell, Charles. 1830–1833. *Principles of Geology.* London: Murray.

Malthus, Thomas. 1798. *Essay on Population.* London: Johnson.

Nordenskiöld, Erik. 1927. *The History of Biology.* New York: Knopf.

Stent, Gunther B. 1971. *Molecular Genetics.* San Francisco: Freeman.

Watson, James D., and Francis H. C. Crick. 1953. "A Structure for Deoxyribosenucleic Acid." *Nature,* **171**:737

Wendt, Herbert. 1976. "Introduction to Paleontology," in H. C. B. Grzimek (ed.), *Encyclopedia of Evolution.* New York: Van Nostrand Reinhold. Pp. 45–58.

2

The Process of Evolution

Chapter 3

The Genetic Basis of Evolution

Humans are a diverse lot of animals, differing in appearance from group to group and person to person. Some of the features in which we differ are influenced greatly by cultural practices. For example, hair can be dyed, curled, cut, or even shaved off completely. Other traits, such as the chemical structure of the red blood cells that circulate through our veins and arteries, are under much more direct biological control. This chapter will begin by sorting out the cultural from the biological influences on our characteristics. After that, attention will be focused on how biological characteristics are transmitted each generation from parents to their children, in the form of cells and the genes that they contain. Last, we will explore some of the pathways from microscopic genes to the visible characteristics of adult humans. The diversity of genetic influence and expression discussed in this chapter represents only a small sampling of all the variation that is already known to exist in human populations. However, the traits discussed will provide an idea of the extent and types of variation on which the forces of evolution can act.

TYPES OF INHERITANCE

In everyday language people use the term *inheritance* to refer to all transfers from one generation to the next. For many purposes this usage meets our needs quite well. Parents may feel that they are

47

responsible for having produced the characteristics of their children without devoting much thought to just how this was done. But to gain a useful knowledge of how human evolution occurs, it is necessary to refine such basic ideas so that the terms used reflect accurately the concepts they are intended to convey.

There are really two rather distinctive types of inheritance: genetic and cultural. In *cultural inheritance,* skills, knowledge, and other non-material attributes can be transmitted from one person or group to another person or group. However, like material possessions (land, money, etc.), the attributes that help determine social status are extra-somatic—outside the body and not part of it. In *genetic inheritance,* only physical units—the gene-containing reproductive cells produced by the parents—are transmitted. There are also other differences between cultural and genetic inheritance. In genetic inheritance, trans-mission is in one direction only, from parents to their offspring. By contrast, in cultural inheritance, transfer can occur from one person to another regardless of their degree of biological relationship. Cultural transmission, moreover, can take place in any direction—even from children to their parents, or by the printed word or recorded sounds, or from someone dead for generations to one now living. Of course, not all biological characteristics are influenced by genes alone. Variations in birth weight from child to child are conditioned more by the different surroundings provided by their mothers' womb than by the children's own genes (Penrose, 1954). However, these maternal influences are just specific types of environmental effects, reflecting levels of nutrition, medical care, and other culturally conditioned factors. Typically they have only a temporary impact on the offspring's body size. Genetically determined variations, on the other hand, often persist for many generations.

Both mechanisms of inheritance play important roles in human evolution. But the general mechanisms of cultural inheritance are reasonably familiar to most people, and their details more properly form the subject matter of other disciplines, from cultural anthropology to law. Because of this, and because in the course of evolution genetic inheritance preceded and still underlies social inheritance, the account here will focus on the physical, or genetic, basis of evolution.

CELLS: UNITS OF LIFE

A cell is the smallest unit of life that can live independently, regulate its own activities, and reproduce complete copies of itself. Some of the simplest forms of life consist of no more than a single cell. Large animals like humans are made up of complexly organized aggregates of millions upon millions of cells. These repeating cellular units, of which all plants and animals are composed, are often said to be the *building blocks* of life. Our understanding of the cellular basis of life is distorted, however, if we think of cells as if they were bricks or boxes. Even the simplest single cells are more like cottages containing families of skilled craft workers and their tools. Just like any functional unit, cells have an internal organization that governs their maintenance and production. In their activities, cottagers take in food and raw materials and turn out finished products. So do cells. The humans who control these craft activities can produce offspring that share their characteristics and abilities, and after a time this new generation of workers can depart, building a new cottage to house themselves. Cells can produce duplicate sets of their genetic control units and then, through the process of division, bud off these key regulatory units into new quarters of their own, giving rise to two cells where just one had been before.

Large, multicellular organisms such as humans are more like giant office buildings or factory complexes than independent cottage industries. The departments, like cells, are attached to one another. In both, there is more communication between units, and there is less independence for each single unit. Though it retains the capability for a diverse range of activities, the gene complex in any single cell of a large organism, like the staff of a department, usually restricts itself to a narrow range of functions. Specialization extends even to reproduction, the biological counterpart of business expansion. In business, the task of expansion is not shared by all units; it is done by only a select few. And the new entity typically incorporates new elements from the outside as well as some from units within. In humans, a new person begins when a sperm cell from the male parent joins with an egg cell from the female parent. These two cells are the only physical bridge from one generation to the next; they carry the genetic instructions needed to produce a new person. To understand how this is possible, we must know more about how cells function.

Cell Structure

Although there are thousands of different kinds of cells, many of them share important features. Figure 3-1 shows an idealized picture of the structures found in the cells of many organisms. The whole cell is enclosed within a *cell membrane,* and the interior is further divided into a number of subunits of various sizes. The most important of these for our purposes is the *nucleus*, which is enclosed by a *nuclear membrane.* The space between the outer cell membrane and the inner nuclear membrane is filled by *cytoplasm.* This is a complex, thick fluid containing many different structures. Those which are of direct interest in the study of heredity are discussed here; some others that play vital supportive roles are shown in Figure 3-1.

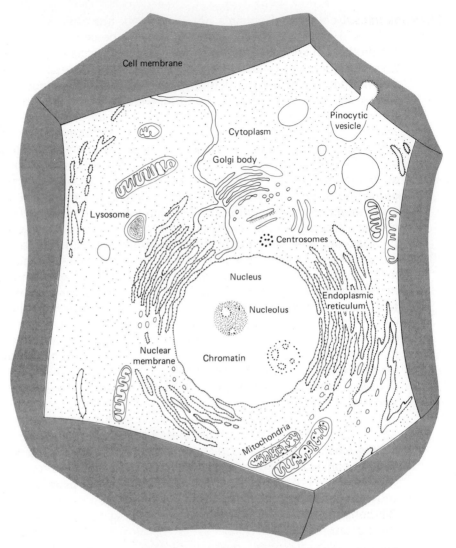

FIGURE 3-1 Composite diagram of an animal cell. Cells in various parts of the body differ greatly in shape; the form shown here would be assumed in uniform aggregates of cells not subject to external deforming forces.
  Source: Adapted from Jean Brachet. 1961. "The Living Cell," *Scientific American*, **205**:50–62 (September).

The cell membrane and nuclear membrane regulate access to the areas they enclose. Protected by its outer membrane, the whole cell carries on an active exchange of materials and energy with its surrounding environment. Various substances also pass in both directions between nucleus and cytoplasm. Both regions are sites of activities necessary for the life of the cell. The genetic material, however, is concentrated in the nucleus.

Chromosomes

Inside the nucleus are the *chromosomes.* A chromosome consists of a linear sequence of thousands of genes, each one of which is an independent unit of biological inheritance. The chemical structure of genes will be explained later in this chapter. Most of the time, chromosomes exist as a tangle of long, thin filaments that are virtually invisible under an ordinary microscope. During cell division, however, the filaments coil into tight spirals. Then the chromosomes appear as short, thickened rods of various lengths. Descriptions of chromosome structure are based on their appearance in this coiled state. Somewhere along the length of the chromosome is the *centromere,* the point of attachment for fibers that control chromosome movement when a cell divides. When many different chromosomes are examined under the microscope, they are seen to differ not only in overall size but also in the position of the centromere. Centromere location and relative arm length allow us to subdivide the whole set of 46 human chromosomes into seven groups. In the last few years, scientists have developed new techniques for revealing the structures of chromosomes in fine detail. Chemical stains, for example, show the existence of many different bands along the length of chromosomes. Being able to determine the number, thickness, and spacing of these bands makes it possible to identify each individual chromosome.

The number of chromosomes in each nucleus is constant for each species. It ranges from as few as two (in *Ascaris megalocephala,* a roundworm that lives as a parasite in the intestine of horses) to as many as several hundred in some invertebrates and plants (one fern, *Ophioglossum petiolatum,* has 1020 chromosomes). Table 3-1 gives the normal chromosome numbers in several species. But the chromosomes in a given cell are not all different. Instead, they occur in pairs, the members of which are alike. Thus, the 46 human chromosomes shown in Figure 3-2 consists of 23 pairs. These chromosomes come from both parents; the 23 chromosomes passed from the father to the child are called *paternal chromosomes;* those contributed by the mother are called

TABLE 3-1   Chromosome numbers in the nuclei of cells of some genetically important organisms

Common name	Scientific name	Chromosome number
Colon bacteria	*Escherichia coli*	1 (haploid)
Bread mold	*Neurospora crassa*	7 (haploid)
Fruit fly	*Drosophila melanogaster*	8
Garden pea	*Pisum sativum*	14
Corn	*Zea mays*	20
House mouse	*Mus musculus*	40
Human being	*Homo sapiens*	46
Chicken	*Gallus domesticus*	78

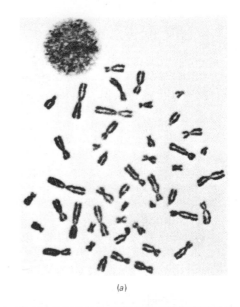

(a)

FIGURE 3-2 The human chromosome complement. (a) Chromosome set found in somatic cell of a human male. The cell has been stopped in the middle of division and treated chemically so that the chromosomes will spread apart for easier observation. (b) Chromosomes shown in *a* have been cut apart, separated, and systematically arranged as a *karyotype,* in which individual chromosomes are paired by size and shape, lined up from longest to shortest, and assigned numbers in sequence. (c) Karyotype of a normal human female. (d) Using chemicals which differentially stain segments of the chromosome makes it possible to distinguish chromosomes with more certainty. The chemicals produce unique banding patterns; here, a schematic representation of results obtained from several banding techniques is shown.

Sources: (a, b, c) A. Redding and Kurt Hirschhorn. 1968. "Guide to Human Chromosome Defects," in D. Bergsma (ed.), *Birth Defects: Original Article Series,* **IV**(4):2, 3. (d) "Paris Conference (1971): Standardization in Human Cytogenetics," in D. Bergsma (ed.), *Birth Defects: Original Article Series,* **VIII**(7):18, 19 (1972).

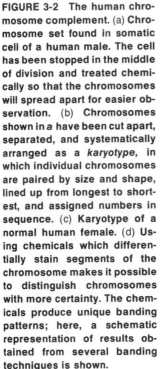

(b)

(c)

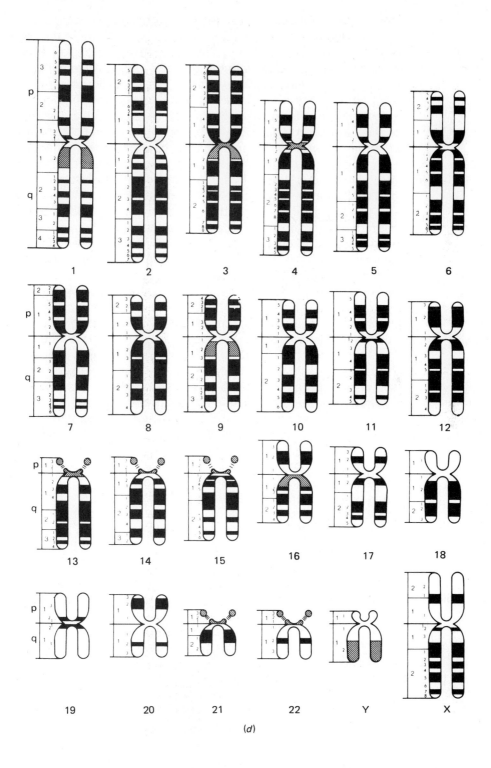

(d)

maternal chromosomes. (These terms refer only to the immediate source of the chromosomes and not to their role in determining the sex of the individual who receives them.) The members of each pair of chromosomes, consisting of one maternal unit and one paternal unit, are called *homologous* (or *like*) *chromosomes.* They are typically identical in overall appearance and in the biological processes they control, and they are different from the other pairs. These identically paired chromosomes are called *autosomes.* The members of one other pair, called *sex chromosomes,* differ in appearance from each other. The sex chromosomes consist of an X (which is long and has a centrally located centromere) and a Y (which is short and has its centromere near one end). In addition to being different from each other in appearance, the X and Y chromosomes pair end to end during the round of cell division that precedes the union of egg and sperm. Other chromosome pairs line up adjacent to each other along their entire length at that time.

Cells in the body of a human female have 22 pairs of autosomes plus two X chromosomes. Body cells of human males have 22 pairs of autosomes plus one X and one Y chromosome. The total in either sex is the same, 46. This is called the *diploid* (2N) chromosome number. The only cells that are not diploid are the egg cells (and their precursors) produced by the female and the sperm cells produced by the male. The term *gamete,* used for both sperm and eggs, highlights their difference from the other cells in the body, which are called *somatic* cells. Each sperm or egg has exactly half as many chromosomes as a somatic cell (23 in humans), and so it is said to be *haploid* (N).

All normal eggs produced by the ovaries of a human female will contain 22 autosomes and one X. Each sperm has 22 autosomes, and on the average half also contain an X chromosome. The remaining sperm carry 22 autosomes plus a Y chromosome. An egg fertilized by an X-bearing sperm gives rise to a female; an egg fertilized by a Y-bearing sperm normally develops into a male (see Figure 3-3). In humans, the chromosomal content of the sperm therefore plays the primary role in determining an individual's sex.

The importance of chromosomes in the study of evolution is discussed in "What Chromosomes Tell Us about Human Evolution," pages 56–57.

Cell Division

In the course of the human life cycle, two distinct types of cell division occur: mitosis and meiosis. *Meiosis* takes place in the reproductive cells of the individuals of one generation and produces the haploid gametes. *Mitosis* takes place in the original fertilized egg (the *zygote*) itself and multiplies the single cell into a multicellular organism. In the resulting adult, the reproductive cells alone remain capable of meiosis—of producing the gametes that can begin the life cycle anew.

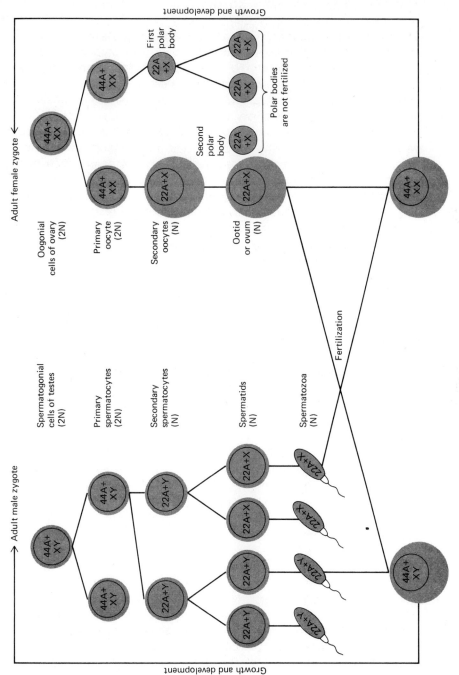

FIGURE 3-3 Chromosomal sex determination in humans.

WHAT CHROMOSOMES TELL US ABOUT HUMAN EVOLUTION

Why study chromosomes? One answer is that knowing the structure of the hereditary apparatus allows us to understand how genes are transmitted from generation to generation. A second answer is more exciting: comparison of the chromosomes of humans and those of other primates can yield direct information about which of these animals are our closest living relatives.

Even before chromosome studies began, many scientists believed, on the basis of comparative anatomy, that our closest relatives were the great apes, a group that includes the orangutans native to Borneo and Sumatra and the gorillas and chimpanzees of Africa. In the early 1960s, even the relatively crude staining techniques then available confirmed that, on the basis of chromosome lengths and centromere locations, there are extensive similarities, despite the fact that we have 46 chromosomes and the great apes have 48.

In 1970, the Swedish biologist Caspersson discovered that a fluorescent dye, quinacrine, produces a series of bright and dull bands along the length of metaphase chromosomes. This and a series of related banding techniques make it possible to show that every chromosome of any species is unique. The techniques are now being used to study human and ape chromosomes in detail. Dorothy Miller (1977) has reported some of the results obtained recently. As might be expected, exactly corresponding banding patterns of human and ape chromosomes are rare, with the X chromosome showing the best match. In contrast, the Y chromosome shows the greatest variation, having the most complex banding pattern in gorillas and the least complex banding

Mitosis and meiosis both take place in the life cycle of all sexually reproducing organisms. The need for both types of division can be seen by first imagining what impossible situations would quickly arise if only one of the two processes operated. Multiplication (mitosis) alone would cause no problems from fertilization through maturity. But if the organism then produced diploid gametes as well as diploid somatic cells, the next generation would begin with fertilization of a diploid egg by a diploid sperm. The second-generation zygote would contain four times the haploid number, or double the diploid number, of chromosomes of the preceding generation. And this doubling would continue in each succeeding generation. Soon the delicate mechanism that regulates cell division would certainly be upset. If it were not, then in only a few generations of repeated doubling the chromosome number would become so unmanageably large that the nucleus would be unable even to hold all the chromosomes. On the other hand, if meiosis

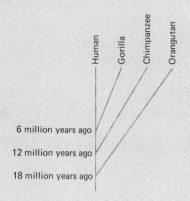

Order of branching of the ancestors of apes and humans, based on evidence from chromosome-banding studies.
Source: Modified from Dorothy A. Miller. 1977. "Evolution of Primate Chromosomes," *Science*, **198**:1116–1124. Fig. 2.

pattern in orangutans and chimpanzees, with humans falling in between. Only the gorilla and human Y chromosomes have a band on the Y chromosome. Another feature shared only by gorillas and humans is the presence on chromosomes 9 and 16 of large blocks of material stained by C banding, another of the new techniques. These and other similarities led Miller to conclude that, among the great apes, the gorilla may be more closely related to humans than the chimpanzee, while the orangutan is the most distantly related to the other great apes and to humans.

Source: Dorothy A. Miller. 1977. "Evolution of Primate Chromosomes," *Science*, **198**(4322): 1116–1124.

occurred without mitosis, the chromosome number would rapidly decline toward zero, and all the genetic material would be lost.

The haploid gametes are therefore a way to maintain a stable chromosome number while providing each individual with genetic material from two different parents. The mixture of genes generates a high probability of new genetic combinations in each generation. This in turn increases the likelihood that some members of the population will be able to survive and reproduce in diverse or changing environments. Chapter 6 gives several examples of the evolutionary flexibility made possible by genetic variation.

Mitosis   Mitosis is the basic process of hereditary transmission and is the simpler of the two types of nuclear division. It is conventionally divided into several distinct stages: interphase, prophase, metaphase, anaphase, and telophase (Figure 3-4). In a preparation of dividing cells

INTERPHASE

1. Chromosomes duplicating but not individually recognizable; nucleus appears to be filled with diffuse chromatin.
2. Single centrosome is visible outside of nuclear membrane.

PROPHASE

1. Chromosomes coil up, becoming visible.
2. Sister chromatids are attached at undivided centromere.
3. Centrosome divides in two.

METAPHASE

1. Nuclear membrane disappears.
2. Spindle appears.
3. Sister chromatids align in equatorial plane of spindle; centromeres attach to spindle fibers.

ANAPHASE

1. Centromeres divide.
2. Sister chromatids move toward opposite poles.
3. Furrow forms in cytoplasm.

TELOPHASE

1. Complete diploid chromosome set reaches each pole.
2. Chromosomes begin to uncoil.
3. Nuclear membrane reappears.
4. Two separate and identical daughter cells exist.

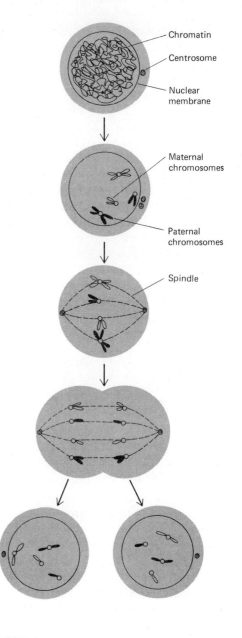

FIGURE 3-4   Mitosis.

seen under a microscope, examples of each of these stages can be found, but so can transitions between them.

Interphase is sometimes described as the resting stage of the cell; actually, during this period the cell is carrying on complex activities, including duplication of all the genes and the chromosomes on which they are carried. While these activities go on, the chromosomes exist as elongated filaments too fine to be individually distinguished except with an electron microscope, so that the nucleus appears to be packed with a mass of tiny threads. *Prophase* begins with the coiling of these filaments; as the coils become progressively shorter, thicker, and denser, the chromosomes become visible. When clearly visible, each chromosome appears to be split lengthwise into two strands that share a single centromere. These strands, called *sister chromatids* (or *bivalents*), result from the duplication during interphase. *Metaphase* is marked by the disappearance of the nuclear membrane and the development of the *spindle.* Spindle fibers radiating from two opposite poles attach to the centromere of each pair of sister chromatids. These pairs align end to end in a ring around the equator of the spindle. During *anaphase,* the centromere of each pair of sister chromatids divides, and the divided centromeres, trailing their attached chromosomal arms, separate and move toward opposite ends of the dividing cell. *Telophase* reverses prophase. After a complete diploid set of chromosomes has accumulated at each pole, the individual chromosomes uncoil again into filaments, and a nuclear membrane reappears to surround each chromosome set. The end of telophase merges into the onset of a new interphase. Two daughter cells now exist, each a precise copy of the single cell that began dividing. The total time required for the whole sequence differs from one type of cell to another, varying from a few minutes to several hours.

Meiosis. Meiosis is the process of cell division in which the chromosome set duplicates once, while the nucleus divides twice. The major ways in which meiosis differs from mitosis are outlined in Figure 3-5. As a result of meiosis, four separate cells are produced, each with a nucleus containing a haploid set of chromosomes.

Results of Cell Division. In discussing mitosis and meiosis, we have been describing the movements of chromosomes in their cellular settings. If we look now at the results of these processes, we can begin to see the consequences of their differences. After mitosis, each daughter cell has the same diploid chromosome set as the other daughter cell and as the original parent cell. Thus, all an individual's somatic cells have the same chromosome makeup (with a few exceptions such as the red blood cells, which lack a nucleus altogether). In contrast, after the meiotic sequence shown in Figure 3-5, the gametes differ from the original reproductive cell: they are haploid rather than diploid. Furthermore, some of the gametes differ from one another; half have the paternal chromosome contribution, and half have the maternal chromosome contribution. Beyond this, a feature of meiosis *not* shown in Figure 3-5 is also very important. At telophase II, the distribution of gametes is not the only one that could have resulted. There are

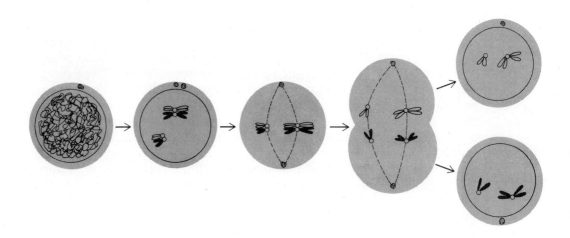

INTERPHASE I	PROPHASE I	METAPHASE I	ANAPHASE I	TELOPHASE I
1. Chromosomes do *not* duplicate; nucleus appears to be filled with diffuse chromatin. 2. Single centrosome is visible outside of nuclear membrane.	1. Homologous chromosomes synapse. 2. Chromosome strands duplicate. 3. Chromosomes coil, making tetrads visible. 4. Centrosome divides in two.	1. Spindle forms. 2. One spindle fiber attaches to centromere of each dyad. 3. Nuclear membrane disappears.	1. Homologous bivalents move toward opposite poles. 2. Cytoplasm begins to furrow.	1. Bivalents reach spindle poles. 2. Nuclear membrane re-forms. 3. Cytokinesis produces two diploid cells.

FIGURE 3-5   Meiosis.

alternative possibilities, even for the same reproductive cell shown at prophase I in Figure 3-5.

The event that determined the outcome shown was the way the tetrads became oriented on the spindle in metaphase I. In the case shown, both maternal chromosomes were nearer to the upper spindle pole, and both paternal chromosomes were nearer to the lower one. The first meiotic division separated the entire maternal set from the entire paternal set.

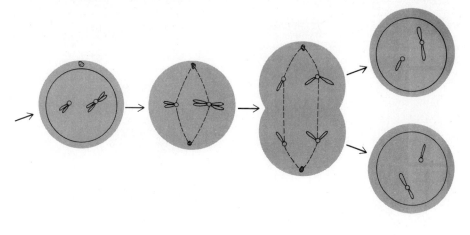

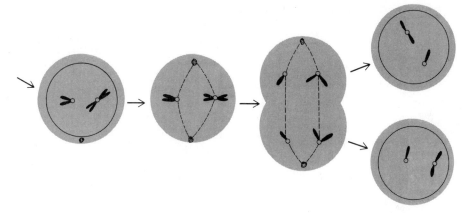

INTERPHASE II –
PROPHASE II

1. Bivalents stay
   compactly coiled.
2. Sister chromatids
   remain connected by
   a single functional
   centromere.
3. Chromosomes do
   not duplicate again.

METAPHASE II

1. Bivalents attach to
   spindle.
2. Centromeres divide.
3. Nuclear membrane
   disappears.

ANAPHASE II

1. One chromatid
   passes to each pole.
2. Cytoplasm begins to
   furrow.

TELOPHASE II

1. Nuclear membrane
   re-forms.
2. Cytokinesis is
   completed.
3. Four haploid
   gametes result.

The second meiotic division merely halved each of these sets. Figure
3-6, an abbreviated form of Figure 3-5, shows a different outcome at
telophase I and telophase II resulting from an alternative arrangement
of chromosomes on the spindle apparatus at metaphase I. From a single
reproductive cell that has just two pairs of chromosomes, it is possible
to produce four different combinations of maternal and paternal chromosomes among the gametes. A human reproductive cell has not just

METAPHASE I
Chromosome alignment
shown in Figure 3-5

TELOPHASE I
Result of alignment
of chromosomes at
Metaphase I

TELOPHASE II
Gametic Chromosome sets
shown in Figure 3-5
(only one gamete of each
type is illustrated)

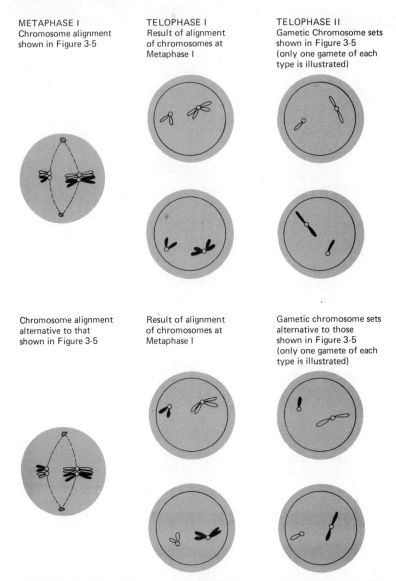

Chromosome alignment
alternative to that
shown in Figure 3-5

Result of alignment
of chromosomes at
Metaphase I

Gametic chromosome sets
alternative to those
shown in Figure 3-5
(only one gamete of each
type is illustrated)

FIGURE 3-6   Alternative gametic combinations produced in meiosis. (Maternal chromosomes are shown in outline, paternal chromosomes in black.)

2 but 23 pairs of chromosomes. Thus, any individual human can produce a large number of different chromosomal combinations among his or her gametes. This diversity is increased by other processes discussed later in this chapter. After these have been presented, it will be possible to get a better idea of the potential for genetic diversity in our species and to consider the significance of this diversity for human evolution.

FROM CELL TO MOLECULE: GENE STRUCTURE AND GENE ACTION

The cell provides a vantage point from which it is possible for us to glance in two directions: up the scale of size from a single fertilized egg to an adult human and down below the reaches of even the most powerful microscope to the molecular level. Since the development of most human characteristics is influenced by events at the molecular level, let us look within these miniature chemical factories before turning back to the people who are the focus of our interest.

Chemical Composition of Chromosomes

The chromosomes provide a good starting point for a detailed examination of how cell components shape appearance, for they are the carriers of genetic instructions from one generation to the next. Since the numbers and gross structures of these minute bodies are the same for all humans, the origin of some human differences must be due to variations too small to be visible under even the most powerful microscope. Knowledge of the chemical composition of chromosomes is needed to understand how genetic transmission controls development. Figure 3-7 shows a human chromosome examined at increasingly higher levels of magnification. It is a complex structure composed chiefly of three kinds of chemicals: *proteins, deoxyribonucleic acid* (DNA), and *ribonucleic acid* (RNA).

Proteins are constructed from units called *amino acids.* Twenty of these are very common; others are found less frequently. All amino acids share a common chemical core but are distinguishable by their unique side groups. Any two amino acids can be linked together by the formation of a *peptide bond,* as shown in Figure 3-8. The resulting molecule would be called a *peptide. Polypeptides* consist of more than 10 amino acids; they are called *proteins* when they contain over 100.

The linear sequence of amino acids determines a protein's *primary structure.* The peptide bonds cause the amino acids in the sequence to assume a screwlike twist in three-dimensional space. This helix, or spiral, is called the protein's *secondary structure.* If no other forces were at work, proteins would look like chains wound loosely into spirals around invisible cylinders. But the various amino acids have side groups of differing sizes and electrical charges. These cause the chain to fold, kink, and coil still further. The result of this additional folding is the protein's *tertiary structure.* In addition, protein molecules may contain more than one chain. For example, red blood cells are filled with a protein, hemoglobin, which is made up of four chains. Such associations of two or more chains make up a protein's *quaternary structure.* The primary, secondary, tertiary, and quaternary structure of collagen, a protein that is a major ingredient of skin, bone, and other connective

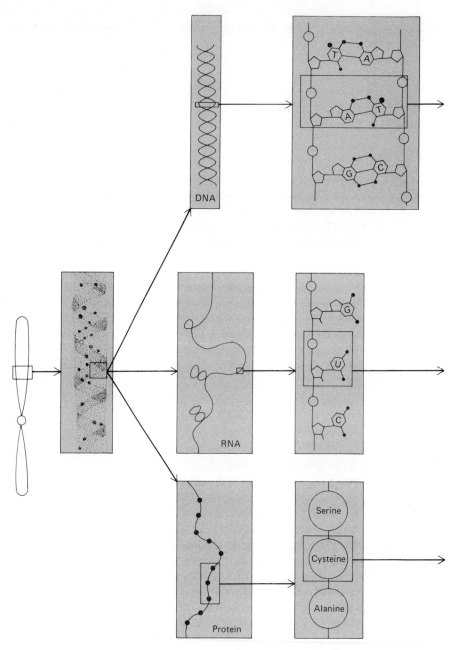

FIGURE 3-7 Chromosome composition. A section of a single human chromosome is subjected here to successively finer levels of resolution. The levels shown reflect observations using a wide variety of physical and chemical techniques, and inferences from data gathered. The precise structural relationships of the components within the chromosome (DNA, RNA, and protein) is uncertain and still under investigation.
Source for first enlargement of chromosome arm segment: After M. W. Strickberger. 1968. *Genetics.* New York: Macmillan.

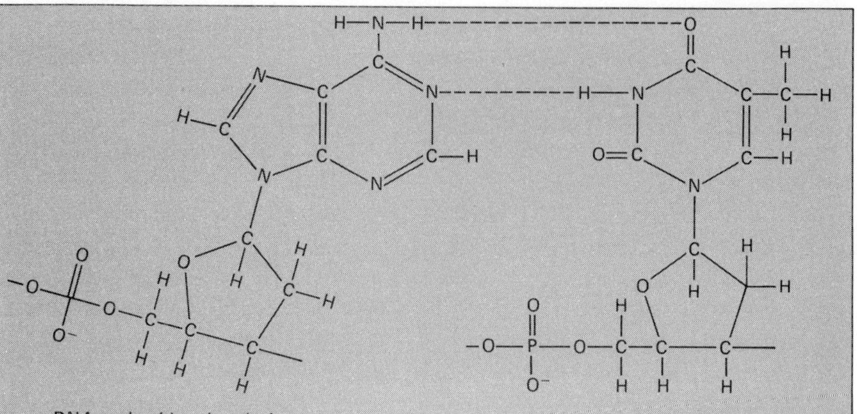

DNA nucleotide pair: adenine   Thymine

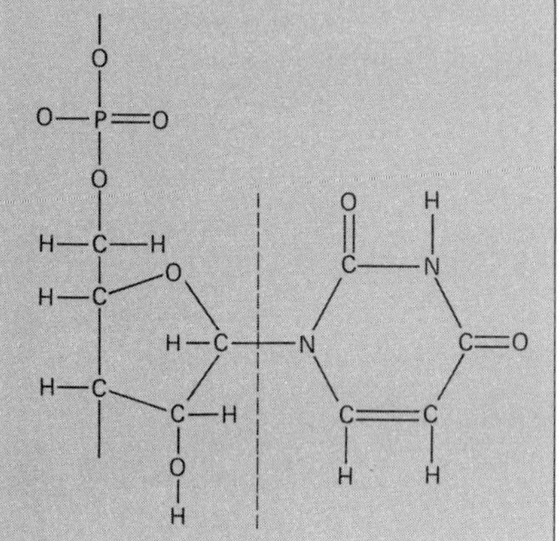

RNA nucleotide: uracil

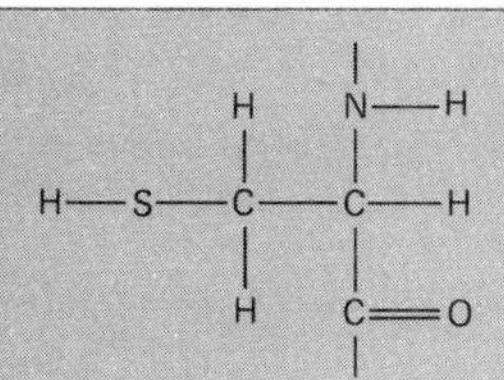

Amino acid: cysteine

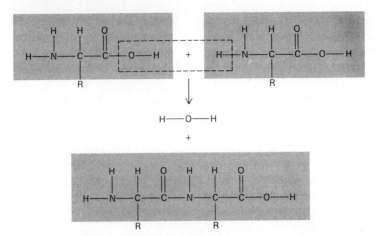

FIGURE 3-8 Amino acids and peptides. Two amino acids (indicated here by their common cores and R, which stands for any side group) can be joined by the formation of a peptide bond. When this occurs, parts of the two amino acid molecules are split off to form a molecule of water.

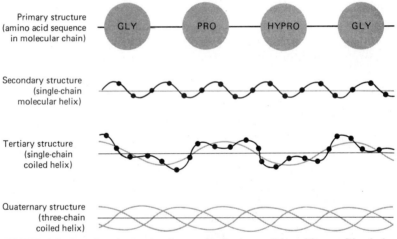

FIGURE 3-9 Levels of structural complexity in proteins. Lines with circles represent the molecular chain of amino acids; smooth lines represent hypothetical axes about which the actual molecular chains are coiled.

tissues, is shown in Figure 3-9. With such great potential for differences in size and shape at a number of structural levels, proteins are quite diverse and difficult to classify.

Perhaps the most basic functional distinction is between *structural proteins* and *regulatory proteins.* Structural proteins are an abundant but not very diverse group. Just one type, *collagen,* makes up about one-third of the substance of the human body. Another, *keratin,* is widely distributed in skin, hair, and nails. Regulatory proteins are much more diverse. Within any single living cell, there are hundreds or thousands of different kinds. Enzymes are regulatory proteins that act within the cells that produce them. As noted briefly in Chapter 2, each enzyme serves as a catalyst that governs the rate at which a specific biochemical process occurs. It is not itself used up in the process and may have an effect even when present in minute amounts. Some enzymes can catalyze the transformation of molecules of one substance into another over 100,000 times per second. Other important groups of proteins include the *antibodies,* which aid in disease resistance, and some *hormones.* Hormones are manufactured in the endocrine glands and are carried in the bloodstream to sites of activity elsewhere in the body. Their function is to regulate activities between, rather than within, cells.

DNA structure shows much less diversity than protein structure. It too forms very long molecules shaped like twisted (helical) ladders made up of alternating sugar and phosphate molecules. The rungs themselves are made up of pairs of chemical bases. Four of these bases occur commonly, and others more rarely. Only two common arrangements will make rungs with halves correctly joined together and of the proper width to fit easily between the side rails. Adenine (usually abbreviated A) must pair with thymine (T); guanine (G) must pair with cytosine (C) (see Figure 3-10). The fundamental unit of DNA is a bit more complex than that of proteins. Instead of a single amino acid, it is a *nucleotide,* the term used for one phosphate + sugar + base segment. DNA molecules from different chromosomes differ chiefly in the sequence and number of these units. *RNA structure* differs from DNA structure in a very few ways. Its sugar is different, but again four bases are commonly found in its molecules. Three of these (adenine, guanine, and cytosine) are identical to those in DNA, but uracil occurs in place of thymine. There is a fundamental structural difference as well: RNA typically occurs as single strands that are shorter than the coiled double helixes of DNA.

Duplication of DNA takes place during the interphase of cell division, as shown in Figure 3-11. Each daughter cell gets a DNA double helix consisting of one old and one new strand. This mechanism maximizes the likelihood of precise continuity of the set of genes from one generation to the next.

Genetic Code

DNA transmits the genetic instructions from one generation to the next. Nearly all DNA is found in the nucleus. Proteins, however, are synthesized in the cytoplasm. RNA is found in both major compartments of the cell. One form, *messenger RNA* (abbreviated mRNA), carries a copy of the nucleotide sequence encoded in DNA out to the cytoplasm. In the

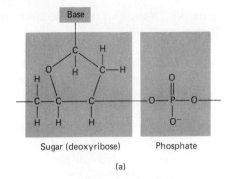

FIGURE 3-10   Components of DNA. (a) Segment of one side rail of the DNA molecular ladder. (b) Bases which form the rungs of the DNA molecular ladder.

cytoplasm the mRNA molecule attaches to a *ribosome,* a tiny particle that moves along the mRNA strand. Another, smaller class of RNA molecules, called *transfer RNA* (abbreviated tRNA), aids in the manufacture of proteins by physically transporting amino acids to their appropriate place in a peptide as it is being formed. The base sequences of the various transfer RNA molecules correspond to the particular amino acid that attaches to each. As the ribosome reaches each successive mRNA triplet, a tRNA molecule with a complementary triplet

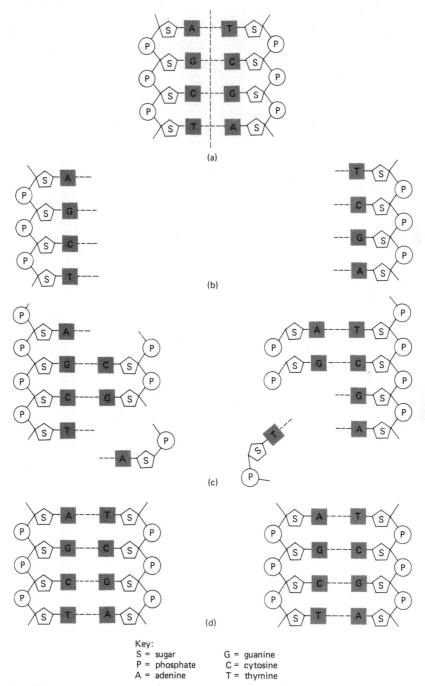

FIGURE 3-11 DNA replication mechanism. (a, b) Double strands separate, and (c) single strands are formed along the exposed single strands of DNA. (d) This produces two double strands, each with the same sequence as the original.

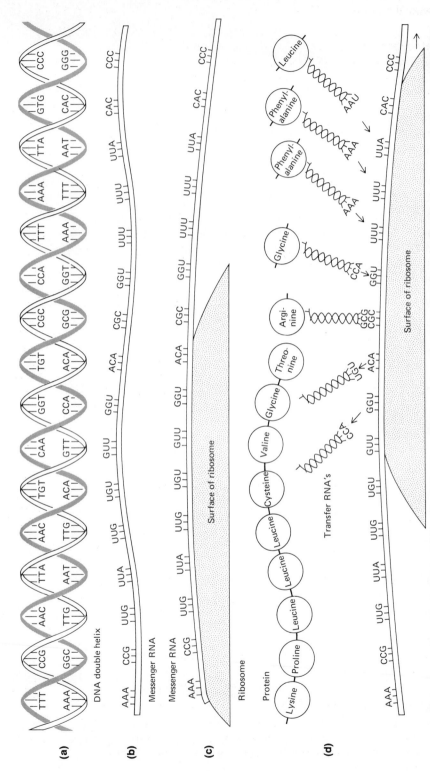

FIGURE 3-12 Protein synthesis. (a) Synthesis begins with the code embodied in DNA. (Although the bases are shown grouped in threes, they are actually spaced evenly along the strands). **(b)** Section of DNA code corresponding to one polypeptide is transcribed into messenger RNA. (The strand copied is shaded.) **(c)** Messenger RNA attaches to one or more ribosomes in the cytoplasm. **(d)** Each ribosome travels along the strand, reading the triplets from the messenger RNA in sequence. Amino acids (labeled circles) are carried to appropriate sites by molecules of transfer RNA. The instructions coded in messenger RNA are translated into an actual polypeptide chain. (The mechanism of recognition between tRNA and mRNA shown here is hypothetical.)

Sources: Modified from M. W. Nirenberg. 1963. "The Genetic Code, II," *Scientific American*, **208**(3):80–94 (March). Diagram on pp. 84, 85. Additional data from (1) F. H. C. Crick 1966. "The Genetic Code, III," *Scientific American*, **215**(4):55–62 (October). Chart on p. 57. (2) J. Hurwitz and J. J. Furth. 1962. "Messenger RNA," *Scientific American*, **206**(2):41–49 (February).

exposed at one point along its length pairs with it. The amino acid sequence in the protein thus reflects the nucleotide sequence in mRNA, which in turn matches that in the DNA.

Protein synthesis is a two-step process:

1. *Transcription.* The first step involves the construction of a particular mRNA molecule along the appropriate section—one gene—of a DNA strand. The length of the DNA segment copied is on the average about 600 nucleotides long, as is the complementary mRNA molecule. The protein coded for by this nucleotide sequence would therefore be 200 amino acids in length.

2. *Translation.* This is the reading of the mRNA triplets by the ribosomes and the lining up of the appropriate sequence of amino acids by molecules of tRNA to form the protein specified. These steps in protein synthesis are diagramed in Figure 3-12.

The relationship of DNA and proteins to evolution is discussed in "What DNA and Proteins Tell Us about Human Evolution," page 72.

GENES: UNITS OF INHERITANCE

From the standpoint of molecular biology, we can now describe what we mean by the word *gene.* Structurally, a gene is a segment of a chromosome containing a length of DNA that codes for a particular protein. This protein can itself serve some function (for example, collagen), or it can combine with another protein. There are about 20 common amino acids in proteins, but only four nucleotides occur with any frequency in DNA. How does the DNA of the cell specify correctly the more complex array of proteins made from the amino acids? An analogy helps us answer this question. Consider the cell's instructions to be in the form of sentences. Each sentence is made up of a meaningful sequence of words (amino acids) that are written with a simple four-letter alphabet (the bases A, C, G, and T). If we assume that the simplest possible system has been used by Nature, the question then becomes: How many letters are needed to spell each word, using combinations of only four letters? There can be only four single-letter words—too few to correspond to 20 amino acids. With two letters per word, there would be four possibilities for the first letter and four for the second; thus 4×4, or 16, different code words could be written—again, too few. With three letters per word, there would be $4 \times 4 \times 4$, or 64, unique words possible; this is more than enough to code for all known amino acids. Several experiments have shown conclusively that a sequence of three nucleotides (constituting a *triplet* or *codon*) in DNA does in fact specify the presence of one amino acid in a given protein. This triplet code is said to be *degenerate* or *redundant,* meaning that in a few cases

WHAT DNA AND PROTEINS TELL US ABOUT HUMAN EVOLUTION

Genes consist of linear sequences of DNA nucleotides, and proteins consist of corresponding linear sequences of amino acids. Study of nucleic acids helps us see how continuity from parents to offspring in genetic traits is maintained, while knowledge of protein structure is basic to understanding how genes convey their messages to influence the developing organism. Comparative studies of these giant molecules in different animals can also help us clarify our degrees of relationship to other species.

Similarities in the nucleic acid sequences of two species can be studied indirectly. The double-stranded molecules of both are separated into single strands and then mixed together to form hybrid sequences. The stability of these hybrid DNAs is then compared with the stability of strands of a single species that have been separated and allowed to rejoin. Such comparisons of chimpanzee and human nucleic acids show that only about 1.17 percent of human DNA differs from that of chimpanzees.

Humans and chimpanzees have identical amino acid sequences in at least five proteins, including three of those found in hemoglobin molecules. In two other proteins each over 140 amino acids long, the human molecules each differed by only one amino acid from the chimpanzee sequence. On the basis of these and a few more indirect studies of protein similarity, Mary-Claire King and A. C. Wilson estimated that for all chimpanzee and human proteins analyzed up to the time of their report (1975), the two species differed by only 19 out of 2633 amino acids. This is a difference of less than 1 percent. Of course, the few molecules studied so far may not correctly represent the overall state of genetic correspondence between chimpanzees and humans. However, should subsequent research support these results, it would seem that despite the striking anatomical and behavioral differences between humans and chimpanzees, at the genetic level we do not differ very much from these great apes. Similar research is still needed to clarify our degree of genetic affinity to gorillas and orangutans.

Source: Mary-Claire King, and A. C. Wilson. 1975. "Evolution at Two Levels in Humans and Chimpanzees," *Science*, **188(4322)**:107–116.

several different triplets can code for the same amino acid. A few other triplets serve as punctuation marks, signaling the beginning and end of each gene's protein message.

Genes are located on chromosomes. In humans and other organisms composed of diploid cells, chromosomes come in pairs. Therefore, genes also come in pairs. But this argument, while logical and correct, does not reflect the way in which genes were first discovered. Mendel, who did his breeding experiments before chromosomes had been seen, *hypothesized* that hereditary particles had to exist. Because his experiments are classic examples of first-rate scientific reasoning, they are summarized in an appendix at the end of this chapter.

The genes that affect a given biological characteristic always occur at a particular place on a given chromosome, which is called a *genetic locus*. Both members of a pair of homologous chromosomes have exactly the same linear sequences of loci along their lengths. Genes of slightly different chemical composition that occupy corresponding loci on homologous chromosomes are called *alleles*. A cell in which both members of a given pair of alleles are the same is said to be *homozygous*. Cells with two different genes at corresponding loci are *heterozygous*.

To simplify further discussion, a few other terms need to be defined. An organism's *genome* consists of all the DNA contained in the set of chromosomes. In the nucleus of each human cell there is roughly 1 m (3.3 ft) of DNA. The function of over 90 percent of this genetic material is still a mystery, despite the large number of enzymes and other proteins that are already known (Lewontin, 1974). The entire genome is present in every cell nucleus, but because cells become specialized for different functions during development (Figure 3-13), each cell expresses only some of its potential characteristics.

Phenotype refers to the characteristics of each individual's appearance, which is the result of gene products' interacting with one another under the influence of the environment. My red hair is one phenotypic characteristic, my brown eyes are another, and my medium stature of 183 cm (5 ft 8 in) is a third. Phenotypic appearance can change during a person's lifetime for various reasons, either environmental or genetic. Should I eat more or exercise less than I do now, my weight would increase to more than its present 68.2 kg (150 lb). Not all phenotypic characteristics are as readily seen as those just listed. My type B blood can be detected only by laboratory tests. Similar tests show that humans vary in hundreds of other biochemical traits, many of which are under more direct genetic control than most visible features.

The *genotype* is an individual's genetic constitution. This is fixed at the time of fertilization, when the father and mother each contribute one gene to every pair in the zygote's genome. Each gene can be represented by some arbitrarily chosen symbol, often a letter of the alphabet. For example, the genotype for red hair could be written as rr. Other examples of genotypes and their corresponding phenotypes are given in the appendix to this chapter. It is theoretically possible to write out any person's entire genotype this way, with one pair of letters representing the genes at each locus: AA BB CC. . . . There are so many

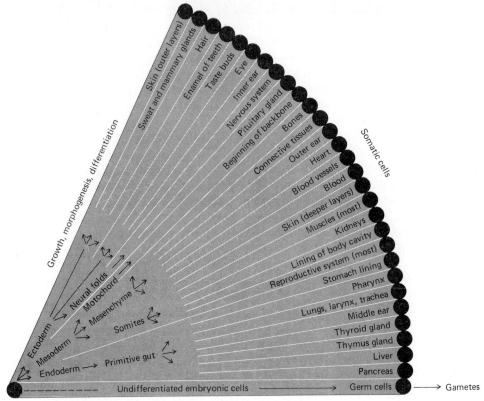

Skin (outer layers)
Sweat and mammary glands
Hair
Enamel of teeth
Taste buds
Eye
Inner ear
Nervous system
Pituitary gland
Beginning of backbone
Bones
Connective tissues
Outer ear
Heart
Blood vessels
Blood
Skin (deeper layers)
Muscles (most)
Kidneys
Lining of body cavity
Reproductive system (most)
Stomach lining
Pharynx
Lungs, larynx, trachea
Middle ear
Thyroid gland
Thymus gland
Liver
Pancreas

Growth, morphogenesis, differentiation

Somatic cells

Neural folds
Notochord
Mesenchyme
Somites
Ectoderm
Mesoderm
Endoderm → Primitive gut

Undifferentiated embryonic cells ────────→ Germ cells ──→ Gametes

Increase in number

FIGURE 3-13   Development: somatic complexity and germinal continuity. All the diverse somatic cells develop—through growth, morphogenesis, and differentiation—from the zygote's single undifferentiated cell. Germ cells originate from the same zygote but remain undifferentiated; they alone provide the gametes which transmit the inherited instructions that control development.

Note: Although this representation is of necessity two-dimensional, cells in an organism have a three-dimensional structure.

genes, however, that this would be unmanageable in practice. Consequently, geneticists usually consider genotypes at only one or two loci at a time.

Blood Groups: Dominance, Lack of Dominance, and Codominance

Of the many examples of Mendelian inheritance known in humans, a great proportion concern inherited differences in components of the blood. Blood is the only tissue routinely sampled for a wide variety of medical procedures as well as theoretical scientific studies. With modern equipment, anthropologists working all over the world can draw, preserve, and ship blood samples to laboratories for intensive study of genetic characteristics.

Blood is a complex fluid in which a variety of cells are suspended. If a sample of blood is placed in a tube and the tube is spun rapidly in a centrifuge, the blood will separate into cells and a yellowish liquid called *plasma*. When the components responsible for blood clotting are removed from the plasma, the result is *serum*, a clear, straw-colored liquid consisting of water, proteins, salts, and other components, as shown in Figure 3-14. Some inherited variants are found in the protein portion of the serum. Others affect the chemical structure of hemoglobin, the protein that colors the red blood cells and transports oxygen and carbon dioxide. Still others cause differences in the chemicals that coat the surface of the red blood cells. These chemicals are called *antigens*, and study of the loci that control their formation gives us much useful genetic information. Red blood cell antigens are part protein and part carbohydrate. Each cell has many antigen molecules on its surface, but not all antigens are alike. Their differences are controlled by a number of separate genetic loci, and different alleles at each locus cause further diversity. The term *antigen* actually refers to a wider category of large, complex molecules, most of them proteins, which have a common function. When introduced into the body of an animal, antigens stimulate the production of *antibodies*, small proteins in the blood plasma that normally protect the body by destroying foreign antigens such as

FIGURE 3-14   Composition of human blood.
Note: One cubic millimeter (1 mm³, about 0.00006 in³) is a drop of blood about the size of the dot on an *i*. Concentration of cells is expressed in cubic millimeters (rather than as percents) because of the great variation in amount and distribution of fluids in the body.

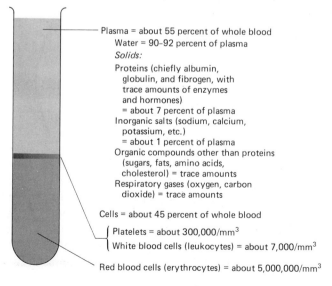

Plasma = about 55 percent of whole blood
  Water = 90–92 percent of plasma
  Solids:
  Proteins (chiefly albumin,
    globulin, and fibrogen, with
    trace amounts of enzymes
    and hormones)
    = about 7 percent of plasma
  Inorganic salts (sodium, calcium,
    potassium, etc.)
    = about 1 percent of plasma
  Organic compounds other than proteins
    (sugars, fats, amino acids,
    cholesterol) = trace amounts
  Respiratory gases (oxygen, carbon
    dioxide) = trace amounts

Cells = about 45 percent of whole blood

Platelets = about 300,000/mm³
White blood cells (leukocytes) = about 7,000/mm³

Red blood cells (erythrocytes) = about 5,000,000/mm³

compounds on the surfaces of viruses or bacteria. When an antigen and its corresponding antibody are brought together, usually some visible effect is produced. *Agglutination,* or the sticking together of red cells and antibodies in a blood sample, is one such effect; *lysis,* or rupture of the cells, is another. When one antibody is mixed with blood drawn from different people, some samples will show no reaction, while other samples will agglutinate or lyse. People who give the same reaction to an antibody share a common antigen and usually belong to the same *blood type.*

Rh Blood Group System.   The first components of this system were discovered in 1939, when Levine and Stetson noted that the blood serum of a woman who had just given birth contained antibodies to an antigen present on her baby's red blood cells. How these antibodies are produced, the effects they may have on the child, and the evolutionary consequences of these effects are discussed in Chapter 6. The name of the system comes from the first two letters of the common name of rhesus monkeys, which are widely used in medical research. When blood from these monkeys was injected into rabbits, the rabbits produced antibodies which reacted positively with cells from the monkeys. The same serum also reacted with blood cells of about 85 percent of a population of New Yorkers of European ancestry, who therefore had the Rh-positive phenotype. The remaining 15 percent of the population had red cells that gave no visible reaction with the antibodies; they had the Rh-negative phenotype.

Studies of offspring from numerous matings showed that people with the Rh-negative phenotype are homozygous for an allele that can be represented by the letter r. People who are Rh-positive, however, may be either RR homozygotes or Rr heterozygotes. Since in heterozygotes the R allele masks the expression of r, R is said to be a *dominant* gene, and r is said to be a *recessive* gene. Table 3-2 shows the proportions expected among the offspring of a mating between heterozygous parents. Further research by Fisher, Race, Wiener, and other geneticists has shown that the Rh system is considerably more complex than outlined here; these complexities will be discussed in a later chapter.

MN Blood Group System.   The MN group shows a different relationship between alleles: *lack of dominance.* There are two antigens in this system, M and N. They are detected in the same way as the Rh system: by injecting the human antigens into the bloodstreams of rabbits. If these antibodies are mixed with samples of blood from a number of people, different patterns of reaction will be seen. The cells in some samples will agglutinate when mixed with serum containing anti-M antibodies, but not when mixed with serum containing anti-N antibodies. The cells in others will do just the reverse: they will react positively with

TABLE 3-2   Proportions of offspring expected from mating of Rh heterozygotes

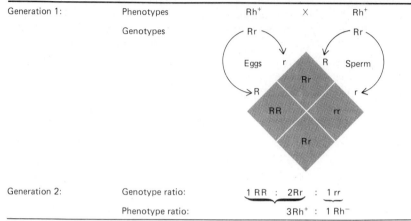

| Generation 1: | Phenotypes | Rh⁺ | X | Rh⁺ |

Generation 2:   Genotype ratio: 1 RR : 2Rr : 1 rr

Phenotype ratio: 3Rh⁺ : 1 Rh⁻

Note: The box shown here is called a *Punnett square* after B. C. Punnett, the English geneticist who introduced this convenient way of representing visually the expected proportions at fertilization.

anti-N but not with anti-M. Some bloods will be agglutinated by both anti-M and anti-N. Thus, these two antibodies can be used to recognize three different MN blood group phenotypes: type M, type N, and type MN.

The presence of each antigen is governed by a single allele. Type M people are genotypically MM; those who produce only the N antigen are NN; and people who have both are heterozygous, with the genotype MN. Since we can distinguish the heterozygote from both homozygotes, no dominance or recessiveness exists, and the alleles m and n are said to be *codominant:* neither gene masks the expression of its allele.

ABO Blood Group System: Multiple Allelism.   ABO blood types are the most familiar of all the human blood groups. This system differs from the MN and Rh in several ways. First, in the ABO system antibodies are not produced in response to foreign antigens that have been introduced into the circulatory system. The blood serum of every person contains one or more of these antibodies. Second, these naturally occurring ABO antibodies can be used to recognize four major red blood cell antigens, and so there are four phenotypes: A, B, AB, and O. These four phenotypes result from combinations of three different alleles: A, B, and O. Whenever a given genetic locus can be occupied by more than two alleles drawn from a larger set of alternative genes, the set is called a *multiple allelic series.* Any one individual can have only two of these alleles. The two genes may be different (genotype AB), or both may be the same (AA or BB). Populations (groups of interbreeding individuals), however, possess all three. Third, genes at the ABO locus interact to produce different patterns of expression. The alleles A and B are

TABLE 3-3   ABO blood group locus

Blood group genotype	Blood group phenotype	Phenotypes: antigens on red blood cells	Antibodies in serum	Reaction when mixed with anti-A	Reaction when mixed with anti-B
AA AO	A	A	Anti-B	Agglutination of cells	—
BB BO	B	B	Anti-A	—	Agglutination
AB	AB	A,B	—	Agglutination of cells	—
OO	O	*	Anti-A, Anti-B	—	—

* Type O individuals have no antigens which react with anti-A or anti-B antibodies. However, they do have a substance, H, which appears to be a precursor of the other antigens.

codominant, and O is recessive to both of them. Thus, as shown in Table 3-3, phenotypes A and B can each be produced by two different genotypes. Phenotypes O and AB, on the other hand, each result from a unique genotype.

Independent Assortment

During meiosis, the genes at different loci pass into gametes independently of each other. This principle was first demonstrated by Mendel's work with peas (see the appendix to this chapter) but can now be illustrated with human genes as well.

In recent years we have learned much more about where specific genetic loci are situated on particular human chromosomes (McKusick and Ruddle, 1977). Present evidence indicates that each of the three blood group loci (Rh, MN, and ABO) is located on a different chromosome pair. As shown in Figure 3-2, the Rh locus is on chromosome pair 1, and the ABO locus is on chromosome pair 9. The MN locus may be on chromosome pair 2. The genes at these three loci should pass into gametes independently. If they do, how many different gametes could result? The answer to this question can be found by looking at Table 3-4. Parents who are heterozygous at one locus can produce two different kinds of gametes. Those who are heterozygous at two loci can produce four different kinds of gametes. Table 3-4 also shows the next step in extending this sequence: eight different kinds of gametes can be produced by an individual who is heterozygous at each of the three blood group loci. A general relationship can be seen from these examples. Each addition of a locus with two alternative alleles doubles the number of genetically different gametes that an individual can

TABLE 3-4   Independent assortment of blood group genes

Number of heterozygous loci:	1	2	3
Genotype	AB	ABMN	ABMNRr

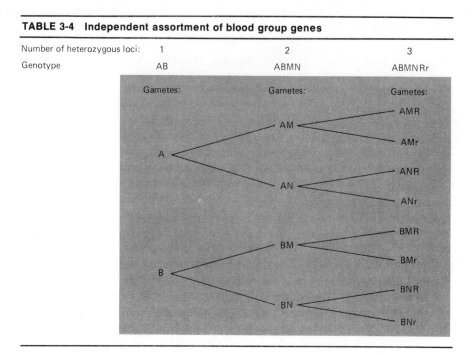

produce because each new locus adds one more pair of alternatives to those already present.

A similar progression can be seen in the diversity of genotypes potentially present among the offspring of a mating between hetero- zygotes. If both parents are heterozygous at one locus, they can produce offspring of three different genotypes (see Table 3-2). Heterozygosity at two loci raises the number of different genotypes to nine, and hetero- zygosity at three different loci gives the possibility for 27 different genotypes among the offspring. You can check this by using the gametes in Table 3-4 to make an 8-by-8 table. Fill it in and then count the number of different genotypes it contains. Table 3-5 summarizes these observations and shows the general relationship inferred from them. As each new independently assorting locus is added, the number of different genetic combinations in the gametes increases, and the diversity of different gametes that can be formed at fertilization increases even more rapidly.

Suppose an individual were heterozygous at just one locus on each of the 23 pairs of chromosomes. How many different gametes could be produced? By the formula in Table 3-5, this would be 2^{23}, or 8,388,608 different gametes—an astonishingly large number. Even so, it consid- erably underestimates the potential for genetic diversity in our species because there are still other processes that lead to the production of new gene combinations in the gametes.

TABLE 3-5   Effect of independent assortment on diversity of gametes and offspring

Number of heterozygous loci	Number of different gametes which can be produced by heterozygote	Number of different genotypes which can result from mating between two heterozygotes
1	2	3
2	4	9
3	8	27
.	.	.
.	.	.
.	.	.
General relationship n	2^n	3^n

Recombination

The genetic diversity that arises from independent assortment of loci on different pairs of chromosomes is increased still further by the exchange of genes between the members of each pair of chromosomes. This process of exchange is called *recombination.* To understand how recombination works, we must look again at chromosome structure.

Each chromosome contains many genes, joined together in specific linear order. It might be expected that the chromosomes and the genes they contain would be transmitted as intact units. They sometimes are, but not always. The exceptions make possible variation above and beyond that provided by independent assortment. Reshuffling of *linked loci*—those located on the same chromosome—results from an actual physical exchange of whole sections of chromosomes at the stage of nuclear division (first meiotic prophase) when all four chromosome strands are still together (Figure 3-15). Figure 3-15 shows the four chromosome strands lined up in the same plane. In the living cell, they form a three-dimensional bundle in which every chromatid of the tetrad is adjacent to all the others (Figure 3-16a). Exchanges can, and do, occur frequently among all of them. Of course, only exchanges between chromosomes carrying different alleles (Figure 3-16b) will be detectable; crossing-over between two maternal or two paternal chromosomes (Figure 3-16c) could have no detectable effect, since identical alleles would change places.

As a result of independent assortment of alleles at *unlinked loci* (those carried on two different pairs of chromosomes), four different types of gametes will be produced (Figure 3-17a). Parental gene combinations and new combinations will be equally frequent. Crossing-over between two linked loci (both on the same pair of chromosomes) also produces four different types of gametes (Figure 3-17b). But in the

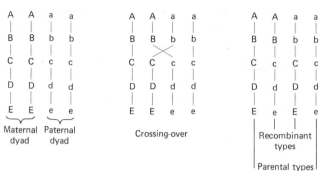

FIGURE 3-15 Linked loci and the new combinations produced by crossing-over.

case of linked loci, gametes carrying new combinations of genes from crossing-over will occur less frequently than the combinations carried by the parental gametes. As a general rule, the farther apart the two linked loci are, the more chances there will be for crossing-over between them, and the greater the frequency of gametes with new combinations.

Linkage

Autosomal *linkage* and X linkage are the two best-known types of linkage in humans. *Autosomally linked* loci are those found on the same pair of autosomes; *X-linked* loci are those which occur only on the X chromosomes. Genes that are X-linked differ in inheritance and expression from those carried on autosomes.

Autosomal Linkage. The Lutheran blood group system is a relatively simple one with two antigens produced by codominant alleles (Lu^a and Lu^b). The locus (which controls the release of ABO blood group antigens into watery secretions such as the saliva) also has two alleles, one of which (Se) is dominant to the other (se). Secretors, people with at least one Se allele, have a water-soluble form of ABO antigens present in their saliva, tears, and other body fluids. The linkage between the Lutheran blood group locus and the secretor locus, discovered over 20 years ago, provided the first known case of autosomal linkage in humans. Our knowledge of the important features of the linkage is even now not very complete. For example, we do not yet know on which chromosome pair the loci are found, although it is known that crossing-over between the two loci occurs frequently enough to be detected.

The Rh locus provides another possible example of autosomal linkage in which the antibodies may be produced by very closely linked loci or even adjacent sites within the same gene. When first discovered, the Rh locus was thought to have two alleles, R and r. This was how it was

(a)

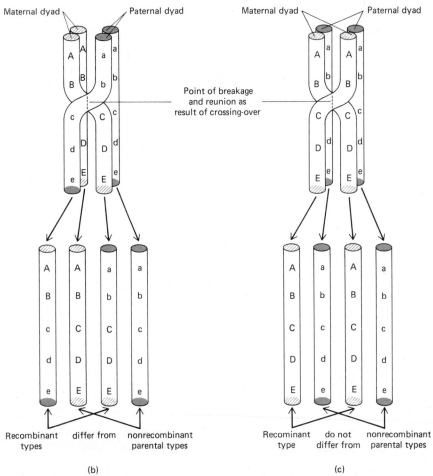

FIGURE 3-16 Results of crossing-over. (a) Tetrad. (b) Crossing-over with detectable effects. (c) Crossing-over with no detectable effects.

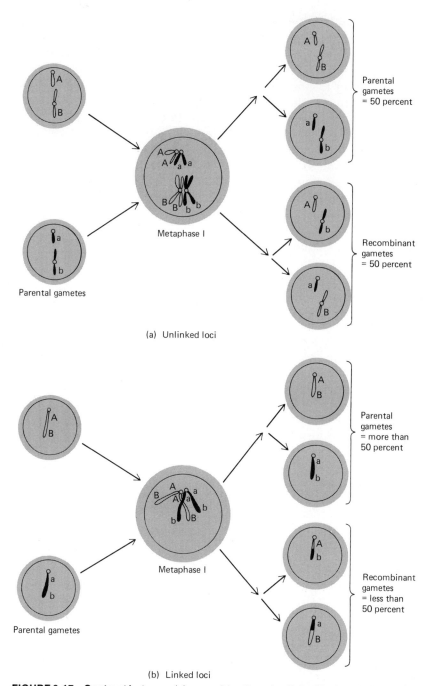

(a) Unlinked loci

(b) Linked loci

FIGURE 3-17   Contrast between (a) recombination of unlinked loci and (b) crossing-over of linked loci.

presented in an earlier section of this chapter. In the course of further research, however, a large number of additional antigens and antibodies have been discovered, and these further subdivide the original two groups. Two conflicting theories now exist to explain the situation. According to one idea, the Rh blood types are controlled by a single locus with a multiple allelic series (with perhaps as many as 40 alleles). The alternative is that the various Rh blood group types are controlled not by one locus but by three (and possibly more) immediately adjacent loci. Each locus is believed to have at least two alleles (C, C^w, and c; D and d; E and e). If the explanation centering on three neighboring loci is correct, it would be an example of the closest type of linkage possible because recombinants are produced so rarely that they have not been detected.

X Linkage. The expression of the many known X-linked genes differs in the two sexes. Females, who have two X chromosomes, may be either homozygous or heterozygous for any allele. Some of these alleles are dominant, and others are recessive. However, the terms *homozygous* and *heterozygous* and *dominant* and *recessive* do not really apply to males. Males have only a single X chromosome and therefore can have only one gene at each locus on it. Any allele, whether it is dominant or recessive in the female, will be expressed in males when present in this *hemizygous* situation. That is, if a male has one gene for, say, color blindness, he will be color-blind, whereas the female can have the same gene but not show the characteristic. Among the hundred or so characteristics known to be controlled by X-linked genes in addition to color vision are one type of muscular dystrophy (wasting away of the muscles), one blood group (Xg), and hemophilia (failure of the blood to clot properly). A pedigree for the inheritance of hemophilia is shown in Figure 3-18.

Genetic Diversity

Independent assortment of genes at just one locus on each of the 23 pairs of human chromosomes allows a multiple heterozygote to produce, potentially, millions of different gametes. Yet each of our chromosomes contains not just a single locus, but many. How many loci are there in the whole human genome? A lower boundary is provided by the number of individual genes known. There are already over 2000 of these, and the total increases each year. This would suggest a minimum of about 100 loci per chromosome pair. Estimates at the opposite extreme are based on measurement of the amount of DNA in the nucleus. There is surprisingly little of this crucial compound, about 3.2×10^{-12} g (1.1×10^{-13} oz) per haploid cell. Dividing this by an even smaller quantity, the weight of a single nucleotide pair, gives an estimate of a rather large

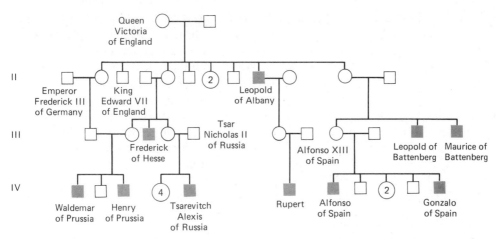

FIGURE 3-18 Pedigree of hemophilia in the royal families of Europe. All of Queen Victoria's children are entered. (Later generations comprise many more individuals than are shown here.)

Source: Curt Stern. 1973. *Principles of Human Genetics* (3d ed.). San Francisco, Calif.: Freeman. Fig. 231, p. 556.

number of nucleotide pairs per haploid cell (2.87×10^9). This estimate can be used to arrive at the number of genes by making a few assumptions. The first is that each locus is responsible for the production of an enzyme or other protein. This is certainly far too simple an approach, since some genes are known to act as switches for others rather than as sites that code for protein production; also, some sections of the genome seem to contain duplicate copies of the same gene. The second assumption is that these proteins have an average length of about 200 amino acids. The third, the most secure, is that each amino acid is coded for by a sequence of three nucleotides. Thus, the average gene would be about 600 nucleotide pairs long. Dividing the total number of nucleotide pairs by this gives an estimate of about 5 million loci in the human genome, or about 200,000 per chromosome. Even if this figure is wrong by several orders of magnitude, it remains impressively large.

The lower and upper estimates of the number of loci per chromosome are, then, about 100 and 200,000, respectively. A conservative estimate in the middle range might be about 10,000 loci per chromosome. Not all these loci would be expected to have pairs of alternative alleles that produce genetic differences in gametes. Many loci probably control essential genetic functions (such as the formation of the deoxyribose sugar that is an essential building block of all DNA) which tolerate no variation, with the result that only one form of a gene exists at the locus. One recent study places the proportion of invariant loci in humans as high as 70 percent (Lewontin, 1974). If only about 1000 loci per chromosome had alternative alleles, what level of diversity could be produced through the genetic reshuffling made possible by independent assortment and crossing-over? The formula in Table 3-5 indicates that a person who is heterozygous at 1000 loci could produce 2^{1000} gametes. A mating between two such heterozygotes could potentially produce 3^{1000} unique genotypes among their offspring. If multiplied out, this would give a number a bit over 1, followed by nearly 500 zeros, a quantity greater than the estimated total of protons, neutrons, and electrons in the entire universe. It utterly dwarfs the size of the world population, which at present is a mere 4 billion. It is scarcely surprising, then, to find that each human is different. From a genetic standpoint, it is quite impossible for more than a small sampling of these unique combinations ever to be born once (with the exception of identical twins)—let alone twice. And as Darwin realized over a century ago, this diversity of hereditary characteristics is the necessary raw material for evolutionary change.

MORE COMPLEX GENETIC SITUATIONS

Until this point, our discussion of how human characteristics are inherited has been limited to traits with simple modes of inheritance. It is useful to begin with these, but it would be extremely misleading to end with them. Let us look at some further complexities.

Genetic Influence versus Genetic Determination

At the MN blood group locus, each antigenic phenotype is controlled by a unique genotype. As a result, this and other blood groups are often said to be genetically *determined.* But a shorthand description like this can be misleading; it is like saying that the presence of light in a room at night is determined by having the switch in the "on" position. This is only partly true. To have light in the room, it is necessary to have the light switch on, but it is also necessary to have an unbroken light bulb in the socket, a properly wired circuit, functioning transmission lines, a properly maintained generating plant with an adequate supply of fuel, and so on. The position of the switch does not determine the existence of the electric power—only its expression. And the expression can be modified by the state of any of the other components. Expression of the hereditary instructions coded in a gene can also be modified by a variety of factors, including the effects of the allele on its homologous chromosome (as in dominance), the influence of genes at other loci, and limiting conditions imposed by the environment.

The starting point for identifying genetic influence on a characteristic is phenotypic differences. In order for a certain characteristic to be affected by alleles at a given locus, the gene must exist in at least two different forms, and the difference must be detectable. If all individuals were of one blood type—say, A—we wouldn't know it. There would be no anti-A serum available because it is produced only by type B individuals. Lactose tolerance is one example of a characteristic showing genetic influence. Virtually all human infants are able to digest lactose, a sugar found in milk, because they produce sufficient quantities of the enzyme needed to break it down to release energy. Some human adults, chiefly northern Europeans and Americans of northern European ancestry, can also digest lactose (Harrison, 1975). But most people over the age of 4 cannot. If given milk, they become bloated with gas, belch, and develop a watery, explosive diarrhea. This difference was first studied by giving milk to individuals and seeing whether digestive disturbances due to an intolerance for lactose developed. More recently, the level of sugar in blood samples has been used to screen for this trait. With this last method, parents and their children can be tested for degree of lactose tolerance. Results from matings of various types suggest that lactose tolerance is transmitted genetically and is dominant to lactose intolerance.

Time of Appearance of Genetically Influenced Traits

The path from genotype to phenotype is sometimes a long one. Although all genes are present from the moment during fertilization when nuclear fusion occurs, not all genes express themselves at the same time. Let us look at some examples.

Polydactyly.   This condition, the occurrence of more than the usual 10 toes or fingers, is typically under the control of a single dominant autosomal allele, and the structures produced vary in size, location, and number. They can be detected by x-rays or inspection of aborted embryos during the first few weeks of development.

Blood Groups.   Infants begin to produce ABO blood group antibodies at the age of 3 to 6 months. The level of these in the blood increases through adolescence and then gradually decreases. Forerunners of the A and B antigens can also be detected quite early in fetal life, although they have not reached their final molecular form.

Eye Color.   Most people have heard that human eye color is controlled by a single pair of genes, with B dominant to b. People with blue eyes are said to be of genotype bb, and those with brown eyes are said to be of either genotype BB or genotype Bb. However, the existence of many different shades of brown and blue eyes, as well as colors (green, gray, hazel, etc.) that are neither blue nor brown, provides ample evidence that the genetic situation is more complex than is usually supposed. At birth, most infants have very light bluish eyes. In some this color is maintained throughout life; in others the eyes darken gradually over a period of time. Some infants, however, have dark eyes when they are born.

Hereditary Baldness.   This condition may appear over a wide range of ages. Most cases occur between the ages of 20 and 50 years, and studies indicate that about 40 percent of all men aged 35 and over in the general American population have some degree of thinning of the hair. Although baldness occurs in both sexes, it is much more common in males. Despite this sex differential, studies of family pedigrees indicate that baldness is not due to a sex-linked gene. With this mode of inheritance ruled out, other explanations must be sought. The age distribution in males provides a clue: the years in which baldness appears are those in which high levels of male sex hormones are produced. These hormones probably affect the expression of genes for the trait, which is therefore said to be *sex-limited.* Further confirmation is provided by a few rarer observations; few eunuchs (castrated males) become bald, but women with tumors of the adrenal cortex have high levels of male-type sex hormones and an unusually high frequency of baldness.

Pleiotropism

Pleiotropism is the term applied to the production of multiple phenotypic effects by a single gene. Some cases of pleiotropism may be traced to a single allele that operates early in embryonic life, and so several different traits can actually be due to the effects of one gene. Thus, many later processes would be indirectly affected. Pleiotropism can also result at any age when an enzyme controlled by a single gene regulates the synthesis of a substance or structure that interacts with many others. Sickle-cell anemia is a particularly dramatic case of pleiotropism. In many African populations, as well as among Americans of African ancestry, people suffer from an inherited anemia that, if untreated, usually causes death before puberty. Family studies show that the anemia is inherited as though it were due to a recessive allele. Therefore, the population can be divided into two phenotypic classes: those who are severely anemic (homozygous recessives) and those who are well (homozygous dominants plus heterozygotes). In slides made from the blood of those suffering from the severe anemia, two-thirds of the red blood cells are the normal flattened disks; the remaining one-third take on a variety of distorted shapes, often resembling the blade of a sickle (Figure 3-19). This is why the disease is usually called *sickle-cell anemia.*

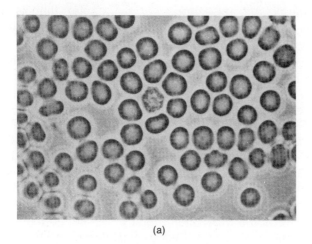

(a)

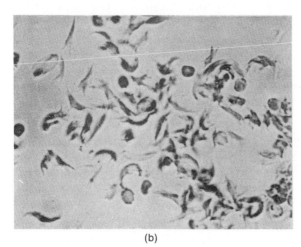

(b)

FIGURE 3-19 (a) **Normal red blood cells from a Hb^A-Hb^A homozygote.** (b) **Red blood cells from an individual with sickle-cell anemia (Hb^{S}Hb^S homozygote).**

Source: Used by permission of Anthony C. Allison, Clinical Research Centre Laboratories, National Institute for Medical Research, Oxford, England.

Most people who appear healthy have none of these altered cells. But slides made from the blood of a few apparently normal people will have some sickle-shaped cells. Using this criterion, the healthy group can be subdivided into two categories: those with no sickle cells must be homozygotes (usually shown as Hb^A Hb^A), and those with a few sickle cells must be heterozygotes (Hb^A Hb^S). Those with many sickle cells must be homozygous for the other allele (Hb^S Hb^S).

Similar results are obtained by a more elaborate method called *electrophoresis.* Blood samples are treated to make the red cells rupture open and release their hemoglobin. These hemoglobin-containing solutions are placed onto absorbent paper to which an electric current is applied. Hemoglobin from a Hb^A Hb^S heterozygote separates into two components, one like that in Hb^A Hb^A homozygotes and the other like that in Hb^S Hb^S homozygotes. This separation occurs because the Hb^A

and Hb^S hemoglobin molecules differ by a single amino acid. At one point glutamic acid in the β chain of hemoglobin A is replaced by valine in hemoglobin S.

Each β chain is 146 amino acids long. Considering just one chain of the hemoglobin molecule, then, the difference between hemoglobin S and hemoglobin A is 1 amino acid out of 146. At the DNA level the difference is even slighter, 1 nucleotide out of 438 (i.e., 146 amino acids $\times$ 1 triplet codon of 3 nucleotides per amino acid). Qualitatively, however, the change from glutamic acid in hemoglobin A to valine in hemoglobin S is radical. At the RNA level it is most likely due to a shift from the triplet GAU to the triplet GUU. This second-position change of A to U produces dramatic effects. When hemoglobin S molecules lose oxygen, as normally happens in the capillaries, they clump together into rods. This distorts the outer membrane of the cell, transforming it into the characteristic sickle shape. Sickle cells interfere with circulation, thus preventing many organs from getting much-needed oxygen and other vital materials. Severe and widespread damage results. Still other changes follow from the body's attempt to compensate for this. From a minute difference at the genetic level—substitution of a single nucleotide—a variety of serious clinical symptoms follow (Figure 3-20). Unless treated, these cause early death.

The sequence outlined here is unusual only in that it is known in such great detail from beginning to end. Many similar cases of genes with pleiotropic effects are also known. This is to be expected. Indeed, in complex organisms it would be surprising to find genes that altered only a single character and affected nothing else.

Quantitative Characteristics

Most of the genetic loci discussed so far govern the appearance of characteristics that can be divided into separate categories. In some cases, there can be variation within one of these categories. For example, two children of the same family might both have polydactyly and yet differ in the number of extra fingers or toes. In such cases, however, the categories themselves—extra digits versus the basic 20—do not overlap. Many familiar traits cannot be conveniently separated into categories. Unlike pea plants, for example, humans are not either very short or very tall. Any large group includes a practically continuous range of stature in which any two individuals may differ only slightly. And not all phenotypic categories are equally frequent. Individuals near the average are most common, and those who deviate in either direction, higher or lower, are seen less often (Figure 3-21). The same is true for weight, skin color, the sizes of various bones in the skeleton, and many other human phenotypic characteristics. The similarities of distribution shared by various quantitative characteristics seem to be due to common

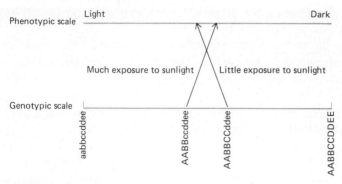

FIGURE 3-23 Norm of reaction.

Norm of Reaction.   Norm of reaction is the range of phenotypic expression that a particular genotype may have under different environmental conditions. Blood group antigens are relatively invariant in molecular form. Given several individuals with the same genotype (such as BB) and conditions that will support life, all will have the same molecular form of B antigen on their red blood cells, though the amount of antigen may vary from person to person. A trait like skin color is far more easily influenced by environmental factors. These include the degree of exposure to sunlight, which stimulates increased production of melanin. Diet can also play a role; excessive intake of certain pigments such as carotene (found in pumpkins and other vegetables) can give a yellowish cast to the skin. Even a small amount of dirt can cause significant darkening. The overall impact of these environmental influences is to blur the phenotypic categories still further and to decrease the correspondence between genotype and phenotype (Figure 3-23).

The diversity of genetic control and expression discussed in this chapter represents but a small sampling of what is actually known to exist in populations of humans and other organisms. However, the traits discussed do provide an idea of the extent and types of variation on which the forces of evolution can act. These forces and their effects are discussed in Chapters 4 through 7.

SUMMARY

1. Cells are the units of life and carry genetic instructions from one generation to the next.

2. Genetic instructions are packaged into chromosomes, which are alike in number and shape in all individuals of each species.

3. In humans, the chromosome set consists of 22 pairs of autosomes and 1 pair of sex chromosomes; one member of every pair is received from each of the parents.

4. Female gametes are all alike in chromosome content (22 A + X), whereas male gametes are of two types (22 A + X; 22 A + Y); therefore, the male's gamete governs the child's sex at the time of fertilization.

5. After fertilization, growth occurs chiefly through cell division; in the human organism, two types of cell division occur: mitosis and meiosis.

6. In mitosis, duplication of chromosomes followed by cell division produces two identical diploid somatic cells; in meiosis, duplication of chromosomes followed by two rounds of cell division gives rise to four haploid gametes with different combinations of genetic material.

7. Chromosomes contain three classes of chemical compounds that are especially important in heredity: deoxyribonucleic acid (DNA), ribonucleic acid (RNA), and proteins.

8. DNA in the cell nucleus carries the genetic code; RNA transmits the message from the nucleus to the cytoplasm, where proteins are made; and proteins form many of the cell's structures and regulate its activities.

9. Alleles that control the same trait segregate or pass into different gametes when the homologous chromosomes on which they are located separate in meiosis.

10. Pairs of genes located on different pairs of chromosomes assort independently of each other when the chromosomes pass into gametes.

11. As a result of segregation and independent assortment, gametes of the same parent may differ in genetic content.

12. New genetic combinations can also result from the exchange of linked genes between homologous chromosomes.

13. For the genetic influence of a locus to be identified, there must be at least two alternative alleles, each having a detectable phenotypic effect.

14. Dominance, recessiveness, the influence of genes at other loci, and pleiotropism are all descriptions of the phenotypic expressions of alleles.

15. Genes at some loci express themselves later in the life cycle than others.

16. Quantitative characteristics are those measured on a continuous scale; they typically represent the cumulative effects of genes at many loci interacting with environmental factors.

APPENDIX TO CHAPTER 3: MENDEL'S DISCOVERIES

Organisms have been scientifically studied in a systematic manner for several centuries. Their cells, and the subunits of these cells such as chromosomes, have been carefully investigated by cytologists for about

100 years. And in the last several decades molecular biologists have refined and extended our knowledge of the appearance and composition of cells, chromosomes, and genes to an incredibly fine degree. But before these details of structure were known, even before chromosomes were suspected to be carriers of genetic instructions, the basic properties of genes had been discovered by Mendel, who inferred their existence and behavior from breeding experiments with plants. The principles governing genetic inheritance have since been demonstrated to hold throughout the many plant and animal species on earth. Because of this, a discussion of Mendel's original experiments with peas gives as good an introduction to the rules governing inheritance as would be provided by examples concerning humans. In addition, it allows a direct contrast of these principles with the alternative ideas that had been put forth earlier to explain inheritance.

In his work, Mendel used pure lines of seeds: those from strains that produced plants of the same appearance generation after generation. Some pairs of strains differed from each other in only a single, clearly visible characteristic: seed color (yellow versus green), plant height (1.83 m—6 ft—versus about 0.3 m—about 1 ft—tall at maturity), and so on. Seven such contrasting pairs were used, and in all cases the results Mendel obtained from crossbreeding were the same. The offspring were all like either one parent or the other. For example, when tall plants were crossed with short ones (by transferring the sperm-bearing pollen from one to the egg-containing ovaries in the flowers of the other), the offspring were all tall. The properties of the other strain disappeared without a trace, regardless of whether they had been introduced from the male or the female parent. Mendel made further matings after the hybrids had matured. When these second-generation plants were used as both parents, some of their offspring were like one parent (tall), and some were like the other (short). The characteristics that had been missing in the second generation reappeared in the third generation.

These results differed from what would have been predicted from the blending theory of inheritance in two ways. First, according to the blending theory, second-generation plants should have been intermediate between the two parents. None were. Something received from a parent of one strain was dominant over something received from the other. That is, it masked the expression of the other strain's characteristic (the characteristic masked is said to be *recessive*). Second, if the blending theory were correct, all plants in the third generation should also have been intermediate. They were not. After considering these results, Mendel reasoned that characteristics such as tallness and shortness are determined by factors which are not blended or otherwise changed while together in the offspring and which he called *hereditary particles*. These stable particles are now known as *genes*. The idea that hereditary particles control biological characteristics is the basis of Mendel's principles.

Segregation: Mendel's First Law

Genes, or hereditary particles, for alternative characteristics (such as tallness or shortness) were together in Mendel's second-generation plants. However, they must have separated, or segregated, as gametes were formed. These segregated units could then recombine via fertilization with those of another individual. Some of the combinations that arose in this manner would be like those which had been present in the parent strains. Mendel's ability to state this principle (now usually called the *law of segregation*) in precise terms resulted from his desire to know exactly how many particles controlled the difference between each pair of characteristics. After noticing that the third generation included individuals like each of the parents, Mendel went on to classify and count all the third-generation plants. In one case there were 1064 third-generation plants in all. Of these, 787 (about three-fourths of the total) were tall, and 297 (about one-fourth of the total) were short. Put another way, there was a ratio of about three tall plants to one short plant.

In the example mentioned above, let the genes which affect the difference in plant height be represented by T and t. Tall plants from a pure line have the genotype TT. Short plants have the genotype tt. Gametes will each carry one gene for plant height, T or t, respectively. After fertilization, second-generation zygotes will also have two genes, one from each parent. Their genotype will be Tt. Unlike the pure-line parents, each of these hybrid zygotes that originated from a cross between two pure lines produced, via segregation, two kinds of gametes. Half carry the T gene, and the other half carry the t gene. To explain the ratios observed in the third-generation zygotes, it is unnecessary to postulate anything more complex than random contact and pairing of eggs and sperm (see Table 3-6).

Independent Assortment: Mendel's Second Law

After determining the results of crosses between strains that differed in a single pair of genes, Mendel went on to see what would occur when several pairs of genes differed. In one case he made a cross between plants grown from seeds that were wrinkled and green and plants grown from seeds that were smooth and yellow. The second-generation hybrid seeds were all smooth and yellow. Plants that grew from them were then used as parents of the next generation, and a total of 556 third-generation seeds resulted. They were of four kinds and were produced in unequal numbers: 315 smooth yellow, 101 wrinkled yellow, 108 smooth green, and 32 wrinkled green. This reduces to a ratio of 9:3:3:1.

Mendel knew from his earlier experiments with single pairs of factors

TABLE 3-6   Particulate versus blending inheritance

	Results expected according to blending theory of inheritance	Mendel's observed results	Explanation of results in terms of particulate inheritance
First-generation gametes	Tall X short	Tall X short	
Second generation	All intermediate	All tall	
Matings between second-generation individuals			
Third generation	All intermediate	3/4 tall: 1/4 short	

that he could expect yellow and green peas to appear in the ratio 3:1. The same was true for crosses between lines that differed only in seed shape: for every three smooth seeds, there was one wrinkled seed. Actually, the same ratios of seed color and seed shape were recurring in this more complex cross. Disregarding shape, 416 out of the 556 seeds were yellow, and the remaining 140 were green; i.e., there were about three yellow seeds for every green seed. Much the same results appear if color is disregarded and only shape is scored: 423 seeds were smooth, and 133 were wrinkled. The ratio is again 3:1. Mendel realized that such results could be explained by assuming that these pairs of factors were behaving as if they had no influence on each other in inheritance. This conclusion is summarized in Mendel's second principle (referred to as the *law of independent assortment*): Members of different pairs of genes pass into gametes independently of each other. An example of the law of independent assortment is shown in Table 3-7.

Physical Basis for Mendel's Laws

Although Mendel did not know it at the time, the abstract rules of genetic transmission he discovered can be explained in physical terms. The key is the realization that genes are located on chromosomes. The chromosomes themselves are distributed according to precise rules

TABLE 3-7   Independent assortment of two pairs of factors

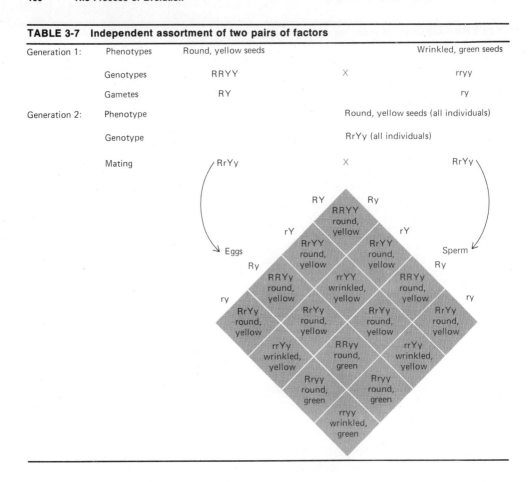

Generation 1:	Phenotypes	Round, yellow seeds		Wrinkled, green seeds
	Genotypes	RRYY	×	rryy
	Gametes	RY		ry
Generation 2:	Phenotype		Round, yellow seeds (all individuals)	
	Genotype		RrYy (all individuals)	
	Mating	RrYy	×	RrYy

during cell division. Two members of the same pair of genes segregate because each is located on one of the members of a pair of homologous chromosomes. As the homologous chromosomes pass into different gametes during meiosis, so do the genes located on them. By extension, members of two pairs of genes located on different chromosome pairs will assort or segregate independently of each other during gamete formation.

SUGGESTIONS FOR ADDITIONAL READING

Stern, Curt. 1973. *Human Genetics*. San Francisco: Freeman. (This is a readily available comprehensive textbook of human genetics for advanced students. However, it is written clearly enough to be read by most undergraduate students without much difficulty.)

REFERENCES

Harrison, Gail G. 1975. "Primary Adult Lactase Deficiency: A Problem in Anthropological Genetics," *American Anthropologist,* **77**(4):812–835.

Lewontin, Richard C. 1974. *The Genetic Basis of Evolutionary Change.* New York: Columbia University Press.

McKusick, Victor A., and Frank H. Ruddle. 1977. "The Status of the Gene Map of the Human Chromosomes," *Science,* **196**(4288):390–405.

Newcombe, H. B. 1964. In M. Fishbein (ed.), *Papers and Discussions of the Second International Conference on Congenital Malformations.* International Medical Congress. Discussion, pp. 345–349.

Ortho Research Foundation. 1960. *Blood Group Antigens and Antibodies.* Raritan, N.J.: Ortho Pharmaceutical Corporation.

Penrose, L. S. 1954. "Some Recent Trends in Human Genetics," *Proceedings of the IXth International Congress of Genetics,* 521–530.

Chapter 4

Populations:
Units of Evolution

In the logic of life there is a scale of organization that goes from simple to complex. A unit that is a functional whole at one stage is usually just a part of some larger system. Chapter 3 began near the middle of the scale, at the level of the single cell, and moved to progressively finer levels of analysis: from the cell to one of its major compartments, the nucleus; then to chromosomes; and eventually to the DNA molecules that make up the genes. This showed us how cells operate to maintain genetic diversity and to translate these genetic instructions into the phenotypic characteristics that we see in people around us. Let us now switch course and move up the scale from cells and individuals to populations.

BREEDING POPULATIONS

Populations, the units of biological evolution on which anthropologists focus, are larger than families and smaller than species. The term *population* must be defined carefully because it is used in a variety of ways. For example, the phrase "the population of the United States" in everyday speech means all residents of the 50 states of our country, without regard to age, sex, or any particular aspect of behavior. In the study of evolution, a population is a group of organisms that breed with one another. In the strictest sense, this *breeding population* (or Mendelian population or deme) is the pool of potential mates, any of whom may be chosen with equal probability. As we will see later in this chapter, each breeding population can also be studied at a more abstract level (which is less realistic but much simpler to work with) as a *gene pool.* A gene pool is the total of all the genes that exist at one locus in the population at the time studied.

Breeding-population boundaries are not absolute: like several teabags in the same pot, Mendelian populations are permeable compartments in a leakproof container, the species. In higher organisms genes can diffuse from one population's compartment to another, but not beyond the confines of the system. In fact, each species can be viewed as the total set of breeding populations capable of exchanging genes.

Each one of us belongs to a breeding population, but given our high level of geographic mobility (the average American family moves once every 5 to 6 years), many humans cannot be assigned to a particular breeding population with any certainty. Other, more readily recognizable breeding populations are all around us, though. All the adult squirrels in a small city park would constitute a breeding population, as would all the cats of reproductive age in a small village and perhaps all the mice in your apartment or dormitory. As these examples suggest, members of most species choose their mates from within a very limited area, and human breeding behavior follows culturally transmitted rules as well.

Limitations on Mate Selection

Distance. The probability of mating drops as physical distance increases, although distance was much more of an isolating factor in the past because travel was so much more difficult. In the villages of Oxfordshire, England, for example, during the seventeenth, eighteenth, and early nineteenth centuries the distance between the birthplaces of spouses never exceeded 16 km (10 mi). But around the middle of the nineteenth century this distance rose dramatically to between 32 and 48 km (20 and 30 mi), chiefly as the result of the increasing availability of mechanized means of travel such as passenger trains (Harrison, 1972). In our own time, travel by car and plane is contributing even more to the decline of distance as a constraint on choice of mate in industrialized societies. For people living in rural villages in developing countries, however, the pool of partners is still very much limited by distance. For example, on the tiny Caribbean island of Saint Barthélemy, which has a surface area of less than 25 km² (9.7 mi²) nearly 100 percent of all mates have come from the island itself for nearly the last century (Benoist, 1964). Saint Barthélemy is thus an extreme example of a *genetic isolate:* a relatively small population that has little or no gene exchange with other populations.

Geographic Features.   Geographic features such as mountains, oceans, and deserts can reinforce the effects of distance. Geographic features are likely to be most effective inhibitors to travel when they divide different ecological zones (regions that differ in climate, rainfall, types of vegetation, and so on). For example, the Alps were long a formidable barrier between the Mediterranean climatic region and central Europe. But these natural features do not inhibit the spread of all species, or even all human populations. Some animals spread relatively rapidly in the face of seemingly formidable barriers (for example, birds and insects cross the Atlantic Ocean in their migrations, and salmon swim far up narrow and shallow streams to spawn); other populations may remain in a limited region even though to a human observer there is little evident reason for this.

In our own species, geographic features, like simple distance, were far more restrictive in the past than they are now. Today, rivers can be spanned by bridges or with boats; mountains can be made passable by roads built across them or tunnels blasted through them; and even the widest oceans and highest mountains can be flown over. The range of ancient hunter-gatherers, however, was much more limited because of these natural obstacles.

Cultural and Behavioral Factors.   Differences in religion, nationality, language, social class, and other customs can further restrict choice of a mate even within the same geographic area. For example, a cultural universal found in all human populations is the *taboo,* or prohibition, against *incest*—marriage with close biological relatives. The incest taboo discourages matings between brother and sister, parent and child, first cousins, and other members of the immediate family. Right now the likelihood of marriage between Arab and Israeli residents of Jerusalem is rather low because of the political situation, which is polarized along religious lines. For centuries it was far more probable that an English or Dutch or Swedish princess would marry a member of the nobility of some other country than a commoner from the city in which she lived.

In India today there is still a rather inflexible caste system. Each Indian is born into one of five major social categories, each of which is further subdivided into still finer levels, and he or she remains in this caste from birth to death. Matings between members of different castes are discouraged by a belief system that marks the higher-caste partner as tainted by the union. For a higher-caste woman the stigma of mating with a lower-caste male is permanent, a system which protects the theoretical purity of descent of the higher-ranked castes.

The strength of cultural influences on mating patterns can be measured by studying the ways in which mates are alike. One commonly used statistical measure of similarity is the *correlation coefficient.* The maximum value that this measure can reach is +1, which indicates that items (such as the same characteristic in both mates) increase or decrease in value together. A correlation coefficient of 0 means that the two items vary independently of each other, and a value of −1 (the lowest possible) means that the items vary in opposite directions (for example, the taller the husband, the shorter the wife). In the last generation in the United States, there was a higher correlation (.75) between brides and grooms for religion than there was for a physical trait such as body weight (.21). From this and other cases like it we can conclude that preferences for culturally determined characteristics must frequently override those for genetically inherited features.

Time and Age.   Age is as important a consideration in choice of a mate as distance or cultural characteristics because there are simple biological limits to the length of the reproductive period. Human gamete production begins during adolescence in both sexes and ends by the early fifties for most women; men usually continue to secrete sex hormones and produce sperm in decreasing amounts into relatively old age. Beyond these fairly broad age ranges, reproduction is not possible.

Most cultures place further restrictions on what is considered a suitable match in terms of age. In the United States, most first marriages occur between people in the same 20- to 24-year-old age range. In other cultures, the age discrepancy may be much greater. For example, among the Tiwi of north Australia (Hart and Pilling, 1961), females are betrothed as soon as they are born, and they must stay married until they die (if the husband dies first, the wife is remarried at the time he is buried). This practice followed logically from the belief common to Australian Aborigines that a woman becomes pregnant because a spirit enters her body; the male is believed to have no role in conception, although every child needs a father for social legitimacy. The right of betrothal belongs to the girl's father, and he uses it wherever it will do him the most good. His choice will usually fall on a friend or ally about his own age (and therefore perhaps as much as 40 years older than the infant girl). With virtually all young girls going to old husbands, very often a young man's first wife is the remarried widow of some older Tiwi male who has just died. For these reasons, it is not at all uncommon for a Tiwi girl's first husband to be several decades older than she and for her last husband to be about as many years her junior, with just the reverse being the case for males.

Tiwi marriages are near the outer extremes of age discrepancy possible in our species. With any significantly greater separation in age, two individuals born into the same locality and appropriately matched in all other respects just could not be mates; by our definition they could not be considered members of the same breeding population, even though they shared as many genes in common as two members of the same generation in the same locality. Membership in a Mendelian population is thus limited in time as well as by location and culture. This point has important implications for the study of long-term evolutionary change because it helps us understand how populations separated by thousands or millions of years of time can look quite different from one another. Genetic differences can build up through time to the point where mating might no longer be possible genetically, even though the populations are related as ancestors and descendants.

Recognizing Population Boundaries

It is easier to define populations theoretically than to recognize their limits in the real world because boundaries are often quite fluid. Populations may increase or decrease in numbers from generation to generation. Certain areas may be occupied only sporadically, and occasionally groups that move into them may be temporarily isolated when nearby populations die out or contract their range. The polar Eskimos are one such population (Murdock, 1934). These people occupy a narrow fringe of the coast of northwest Greenland, hundreds of miles from their nearest neighbors and only 1600 km (1000 mi) from the North Pole itself. When they were first discovered by John Ross in 1818, they had lived in isolation for so long that they believed themselves to be the only people in the world, and no other Eskimo group knew of their existence. For one small group to go on living in isolation from its nearest neighbors is rare, particularly in mainland areas. Much more frequently a group that has declined in numbers as a result of disease or because many group members have been killed by raiders will be absorbed by another population.

Time boundaries are also fluid. A human *chronological* generation, for example, is usually given as 20 years. A human *pedigree* generation includes all the children born to the same pair of parents. People who belong to the same pedigree generation may differ from one another in age by more than 20 years, and those in different pedigree generations may differ by much less than 20 years (see Figure 4-1). As a result, people often choose mates who are of a generation different from their own, though not very different in age. Geographic and cultural boundaries also blur and change, as we saw above.

Appearance is sometimes a way to recognize members of a Mendelian population. Those in one local population may look very much like one another and unlike members of neighboring populations. A species that has such populations is said to be *polytypic.* Domestic dogs are one of the most familiar examples of a polytypic species. Each registered breed is a breeding population kept artificially closed by the refusal of registration to offspring of any mating that includes an unregistered parent. Therefore, any two dachshunds resemble each other more than

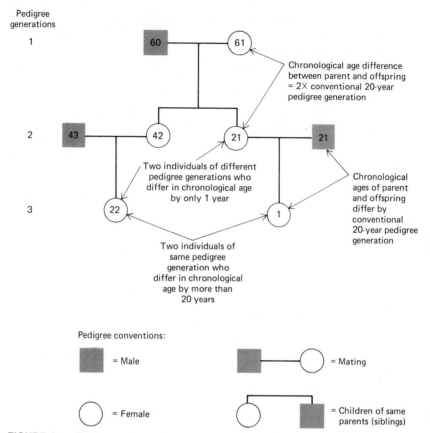

FIGURE 4-1 **Pedigree generations versus chronological generations.**

either one resembles a Great Dane or even a basset hound. But even a single Mendelian population may be *polymorphic;* it may include different phenotypes that do not grade into one another. Common ladybird beetles, for example, come in two distinct color patterns: an orange-red background with black spots and a black background with reddish spots. The color reversal is controlled by a single genetic locus. The distribution of phenotypic variation in populations is discussed in greater detail in Chapter 7. Here it is enough to note that phenotypically similar individuals can belong to different populations (for example, dark skin color is found in some Pacific Islanders as well as in Africans) and that individuals differing significantly in appearance (for example, blonds and brunettes) may belong to the same population.

Although their boundaries are difficult to locate and are by no means absolute, populations are real units of evolution. Each population represents the visible, if temporary, expression of a particular gene pool. Since each population is partially isolated from all others, it can build up a gene pool that is particularly suited to local conditions. At the same time, since it is partially open, its gene pool can incorporate advantageous alleles that arise in other groups. Because of these features, populations represent highly flexible units of evolutionary change.

BREEDING POPULATIONS AND EVOLUTIONARY CHANGE

Population boundaries are not rigid; matings can and do take place in each generation between members of different local groups. These subdivisions are thus said to be *open populations.* A species, in contrast, is a *closed population.* As a rule, few if any productive matings can take place between members of different species, despite persistent folk myths to the contrary. There is no documented case of a human mating with any other animal and producing offspring. The Minotaur (half man and half bull) of Greek legend is a product of the human mind rather than of the reproductive organs. On a more mundane level, ring-tailed Maine coon cats have resulted not from matings between cats and raccoons (which have very different chromosome numbers as well as courtship patterns) but from genetic variations that have arisen within cats. The same is true for short-tailed cats, which have not come from mating between cats and cottontail rabbits. These variations, in other words, happen because of gene exchanges and changes over time within and among the breeding populations within a particular species. The large-scale changes in appearance through time that occur in most species—for example, the roughly threefold expansion in the size of the human brain over several million years—are chiefly a result of genetic processes. Change through time—evolution—takes place whenever there are heritable alterations in the genetic material of a breeding population. In the strictest sense of this definition, a single nucleotide substitution or change in chromosomal structure would qualify as an evolutionary event. But from a practical point of view, most such changes are significant only when they increase or decrease in frequency in a population.

Population genetics is the discipline that deals with these shifts in *gene frequency* or *allele frequency.* The frequency of a gene is its fraction of the total alleles at a given locus in a population, usually expressed as a decimal. In some cases changes in frequency can be measured directly—for example, by comparing the allele frequencies at a blood group locus in two successive generations of a single population. In other instances, as in the case of changes in quantitative characteristics, changes in gene frequency must be inferred from the distributions of phenotypes. If we want to detect changes in gene frequency through time, we can look at populations as collections of genotypes. But because sexual reproduction breaks up the diploid genotype combi-

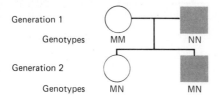

Generation 1

Genotypes   MM   NN

Generation 2

Genotypes   MN   MN

FIGURE 4-2   Gene and genotype frequencies in two successive generations. The numbers of each allele (2M + 2N) are the same in the two generations; but as a result of meiosis and fertilization, the genotype frequencies are completely different.

nations in each generation (Figure 4-2), the *distribution* of genotypes can change in each generation, even though the types and numbers of alleles stay the same. To account for this in the study of evolution, a population can be treated as a *gene pool,* which is the total of all the genes that exist at one locus in the population at a particular time (Figure 4-3).

Changes in Gene and Genotype Frequency

The basis for the study of changes in gene frequency was provided by Darwin's realization that the frequency of phenotypic characteristics could change through time and by the rediscovery, at the turn of the century, of Mendel's laws governing the inheritance of genes that produce the phenotypic characteristics. During the first decade of the twentieth century there was widespread interest in Mendel's principles, but there was also disagreement about how generally applicable these laws were to various organisms, including humans. At one popular lecture on Mendelism, a critic pointed out a puzzling problem. People in some populations have abnormally shortened fingers, a trait known as *brachydactyly*. It occurs in individuals who have at least one B gene,

FIGURE 4-3   Populations as pools of genotypes and genes.

Population boundary

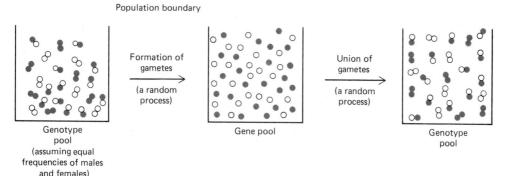

Formation of gametes

(a random process)

Union of gametes

(a random process)

Genotype pool (assuming equal frequencies of males and females)

Gene pool

Genotype pool

For one locus:

O, ● = Alternative alleles

OO, O●, ●● = Diploid combinations

$N = 24$

TABLE 4-1   Brachydactyly: Expected distribution of offspring from a mating of two Bb heterozygotes

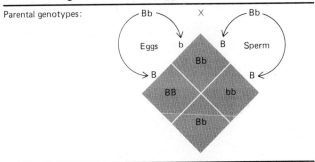

---

which is dominant to its allele b. From Mendel's principles it would be predicted that, in matings between two Bb heterozygotes, genotypes among the children should appear in the ratio of 1BB:2Bb:1bb (Table 4-1). Three out of every four children should have brachydactyly. Thus, it was argued, if in each generation three individuals with brachydactyly were born for every unaffected one, the whole population would gradually approach these proportions. Yet this had not happened. Brachydactyly had been known for a long time, but it had not spread widely through the population. This apparent paradox could be resolved by assuming that Mendel's principles did not apply here.

An English mathematician, G. H. Hardy, proposed an alternative solution to the problem. He realized that the paradox resulted from a confusion between dominance (a gene's mode of expression) and predominance (or, more simply, frequency) in the population. Hardy agreed that a 3:1 ratio of affected to normal would occur in offspring from matings of two Bb individuals. But since brachydactyly is quite a rare condition, very few matings of this type would ever take place. The only way to determine what would happen to the relative frequencies of the B and b alleles would be to consider the results of all possible matings in the population (the basic concepts of probability, which help here, are discussed briefly in "Probability").

Consider a locus (such as that which determines the presence or absence of brachydacytyly) at which there are just two alleles, B or b. Their frequencies in the population can be represented by the two letters p and q, respectively. Thus, p is the frequency of B, and q is the frequency of b. Since these are the only alleles at this locus in the population, the sum of $p + q$ must total 100 percent. The values of p and q are usually expressed not as percentages but as their decimal equivalents, and so $p + q = 1$. For example, if 1 percent (or 0.01) of the alleles at this locus are B, then the remaining 99 percent (or 0.99) must be the alternative b alleles. With two alternative alleles at a locus, three genotypes are possible: BB, Bb, and bb (Table 4-1). Any individual in

PROBABILITY

Behind the statistical concept of probability is the everyday idea that any given event has a certain chance or likelihood of occurring and that these likelihoods may differ from one event to another. It is completely certain that I will die someday, and it is clearly absolutely impossible that I could swim across the Atlantic Ocean. The probabilities of these events could be taken as the end points of a *probability scale* (shown below) with $p = 0$ (absolute impossibility) as the lower boundary and with $p = 1$ (absolute certainty) as the upper boundary. Between them lie the probabilities of many other events. For example, there is 1 chance in 2 that a penny tossed in the air will come down heads, and so $p = .5$; there is 1 chance in 6 that a die rolled on a table will show six spots on its upper surface, and so $p = .167$; and so on.

How are these estimates of probability derived? Even without tossing any coins we might intuitively feel that, for all practical purposes, a coin can land in only one of two ways: heads up or tails up. Thus there is 1 chance in 2 that the coin will come up heads, and so $p = \frac{1}{2}$, or .5. This intuitive estimate is sometimes referred to as the *a priori probability* or *mathematical probability*. Another approach is possible, however. If a very large number of pennies were flipped, we would find that half of them (or very nearly) would come up heads. Thus $p = 100/200$, again 0.5. This experimental approach leads to a working definition of what

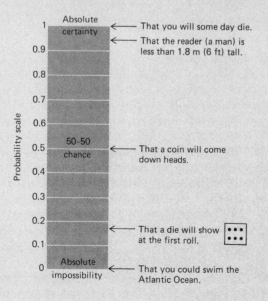

The probability scale.

Source: M.J. Moroney. 1962. *Facts from Figures*. Baltimore, Md.: Pelican. Fig. 1, p. 5.

is sometimes referred to as *actual* or *empirical probability*. Here,

$$\text{Probability} = \frac{\text{total number of occurrences of one alternative}}{\text{total number of trials}}$$

It is useful in cases where we have little information on which to base an a priori estimate. Thus a gardener with a batch of new seeds may not know what percentage of those planted will germinate. But if 200 are planted and 90 of them sprout, the gardener can estimate that the probability of germination in the rest of the batch is about 90/200, or .45.

For the concept of probability to be applicable to a problem, several conditions must be met:

1. Events, experiments, or observations must be repeatable, and each trial must have alternative outcomes that are clearly distinguishable (a coin must land either heads up or tails up, a die must land on one of its six faces, a given birth must produce either a boy or a girl, and so on).

2. It must be possible for the alternative outcomes to result even if the conditions under which the trials are performed are highly uniform (that is, even if a coin were tossed precisely the same way each time, it should be possible for it to land in either of two ways).

3. The ratio of the occurrence of one alternative to the total number of trials (sometimes called the *frequency ratio*) should approach some constant as the number of trials becomes large (if we toss a few pennies, the proportion that turn up heads may deviate substantially from ½; if we toss 2 tons of pennies, 1 ton should turn up heads, and 1 ton should turn up tails).

With these concepts as a foundation, it is possible to formulate rules to deal with more complex cases.

Multiplication Rule. If the occurrence of one event does not affect the outcome of another (simultaneous or subsequent) event, the two events are said to be *independent*. The probability that two or more independent events will occur is the product of the probabilities of each event's happening separately. For example, if the probability that one coin will land heads up when flipped is ½, then the probability that two coins will both land heads up is ½ × ½, or ¼.

Addition Rule. If the occurrence of one event precludes the occurrence of another, the two events are said to be *mutually exclusive*. The probability that an outcome will result in one of several mutually exclusive ways is the sum of the probabilities of occurrence of the several different possible ways. For example, consider the case represented by "Heads I win, tails you lose." Here the outcome is winning from the toss of a single coin. Heads and tails are the mutually exclusive events. I will win if the coin turns up heads, with $p = \frac{1}{2}$. You will lose (and hence I will of necessity win) if the coin turns up tails, also with $p = \frac{1}{2}$. Adding the two probabilities together, we see that the total probability of my winning is $p = \frac{1}{2} + \frac{1}{2} = 1$. Thus my winning is certain. Probability theory is appealing and useful because it allows us to proceed from a few simple assumptions and definitions to some reasonable predictions about the outcome of intrinsically variable events. The principles of probability do not allow us to reach conclusions that are certain, but these principles do place limits on the degree of our uncertainty.

the population who is of genotype BB must have received a B gene from his or her father. The probability of this is p, which is simply the frequency of the B allele in the population. Such an individual must also have gotten a B gene from his or her mother, again with a probability of p for the same reason. Thus the probability that any randomly chosen individual in the population is of genotype BB is the same as the frequency of that genotype in the population. This is $p \times p$, or p^2. The same process of reasoning leads to the conclusion that the frequency of the bb genotype in the population would be q^2.

Finding the frequency of Bb heterozygotes is a bit more complicated. The key is that an individual can become genotype Bb in two ways. He or she can inherit a B gene from the father (with a probability of p) and a b gene from the mother (with a probability of q). The probability of getting the genotype Bb this way is thus $p \times q$, or simply pq. The same genotype can result in just the opposite way as well, that is, if b came from the father and B came from the mother. This would also have a probability of pq. The frequency of the Bb genotype in the population is therefore the total of these two probabilities: $pq + pq$, or $2pq$. Since there are just three genotypes at this locus in the population, their frequencies must add up to 100 percent. That is, p^2 (the frequency of BB) + $2pq$ (the frequency of Bb) + q^2 (the frequency of bb) equals 1 (Table 4-2). A numerical example is given in Table 4-3. The total frequencies of offspring of each genotype can be obtained by adding up the columns in Table 4-4, which shows all possible matings. These totals, simplified by a bit of basic algebra, are the same as the offspring frequencies in the generation of the parents.

TABLE 4-2 Brachydactyly: Genotypes and their expected frequencies in one generation

Genotypes	Genotype frequencies
BB	p^2
Bb	$2pq$
bb	q^2
Total	$p^2 + 2pq + q^2 = 100\%$, or 1

TABLE 4-3 Relationship between gene and genotype frequencies in a population existing under Hardy-Weinberg conditions*

	Sperm	
	A $p = .3$	a $q = .7$
A $p = .3$	AA $p^2 = .09$	Aa $pq = .21$
a $q = .7$	Aa $pq = .21$	aa $q^2 = .49$

Eggs

* This table is merely a way of graphing (for a particular set of values, $p = .3$ and $q = .7$) the binomial expansion $(p + q)^2 = p^2 + 2pq + q^2$. The Mendelian ratio among offspring of parents both heterozygous for a single pair of alleles is just a special case of this where $p = .5$ and $q = .5$ (since the two types of gametes are produced in equal frequencies).

Constancy of Gene and Genotype Frequency: Hardy-Weinberg Law

When there is no change in gene or genotype frequency from one generation to the next, a population is in *equilibrium*. In gene pools this equilibrium is a dynamic balance, not a static one, because genotype combinations do not stay intact as they are passed from generation to generation; instead, these combinations are reshuffled by crossing-over and independent assortment during meiosis. Two brown-eyed parents can give birth to a blue-eyed child; two parents of medium stature can produce a tall son or daughter. But as long as no genes are lost or gained in the reshuffling, the nature and extent of variation in the population, as measured by gene and genotype frequencies, remain the same.

Hardy's discovery of the conditions necessary to maintain the constancy of genotype frequencies in successive generations was duplicated independently at just about the same time by a German physician,

TABLE 4-4 Brachydactyly: Expected frequencies of all possible matings and resulting offspring genotypes in the population

Genotype frequencies in the parents: p^2 BB + $2pq$ Bb + q^2 bb

Mating		Frequency of mating	Offspring genotypes		
Males	Females		BB	Bb	bb
BB	× BB	$p^2 \times p^2 = p^4$	p^4	—	—
BB	× Bb	$p^2 \times 2pq = 2p^3q$	p^3q	p^3q	—
BB	× bb	$p^2 \times q^2 = p^2q^2$	—	p^2q^2	—
Bb	× BB	$2pq \times p^2 = 2p^3q$	p^3q	p^3q	—
Bb	× Bb	$2pq \times 2pq = 4p^2q^2$	p^2q^2	$2p^2q^2$	p^2q^2
Bb	× bb	$2pq \times q^2 = 2pq^3$	—	pq^3	pq^3
bb	× BB	$q^2 \times p^2 = p^2q^2$	—	p^2q^2	—
bb	× Bb	$q^2 \times 2pq = 2pq^3$	—	pq^3	pq^3
bb	× bb	$q^2 \times q^2 = q^4$	—	—	q^4

BB column:
$$= p^4 + p^3q + p^3q + p^2q^2$$
$$= p^4 + 2p^3q + p^2q^2$$
$$= p^2(p^2 + 2pq + q^2)$$
$$= p^2 (1)$$
$$= p^2$$

Bb column:
$$= p^3q + p^2p^2 + p^3q + 2p^2q^2 + pq^3 + p^2q^2$$
$$= 2p^3q + 4p^2q^2 + 2pq^3$$
$$= 2pq (p^2 + 2pq + q^2)$$
$$= 2pq (1)$$
$$= 2pq$$

bb column:
$$= 2p^2q^2 + pq^3 + pq^3 + q^4$$
$$= 2p^2q^2 + 2pq^3 + q^4$$
$$= q^2 (p^2 + 2pq + q^2)$$
$$= q^2 (1)$$
$$= q^2$$

Genotype frequencies in the offspring: p^2 BB: $2pq$ Bb: q^2 bb

Wilhelm Weinberg. Both men are therefore given credit in the name *Hardy-Weinberg law*. The discovery of this law represents the extension of Mendel's principles to the population level. In each family the genotypes of the parents determine the genotypes of the children; likewise, the genotypes of all the parents in the population determine the genotypes of all the children in the population's next generation, provided that some specific conditions are met. The demonstration of this point provides the basis for understanding and measuring evolutionary change. The discovery by Hardy and Weinberg describes the basis for the existence of a special kind of equilibrium. A population should be in *Hardy-Weinberg equilibrium* whenever genotype and gene frequencies stay constant from one generation to the next because certain conditions exist: (1) Mating must be random, (2) no mutations can occur, (3) the population must be infinitely large, (4) no gene exchange can take place with other populations, and (5) persons of all genotypes must produce the same average of offspring. Deviation from random mating will bring about a change in the proportions of genotypes, but it will not cause a change in gene frequency. The types of nonrandom mating will be discussed later in this chapter. If any of the other ideal conditions (no mutation, an infinitely large population, no gene exchange, and production of the same number offspring by persons of all genotypes) are not met, there can be a change in gene frequency. The effects that mutation and gene exchange between populations have on gene frequency will be covered in detail in Chapter 5; the consequences of finite population size and differential production of offspring by persons of various genotypes will be explored in Chapter 6.

The Hardy-Weinberg law shows that the frequencies of genotypes in a population are determined only by the frequencies of alleles in the gene pool. If genotype frequencies (based on a large enough sample to be representative of those in the population and calculated as shown below) deviate from those which are predicted by the Hardy-Weinberg formulation, we are alerted that something must be occurring to upset the expected equilibrium. Therefore, whenever the Hardy-Weinberg equilibrium is upset, evolutionary changes must be taking place.

Determining Gene Frequency

Gene frequency may be determined in two ways: by the *gene-counting method* and by the *square-root method*. The gene-counting approach can be demonstrated most easily with an example. Suppose that blood samples are drawn from a small group of 100 people and then are tested for the presence of MN blood group antigens (the results are shown in Table 4-5, page 118). The gene-counting method depends on our being able to distinguish, at the locus being studied, every different genotype in the population, which is possible only for codominant alleles. Homozygotes have, by definition, two genes of the same kind. Hetero-

THE HARDY-WEINBERG LAW

It is doubtful that any or all of the conditions necessary for the Hardy-Weinberg equilibrium are ever exactly satisfied or even very closely approximated in humans. If this is the case, then what is the value of the Hardy-Weinberg equilibrium? In answering this, it is best to point out that the Hardy-Weinberg law is not like most "laws" in science. A much more typical example of these is Bergmann's rule: Within any single species, the average size of individuals tends to be smaller in warmer climates and larger in colder climates. Formulation of Bergmann's principle was preceded by extensive observations of many populations of mammals all over the world. Then on the basis of these observations, the rule—a statistical generalization with few exceptions—was formulated. *Inductive reasoning* is the term applied to such an attempt to draw a general conclusion from analysis of many bits of data.

The Hardy-Weinberg law was not arrived at in this manner. Neither Hardy nor Weinberg actually observed what occurred in populations and then proceeded to a generalization. Instead, the process was *deductive:* each of these scientists began with what was known about Mendelian genetics and then stated what the consequences would be in a population with certain ideal characteristics (the five conditions listed on page 116). Genotype and gene frequencies will remain constant generation after generation as long as these conditions are met (or, in a practical sense, reasonably approximated). The Hardy-Weinberg principle does not describe the real world, but rather is a simplified or ideal model of it.

zygotes have one allele of each type. Thus each MM homozygote would add two M genes to the total for this allele in the population, while each MN heterozygote would add only one. The total number of N genes in the population can be seen as resulting in the same way: two N alleles from each NN homozygote and one N allele from each MN heterozygote. These conclusions can be stated more concisely as follows: frequency of allele M = frequency of genotype MM + 1/2 frequency of genotype MN; and frequency of allele N = frequency of genotype NN + 1/2 frequency of genotype MN. An alternative formula based on the same reasoning is presented in Table 4-5 (page 118).

The square-root method must be used when not all genotypes can be distinguished, as in the case of dominance, when a person showing a particular phenotype can be either homozygous or heterozygous.

An interesting example of a trait showing dominance and requiring the use of the square-root method is provided by differences in the

---

TABLE 4-5   Determination of gene frequency: Gene-counting method

Phenotypes	M	MN	N	Totals
Genotypes	MM	MN	NN	
Number of individuals	16	48	36	100
Numbers and types of alleles	32M	48M + 48N	72N	200
	80M		120N	200

$$p = \text{frequency of M alleles} = \frac{\text{number of M alleles}}{\text{total of M + N alleles}} = \frac{80}{200} = 0.4$$

$$q = \text{frequency of N alleles} = \frac{\text{number of N alleles}}{\text{total of M + N alleles}} = \frac{120}{200} = 0.6$$

---

ability of people to taste a compound called *phenylthiourea* or *phenylthiocarbamide* (abbreviated PTC). Human populations can be divided into two major categories. Most people are tasters, but a significant minority are nontasters. Although closer studies have shown that tasters differ widely in their ability to detect PTC (some can taste it only in concentrated solutions, while others can reliably attest to its presence even in very dilute solutions), inability to taste it is under genetic control to a substantial extent. Family studies show that most tasters are either homozygous dominants (TT) or heterozygotes (Tt). Nontasters appear to be homozygous recessives (tt).

About 91 percent of American blacks are tasters, and the remaining 9 percent are nontasters. What are the frequencies of the T and t alleles? The relative proportion of tasters who are heterozygotes rather than homozygotes cannot be determined from this information, but all the nontasters must be of genotype tt. From the Hardy-Weinberg principle we know that their frequency in the population, 9 percent (or 0.09), must equal q^2. If $q^2 = 0.09$, then taking the square root of both sides of this expression will give the numerical value of q, the frequency of the t allele. We know that $p + q = 1$, and thus p, the frequency of T, is simply $1 - q$ (see Table 4-6).

TABLE 4-6   Determination of gene frequency: Square-root method

	Taster	Nontaster
Phenotypes		
Phenotype frequencies	.91	.09
Genotypes (mode of inheritance known from family studies)	TT + Tt	tt
Genotype frequencies (predicted by Hardy-Weinberg law)	$p^2 + 2pq$	q^2
Determination of gene frequency:		
Step 1		$q = \sqrt{.09} = .3$
Step 2	$p = 1 - q = .7$	

DEVIATIONS FROM HARDY-WEINBERG EQUILIBRIUM: NONRANDOM MATING

Any of the deviations from random mating described in this section will alter genotype proportions in the population, but by themselves they will have no effect whatever on gene frequency. Furthermore, for one trait, any deviation from Hardy-Weinberg proportions can be reversed with a single generation of random mating. Departure from any of the other four conditions necessary for the existence of the Hardy-Weinberg equilibrium, however, can cause a shift in the frequency of alleles in the population. By one definition, any such shift in gene frequency is an evolutionary change. The causes of these deviations, called *forces of evolution* (Table 4-7), are discussed in Chapters 5 and 6. But although deviations can serve as indicators that one or more forces of evolution are operating, the fact that frequencies in a population agree with the Hardy-Weinberg predictions does not automatically indicate equilibrium. The very existence of several different forces of evolution, each capable of increasing or decreasing gene frequency, suggests that occasionally one could cancel out another's effects. This and other complications will be taken up in more detail in Chapter 6.

Random mating (or *panmixis*) exists when the different genotypes in the population are formed in proportion to the frequencies of alleles in the gene pool, as they will be when two random processes actually occur in sequence: formation of gametes in meiosis and union of gametes in fertilization (Figure 4-4). The first of these two processes is not affected by the mating system, but the second is. Mating probably occurs at random for a large number of genetic characteristics because many traits, such as blood groups, are not readily apparent. But mating can be simultaneously random for these traits and nonrandom for others such as skin color. For this reason, mating patterns, random and nonrandom, are locus-specific. This is made possible, as you now know, by the great variety of genotypic combinations generated by crossing-over and independent assortment.

Any departure from mating in proportion to genotype frequencies is by definition *nonrandom*. The two major categories of nonrandom mating, *assortative mating* and *inbreeding*, are patterns that reflect

---

TABLE 4-7   Hardy-Weinberg conditions and results of deviations from them

Condition	Result of deviation
1. Random mating	Change in genotype frequency
2. No mutations	Mutation
3. Infinitely large population	Random genetic drift
4. No gene exchange with other population	Gene flow
5. People of all genotypes produce same average number of offspring	Selection

(Mutation, Random genetic drift, Gene flow, Selection) } Forces of evolution

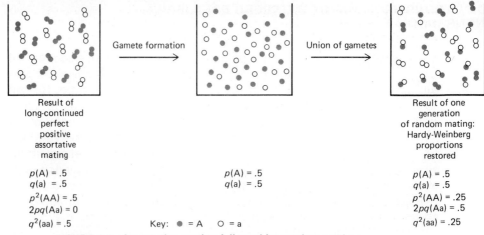

Result of
long-continued
perfect
positive
assortative
mating

$p(A) = .5$
$q(a) = .5$
$p^2(AA) = .5$
$2pq(Aa) = 0$
$q^2(aa) = .5$

$p(A) = .5$
$q(a) = .5$

Key: ● = A   ○ = a

Result of one
generation
of random mating:
Hardy-Weinberg
proportions
restored

$p(A) = .5$
$q(a) = .5$
$p^2(AA) = .25$
$2pq(Aa) = .5$
$q^2(aa) = .25$

Gamete formation

Union of gametes

FIGURE 4-4 Assortative mating followed by random mating.

human values and behavior. They may have important biological, social, and cultural causes as well as consequences.

Assortative Mating

When choice of a mate is based on preference for a particular characteristic, *assortative mating* occurs. Many of the selected characteristics have a genetic basis, although the selection process may actually be based on phenotypic criteria such as body size, hair color, skin color, intelligence, personality, some complex blend of all these, or some purely culturally determined feature. Aside from the characteristics on which it is based, assortative mating can be in either of two opposite directions and can take place to varying degrees. *Positive assortative mating* occurs when people mate with those of similar phenotype more frequently than would be expected from their frequency in the population. For example, tall women often marry tall men, and short people usually marry others near their height. Such phenotypic resemblance usually reflects similarity in the underlying genotypes. But phenotypic similarity does not always indicate genetic similarity; in some human groups the heritability of stature is near zero. Furthermore, choice of a mate can be influenced by phenotypic traits in which variation is due largely or even entirely to social or cultural causes. Regardless of its causes, positive assortative mating is indicated by the finding that mates resemble each other more than would be expected from phenotypic frequencies in the population. The results of some systematic studies are shown in Table 4-8.

When mates have fewer phenotypic characteristics in common than

TABLE 4-8 Assortative mating for some physical, psychological, and sociological characteristics in various United States population samples

Characteristic	Correlation*
Weight	.21
Stature	.38
Eye color	.42
Intelligence (Stanford test)	.47
Memory	.57
Age	.94
Religious affiliation	.75
Drinking habits	.55
Years of education	.40

* Degree of similarity between mates is measured on a scale from +1 for perfect positive assortative mating, through 0 for random association, to −1 for perfect negative assortative mating. It should be remembered that these figures are specific for the populations in which they were gathered and might differ at other times or in other places.

Sources: (1) J. N. Spuhler. 1962. "Empirical Studies on Quantitative Human Genetics," *Proceedings of the UN/WHO Seminar on Use of Vital and Health Statistics for Genetic and Radiation Studies, 1960.* Geneva: World Health Organization. (2) J. N. Spuhler. 1969. "Assortative Mating with Respect to Physical Characteristics," *Eugenics Quarterly,* **15**(2): 128–140.

would be expected by chance, the explanation may be *negative assortative mating*, in which individuals prefer mates less similar to themselves than would be expected from their frequency in the population. This is not common; it has been suggested that people with red hair choose mates with red hair less frequently than would be expected by chance, but no conclusive studies have been done.

Inbreeding

Inbreeding exists when an individual's parents share one or more common ancestors. Marriages between biologically related people are said to be *consanguineous* (literally, "of like blood"), and the children that result are said to be *inbred*. Because these parents share common ancestors, identical copies of any gene present in a common ancestor may have been passed down to both of them. Thus, at any locus their children could get two copies of what was, to begin with, the same gene (Figure 4-5). Assortative mating may increase the similarity of mates and their offspring for particular characteristics and the loci that control them; inbreeding increases the probability of homozygosity at *all* loci.

The total inbreeding present in a given population can result from several quite different causes. One of these is *random inbreeding*, which takes place because of the greater chance of marrying a biological relative in a population of limited size. Choice of mates, for example,

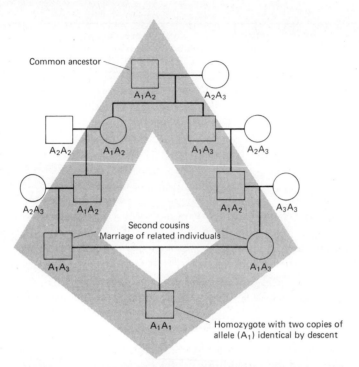

Common ancestor

A_1A_2 A_2A_3

A_2A_2 A_1A_2 A_1A_3 A_2A_3

A_2A_3 A_1A_2 A_1A_2 A_3A_3

Second cousins
Marriage of related individuals

A_1A_3 A_1A_3

A_1A_1 Homozygote with two copies of
allele (A_1) identical by descent

FIGURE 4-5 Inbreeding: pedigree showing descent of the same allele through two lines of relatives. Shaded area includes individuals in the path of descent of the shared allele (A_1).

might be quite limited on a small island or in an isolated village; in both cases, geographic factors and cultural limitations might make it likely that the pool of potential mates of appropriate age would include some relatives. Inbreeding can also be due to conscious choice. Members of some social classes marry relatives more frequently than might be expected from their frequency in the population, in order to concentrate wealth, privilege, or power. This results in *nonrandom inbreeding*, which is really assortative mating on the basis of family relationship. The cultural rules governing such matings may be either *prescriptive* (describing which relatives *should* be chosen as mates) or *proscriptive* (specifying who *should not* be married). Prescriptive mating rules are found in all cultures, but only a minority of the actual marriages may be of the prescribed type because of the limited availability of the ideal mates in real populations of finite size. Prescriptive and proscriptive rules may differ from society to society, but one proscriptive rule is common to all: the *incest taboo*. Incest rules forbid matings between certain close relatives such as parents and children or brothers and sisters, and they are more generally followed than most prescriptive rules.

Regardless of its causes, the direct effect of nonrandom inbreeding on a population is to decrease the frequency of heterozygotes. This in turn automatically increases the frequency of both homozygotes to the same degree. The extent to which this redistribution of genotype frequencies takes place is measured by the *inbreeding coefficient* (often represented by the letter *F*), which is the probability that the two alleles in a homozygous individual were both inherited from a common ancestor. Since it is a measure of probability, *F* can range from 0 to 1, where 0 = complete absence of inbreeding (i.e., random mating in a finite population) and 1 = complete inbreeding. The inbreeding coefficient measures the net effect of all types of marriages among relatives. The frequency of marriages between relatives varies greatly among human societies (Table 4-9). A study in one part of the United States (Baltimore, Maryland) showed that only about 0.05 percent of marriages were between first cousins; but in one rural province of India, the corresponding proportion of first-cousin marriages was 33.3 percent. As is noted in "We're All Inbred" (pages 124–125), people in all populations have some ancestors in common, though all of these are rarely known.

Inbreeding also has genotypic and phenotypic effects on individuals. As homozygotes become more common, many recessive genes that were formerly concealed in heterozygotes will be expressed. Consequently, children with some recessively inherited conditions such as albinism (absence of skin pigmentation) are as much as 10 times more frequent in inbred families as in the general population. Since many (though by no means all) recessive genes are harmful, individuals who are homozygous for them often seek medical treatment or genetic counseling. When the family history is taken, it is often found that the affected person is the child of a consanguineous marriage. Some population studies also point to the harmful effects of inbreeding on individuals. A survey conducted in two rural French provinces indicated

TABLE 4-9   Average inbreeding coefficients of selected human populations

Population	Inbreeding coefficient*
Netherlands	Less than .00001
U.S. (Roman Catholics)	.00001–.0001
U.S. (Mormons)	.001–.01
U.S. (Ramah Navajo)	.001–.01
India (Bombay)	.001–.01
India (rural province)	.01–.05

*The inbreeding coefficient *F* is defined as the probability that at a given locus an individual has two identical copies of an allele present in an ancestor common to both the individual's parents.

Source: Russell M. Reid. 1973. "Inbreeding in Human Populations," in M. H. Crawford and P. L. Workman (eds.), *Methods and Theories of Anthropological Genetics.* Albuquerque: University of New Mexico Press. Pp. 83–116.

WE'RE ALL INBRED

Despite the problems that may arise from very close marriages, all of us are inbred to some extent. This can be seen by counting the number of ancestors that any individual would have in each generation back through time. There would be two parents, four grandparents, eight great-grandparents, and so on. After the passing of a mere 20 generations, an individual has, in theory, over 1 million direct ancestors: 1,048,576, to be exact.

The qualification "in theory" is an important one, however. The reason can best be illustrated with a specific case. From the time of the Norman Conquest of England until the present, about 900 years have elapsed. Allowing 20 years per generation, this is equivalent to about 45 generations. Each living resident of England would now be expected to have 2^{45} different ancestors. But this number (over 35 trillion) is many times greater than the actual total population of England (estimated to have been about 1.5 million) at that time. Thus beyond a certain point back in time, the number of ancestors is finite. All descendants will share the same pool of ancestors in common.

Precise family relationships are rarely known beyond a few generations back except among the nobility. Even for royalty the family tree would resemble a rank bush unless it were carefully pruned by the (perhaps selective) forgetting of nonillustrious ancestors (see illustration opposite). In reality, virtually all persons of English descent can claim to be descended from nearly everyone who was alive at the time of the Norman Conquest. It's the rare person of English ancestry who may *not* be descended from William the Conqueror. This ancestry probably doesn't have much in the way of genetic consequences, however. The chance that any given person would have received one particular gene from William the Conqueror is vanishingly small, since there's only 1 chance in 2 that it would have been transmitted from each generation to the next and since such a transfer would have had to occur 45 times in succession. This is rather unlikely since $(1/2)^{45} = 2.84 \times 10^{-4}$. In fact, it is quite possible that one of William's descendants might not have received even a single gene from this illustrious ancestor. The chance that at least one gene was transmitted

that about 25 percent of children whose parents were cousins died before reaching adulthood. In a comparable group of children of unrelated parents, about 12 percent died before adulthood. In the United States, about 23 percent of the children from cousin marriages died early, compared with only 16 percent from parents who were not biologically related (Bemiss, 1958). If all these differences are related

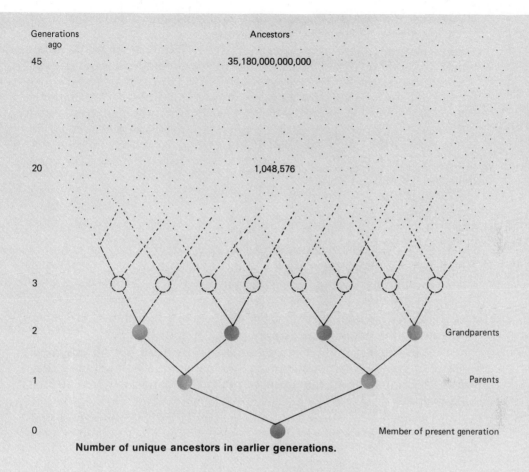

Generations ago	Ancestors
45	35,180,000,000,000
20	1,048,576
3	
2	Grandparents
1	Parents
0	Member of present generation

Number of unique ancestors in earlier generations.

from William the Conqueror to a given Englishman or Englishwoman is just the figure above for one locus multiplied by the total number of loci in the human genome (estimated to be about 200,000); this would be $(2.84 \times 10^{-4}) \times (200,000) = 5.6 \times 10^{-9}$, which is still quite low. Nevertheless, the chance that an English couple share a common ancestor of William's *generation* is virtually a certainty.

to the effects of inbreeding, then mating patterns take on considerable importance. Some caution is needed in interpreting these different findings, however, because at least some of the differences between studies might be due to the physical and cultural environments in which the populations lived. Marriages between cousins could be more frequent in relatively isolated rural areas which are also depressed economically, and thus some health problems might result from poor diet and little health care. Also, diseases like diabetes (which has a heritable component and tends to cluster in families) could be lethal before adulthood where insulin is unavailable but only mildly detrimental where insulin can be obtained.

Inbreeding increases the frequency of homozygotes in populations, thus setting up the conditions under which some people, those who have two copies of some potentially harmful recessive allele, could be selectively eliminated. This would bring about a change in gene frequency. Other possible causes of a change in gene frequency are taken up in Chapters 5 and 6.

SUMMARY

1. Each of us belongs to a breeding population or pool of potential mates; these populations are the units in which evolution takes place.

2. Evolutionary change is due to changes in gene frequency, that is, shifts in the balance of alleles within a population.

3. The Hardy-Weinberg law describes a stable equilibrium for genotype and gene frequencies if certain ideal conditions exist.

4. These ideal conditions are: (a) random mating, (b) no mutation, (c) an infinitely large population, (d) no mating outside the population, and (e) production of the same number of offspring by people of all genotypes.

5. Deviations from random mating include assortative mating and inbreeding; both cause a change in genotype frequency but not in gene frequency.

6. Failure to satisfy the other four conditions will cause a change in gene frequencies; consequently, the resulting changes are called *forces of evolution.*

SUGGESTIONS FOR ADDITIONAL READING

Haldane, J. B. S. 1964. "A Defense of Beanbag Genetics," *Perspectives on Biology and Medicine,* **7:**343–359. (This brief paper gives a logical and lively explanation for the use of simplifying assumptions such as the idea that populations of complex organisms can be represented for some purposes as pools of genes that assort and recombine at random.)

Jameson, D. L. (ed.) 1977. *Benchmark Papers in Genetics.* Vol. 8. *Evolutionary Genetics,* New York: Dowden, Hutchinson, and Ross. (In this volume Jameson has collected a number of classic papers in the field of population genetics, including those by G. H. Hardy and W. Weinberg, along with editorial comments on the significance of each publication.)

REFERENCES

Bemiss, S. M. 1858. "Report on the Influence of Marriages of Consanguinity on Offspring," *Transactions of the American Medical Association,* **11**:319–425.

Benoist, J. 1964. "Saint-Barthélemy: Physical Anthropology of an Isolate," *American Journal of Physical Anthropology,* **22**(4):473–487.

Harrison, G. A. 1972. *The Structure of Human Populations.* London: Oxford University Press.

Hart, C. W. M., and A. R. Pilling. 1961. *The Tiwi of North Australia.* New York: Holt, Rinehart and Winston.

Moroney, M. J. 1962. *Facts from Figures.* Baltimore: Pelican Books.

Murdock, G. P. 1934. *Our Primitive Contemporaries.* New York: Macmillan.

Plumb, J. H. 1970. "Genealogy: The Well-pruned Family Tree," *Horizon,* **12**(1): 118–120.

Chapter 5

Sources of Genetic Variation: Mutation and Gene Flow

Genetic variation is the basis of evolutionary change. Without it, as Darwin clearly realized over a century ago, evolution must necessarily stop. By now you have a basic knowledge of the factors that make it possible for evolution to continue. Chapter 3 described the nature of the physical units (genes and chromosomes) that remain distinct as they are passed from parent to offspring, and Chapter 4 discussed the structure of the populations in which mating and genetic transmission take place. As genes are transmitted from generation to generation, changes in frequency sometimes take place. These changes can be due to mutation, gene flow, genetic drift, and natural selection. Since the two forces of natural selection and genetic drift (Chapter 6) continually reduce the variation originally present in populations, there must be some countervailing forces, some other deviations from Hardy-Weinberg equilibrium. These are mutation and gene flow, processes that replenish the supply of genetic variations in populations of humans and other organisms. A *mutation* is any heritable change in the structure or amount of genetic material in a cell. Although rare, mutations are important, for all genetic differences can be traced to this one source. *Gene flow* is the transfer of alleles from one gene pool to another through interbreeding between members of two populations. Gene flow can therefore supplement mutation in introducing genetic variation into a population.

MUTATION

The heritable genetic changes called *mutations* can take place in any cell. After a somatic cell mutation occurs, an individual is a *mosaic* and consists of blocks of cells with different genetic contents. The effect of a somatic mutation may be large or small, depending on when it happens and on the particular cell group in which it occurs. Some forms of cancer, for example, may be caused by somatic mutations that increase the rate of cell division, whereas a somatic mutation for increased pigmentation in one skin cell may produce only a freckle. Fortunately, the direct effect of the somatic mutation is always limited to the individual in whom it occurs. In contrast, mutations in gametes and the reproductive cells that produce them can be transmitted by sexual reproduction to later generations. All new genes must have arisen from old ones by mutations of this sort.

Process of Mutation

Mutations are often said to occur at random, but this does not mean that absolutely anything can occur within a cell's genome (the whole set of genes contained in the nucleus); it means that we do not yet know the rules. Because of the chemical composition of the genetic material, we know that only certain changes are possible and that not all possible changes are equally likely. But there are many nucleotides in the genes of each cell nucleus, and we do not know enough about the chemical events that take place at this submicroscopic level to predict where in the genome the next mutation will occur, when it will arise, or what its phenotypic effect will be. We can, however, predict the probabilities that certain mutations will occur. We also know, from the results of numerous studies, that no particular environmental agent seems to produce, other than rarely and by random chance, mutations that help suit the gene pool of a population to its environment. For example, mutations caused by a rise in temperature do not increase an organism's ability to tolerate heat, and mutations caused by radiation are more likely to cause death or genetic damage than to increase resistance to high-energy subatomic particles.

The process of mutation, or change in the chemical structure of DNA molecules, can be spontaneous or induced. *Spontaneous* mutations are those which take place without human intervention and for which no cause is known. *Induced* mutations are those which are actually promoted by certain human activities. Agents known to produce mutations or to raise the frequency of mutations above the spontaneous level are called *mutagens.* Artificial mutagens include ionizing radiation such as x-rays, high and low temperatures, viruses, and a variety of chemicals from the mustard gas formerly used in warfare to the caffeine in coffee. Most of these mutagens have other biological effects in addition to the ability to modify genetic material: caffeine is a stimulant, and large doses of radiation can disrupt activities throughout the body, but mutation is not just a side effect of these large-scale changes. Relatively small doses of radiation, for example, can have great effects. Just one-twentieth of the amount of energy necessary to kill a person would be enough to cause, on the average, a break in one chromosome in each exposed cell. All the artificial mutagens mentioned here have counterparts in the natural environment. Background radiation, for example, comes from various sources and differs from place to place (Table 5-1). But we do not yet know the relationship between naturally occurring mutagens and rates of spontaneous mutation in populations; most of our knowledge of mutation comes from the study of artificially induced changes.

Mutagens appear to act not by a direct physical process but by altering the chemical environment of the cell. All mutations, whether spontaneous or induced, involve changes in the chemical structure of the DNA molecules. These changes can originate in various ways. Radiation, for example, can cause mutations by three different paths. First, a high-energy subatomic particle passing through a cell can score a direct hit on a nucleic acid molecule, causing a rearrangement such as loss of a nucleotide's side chain. Second, radiation striking other atoms in a cell can cause *ionization* by releasing electrons. In this way, previously stable molecules become able to react chemically with other compounds, including nucleic acids. Third, radiation passing through cells can produce toxic chemical by-products that can interfere with DNA structure and function. Chemical mutagens bring about changes in a different manner. In addition to adenine, guanine, cytosine, and thymine, there are less frequent naturally occurring nucleic acid bases. These, as well as the four frequent bases, and *base analogs* (synthetic

TABLE 5-1   Background radiation: Some sources and amounts*

Altitude	Composition of surface		
	Open ocean	Ordinary granite	Sedimentary rock
Sea level	53	143	76
1524 m (5000 ft)	—	150	83
3048 m (10,000 ft)	—	190	123
4572 m (15,000 ft)	—	270	203
6096 m (20,000 ft)	—	414	347

* Figures refer to total radiation dose received from cosmic rays plus the materials composing the surface. The dose is given here in milliroentgens (1 mr = $1/100$ roentgen). The *roentgen* is one of several measures of the intensity of radiation; it is a measure of the degree of ionization caused by a given amount of energy. For living tissues it is roughly equivalent to the *rad*, which is the quantity of radiation which will result in energy absorption of 100 ergs per gram of irradiated material.

Source: 1962. *Report of the United Nations Scientific Committee on the Effects of Atomic Radiation.* New York: United Nations.

chemicals that mimic naturally occurring bases) can undergo slight alterations in chemical structure that can cause permanent changes in base-pairing relationships. If such a change occurs during DNA replication, a different base can be incorporated into the new strand being formed, thus producing a permanent genetic modification. Let us look more closely at the various types of mutations before considering their effects on individuals and on populations.

Types of Mutations

In classical genetics, three types of mutations are recognized (Vogel and Rathenberger, 1975): gene mutations, chromosome mutations, and genome mutations. Gene mutations cannot be detected under the light microscope because they are due to changes at the molecular level and often affect only a single nucleotide. Chromosome mutations are rearrangements in chromosome structure that are visible under the light microscope. Genome mutations alter chromosome number, from the gain or loss of a single chromosome to the gain or loss of entire haploid chromosome sets. These three categories are a useful framework for organizing our discussion, but actual mutation processes are of course not so easily differentiated. Physical and chemical changes in the genetic material take place on a continuum, with effects that range from slight to great.

Gene Mutations.   *Gene mutations,* which are also called *point mutations,* take place, as we have said, at the molecular level. There are two immediate causes of these mutations: (1) base substitutions; (2) frame shifts due to the gain or loss of one or more bases.

Base substitutions arise whenever a new DNA base is substituted for the previously existing one. The most important feature of base-substitution mutations is their limited impact on the primary structure of proteins. Usually only the single corresponding RNA triplet mirrors this change. At the protein level, just the one amino acid coded for by this triplet may be different. Figure 5-1 shows how this works and how phenotypic effects may vary. The RNA triplet GUU, for example, calls for the presence of the amino acid valine. A change from U to A in the third position, giving GUA, would have no effect because GUA also codes for valine. The mutation from GUU to GUA would be *silent,* that is, not phenotypically detectable. But the same change from U to A in the second position of one triplet in the RNA molecule that codes for the β chain of hemoglobin produces sickle-cell anemia—with enormous phenotypic consequence (see Figure 3-20).

Frame shifts are due to additions or deletions in the DNA molecule. Although only a single base may be gained or lost, the effects would not be limited to the triplet in which the change occurred. This is so

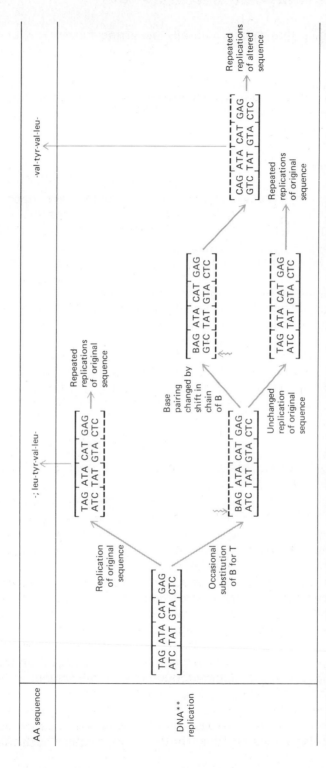

FIGURE 5-1 Base substitution, following replication with trace amounts of a base analog present. The base analog is 5-bromouracil (abbreviated B), which usually pairs with G. Solid line indicates the DNA strand which serves as a template; dotted line indicates the newly synthesized strand.

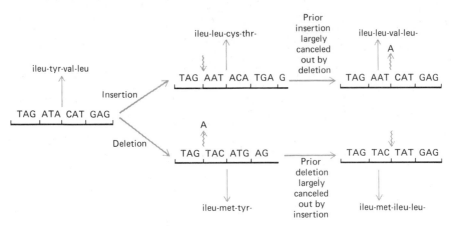

FIGURE 5-2   Frame-shift mutations.

because bases in the RNA strand are read in sequence, one triplet at a time, in much the same way that you read words from left to right across this page. Deletion of one base in a triplet means that one base from the next triplet will be "moved over" to make three, and so on through the sequence. It is as though we removed one letter in a word and the sentence changed from "The book is on the table" to "[T]heb ooki so nt het able." The reading of the triplets changes in the same way by the insertion of one extra base. After an insertion, the added base becomes the first letter of the next triplet. This shift in the sequential reading of triplets is maintained until the end of the gene; thus an insertion can sometimes cancel out much of the effect of an earlier deletion, and a deletion can reverse the effect of a previous insertion. In this case, the only part of the message that will remain altered is the segment between the two mutations. Figure 5-2 illustrates all the changes that can occur through frame-shift mutations.

Chromosome Mutations.   All structural changes that can be seen when chromosomes of a cell are viewed under a microscope are called *chromosome mutations,* or sometimes *chromosome aberrations, chromosome abnormalities,* or *chromosome variations.* These larger-scale changes seem to occur less frequently than point mutations and are often associated with substantial phenotypic effects. The structural changes appear to result from breakages or other events that lead to rearrangements of chromosome segments. Types known in humans include deletions, duplications, inversions, and translocations (see Figure 5-3). Single or multiple chromosome breaks can occur within cells, and broken pieces can join together in a great variety of ways. Whenever a broken piece does not reattach to a centromere, it will be lost in cell division. If a chromosome fragment attaches to a new

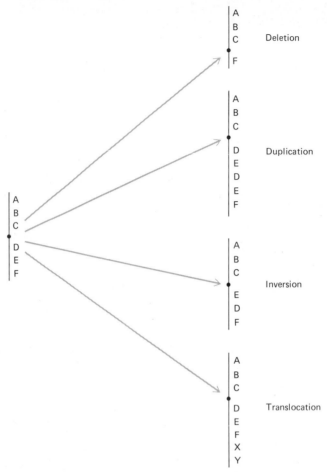

**FIGURE 5-3 Types of structural rearrangements in chromo-
somes.**

chromosome, some loci may be present more than once. Either loss or
gain of loci can cause disruptions in development, but deletions seem
more likely to be harmful.

Deletions can result from a single chromosome break that is not
repaired before cell division. The piece lacking a centromere will be
unable to attach to a spindle fiber and consequently will be lost.
Deletions can also occur through unequal crossing-over during meiosis.
An example is provided by an unusual type of abnormal hemoglobin
called *hemoglobin Lepore.* Like normal hemoglobin, the hemoglobin
Lepore molecule has two alpha (α) chains. But in place of the two beta
(β) chains are shorter polypeptides. These appear to have a hybrid
origin, as shown in Figure 5-4. More than half of the molecule has the
same amino acid sequence as a normal β chain. But in place of the first

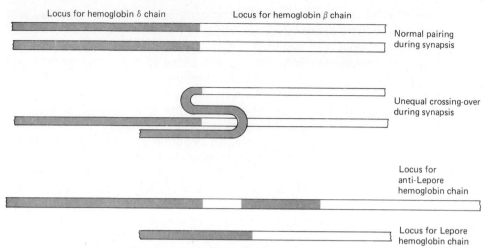

FIGURE 5-4   Chromosome deletion: hemoglobin Lepore locus. The loci which produce the β and δ hemoglobin chains are closely linked on the same pair of chromosomes. Unequal crossing-over produces a new hybrid genetic locus consisting of parts of the former β and δ loci. Anti-Lepore hemoglobin, produced by the chromosome segment bearing a partial duplication, has also been found.

Source: Adapted from J. S. Thompson and M. W. Thompson. 1973. *Genetics in Medicine.* Philadelphia, Pa.: Saunders. Fig. 5-4, p. 99.

part of the β chain is a short segment which is just like the beginning of a normal delta (δ) chain. The delta is produced by a different locus from that which produces either the α or the β hemoglobin chains, and it replaces the β chains in hemoglobin A_2. This A_2 variant makes up about 2 percent of the hemoglobin of normal adults. Homozygosity for the gene that produces hemglobin Lepore causes one form of hereditary anemia and symptoms such as abnormal red blood cells and physical weakness. Apparently in this case the mutation causes some harm, but does not affect survival.

In *duplication,* the length of a chromosome is increased by the repetition of a section. Duplications are thus the opposite of deletions also resulting from unequal crossing-over. *Inversions* change the order of loci along part of the chromosome's length, but they do not alter the amount of genetic material present. For an inversion to occur, two breaks must take place, and the chromosome segment between them must be turned around a full 180°. A chromosome bearing an inverted section may not pair properly with its homolog during meiosis, which in some cases can lead to formation of abnormal gametes.

Translocation is the shift of a piece of chromosomal material to a nonhomologous chromosome. A broken section of one chromosome may become attached to the end of a member of another pair. Alternatively (and more frequently, at least among known survivors), breaks may appear in two different chromosomes at the same time, and sections become exchanged. The effect of the translocation on offspring depends on what then happens in meiotic division. If both chromosomes with translocated sections go into the same gamete, the genome will contain the full normal haploid set of genes. If the chromosomes that have exchanged sections enter different gametes, each of those cells will have an abnormal genome, with some loci missing and others represented in duplicate. In effect, both a duplication and a deletion will have resulted. Translocation can produce abnormalities of development. One example is *Down's syndrome,* or mongolism, which occurs in about 1 out of every 600 births. Symptoms commonly include modification of the skin around the eyes (giving a facial appearance like that found among people of Asian ancestry), mental retardation, short stature, a thickened tongue, and an unusual pattern of skin ridges on the hands and feet. Not all translocations are equally damaging. The present human diploid chromosome number of 46 may have been reached through a translocation. Our nearest living relatives, the great apes (gorilla, chimpanzee, and orangutan), all have a diploid chromosome number of 48. What is more, their chromosome sets include more pairs of acrocentric chromosomes (those with only one arm) than ours. A reciprocal translocation between members of two pairs of acrocentric chromosomes could have reduced a chromosome number of 48 to 46 with little or no loss of genetic material (Figure 5-5).

Genome Mutations. *Genome mutations,* or changes in chromosome number, result in even larger-scale changes in the karyotype than most changes in chromosome structure, and they can cause correspondingly greater disturbances in the phenotypes of individuals. Most cases of Down's syndrome, for example, are due to the presence of an extra copy of chromosome 21 (hence the term *trisomy-21,* which is sometimes used for this abnormality). Trisomy occurs when two chromosomes fail to separate during meiosis. In a few cases, Down's syndrome is produced by translocation of most of chromosome 21 onto a member of another pair. A common sex chromosome trisomy, XXY, gives rise to *Klinefelter's syndrome,* which affects about 1 out of every 400 male babies alive at birth. Symptoms usually include above-average height, enlarged breasts, and failure to produce sperm. Mental defects are also not unusual. A different chromosomal situation (XO) is represented in *Turner's syndrome,* which afflicts about 1 in every 2500 newborn females. Some of the physical features are dramatic opposites to those seen in Klinefelter's syndrome. Females with Turner's syndrome are of less than average stature—frequently under 152 cm (less than 5 ft)—and have broad chests with poorly developed breasts. The external genitals are juvenile in appearance, and the internal sex organs, including the ovary, are rudimentary. Failure to menstruate is common.

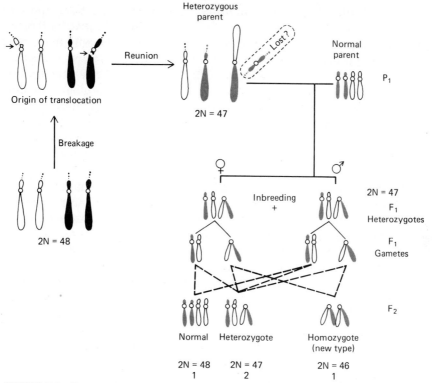

FIGURE 5-5   Translocation leading to a reduction in chromosome number. Reciprocal translocation between two nonhomologous acrocentric chromosomes leads to formation of one metacentric chromosome, which may become established in the population, eventually replacing the two acrocentrics. The human 46-chromosome complement may have originated in this way from a 48-chromosome ancestor.

Source: Modified from J. L. Hamerton and H. P. Klinger. 1963. "Chromosomes and the Evolution of Man," *New Scientist,* **341**:483–485. Figs. 4, 5, p. 485.

IMPACT OF MUTATION: INDIVIDUALS AND POPULATIONS

In terms of their impact on individuals, mutations can be classed by the amount of genetic material affected (as we did above), by the change they make in an allele's mode of expression, by their effect on the phenotypic appearance of the individual who inherits the allele, and by their impact on that individual's survival, reproduction, or both. It is commonly assumed that the effects are proportional to the scale of the changes. That is, a large-scale change in genetic material is expected to produce a highly visible phenotypic change that will exert a great effect on survival and on transmission of the mutation to future generations. But this is not always true: a few females with trisomy-21 (a relatively large-scale genetic change) have shown relatively mild forms of the usual symptoms of Down's syndrome and have had children. On the other hand, a point mutation affecting the activity of a single enzyme can produce an albino. Albinos, who are homozygous for the recessive gene, lack melanin pigment in the skin, hair, and eyes and have increased susceptibility to skin cancer. Homozygosity for the gene that produces albinism also has a measurable effect on future generations: albinos produce on the average only about 60 percent as many offspring as normally pigmented individuals.

Impact on Individuals

Mode of Expression.   A *recessive mutation* is one that changes a gene's mode of expression from dominant (detectable even in heterozygotes) to recessive (detectable only when homozygous). A *dominant mutation* transforms a gene that has been recessive to one that shows a dominant mode of expression. Dominant mutations are, of course, visible in the first generation of heterozygous offspring to which they are transmitted. If harmful, they can be eliminated or at least kept to a very low frequency by natural selection. Recessive mutations, on the other hand, can build up in heterozygotes for many generations before becoming frequent enough to give rise to significant numbers of homozygotes—that is, before becoming apparent. Tay-Sachs disease, for example, is due to homozygosity for a recessive gene. Heterozygotes are phenotypically normal, but they show a lower level of one enzyme than those who are homozygous for the dominant allele; they are *carriers* of a potentially harmful gene but do not themselves suffer from the gene's effect. In homozygotes, the buildup of complex fatty compounds in cells of the central nervous system leads to blindness, mental deterioration, muscular degeneration, and paralysis, usually terminating in death during the second or third year of life (Fredrickson and Trams, 1966). The frequency of this disease in most populations is about one case per every one-half million people of normal phenotype. But in one population, the Ashkenazy Jews of central Europe, the frequency of Tay-Sachs disease is markedly higher (about 1 affected child in every 6000 births). One theory to explain the buildup is that heterozygous carriers of the gene produce slightly more surviving offspring than dominant homozygotes; this possibility will be taken up again in Chapter 6.

Phenotypic Appearance.   Phenotypic effects can be defined at a number of levels. Some may be so slight that they are, for practical purposes, invisible. These mutations produce changes within the range of phenotypic characteristics considered normal, and therefore they do not attract attention. For example, a mutation from low to high pigment production at one of the five or so loci that influence skin color would rarely be noticed. Another genetic change could produce a noticeable phenotypic effect and be classed as a *visible mutant* such as brachy-dactyly. A mutation might also produce no visible change in external characteristics such as weight, skin color, or stature and yet give rise to an enzyme different in one or more amino acids from the nonmutant polypeptide. But because such differences may be detectable only with the aid of sophisticated research equipment, or may not be detectable at all because the mutation causes death before the defect comes to anyone's attention, we may not even know of the occurrence of variations outside a certain range. Recent work done in molecular genetics (Lewontin, 1974) thus raises important problems for evolutionary studies. Molecular geneticists have found that variations at the molecular level seem to occur more frequently than readily observable changes in external characteristics. They have also found that some amino acid changes are tolerated more readily than others. Enzymes seem to have *active sites* that are essential to proper functioning. Mutations within these regions might be more damaging than those elsewhere in the molecule and might cause death long before the variation could become apparent. Therefore, only those mutations within a certain range of phenotypic effects are likely to be noticed; those with very great or very slight influences may frequently escape detection.

Survival and Reproduction.   Mutations are usually grouped into three classes on the basis of their probable effects on a possessor's survival and reproduction: lethal mutations, sublethal mutations, and neutral mutations.

Lethal mutations are those which can cause their possessor's death. Recessive lethal genes cause death only when they are present in the homozygous condition, and so some of these genes may exist in heterozygotes. Dominant lethal genes, in contrast, cause the death even of heterozygotes. In the strict sense, any newly detected dominant lethal gene must have arisen in the reproductive cells of the affected individual's parents. There is another possibility: the gene may be variable in its expression—it may usually be lethal, but not always. These mutations can then range to *sublethal;* they can cause some reduction in an individual's ability to survive and lead a normal life, including the ability to reproduce. A good example of this is Down's syndrome. Most individuals who inherit this chromosomal abnormality die in their teens or early twenties, but a few trisomic women have given birth to children (some of whom also had the disease).

Many new mutations are lethal or subvital. This might seem puzzling at first, since we tend to equate evolutionary changes with progress. But mutations are random changes acting on the structure of nucleotides that have often survived for millions of years. Moreover, at least some of the genes we have inherited are shared in common with even earlier nonhuman ancestors. Any genes that have survived unchanged for such long periods must have proved useful in a variety of genomes subjected to quite different environments. The chance that any one random change could suddenly improve one of these nucleotide sequences is relatively remote, but the probability that such undirected change could disrupt the function of the polypeptide coded for by such a sequence is enormous.

Neutral mutations are those which have no influence on the probability of survival or reproduction of their possessor. Whether or not absolutely neutral mutations exist is currently a subject of considerable debate among evolutionists, as is their relative frequency. The case in support of the existence of neutral genes comes chiefly from molecular evidence. In some cases, the substitution of certain amino acids for others does not seem to affect either the chemical properties of proteins (such as electrical charge or three-dimensional shape) or their metabolic function. The argument against the existence of these mutations is based on theory. Each gene produces a chemical product that in most cases does not act directly and by itself to produce a single visible trait. Rather, in the course of development this gene product interacts with the products of other genes. As a result, it may influence a large number of different internal reactions and external characteristics, which is why pleiotropism exists. The basic question is whether any given gene could be completely without influence on the probability of the individual's survival or reproduction. Whether we can measure this influence is a different question which will be dealt with later.

Beneficial mutations may also exist. Each of us carries, on the average, several new mutant genes that originated in the reproductive cells of our parents. Vogel and Rathenberger (1975) have estimated that there may be as many as 17.5 new mutations per haploid gamete, or about 35 per individual. It is rare for these new mutations to become established in populations, perhaps because of their tendency to disrupt rather than improve the functions controlled by existing loci. Mutations that are beneficial, that confer above-average chances of survival, do arise from time to time. But whether a given change is helpful or harmful, and to what degree, depends on the organism's internal and external environments.

Role of Internal and External Environments

The enzymes, hormones, and other regulatory compounds produced by some genes create an *internal environment,* a physical and chemical setting that influences the expression of other loci. As a result, the rest of the genome provides a background against which any mutation is expressed. A base deletion could produce a frame shift which disrupts a polypeptide's sequence in one case but which restores an enzyme's function by canceling out a disruption caused by a previous insertion in another case. Interactions with enzymes produced by genes at other loci can further alter the expression of any mutant gene.

The external environment can also give one gene an advantage over others. Many microorganisms can live and grow in simple solutions containing a few sugars, a nitrogen source, a vitamin such as biotin, and a few other simple compounds. These raw materials are transformed by the organism's enzymes into substances needed for its functioning and survival. Any mutation that destroys an enzyme's ability to transform a given raw material into a needed product would be disadvantageous because it would interfere with the individual's ability to survive and reproduce. Individuals without the mutation would continue to reproduce as before. The situation is reversed when the solution is supplemented with the raw material needed by the mutant. Then, not only can the mutant strain survive and reproduce, but sometimes it can also do so at a faster rate than the one with the normal allele. This may be due to the fact that the mutant has eliminated several *biosynthetic steps* (each step involving the transformation of one compound into another) and saves energy by doing so (Zamenhof and Eichhorn, 1967). To use an analogy from everyday life, although nearly everyone can go through the steps of cooking a meal, it is quicker to eat one made by someone else and this frees time and energy for other tasks.

This experimental work suggests that genes not needed by the organism are just so much excess baggage—useless or even burdensome. To carry the above analogy a bit further, retaining genes that are no longer needed is like carrying hunting and camping equipment on a walk through a large city. It is very unlikely that the things could be used, and they would certainly be a nuisance. This idea that unneeded genes are a burden has recently been used to explain one inborn error of metabolism that is common to our species and a few others (Jukes and King, 1974; Pauling, 1970). Humans, monkeys, guinea pigs, the Indian fruit bat, and some species of birds all lack the ability to synthesize ascorbic acid (vitamin C). If this chemical is absent from the diet for long periods, the result is scurvy, a disease marked by weakness, loosening of the teeth, pain in the joints, bleeding in various tissues, and eventually death.

Vegetable foods contain abundant amounts of ascorbic acid. Since guinea pigs and many of the primates eat large amounts of vegetation, they would not need to make ascorbic acid. Thus the gene for the production of ascorbic acid could have been lost through mutation, with no cost and perhaps even a slight gain in efficiency. Comparative studies of the abilities of different primates to synthesize ascorbic acid could help establish the stage in primate evolution at which this genetic change took place. One thing seems certain: humans who eat a diet low in vitamin C, which is very different from the diet of their primate ancestors, will now pay the price, in impaired health, of a possible earlier gain in efficiency.

Rates of Mutation

Known Rates.   Mutations are very infrequent events. As Table 5-2 shows, the numbers range from about 10^{-4} (1 mutation per 10,000 gametes) to about 10^{-6} (1 mutation per 1,000,000 gametes). Most fall in the middle of this range, and thus the average rate of mutation for human traits studied so far is on the order of 1 gene in every 100,000 gametes in each generation. Mutation rates (abbreviated μ) are expressed in terms of the number of mutated genes which appear in a large sample of gametes in a single generation. Estimates of about the same magnitude have been found in some other organisms, such as the fruit fly. So far, human mutation rates have been calculated mostly for harmful and easily detected traits. The range and average may therefore not be representative for all human loci, and mutation rates could conceivably range from much higher to considerably lower than those now known.

Many factors determine whether a given type of genetic change will be detected and studied. Some mutations, though serious, may occur so rarely that they never come to a scientist's attention. But, as we saw earlier, a mutation's effect on phenotypic appearance could also play a role in whether or not even a common change is detected. Slight changes may produce a phenotype within the "normal" range of expression and so go unnoticed. Mutations that produce major changes can also be missed. It has been estimated that one-fourth to nearly one-half of all human embryos are spontaneously aborted early in pregnancy. These losses are often neither noticed by the woman nor studied by a doctor. One study discussed by Vogel and Rathenberger (1975) has shown that among spontaneously aborted embryos large enough to be seen, about one in every four or five has a detectable chromosome abnormality. This process might therefore provide a very effective screen against the birth of individuals with gross disturbances. Mutation-

---

TABLE 5-2   Mutation rates in humans

Trait	Type of mutation	Population	Mutation rate
Achondroplasia (skeletal abnormality causing dwarfism)	Autosomal dominant	Denmark	1×10^{-5}
Retinoblastoma (tumors of the retina of the eye)	Autosomal dominant	Michigan	8×10^{-6}
Neurofibromatosis (abnormal pigment spots on the skin and tumors in the nervous system)	Autosomal dominant	Michigan	1×10^{-4}
Hemophilia A (failure of the blood to clot properly)	Sex-linked recessive	Finland	3.2×10^{-5}

Source: Friedrich Vogel. 1971. "Mutations in Man," in L. N. Morris (ed.), *Human Populations, Genetic Variation, and Evolution*. San Francisco: Chandler. Table 1, pp. 134–135.

rate estimates are thus true for traits that arise frequently enough and with effects that are large enough to attract attention, but not so great as to interfere too early with survival.

There are other complications as well. Phenotypic traits may be heterogeneous. Several individuals may appear to have the same trait—the same disease or the same set of characteristics. But some cases may be caused by quite different loci from those which cause others. Hemophilia (failure of the blood to clot quickly after a cut) was thought to be caused by a defect in a single sex-linked gene. Now two different sex-linked types are known, each controlled by a different locus. *Phenocopies*, traits caused by environmental agents that mimic the result usually produced by mutations, may be present. Abnormally slow clotting of the blood, for example, can also be due to a deficiency of vitamin K, and it can be produced by the administration of drugs such as heparin.

Calculating Mutation Rates   Mutation rates for specific traits can be determined by *direct* and *indirect* methods. The direct method can be used only for dominant mutations, but it is simpler and so will be discussed in detail here. The indirect method can be used for either dominant or recessive mutations. Thus it is possible, at least for dominant mutations, to determine rates in two ways to check on the consistency of results. The indirect method is mathematically more complex and depends on an understanding of certain topics that are more properly discussed in the chapter on selection; see Stern (1973) for a good presentation of this method. In using the direct method, two major assumptions are necessary: (1) that in the group all mutations of the type under study are detected and (2) that the trait is always produced by the same dominant gene, and not by recessive alleles, genes at other loci, or phenocopies.

An early (1941) and now classic example of direct determination of mutation rates was done by the Danish geneticist E. T. Mørch. He used records on 94,075 births at the Lying-in Hospital in Copenhagen, among which were 10 children with achondroplasia. This condition is due to a defect in bone growth that produces very short arms and legs on a body of usual size. It is caused by a dominant gene; individuals of normal phenotype are dd, and achondroplastic dwarfs are Dd. Genotype DD is a theoretical possibility from a marriage between two dwarfs. Two of the ten affected newborn infants had a parent who was also a dwarf. By overwhelming probability, therefore, these two cases were due to normal inheritance of the parent's dominant gene, and not to new mutations. But in the remaining eight cases both parents were normal; each of the affected children was the inheritor of a new dominant mutation to achondroplasia.

The information on the incidence of the condition can be used to calculate the mutation rate. Each of the 94,075 children had two alleles at this locus; thus, a total of 188,150 genes is represented. Of these, eight are the result of new mutants. Thus

$$\mu = \frac{\text{new mutant genes}}{\text{total alleles}} = \frac{8D}{188,150} = 0.000043 \cong 4 \times 10^{-5}$$

In this population, then, the mutation from d to D occurred about 4 times in every 100,000 genes.

Mutation as a Force of Evolution

Mutations are the ultimate source of new genetic material for species in their evolution. But with rates as low as those estimated above, is mutation of any quantitative significance? The question becomes more interesting if we put it in the context of events that have a direct influence on our lives.

It has been estimated from both mice and fruit flies that the amount of radiation necessary to cause a doubling of the mutation rate is on the order of 40 rads. The International Commission on Radiological Protection has suggested that in the future the maximum radiation exposure of human populations from peaceful uses of atomic energy be an average of about 5 rads per person in each generation. To this we can add a probable additional 3 rads for medical radiation. This total of 8 rads would be about 20 percent of the doubling dose and might be expected to increase the mutation rate by a proportionate amount each generation. Studies of human populations indicate that about 8 individuals out of every 100 carry one new lethal gene or its equivalent (several sublethals, which add up to the same total effect). Increasing this by 8 percent would add another 64 lethal genes per 10,000 births.

In 1973 there were about 3,141,000 live births in the United States. The allowable figure for radiation exposure given above would produce in the American population each year over 20,000 additional mutations serious enough to cause death. This is a staggering figure. How many deaths would have occurred if alternative sources of energy had been used? And what is the significance of this figure when weighed against the 56,590 deaths from automobile accidents in the United States in the same year? There are no simple answers to questions like these, but it does help to have the perspective that wider knowledge can provide. Mutation has always been part of the price that must be paid for the survival of populations during evolution. The major difference between the present and the past is the extent to which this force of evolution can be brought under conscious and rational control.

GENE FLOW

Gene flow is the transfer of alleles from one group to another through matings that take place across population boundaries. Unlike mutation, gene flow *decreases* differences between populations, making them more alike genetically. This occurs because as an allele increases in frequency in one population, the chances are increased that it will be present in someone who goes to a second group as a result of mate exchange. The allele can then be transmitted to children born in the second group. In this way the allele would decrease slightly in frequency in the group from which the mate was drawn and would increase in frequency in the group which he or she joined. Intergroup marriages of this type have produced a slow trickle of genes from population to population that has helped to maintain the unity of our species over a very long span of time. Although *migration* is commonly used as a synonym for *gene flow*, it really refers to the mass movement of people (see "Major Migrations in the History of Europe," pages 158–159) rather than to the transmission of genes. Migration is often followed by *hybridization*, large-scale intermarriage that melds two previously distinct populations into a single new group. Migration and gene mixture can occur independently of each other. For example, in many large cities, immigrant populations may remain at least partially *endogamous* (choosing mates from within their own groups), as opposed to being *exogamous* (bringing mates in from other groups), for several generations. The settlement of the New World provides an even clearer example of the spread of people without genetic mixing, though as defined above the term *migration* is not entirely appropriate to this case. Over 10,000 years ago, small hunting bands wandered across the Bering Straits into North America. These ancestors of the American Indians encountered no other human inhabitants; thus as they traveled across the continent, they remained genetically unmixed until the European migrations began in the sixteenth century. Gene flow, in contrast, often takes place with little or no population movement. For example, in each generation about 15 percent of marriages among Australian Aborigines take place between members of adjacent tribes. Although the exact percentage may differ from group to group and from time to time, this trickle of genes from one population to another can serve as a substantial homogenizing influence on the gene pool of a species and can counter the differentiating effects of other evolutionary forces.

Role of Gene Flow in Evolution

Large-scale movements of populations are unlikely unless the people involved have two important items in their material culture. The first is an efficient means of transportation, and the second is a supply of food

MAJOR MIGRATIONS IN THE HISTORY OF EUROPE

Barbarians as an epithet for the peoples who settled in the Roman Empire in the fourth to the sixth centuries after Christ is no more satisfactory than *immigrants* as a term to describe everybody, from William Bradford to Al Capone, who settled in America. The barbarians did not speak a common language, nor did they all migrate simultaneously, nor did they regard themselves as one people, any more than Lord Baltimore's descendants would have felt deep kinship with nineteenth-century Irish Catholic immigrants—even though the point of origin was the same. The Visigoths and Ostrogoths (see the arrows on the map opposite) had moved out of Scandinavia by the second century after Christ; they settled down along the Danube close by the Alans, and eventually all these people turned westward. The Visigoths, after a long journey through Greece and the Balkans, sacked Rome in A.D. 410 and during the same century settled in Spain. As allies of the Romans they fought in Spain against the Vandals and farther north against the Huns and Franks. The Huns, a vigorous Asiatic people, entered Europe around A.D. 373; their career lasted only about 75 years (by way of contrast to that of the enduring Goths). The Vandals, too, had an energetic but brief life as a nation. Joined by the Suevians, they left Germany around A.D. 400, swept into Spain 10 years later, settled in North Africa 30 years after that, and themselves sacked Rome in A.D. 455, only to be annihilated by the eastern Romans 80 years later. The trajectories of the major barbarian tribes can all be traced on the map: Jutes, Angles, and Saxons invading Britain; Vandals in north Africa; Huns apparently exploding and then vanishing; Goths everywhere; Alans traveling slowly westward from north of Persia; and Franks, Burgundians, and Lombards making briefer journeys but more indelible marks. What no map can show is how these newcomers mingled with one another and with the Romans or how they eventually settled down among the ruins of the classical world to build a feudal civilization and slowly to create all the Romance and Germanic languages spoken in the modern world.

Source: Adapted from Robert Winston. 1970. "The Barbarians," *Horizon,* **12**(3):66–81, and map on pp. 76, 77.

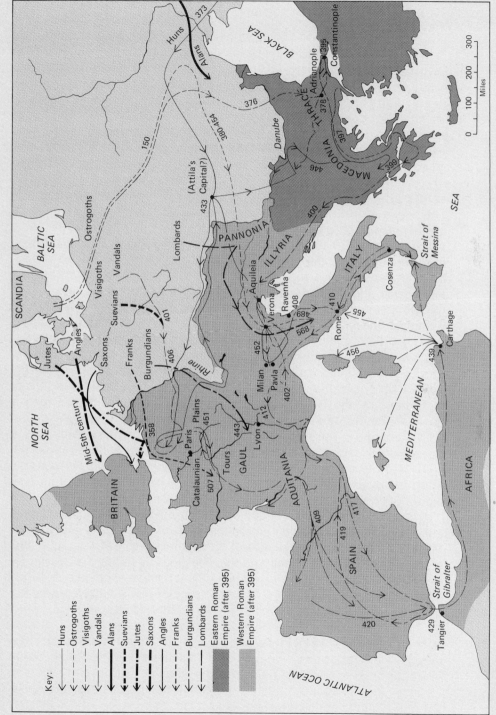

Migrations of barbarian peoples.

Key:
- Huns
- Ostrogoths
- Visigoths
- Vandals
- Alans
- Suevians
- Jutes
- Saxons
- Angles
- Franks
- Burgundians
- Lombards
- Eastern Roman Empire (after 395)
- Western Roman Empire (after 395)

SCANDIA

BALTIC SEA

NORTH SEA

Jutes
Angles
Saxons
Suevians
Visigoths
Vandals
Ostrogoths
Huns

373
150
376
380-454
433 (Attila's Capital?)

PANNONIA
ILLYRIA
THRACE
MACEDONIA
Adrianople 378
Constantinople 395
BLACK SEA

Lombards
Franks
Burgundians
401
406
Rhine

Mid-5th century
358

BRITAIN
Paris
Catalaunian Plains 451
Tours 507
GAUL
Lyon 443
412
AQUITANIA

Aquileia
Verona
Ravenna 408
Milan 452
Pavia
402
489
493
410
400
446
397
399

ITALY
Cosenza
Rome 455
456
Strait of Messina

Carthage 439

MEDITERRANEAN SEA

SPAIN
409
419
417
420

AFRICA

Strait of Gibraltar
Tangier 429

ATLANTIC OCEAN

Miles
0 100 200 300

that can sustain the population on its journey. A large group moving through unfamiliar territory could not be sure of getting enough food by hunting or collecting. At least part of its food supply would have to be brought along and would therefore have to be nonperishable and easily transported. The problem could be solved by carrying concentrated high-energy cereal grains such as wheat, by having herds of animals that could be driven along, or even by raiding the supplies of settled populations through whose territory the migrants passed. These two requirements for transportation and food make it probable that mass migrations were relatively rare before the agricultural revolution that occurred in the Old World during the Neolithic period, about 8000 to 10,000 years ago. (The Neolithic period is formally defined as the time when ground or polished stone replaced the chipped stone tools of the preceding Paleolithic and Mesolithic periods.) Some of the same cultural conditions that brought about the domestication of plants and animals also led to the invention of writing a few thousand years later. For these reasons, in the last few thousand years of recorded history large-scale migrations occurred with far greater frequency than during the preceding 3 million years of human evolution.

Many of these migrations struck like waves on the human populations already settled in each region, leaving marks on cultural and historical records as well as churning up gene pools. In the history of western Europe, the shifts of various Asian and Germanic tribes became particularly frequent beginning in the fourth century after Christ (see "Major Migrations in the History of Europe") and are remembered largely because the invaders contributed to the fall of Rome toward the end of the fifth century. The frequency of these major population shifts during the period of recorded history, and their impact on political and other cultural developments, may explain why migration has been used so often as an explanation for earlier patterns of human evolution (see Figure 5-6 for an example reprinted from an earlier text).

Archeological evidence suggests that, prior to the Neolithic period, most humans lived in small bands that occupied relatively fixed areas. Most gene flow probably occurred through matings between members of immediately adjacent populations. Present-day hunter-gatherers, for example, have systematic arrangements for mate exchange between adjacent bands to make up for the shortage of mates within one group or to ensure peaceful relations between groups. Most hunter-gatherers are *patrilocal*, which means that the males stay in the territory of their fathers, while the women are given as mates to surrounding groups, though there are many variations on this basic theme. Some South American tribes conduct raids for the specific purpose of stealing women. Conflict between other American Indian groups has sometimes led to the annihilation of the males and the assimilation of the females into the victorious group.

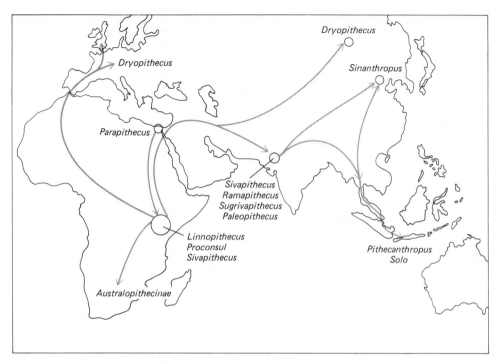

FIGURE 5-6   Migration of Miocene apes from east Africa. Groups named on this map are nonhuman primates or primative human hunter-gatherers. The migrations shown are highly conjectural.

Source: Bertram S. Kraus. 1964. *The Basis of Human Evolution.* New York: Harper and Row. P. 259. Copyright © 1964 by Bertram S. Kraus; used by permission of Harper and Row, Publishers, Inc.

In still earlier stages of human evolution, before the development of language, human populations may have mated with members of their own group more frequently than they do now. But even without culturally determined rules for mate exchange, nonhuman primates and other animals maintain gene flow among subpopulations, and thus it is reasonable to believe that our early human ancestors did so as well. Gene flow through mate exchange among adjacent bands or groups has probably been more typical for most of human evolution than either complete isolation or large-scale migration and hybridization.

How Gene Flow Works

Rate of Gene Flow.   The rate of gene flow depends on the size of the population and the proportion of it that is exchanged with surrounding groups each generation. If the size of the breeding population N is 200 and the number of mates m exchanged with neighboring groups is 20, then $m/N = {}^{20}\!/_{200}$, for a gene-flow rate of 10 percent in this generation.

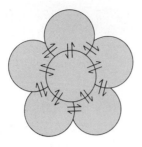

FIGURE 5-7
Spatial distribution of small human populations.

It is unlikely that all the gene exchange will be with just one group, however. Most small human populations are surrounded by several adjacent groups among whom gene flow will be divided, as shown in Figure 5-7. Thus, if the total gene-flow rate is 10 percent and this is divided evenly among the five groups, the rate of gene flow into any single group will be about 2 percent.

Gene Flow through Time. Genes flow from population to population each generation. If an allele is present in just one population in an area, during one generation it can diffuse to the ring of surrounding populations, but not until the next generation can it flow from any one of these to more distant groups. Furthermore, unless something has occurred to increase the frequency of the allele within each group, its frequency will be diluted still further in each of the more distant groups to which it is transmitted (Figure 5-8). By this gradual, generation-to-generation process alone it is possible for a new gene to spread over a very wide area, though the rate is very slow.

A low but relatively constant flow from one region to another is one influence (others are discussed in Chapter 6) that can produce a genetic *cline* (Glass, 1954), a progressive change in the frequency of a char-

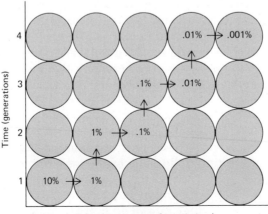

FIGURE 5-8   Gene flow in time and space. As a gene is transferred through five populations over the course of four generations, its frequency can be reduced from a relatively high level (about 10 percent) to a level so low that it will all but vanish.

acteristic from one population to the next in a geographic area. One example of a cline thought to be due to gene flow is the gradual decrease in the frequency of the B allele of the ABO blood group system from about .25 to .30 in central Asia to nearly 0 in parts of western Europe. Candela (1942) has suggested that this cline may be due to the westward expansion of Mongolian populations between A.D. 500 and 1500. A very different spatial distribution of gene frequencies would result from a long-distance migration by a group that did not mate with populations between its place of origin and its destination. In such a case, frequencies would be expected to show a sharply discontinuous pattern (Morris, 1972). In the area of resettlement the result is often a hybrid population, distinctive in gene frequency from surrounding groups as well as from the migrant ancestral group.

Migration and Hybridization

Although large-scale migration and hybridization of populations is a relatively recent occurrence in the history of our species, it is and has been, as we have seen, frequent and well recorded. Let us look at some cases of actual and probable hybridization.

Hawaii.   This state consists of a group of islands that lie in the Pacific Ocean almost at the midpoint of a straight line drawn between California and Australia. The first humans reached Hawaii from other islands in Polynesia sometime within the last 2000 years, possibly not much earlier than A.D. 1000 or 1200. From a western perspective, the islands and their inhabitants (one-third of a million people) were discovered by Captain James Cook in 1778. By 1810 a whaling station had been established. When ships stopped for supplies, interbreeding between sailors and native women was an inevitable result. Traders and missionaries arrived a decade later, contact between the two peoples now became wider and more permanent. Unfamiliar infectious diseases such as measles soon took a subtantial toll of native Hawaiians. But beginning in the 1850s, laborers from China, Japan, Korea, the Philippines, and elsewhere were brought in to work on the large plantations established by American entrepreneurs. By the middle of the twentieth century, a federal census of Hawaii (Table 5-3) showed that people of nearly pure or predominantly Polynesian ancestry had decreased in number from over 300,000 at the time of Cook's contact to fewer than 65,000 but that the total population of the island had grown by over 50 percent. What is more, all these diverse human elements had been united into a new stock called *golden people* by some writers.

The Basques.   Hawaii provides perhaps the most visible example of a human blend. The case of the Basques of France and Spain is a bit

TABLE 5-3   Population composition of Hawaii

Population	Numbers
Japanese	156,000
Caucasians	115,000
Filipinos	100,000
Chinese	28,000
Koreans	6,000
Part Hawaiian	50,000
Pure Hawaiian	14,000
Total	469,000

Source: Federal census of 1940.

different. These people are still isolated, but we are able to predict some of the changes in gene frequency that would result if they began to interbreed with adjacent groups.

The Basques, a genetic isolate, live on the border between Spain and France, from the Pyrenees down to the southeast coast of the Bay of Biscay. They are isolated partly by geography and partly by culture. The most important cultural difference is their language, which, unlike Spanish and French, is not an Indo-European tongue. The barriers are not complete, since on the edges of their territory Basque gene frequencies grade into those of the surrounding populations. Nevertheless, some of their characteristics are unusual for Europe. One example occurs at the Rh blood group locus. Among Basques, the Rh-negative allele has a frequency of .53. Elsewhere in Europe it is unusual for this allele to exceed a frequency of .40; in the non-Basque population of Spain it is about .35. What would happen if at some point in the future the Basques interbred freely with Spanish people?

We can predict the gene frequency at any locus following hybridization if we know the proportions that each of the formerly separate populations contributes to the hybrid group and if we know the gene frequency in each parent population. The formula used here is

$$q_H = m_1 q_1 + m_2 q_2$$

where

m_1 = percent of the hybrid population contributed by population 1
m_2 = percent of the hybrid population contributed by population 2
q_1 = frequency of one allele in population 1
q_2 = frequency of the same allele in population 2
q_H = frequency of the same allele in the hybrid population

In 1970 there were about 750,000 people of Basque descent in Spain and 33,250,000 non-Basques, a total of about 34,000,000. The Basque proportion of the total (m_1) is 750,000/34,000,000, or about 0.02 (2 percent). By subtraction, the non-Basque proportion (m_2) is 0.98 (98 percent) of the total. The Rh-negative allele frequencies are $q_1 = .53$ and $q_2 = .35$. Combining this information, we find that

$$q_H = (.02)(.53) + (.98)(.35) = .0106 + .3430 = .3536$$

This predicted frequency is only a tiny fraction higher than the actual frequency of the Rh-negative allele in the non-Basque Spanish right now. In this case, before hybridization there would be a substantial difference in gene frequency between the two parent populations. Afterward, the contribution of the smaller group would be swamped—diluted so much that it could no longer be detected—by the genes derived from the larger group.

American Blacks   Anthropologists are not often asked to predict what the gene frequency will be in a potential hybrid population; far more commonly they are asked to help reconstruct processes that went on in the past, such as the extent to which differential ancestral populations contributed to some present hybrid group. American blacks provide a good example of such a situation. Can we reconstruct the proportions of genes that this population has received from various sources?

From the colonial period to the Civil War, the ancestors of today's American blacks were brought here as slaves, chiefly from west Africa. Their gene pool has been augmented by alleles from a white population of predominantly European ancestry. Some American Indians may have added their genes to the hybrid population, but their contribution is thought to have been very slight (Glass, 1955). Most anthropologists assume that west African blacks and American whites of European ancestry are the major ingredients of the hybrid American black population. Another major assumption in estimating the sources of alleles in the black gene pool is that gene frequencies in west African blacks and American whites are the same today as they were when hybridization began. There is no way of directly testing the accuracy of this assumption, but it seems reasonable in light of what is known of the history of the populations. A more complete list of the assumptions used in hybridization studies is given in Table 5-4.

In theory, any allele that differs in frequency between the two populations could be used to estimate the extent of hybridization. In practice, it is best to choose a gene for which the ancestral populations differ greatly in frequency. Some alleles at the Rh blood group locus meet this requirement. For example, the $R°$ allele is common ($q_1 \cong .55$) in west African blacks and infrequent ($q_2 = .03$) in American whites. In American blacks, it has an intermediate frequency ($q_H \cong .45$).

The formula by which this information can be used to measure the relative genetic contributions of one of the ancestral groups to the hybrid population is

$$m_1 = \left| \frac{q_H - q_2}{q_1 - q_2} \right|$$

TABLE 5-4   Assumptions made in estimating the composition of hybrid populations

1. The exact composition of the populations ancestral to the hybrid population is known.

2. The allele frequencies used are based on representative samples from the correct populations (those actually ancestral to the hybrid population).

3. There has been no change in frequency of the allele studied in the ancestral populations from the time at which hybridization began to the present.

4. Interbreeding between the ancestral populations is the only factor which has determined gene frequency in the hybrid population; that is, no mutation, genetic drift, or selection has occurred.

Source: Adapted from T. E. Reed. 1969. "Caucasian Genes in American Negroes," *Science,* **165:** 762–768. Reprinted in Laura Newell Morris (ed.). 1972. *Human Populations, Genetic Variation, and Evolution.* San Francisco: Chandler.

---

where m_1 is the proportion added by west African blacks and q_1, q_2, and q_H are the frequencies of the same allele (here R°) in each of the populations. The vertical lines enclosing the right side of the formula indicate that only the absolute value of the expression is important here; that is, the sign may be disregarded. This is logical, since a group cannot make a negative contribution.

In the example here,

$$m_1 = \left| \frac{.45 - .03}{.55 - .03} \right| \cong .8$$

This calculation suggests that approximately 80 percent of the R° alleles in the American black population came from their west African ancestors. Results similar to these have been obtained using data from several other genetic loci. Of course, any single locus gives only an estimate of the total amount of hybridization, which is really a property of the whole gene pool. But estimates from other loci generally indicate that about 20 to 30 percent of the alleles in the gene pool of American blacks are the result of gene flow from American whites.

This gene flow probably took place as a generation-by-generation trickle. Before the Civil War, gene flow between the two groups was predominantly in one direction, from matings between white men and black women. Later, matings between pairs of hybrid parents produced some children with phenotypic traits within the range of the highly variable white population, a range defined by society and based on appearance and ancestry. Those whose hybrid ancestry could not be detected from their appearance could and often did enter the white population, and so gene flow was no longer only unidirectional. In recent years the remaining barriers to gene flow have substantially decreased, though perhaps not symmetrically. Dyer (1976) reports that since the Civil War matings between black males and white females have been more frequent than the opposite combination. Over the long term, cultural barriers to gene flow between socially defined blacks and whites may well disappear completely. If completely random mating occurred between the two groups, at equilibrium the resulting population would look much like the present white population. It would take a number of generations for equilibrium to be reached if these two semi-isolated gene pools were to become blended into one. In fact, at the present time there is not a uniform distribution of alleles from African and European ancestors even within the American black population. The figure of 20 percent white admixture given above is only an average; it has been estimated that about one-fourth of all American blacks have more than 50 percent European ancestry. Fewer than about 1 in 20 have virtually unmixed African ancestry, and some people socially defined as black have received over 90 percent of their genes from European ancestors.

Pygmies and Negritos.   Genetic methods are powerful tools for teasing out the intertwined strands of ancestry from a population's evolutionary line. But these tools are useless without historical records; it is just not possible to use genetic traits alone to reconstruct the ancestry of a population. The Pygmies and Negritos of Africa and southeast Asia, respectively, illustrate the dangers of trying to use genetic or phenotypic data alone to reconstruct ancestry. Both groups are dark-skinned and short in stature (males average less than 152 cm—under 5 ft—and females are shorter still), and both have relatively large heads with tightly curled hair. These shared features led some anthropologists to suggest a common ancestry for the two groups, although they are separated by 9600 km (6000 mi). A frequent explanation was that the Pygmies and Negritos had a common origin at some time in the distant past and were the earliest inhabitants of the areas in which they are found at present. Any differences between the two (for instance, in frequencies of blood group alleles) were said to be due to hybridization with populations thought to have entered Africa and southeast Asia later.

But not all differences between Pygmies and Negritos can be explained by hybridization, as Table 5-5 shows. For example, Pygmies of the Ituri region of the Congo differ substantially from the short-statured Semang Negritos of Malaya in frequency of the B allele at the ABO locus (21.9 versus 9.0). The high frequency of the B allele in Ituri Pygmies cannot be accounted for by gene flow from blacks of full stature in their own region because they have a much *lower* frequency of this gene. Nor can the low frequency of the B allele in the Semang be explained by gene flow, since neighboring Malayans have a *higher* frequency of the B gene than the Semang. Similarity of appearance, therefore, does not always

TABLE 5-5   Contrast between stature and ABO gene frequencies as measures of similarity

| | Populations | | | |
| | Africa | | Southeast Asia | |
Alleles	Normal stature (Ituri region); N = 150	Pygmies (Ituri region); N = 8000	Negritos, Semang; N = 119	Normal-stature Malayans; N = 1963
A	21.3	22.7	8.3	18.7
B	7.7	21.9	9.0	18.1
O	71.1	55.3	83.0	63.2

Source: Based on Ilse Schwidetzky. 1962. "Neuere Entwicklungen in der Rassenkunde des Menschen," in Ilse Schwidetzky (ed.), *Die Neue Rassenkunde.* Stuttgart: Gustav Fischer Verlag. Table 12, p. 81; table 14, p. 84.

indicate closeness of genetic relationship. There are many other explanations for patterns of genetic and phenotypic resemblances. Gene pools that were originally different can be made more similar by the operation of common selective factors, as outlined in Chapter 6. Even in the absence of evolutionary changes, different genotypes can be molded into similar phenotypes, as demonstrated by the work of Boas and others on immigrants to the United States, taken up in Chapter 7. Without historical information, we could not pick any one explanation with any certainty of being able to prove it.

SUMMARY

1. Mutations, the source of all genetic differences, are random heritable changes that can occur in any cell; only gamete mutations, however, can be passed on to future generations.

2. These inherited rearrangements occur on a wide scale from point mutations (which include base substitutions and frame shifts due to the addition or deletion of bases) through structural variations in chromosomes (which include larger-scale deletions, duplications, inversions, and translocations) to changes in chromosome number (the gain or loss of single chromosomes or whole haploid sets).

3. For individuals, the phenotypic effects of different mutations range from invisible to very noticeable, and their impact on the ability to survive and reproduce ranges from apparently none to either a lethal or a positive advantage.

4. Most new mutations are harmful because they disrupt DNA sequences built up over millions of years of evolution; however, rare mutation can confer a benefit on individuals or populations.

5. The relative advantage or disadvantage of a given mutation also depends on the genetic background of the individual and on the external environmental conditions. These factors will determine whether a mutation is lost from the population or increases in frequency in later generations.

6. Frequencies of both dominant and recessive mutations can be calculated.

7. Rates of mutations are low (averaging about 1 gene per 100,000 gametes per generation), but the known range is wide (about an order of magnitude above or below the average), and there are reasons to believe it might be even wider.

8. Gene flow is the addition of foreign alleles to a gene pool as a result of interbreeding between two populations.

9. The terms *migration* and *hybridization,* which are sometimes used as synonyms for *gene flow,* instead describe mass population movements and any resulting interbreeding on a large scale.

10. For over 99 percent of human evolution, gene flow has probably operated at a low but relatively constant rate, helping to maintain the unity of the species.

11. The gene frequency in a potential hybrid population can be predicted if the numbers and gene frequencies of component groups are known; these predictions could help estimate the frequency of inherited traits, including genetic diseases, in future populations.

12. If gene frequencies are known, the proportions contributed by known ancestral populations to a hybrid population can be estimated, as in the case of American blacks.

SUGGESTIONS FOR ADDITIONAL READING

Boyd, William C. 1963. "Four Achievements of the Genetical Method in Physical Anthropology," *American Anthropologist,* **65**:243–252. (This brief article is written by one of the earliest active campaigners for the application of modern genetic methods to traditional anthropological problems. Boyd describes how these methods have helped clarify the ancestry of the Gypsies, American blacks, the Lapps, and the Pygmies.)

Morris, Laura Newell. 1971. "Mutation," sec. III and associated readings, in L. N. Morris (ed.), *Human Populations, Genetic Variation, and Evolution.* San Francisco: Chandler. Pp. 105–120. (These provide a relatively recent selection of basic references on the subject, useful at an intermediate level.)

Novitski, Edward. 1977. In *Human Genetics.* New York: Macmillan. Chaps. 10–16. (These chapters provide good, clear, detailed descriptions of the mechanisms and causes of mutation in humans. The discussion is at a level well suited to undergraduate students.)

Ohno, Susumu. 1970. *Evolution by Gene Duplication.* New York: Springer-Verlag. (This is a speculative but fascinating overview, at an advanced level, of the role that mutation may have played in the origin of major evolutionary advances.)

REFERENCES

Candela, P. B. 1942. "The Introduction of Blood-Group B into Europe," *Human Biology,* **14**(2):413–443.

Crow, James F., and M. Kimura. 1970. "Populations in Approximate Equilibrium" and "Distribution of Gene Frequencies in Populations," in *An Introduction to Population Genetics Theory.* New York: Harper and Row. Chaps. 6, 9.

Dyer, K. F. 1976. Patterns of Gene Flow between Negroes and Whites in the U.S.," *Journal of Biosocial Science,* **8**(2):309–333.

Fredrickson, D. S., and E. G. Trams. 1966. "Ganglioside Lipidosis: Tay-Sachs Disease," in J. B. Stanbury, J. B. Wyngaarden, and D. S. Fredrickson (eds.), *The Metabolic Basis of Inherited Disease.* New York: McGraw-Hill. Pp. 523–528.

Glass, B. 1954. "Genetic Changes in Human Populations, Especially Those Due to Gene Flow and Genetic Drift," *Advances in Genetics,* **6**:95–139.

Glass, B. 1955. "On the Unlikelihood of Significant Admixture of Genes from the North American Indians in the Present Composition of Negroes in the United States," *American Journal of Human Genetics,* **5**(4):1–20.

Jukes, T. H., and J. L. King. 1975. "Evolutionary Loss of Ascorbic Acid Synthesizing Ability," *Journal of Human Evolution,* **4**(1):85–88.

Lewontin, R. C. 1974. *The Genetic Basis of Evolutionary Change.* New York: Columbia University Press.

Mørch, E. T. 1941. "Chondrodystrophic Dwarfs in Denmark," *Opera ex Domo Biologiae Hereditairiae Humanae Universitatis Hafniensis,* **3**:1–200.

Morris, Laura Newell. 1972. "Gene Flow," in L. N. Morris (ed.), *Human Populations, Genetic Variation, and Evolution.* San Francisco: Chandler. Pp. 409–426.

Pauling, L. 1970. "Evolution and the Need for Ascorbic Acid," *Proceedings of the National Academy of Sciences, U.S.A.,* **67**(4):1643–1648.

Reed, T. E. 1964. "Caucasian Genes in American Negroes," *Science,* **165**(3895):762–768.

Stern, C. 1973. *Principles of Human Genetics.* San Francisco: Freeman.

Vogel, F., and R. Rathenberger. 1975. "Spontaneous Mutation in Man," *Advances in Human Genetics,* **5**:223–318.

Zamenhof, S., and H. H. Eichhorn. 1967. "Study of Microbial Evolution through Loss of Biosynthetic Functions: Establishment of 'Defective' Mutants," *Nature,* **216**(5114):456–458.

Chapter 6

Structuring Genetic Variation: Genetic Drift and Selection

Mutation and gene flow introduce alleles into populations, serving as sources of the genetic variation that is the raw material for evolution. Gene flow also decreases differences between populations, and mutation often has much the same effect. This is so because mutations are similar in kind and frequency in human populations of reasonably large size, while in small populations a mutation should be so rare that by itself it could do little to change gene frequency. By the process of elimination, genetic drift and natural selection remain as the chief forces of evolution that can produce genetic differences between populations and the major shifts in genotypic and phenotypic characteristics in the course of human evolution.

Genetic drift is random change in gene frequency due to chance in a small population. It occurs because the gene frequency among the offspring in a population may not be exactly the same as the gene frequency at the same locus in the generation of their parents. This change takes place simply as a result of accidents. The smaller the population, the greater the chance of *sampling error,* the chance that a small group (here, the offspring actually born) is not representative of a larger one from which it is drawn (in this case, all the offspring who might possibly have been born). *Selection* takes place when some genotypes produce more surviving offspring than others. Both drift and selection can change gene frequencies, but they do so in different ways.

Selection can lower the frequency of an allele by reducing the chances of survival or reproduction of an individual who has it. Drift—accidental change—can shift the frequency of an allele regardless of whether the gene benefits or harms the organism.

GENETIC DRIFT AND HUMAN POPULATIONS

The sampling error that produces genetic drift happens at several points in the reproductive process. First, some but not all members of the population have the opportunity to reproduce. Then, sampling occurs again as combinations of genes are packaged into gametes, and yet again as one sperm from among the millions produced by a man fertilizes the single egg that happens to be released in a given month by a woman. The chance for error exists in a small population because each generation produces many more gametes than are used in reproduction. If only a few offspring are produced, they may not display the whole range of characteristics present in the parents, and these characteristics will then be lost to that population. With a large population and many offspring, the chances that this will happen are far less. Although we are now worried about overpopulation, for most of the evolutionary history of our species human populations have been small, and genetic drift has therefore been an important force.

Hunter-Gatherers: The Unit of Human Evolution

The present world population of about 4 billion people has been made possible in part by the industrial revolution, which began several hundred years ago in some regions and is continuing to spread even now. The industrial revolution, in turn, rests on agriculture, which originated only about 10,000 years ago. For 99 percent of the last 3 million years of human evolution, then, human groups depended for survival on hunting wild animals and gathering uncultivated food plants. Lack of a fixed, controllable food supply and the technology to expand it kept these groups small and therefore subject to the effects of genetic drift.

A few hunting-and-gathering groups still survive today, and other groups lived by this subsistence pattern recently enough for information on their way of life to have been recorded; thus we have a basis for looking at, and generalizing about, the characteristics of these populations in terms of evolution. Of course, recent hunter-gatherers may not be identical to such groups in the past. For one thing, virtually all recent and contemporary groups occupy marginal habitats, having been displaced from richer areas by agricultural, herding, and industrial populations. Some items of material culture, including tools and weapons, have changed over the course of time, and some aspects of behavior and social organization may have changed as well. Certainly environments have been altered by changes in climate, which also affect the richness of the food resources available. Some hunting-and-gathering groups, for instance, depended on seasonal migrations of large game animals, and others foraged for smaller game and vegetable foods. Just as there have been differences through time, there are variations now. Not all recent hunting-and-gathering groups are alike, and some differ from the norm in ways that influence population size and structure. With these cautions in mind, it is still possible for us to use the available information with some confidence, for all hunting-and-gathering groups share some important features in common, and they differ as a category from peoples who have a different means of subsistence.

Density.   Compared with the agricultural and industrial societies that are much more familiar to us, hunting-and-gathering groups are spread quite thinly over the regions in which they live. *Population density,* the number of individuals living on a given unit of land area, is the measure used by anthropologists and other scientists to express this distribution. Given a measure of population density and an estimate of the total land area inhabited, the size of the human population in any region can be reconstructed. When such an estimate of size is combined with knowledge of some common features of human behavior, it also becomes possible to create a picture of *population structure:* the ways in which breeding groups are distributed over the regions they inhabit and the consequences of the distribution for systems of mating.

Population densities of recent hunting-and-gathering peoples vary over a wide range. The lowest levels occur where the climate is harsh. Aborigines living in desertlike areas of south Australia, where rainfall is less than 25.4 cm (10 in) per year, have densities of 1 person for each 207 to 233 km² (80 to 90 mi²). Slightly lower densities of about 1 person per 324 km² (125 mi²) are recorded for the Caribou Eskimos, who live in the barren, cold area west of Hudson Bay. If we use these cases as a standard, 1 person per 259 km² (100 mi²) is a fair estimate of the lowest human population density. The highest densities occur in areas where food resources are abundant. Before the Europeans came, as many as 10 people per 2.6 km² (1 mi²) lived in parts of California where pine nuts, acorns, and a variety of other seeds were in rich supply. The northwest coast of the United States and Canada, with abundant fish and sea mammals and high rainfall producing level vegetation, may have supported even higher densities. Some islands off the coast of Australia had densities as high as 2 or 3 people per 2.6 km² (1 mi²), but there most food came from the tidal regions along the shore; 1 person per 2.6 km² (1 mi²) is probably close to the highest density for hunter-gatherers in inland areas.

Group Size. Population density is an important factor in determining the size of human breeding populations, but there are other influences as well. Humans are social animals who live in groups. Among hunter-gatherers, three types of groups commonly occur: the family, the band, and the tribe. The smallest and most fundamental unit is the *family.* This may be either a *nuclear family* (which includes a man, one or more wives, and their unmarried children) or an *extended family* (which may consist of a father and his married sons and their families or of a group of brothers and their families). Nuclear families are small social and economic units with relatively high mobility and low ecological impact. When food is not abundant in any single place, families may scatter over the available territory and survive without overtaxing the local resources. At other times, several families may come together to form a *band* of 20 to 60 individuals. The family-based bands are really quite fluid in composition, and each year they may combine with one to five other bands.

Most bands of hunter-gatherers are *patrilineal;* relationships are traced through the male line of descent. At the present time bands are *exogamous;* females come in as mates from nearby bands. In the more distant past, mating may have been more *endogamous,* with the majority of mates chosen from within the group, as is still the case in some nonhuman primate groups. The *tribe,* the largest social unit, is composed of several bands whose members share a common language and occupy adjacent areas. Tribes are also partly exogamous; in each generation, about 15 percent of the members of any Australian tribe marry people from surrounding tribes (Tindale, 1953). Tribal size varies from place to place, but it averages about 500 members in most regions. The evidence from Australia suggests that this size results from a balance among several environmental and cultural factors. Members of a small tribe with a relatively high population density would have frequent contact with surrounding tribes and would thus run the risk of losing their linguistic identity. A group of the same size living in an area of poorer resources would occupy a larger territory; each individual would then have less frequent contact with members of other tribes.

The upper limit to group size in most environments would be set by the type of transportation available. The fact that all travel must be done on foot places a limit on the size of the area over which linguistic contact can be maintained, for beyond a certain distance dialects can lead to differentiation. These upper and lower limits on size have important implications for evolution because a tribe is only partly exogamous and is therefore a *genetic isolate,* a group whose members tend to breed more frequently with one another than with members of other groups.

Group Structure.   Populations on the order of 500 are small enough to allow genetic drift to occur. But total size alone does not tell us the potential for random changes in gene frequency. Not all individuals in a population at any one time participate equally in reproduction. Some are too young, and others are too old; and even among those of childbearing age, some are less fertile than others. For purposes of measuring evolutionary change, then, anthropologists and geneticists are less interested in the total number (N) of people in the group than in what is called the *effective population size* (usually abbreviated N_e). The value of N_e will approach N most closely when there are no individuals above or below reproductive age, when there are equal numbers of male and female parents, and when these pairs of parents all produce the same number of surviving children. Deviation from any of these conditions will make N_e lower than N. Table 6-1 shows total size and effective size of some recent human populations.

How Drift Works

Drift occurs in human populations because a small sample of gametes is not likely to contain the range and proportions of genes present in the larger population from which the gametes were drawn. This effect can be seen by contrasting the results of one generation of reproduction

---

TABLE 6-1   Total size and effective size in some human populations

Population	Total size	Effective size*
North America (Arizona)		
Ramah Navajo		
Havasupai	614	129
Caribbean		
Providentia Island	177	39
Central America		
Paracho	4593	967
South America (Brazil)		
Camayura	110	23
Europe		
Swiss Alpine village	250	96
Africa (Sudan)		
Dinka village	375	109

  * There is a formula for calculating the effective population size. It is $N_e = 4N_mN_f/(N_m + N_f)$, where N_m is the number of males of reproductive age and N_f is the number of females of reproductive age.
  Source: Modified from J. Buettner-Janusch. 1973. *Physical Anthropology: A Perspective.* New York: Wiley. Table 11-6, p. 355.

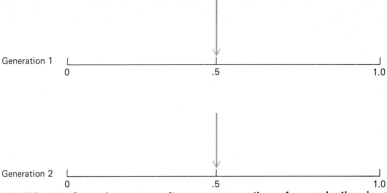

FIGURE 6-1 Gene frequency after one generation of reproduction in a population where $N = \infty$.

in two populations that differ widely in size. Both populations have a single locus with alternative alleles, A and a. These alleles are of equal frequency, and thus p (the frequency of A) and q (the frequency of a) both equal .5.

If the first population contains an infinite number of individuals in the first generation, these parents produce an infinite number of gametes that unite to produce an infinite number of children. In this extreme case, the entire population is the sample. With a population of infinite size it is impossible for the gene frequency to change (see Figure 6-1), as long as other forces of evolution do not operate.

For contrast, let the second population be the smallest possible one for a sexually reproducing species: one male and one female. Gene frequencies of A and a will be equal to .5 if both parents are heterozygotes, Aa. If population size remains constant, what will the gene frequency be in the second generation? There is no single answer. These two parents can produce children of several different genotypes, as shown in Table 6-2. The gene frequency in the next generation will depend on this combination of offspring genotypes. The columns in Table 6-3 giving gene frequency in the second generation show that there are five different possible frequencies of the A allele in this

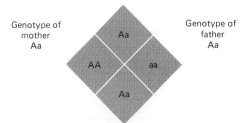

TABLE 6-2 Genotypes possible from a mating of two heterozygous parents

TABLE 6-3 Genotype frequencies resulting from possible combinations of two offspring from two parents of genotype Aa

Genotype of first child		Genotype of second child	Gene frequency in second generation		Probability of this gene frequency in second generation
			A	**a**	
AA		AA	1.0	0.0	$\frac{1}{4} \times \frac{1}{2} = \frac{1}{16} = .0625$
AA	or	Aa	.75	.25	$\left. \begin{array}{l} \frac{1}{4} \times \frac{1}{2} = \frac{2}{16} \\ \frac{1}{2} \times \frac{1}{4} = \frac{2}{16} \end{array} \right\} = \frac{4}{16} = .25$
		AA	.75	.25	
AA		aa	.50	.50	$\left. \begin{array}{l} \frac{1}{4} \times \frac{1}{4} = \frac{1}{16} \\ \frac{1}{4} \times \frac{1}{4} = \frac{1}{16} \\ \frac{1}{2} \times \frac{1}{2} = \frac{4}{16} \end{array} \right\} = \frac{6}{16} = .375$
aa		AA	.50	.50	
Aa		Aa	.50	.50	
Aa		aa	.25	.75	$\left. \begin{array}{l} \frac{1}{2} \times \frac{1}{4} = \frac{2}{16} \\ \frac{1}{4} \times \frac{1}{2} = \frac{2}{16} \end{array} \right\} = \frac{4}{16} = .25$
aa		Aa	.25	.75	
aa		aa	0	1.0	$\frac{1}{4} \times \frac{1}{4} = \frac{1}{16} = .0625$

generation, as illustrated in Figure 6-2. Furthermore, not all these possible outcomes are equally likely, as indicated in the right-hand column of Table 6-3. In the second generation the frequency of the A allele may remain the same as in the previous generation, 50 percent. But this will happen less than half the time (6 times out of 16, to be exact). The frequency of the A allele can increase to 100 percent 1 time in 16. There is the same chance (1 in 16) that the frequency of A can drop to 0. In either of the last two cases, one of the alleles, a or A, will be completely lost from the population.

Now let us look at these results in terms of human evolution. Suppose that each of 16 islands was settled by a different pair of parents, both of genotype Aa. If each pair of parents produced two children, in the next generation we should not be surprised to find that the frequency of the A allele remained unchanged (at .50) on only six of the islands. Nor should it be disturbing to find that the frequency had shifted to .25

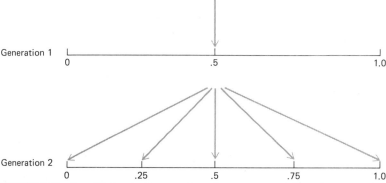

FIGURE 6-2 Possible gene frequencies after one generation of reproduction in a population where N = 2.

on four islands, to .75 on another four, to 0 on one, and to 1.0 on another. In fact, a distribution of this sort is what we would expect from the chance effects of sampling in such small populations. This kind of random shift in gene frequency is sometimes given a specific label: *generation change* or *generational drift*. The hypothetical populations discussed above are outside the practical size limits encountered by anthropologists in real human populations, but they show in a dramatic way the relationship between size and the potential for generational change in gene frequency.

Population Size and Probability of Drift

A greater variety of gene frequencies can exist in populations of more than two persons, since gene frequencies are no longer limited to the categories of 0, .25, .50, .75, and 1.0. Larger populations can have the same categories of gene frequency as smaller populations, plus additional categories intermediate between them, as Figure 6-3 shows. The three bar graphs in Figure 6-3 are called *histograms*. In these bar charts used by statisticians, the areas of the rectangular blocks indicate relative frequencies of various classes of items. Along the bottom of each one used here are the various classes of gene frequency in the population. The height of each of the vertical bars indicates the probability of occurrence of each gene frequency given. The symmetrical, stepwise arrangement of the bars in these graphs shows that in each case we are looking at a *binomial* distribution. Some form of this distribution will be seen whenever there are two alternatives (such as two alleles, A and a) that can occur in different combinations (AA or Aa or aa in one individual or Aa and Aa versus Aa and AA in two individuals). A simple mathematical formula will tell us how frequently each alternative can occur (Table 4-3).

For our purposes, the type of distribution is important because of some consequences that follow from it. The number of vertical bars in the graph of a binomial distribution increases with N (the population size). As the number of bars increases, the difference in height between adjacent bars decreases. To see why this is so, picture a flight of stairs. If the distance between the bottom and top of the stairway remains the same, but more and more steps are added, the height of each step decreases. If more intermediate gradations are added, the stairs will come to resemble a corrugated ramp and then a smooth inclined plane. The same sort of process takes place with binomial distributions. As N becomes very large, the distribution comes to look less like a double set of steps and more like a smoothly sloping mound. Actually, as Figure 6-4 shows, a normal curve can be fitted to almost any binomial distribution. But the more vertical bars there are in the binomial, the better the fit becomes; that is, there is less of a gap between the curve and the steps (see "Characteristics of the Normal Curve" on pages 170–171).

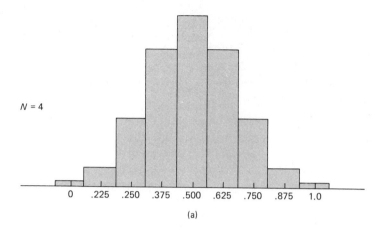

(a)

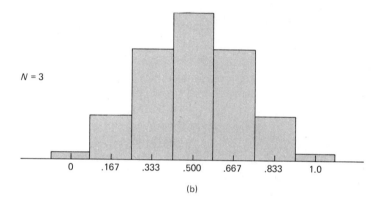

(b)

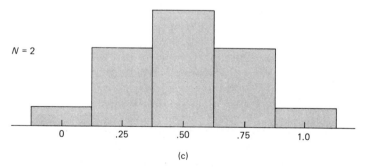

(c)

FIGURE 6-3 Probabilities of various gene frequencies after one generation in (a) a population where $N = 4$, (b) a population where $N = 3$, and (c) a population where $N = 2$. Height of bars = probability of each gene frequency.

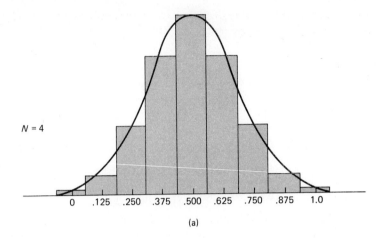

(a)

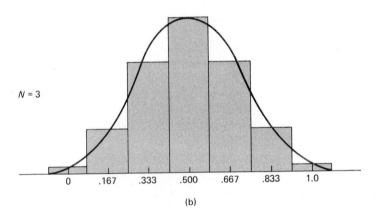

(b)

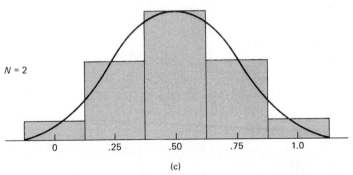

(c)

FIGURE 6-4 Fit of normal curve to binomial distributions of gene frequency (a) where *N* = 4, (b) where *N* = 3, and (c) where *N* = 2.

The Normal Curve and Its Properties.   In technical terms the steplike binomial distribution can be more closely approximated by a *normal curve*.

The bell shape of the normal curve depends on two things: (1) the frequencies (*p* and *q*) of the alleles (respectively, A and a) in the population and (2) the population size (*N*). If these are known, the value of one (or two, three, or more) standard deviations can be calculated.

The formula used here is $\sigma = \sqrt{\dfrac{pq}{2N}}$. (It is not necessary to understand the origin of this formula in order to use it, but a derivation of it is given in Crow and Kimura, 1970, p. 327.)

The baselines of the normal curves used here represent scales of gene frequency. The gene frequency in the previous generation gives the location of the mean of the curve; the value of 1σ tells how far from the mean the frequency could have shifted as a result of sampling error alone; and the areas under different parts of the curve give the probabilities of these different changes in gene frequency ($+1\sigma = 34.1$ percent, $+2\sigma = 34.1$ percent $+ 13.6$ percent, $+3\sigma = 34.1$ percent $+ 13.6$ percent $+ 2.2$ percent). Let us look at some examples.

Example 1. $N = 50, p = .5$. What are the chances for changes in gene frequency in the next generation? First it is necessary to calculate the value of one standard deviation:

$$\sigma = \sqrt{\frac{pq}{2N}} = \sqrt{\frac{(.5)(.5)}{2(50)}} = \sqrt{\frac{.25}{100}} = \frac{.5}{10} = .05$$

Figure 6-5 shows the meaning of this number. There is a chance of 34.1 percent that in generation 2 the frequency of A will be between .50

FIGURE 6-5   Probability of drift when $N = 50, p = .5$.

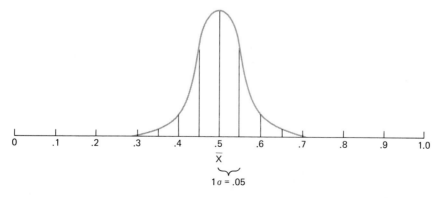

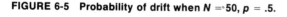

CHARACTERISTICS OF THE NORMAL CURVE

In technical terms, the steplike binomial distribution can be more closely approximated by a *normal curve.*

The normal curve has a number of distinctive characteristics, as follows:

1. It is always symmetrical about the *mean value* (which can be represented by a line drawn vertically from the base to the highest point of the curve).

2. It is always bell-shaped (rounded at the top and flared out at the bottom).

3. There is an area on each side of the mean value at which the curve changes direction; this point is called the *point of inflection.*

4. A line can be drawn downward from the point of inflection and perpendicular to the baseline. The distance from the spot at which this line meets the base to the mean value is called *1 standard deviation* (1σ).

5. The area between the mean value and the vertical line at 1σ is equal to about 34.1 percent of the total area under the curve.

6. If we take the length of 1σ and measure it off again along the baseline, we get 2σ. The area under the second standard deviation is equal to about 13.6 percent of the total area under the curve.

7. If the distance of one more standard deviation is measured off

and .55; there is an equal probability that the frequency of A will be between .50 and .45. There is a chance of 13.6 percent that in generation 2 the frequency of A will be between .55 and .60; there is an equal probability that the frequency of A will be between .45 and .40. There is a chance of 2.2 percent that in generation 2 the frequency of A will be between .60 and .65, and there is an equal probability that the frequency of A will be between .40 and .35.

These separate probabilties can be combined. Thus, there is a probability of 47.7 percent (34.1 percent + 13.6 percent) that in generation 2 the frequency of A will be between .50 and .60 or between .50 and .40. Similarly, the chance that from generation 1 to generation 2 the frequency of A will have changed 1σ in either direction (from .45 to .55) is 68.2 percent (34.1 percent + 34.1 percent). Finally, it is statistically possible for the gene frequency to have reached some value beyond 3σ, but for all practical purposes such a chance is so slight that it is usually ignored.

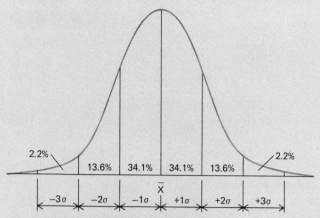

2.2%   13.6%   34.1%   34.1%   13.6%   2.2%

$\overline{X}$

-3σ -2σ -1σ $+1\sigma$ $+2\sigma$ $+3\sigma$

The normal curve and its properties.

again, this will give us 3σ. The area under the third standard deviation is about 2.2 percent of the total.

8. The area including $3\ \sigma$ on one side of the curve will thus include $\sim$34.1 percent + $\sim$13.6 percent + $\sim$2.2 percent $\cong$49.9 percent of the total.

9. Since the curve is symmetrical, the area including 3σ on the other side of the curve will also include about 49.9 percent of the total.

10. Therefore, the region between -3σ and $+3\sigma$ includes nearly the total area ($\sim$99.8 percent) under the entire normal curve.

Example 2. $N = 5000, p = .5$. The gene frequencies are the same here as in example 1, but the population size is much larger. The approach remains the same:

$$\sigma = \sqrt{\frac{pq}{2N}} = \sqrt{\frac{(.5)(.5)}{2(5000)}} = \sqrt{\frac{.25}{10,000}} = \frac{.5}{100} = .005$$

The resulting distribution of gene frequencies is shown in Figure 6-6. As expected, with a larger population the potential deviation due to drift is much smaller. It is now possible, moreover, to see precisely how much less. The total practical range (from -3σ to $+3$) of potential change in gene frequency is from just .485 to .515; this is less than the range of 1σ with the smaller ($N = 50$) population.

This conclusion can be supported with a hypothetical example. Suppose an anthropologist discovers that two groups differ greatly in size (one with $N = 50$ and the other with $N = 5000$) but have the same gene frequency of A ($p = .5$). Imagine further that a similar chance

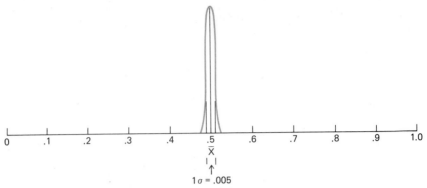

0   .1   .2   .3   .4   .5   .6   .7   .8   .9   1.0

$\overline{X}$

$1\sigma = .005$

FIGURE 6-6   Probability of drift when $N = 5000$, $p = .5$.

event has taken place in each: a boat has overturned, drowning all five people aboard, or an automobile has crashed, killing the driver and four passengers. If in both cases all five individuals were of genotype AA (as would be possible for a group of close relatives), then 10 A alleles would have been lost from each population. The parallel events have, however, surprisingly different genetic consequences for the two populations. In the smaller population, after the accident the frequency of the A allele would be

$$\frac{40A}{40A + 50a} = \frac{40}{90} = .444$$

From the previous frequency of .50 this is a reduction of .056, or an 11.2 percent change (.056/.50) in gene frequency attributable to the accident. In the larger population, the postaccident frequency of A would be

$$\frac{4990A}{4990A + 5000a} = \frac{4990}{9990} = .499$$

This is a reduction of only .001 from the previous .50, or a 0.2 percent change (.001/.50) in gene frequency.

This numerical demonstration of the different impacts of what seem to be the same event actually fits very well with our own day-to-day experience. Someone who has just $1 and loses it is far worse off than someone who loses $1 from a total of $100.

Example 3. $N = 18$, $p - .1$. In this small population, the frequencies of the A and a alleles are unequal. The method remains the same, however, beginning with calculation of the value of one standard deviation:

$$\sigma = \frac{pq}{2N} = \sqrt{\frac{(.1)(.9)}{2(18)}} = \sqrt{\frac{.09}{36}} = \frac{.3}{6} = .05$$

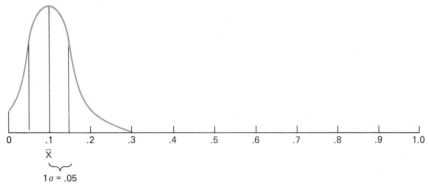

FIGURE 6-7   Probability of drift when N = 18, p = .1.

Figure 6-7 indicates that in the next generation there is about a 2.3 percent chance that gene frequency will be 0, and thus the population may have lost the A gene entirely. When only one gene is left at a locus in a population and all its alternative alleles have been lost, *fixation* has occurred. Random genetic drift, then, can lead to the loss of one or more alleles. Once lost, a gene can be reintroduced only by mutation or by gene flow from other populations. Figure 6-8 (page 174) shows a schematic view of generational drift in one small population.

Founder Effect

Generational drift, although important, is not the only kind of genetic drift. Another type of random change can result when new populations are begun by small groups that leave the parent population and move to a new, previously uninhabited area (Figure 6-9, page 175). There is a very good chance that the migrating group will not have a gene frequency representative of the parent population. This type of nonrandom sampling, called the *founder effect*, must have happened many times in human evolution, although we know of only a few instances.

GENETIC DRIFT IN HUMAN POPULATIONS

Potential for Drift: The Pitcairn Islanders

On December 23, 1787, a British ship, the *Bounty*, set sail from England; the ship was bound for Tahiti, where specimens of the breadfruit tree were to be collected, and then for the West Indies, where it was hoped the breadfruit would provide cheap food for the plantation slaves of the Caribbean islands. The ship reached Tahiti, the plants were gathered and loaded, and the *Bounty* set out. But on April 23, 1789, there was

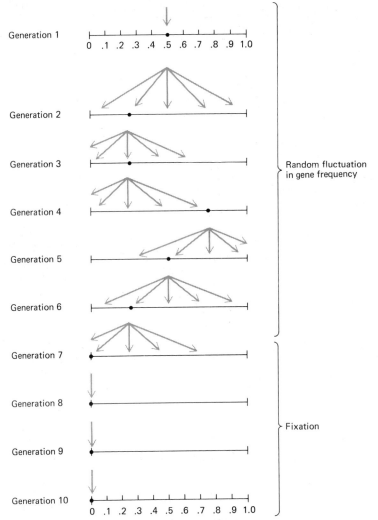

FIGURE 6-8 Generational drift in a small population.

mutiny on board. The captain and some of his supporters were set adrift, and 25 mutineers took the ship back to Tahiti. Sixteen stayed there; the other nine, more fearful of British justice, sailed on, taking with them nine Tahitian women as wives and six Tahitian men as servants. The small band sailed 4000 km (2500 mi) southeast of Tahiti and early in 1790 landed, by chance or design, on Pitcairn Island. This deserted bit of land, only slightly over 3.2 km (2 mi) long, was the scene of recurrent violence over the next 10 years. By 1799 only the women and two of the Englishmen had survived from among the founding group. But despite losses among the adults, the colony had already begun to grow as children were born on the island.

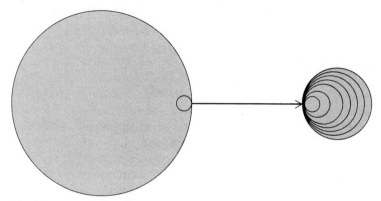

FIGURE 6-9   Founder effect.

Contact with the world beyond Pitcairn was renewed in 1808, when the island was visited by an American vessel searching for water and seals. At that time, the colony consisted of about thirty-five people: one surviving mutineer, eight or nine Tahitian women, and perhaps twenty-five children. Table 6-4 summarizes the further increase in the size of the Pitcairn population from 1808 to 1856, at which time the entire group moved to the larger area of Norfolk Island. After this point, the history of the population becomes more complex. In 1858, 16 of those on Norfolk Island became homesick and returned to Pitcairn. They were joined in 1864 by 26 more returnees. By 1934, the Pitcairn group alone

TABLE 6-4   Population growth for the Pitcairn colony

Year	Total	Males	Females
1808	35		
1814	40		
1825	66	36	30
1839	106	53	53
1840	108	53	55
1841	111	54	57
1842	112	53	59
1843	119	59	60
1844	121	60	61
1845	127	65	62
1846	134	69	65
1847	140	72	68
1848	146	74	72
1849	155	76	79
1852	170		
1853	172	85	87
1855	187	92	95
1856	193	94	99

Source: H. L. Shapiro. 1962. *The Heritage of the Bounty.* Garden City, N.Y.: Anchor Books, Doubleday. P. 209.

had grown to about 200, and an additional 25 had left to scatter over the world. The colony on Norfolk Island, meanwhile, had increased to over 600, although this total included a few recently arrived foreigners. In all, at least 800 and perhaps as many as 1000 offspring were descended from the original mutineers and their companions, and this multiplication had occurred over only 145 years. (The population history summarized briefly is taken from *The Heritage of the Bounty*, by Harry L. Shapiro, the anthropologist who studied the group, Shapiro, 1962.)

When a new group begins by budding off from a larger, established population, shifts in gene frequency can occur in several mutually reinforcing ways. There is the founder effect itself, the chance that the small group will not reflect allele frequencies in the larger population from which it was drawn. This effect should be particularly pronounced in the case of rare genes. In the example of the Pitcairn Islanders it is unlikely that a gene with a frequency of 1 percent in the English population would be represented among the nine mutineers. But suppose such a rare gene were present in just one of them. In such a small group, that gene would immediately have a much higher frequency. If its carrier were a heterozygote, the gene would be one of the 18 at this locus in the nine mutineers (thus $1/18$ = a frequency of 5.6 percent). Even when diluted in half again by the addition of the nine Polynesian women (assuming that none of them carried the gene), the frequency would drop to 2.8 percent. This is still nearly three times the frequency of the gene in the English population.

The chance deviation could have been magnified if some of the Polynesians had been related and thus shared some genes inherited from common ancestors. The number of independent genomes would then have been lower than the number of individuals; N_e would have been lower than N. Effective population size would have been reduced still further if some individuals had left more offspring than others. For example, there is a strong suggestion that all or most of the children on Pitcairn were sired by just six of the Englishmen; the other males, although present, made little or no genetic contribution to future generations.

The founder effect and generational drift commonly occur together. The genetic bottlenecks mentioned above do not disappear immediately, even if the founder population increases rapidly in size. The gene pool of the larger population would consist only of more copies of the same genes. There might be many people, but little genetic diversity.

Evidence of Drift: The Dunkers

Cases like that of the Pitcairn Islanders suggest that the founder effect and generational drift can occur in human populations. But for genetic drift to be more than an intriguing theoretical possibility, more positive evidence is needed from populations in which actual shifts in gene frequency could be measured and in which forces of evolution other than genetic drift could be ruled out. Good candidates for such studies would be groups like the Dunkers of the United States, who are small in number and reproductively isolated and yet similar in past ancestry and present environment to the larger population surrounding them. This group was first studied in the 1950s (for details, see Glass, Sacks, John, and Hess, 1952). The sect originated in 1708 in the Rhineland region of Germany. In 1719 there was a migration of 28 people to Germantown, Pennsylvania, and over the years they were joined by several hundred others, most from the same general area. In 1881 the sect split into three groups, with the Dunkers remaining the smallest and most conservative. At the time they were studied, the group was made up of 3500 individuals, divided among 55 communities. Genetic information was collected on one community that has consisted of about 100 individuals for at least three generations. Some of this is summarized in Table 6-5, which also gives information on frequencies of the same alleles in Germany and the United States. Germany is taken to represent the gene pool from which the founder stock of the Dunkers was drawn, and the United States is the group that surrounds it at the present time.

Allele frequencies in the Dunker isolate differ substantially from those in Germany and the United States. Why do these differences exist? Dunkers dress distinctively, but in most other respects their way of life is quite similar to that of the surrounding American population; natural selection, therefore, does not seem to explain the differences in gene frequency. Mutation can be ruled out as a source of genetic differences even more easily: mutation rates would have to have been enormously higher than any yet observed to have produced the differences seen in the time available. Some gene flow (about 10 to 15 percent per generation) occurs between the Dunkers and neighboring groups, but this would provide an explanation only if the Dunkers' frequencies were intermediate between those in Germany and those in the United States. This is not the case; the frequencies of A and M are higher, and those of B, O, and N are lower, in the Dunkers than in either Germans or

TABLE 6-5 Gene frequencies in the Dunker isolate, in Germany and in the United States

Populations	ABO locus				MN locus		
	Alleles:	A	B	O	Alleles:	M	N
Germany		.29	.07	.64		.55	.45
United States		.26	.04	.70		.54	.46
Dunkers		.38	.02	.60		.66	.34

Americans. Although some of the gene frequencies in the Dunkers are a bit high to be accounted for by drift alone, in view of the small population size of this isolate, drift remains the simplest single explanation for all the divergences.

Several other characteristics were also examined in the Dunker study: amount of middigital hair (the hair on the middle section of the fingers), distal hyperextensibility of the thumb (a type of double-jointedness), shape of the earlobes, and handedness (the tendency to use one hand rather than the other). The transmission of these traits is less well understood than that of the blood group antigens, and there is a good chance that more than one locus determines the expression of each (that is, they may be polygenic characteristics). The results obtained were therefore not as clear-cut as in the case of the blood groups because, among other things, frequencies of the polygenic traits were not available for the German population and so could be studied only in the Dunkers and in the adjacent American population. Even with these limitations, amount of middigital hair, earlobe shape, and degree of thumb extensibility of the Dunkers were strikingly different from those of American whites. Among the complex traits, only the frequency of right- and left-handedness was the same in the Dunkers and their neighbors.

The interpretation of evolutionary patterns in complex characteristics is difficult because each of the loci contributing to a polygenic characteristic could be affected differently by drift. Thus the results at one locus could cancel out those at another, and the expression of the characteristic would not be altered at all. One of the problems in detecting drift in long-term studies is that, beyond a few generations, all our reconstructions of human evolution are based almost completely on skeletal characteristics, which are determined by an interaction of multiple loci and environmental influences. As a result of this limitation, most of our conclusions about the effect of drift on human evolution must come from the study of contemporary populations.

Parma Valley Villages

The Parma Valley in Italy, recently the object of a rather elegant study in population genetics by Cavalli-Sforza (1969) and his colleagues, provides another example of drift as a significant determinant of genetic variation in human populations. The Parma Valley in north-central Italy has been settled since prehistoric times and has received no major immigrations since about the seventh century before Christ. Because of this fact, demographic and genetic equilibrium has probably been reached. The geology and topography of the valley, which is 90 km (56 mi) in length, have shaped the pattern of human settlement and created a natural laboratory for the study of the causes of variation in human

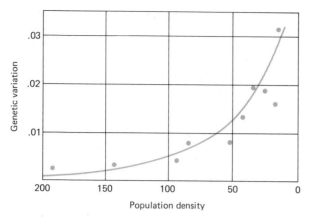

FIGURE 6-10 Genetic variation as a function of population density. Genetic variation is lowest in the relatively densely populated plains, which include the city of Parma; as population density decreases through the hills and into the mountains, genetic variation rises steadily.

gene frequency. Small villages of 200 to 300 residents are scattered about in the steep upland areas, larger villages are found in the lower hills, and the city of Parma is situated on the lowland plains.

The geneticists expected that the effects of drift would be greatest in the small and relatively isolated mountain villages, least in the larger and more accessible settlements on the plains, and intermediate in the hill communities. Their findings, shown in Figure 6-10, supported this hypothesis. Moreover, it was possible to rule out gene flow and selection as explanations for the village-to-village genetic differences. These more systematic forces of evolution would not be expected to cause the same degree of variation for all loci, whereas drift would. And the data gathered showed that variations in gene frequency from one village to another were the same for all genes.

Drift as an Explanation for Human Evolution

At the present time there is some difference of opinion among population biologists concerning the relative role of genetic drift in evolution. Some believe that drift should be offered as an explanation only when other forces of evolution can be ruled out, which would mean in all cases except those in which the history of the population is known in detail, as with the Dunkers and the villagers of the Parma Valley. Another viewpoint is based on the hypothesis that many genes are selectively neutral, giving their possessors neither an advantage nor a disadvantage in survival or reproduction (Kimura, 1968), and that drift should be the major determinant of the frequency of these neutral genes. Rates of change in gene frequency, and the replacement of one allele by another, would then be determined almost exclusively by chance, depending on population size. Differences between populations for such genes would be directly proportional to the time since their divergence from a common ancestor. Given these assumptions, data on frequencies of genes assumed to be neutral can be used to reconstruct patterns of relationships between human populations and between organisms that are more distantly related. In recent years, such studies have been a major topic of research by population biologists.

SELECTION

Selection occurs whenever there is a *causal* relationship between genotype and probability of survival or reproduction. Owing to selection, alleles that enhance the probability of survival or reproduction or both will be retained in the population; alleles that cause a genotype to produce fewer surviving offspring will decline in frequency and may ultimately be lost. When presented this way, selection sounds like a sieve that separates nuggets of ore from sand—a simple and uncreative process. It is, however, much more complex and interesting. For one thing, although genetic loci can be thought of in isolation, they do not exist separately. Each one is part of a genotype with a given phenotypic expression. To extend the sieve analogy, it is as if some nuggets or lumps stay together instead of breaking up and passing through the sieve because certain of their particles add stability. In genetic terms, one genotype is superior to another because its component genes are coadapted to produce an organism more suited to its environment. Whether the process which retains the rare adaptively superior phenotypes is a creative force or not is a fundamental philosophical question that cannot be done justice here. But it is a fact that selection has made a major contribution to the diversity and efficiency of the living creatures on our planet today.

Natural selection was the mechanism Charles Darwin advanced in 1859 in *The Origin of Species.* Darwin had observed variations in nature, and he had also studied the results of selection in animal and plant breeding. As a result, he came to believe that artificial and natural selection work by the same mechanisms. The only difference is the agent of change—human beings or conditions in the natural world. To Darwin, artificial selection made a suitable model for natural selection, as is evident in the following passage from *The Origin of Species:*

In order to make it clear how, as I believe, natural selection acts, I must beg permission to give one or two imaginary illustrations. Let us take the case of a wolf, which preys on various animals, securing some by craft, some by strength, and some by fleetness; and let us suppose that the fleetest prey, a deer for instance, had from any change in the country increased in numbers, or that other prey had decreased in numbers, during that season of the year when the wolf was hardest pressed for food. Under such circumstances the swiftest and slimmest wolves would have the best chance of surviving and so be preserved or selected—provided always that they retained strength to master their prey at this or some other period of the year, when they were compelled to prey on other animals. I can see no more reason to doubt that this would be the result, than that man should be able to improve the fleetness of his greyhounds by careful and methodical selection, or by that kind of unconscious selection which follows from each man trying to keep the best dogs without any thought of modifying the breed. I may add, that, according to Mr. Pierce, there are two varieties of the wolf inhabiting the Catskill Mountains, in the United States, one with a light greyhound-like form, which pursues deer, and the other more bulky, with shorter legs, which more frequently attacks the shepherd's flocks.

Darwin believed that the characteristics which have become established in populations are those which increased the chances of their possessor's leaving behind more offspring. But to attribute evolution to selection in this manner is really just to make a judgment after the fact. The conclusion may be correct, but the explanation does not prove it. Darwin's formulation did, however, lead to the next stage of thought about selection. Once it was accepted that traits were supposed to confer benefits, people began to ask just why they did. In most cases, the answer to this question involved pairing an obvious characteristic with an equally obvious explanation. Humans have larger brains than other animals. One reason given for this differential was that larger brains allow the more flexible behavior needed to deal with the greater complexities of human existence. This approach toward an explanation of how selection operates states a correlation and implies—but again does not prove—a cause. It also suggests where such a cause might be sought. On a still less abstract level, scientists deal with selection as part of the genetic mechanism of evolution. If a characteristic does give a higher rate of survival or reproduction, then in what direction and how rapidly will a change in gene frequency take place? Since this third level is the one that is best understood at the present time, it is the best place to begin a study of how selection operates.

How Selection Works

Darwin said that selection was caused by "survival of the fittest." By the "fittest" organisms he meant those which were best able to carry out the functions of existence, those *adapted* best to some environmental situation or way of life. A "fit" deer was one that could run swiftly and escape from predators. In evolutionary studies today, *fitness* means only relative reproductive ability. The most fit genotype is the one that leaves the largest number of surviving offspring, compared with other genotypes. This comparison has to be done in retrospect. All the people alive 100 years from now will be our descendants, but not all of us will be their ancestors. From a vantage point sometime in the future, it would be possible to give each of us a rating according to the relative contribution we made as the ancestors of those future people. How would these scores be assigned?

Suppose that in a population there are three genotypes: AA, Aa, and aa. To each of these we can give a corresponding fitness value: W_1, W_2, and W_3. Imagine that we can measure the average number of offspring born to each genotype. We might find that a typical AA homozygote produced five children; an Aa heterozygote, four; and an aa homozygote, three. The optimum genotype, here AA with five offspring, is arbitrarily assigned a fitness value of 1. Individuals of genotype Aa have four children rather than five, and so their fitness value is four-fifths that of genotype AA, or 0.8. Similarly, genotype aa has a fitness value three-fifths that of AA, or 0.6.

The *selective coefficient* is the converse of the fitness value. It tells how much lower the reproductive rate of one genotype is compared with that of the genotype with optimum fitness. In the example just given, the selective coefficient of the AA genotype (S_1) is 0, that of the Aa heterozygote (S_2) is 0.2, and that of the aa homozygote (S_3) is 0.4. The relationship between the two measures, then, is: fitness = 1 − selective coefficient. The reverse is also true: $S = 1 - W$.

Selection at the Genotypic Level

Selection at a single locus can occur in a number of different ways, depending on which genotypes are selected for or selected against. These possibilities are summarized in Table 6-7 (see Appendix to Chapter 6, page 199), and a simple visual summary of the various effects of selection on single-locus gene frequencies is presented in Figure 6-11.

Search for Natural Selection

The section on selection at the genotypic level presented in the appendix to this chapter consists mainly of a deductive, mathematical approach in which the basic idea is that if some genotypes have different fitness values from those of other genotypes, the relative genotype proportions will change in the next generation, as will the gene frequencies that are derived from them. However, there is a large gap between our theoretical notion of how selection could work and our ability to demonstrate that it actually does operate this way in real populations. Selection is hard to demonstrate because of several limitations: (1) we do not have

Selection against:

1. None of the genotypes
   (stable equilibrium,
   stable polymorphism)

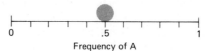

2. One homozygote (aa)
   (directional change in
   gene frequency, transient
   polymorphism)

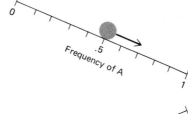

3. One homozygote and one
   heterozygote (AA and Aa)
   (directional change in
   gene frequency, transient
   polymorphism)

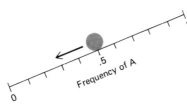

4. Heterozygote only
   (unstable equilibrium,
   indeterminate duration
   of polymorphism)

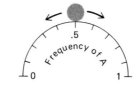

5. Both homozygotes
   (stable equilibrium,
   balanced polymorphism)

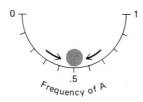

 = Gene frequency in population at beginning of selection

FIGURE 6-11   Effects of selection on gene frequencies at a single locus.

enough generations available for study, (2) we need to study large numbers of individuals in any generation, and (3) it is difficult to identify each individual's genotype correctly.

Number of Generations.   In order to detect a change in the frequency of an inherited characteristic, it is necessary to have comparable information from at least two generations. But detection of a change in gene frequency does not prove that selection has been operating, since other forces of evolution can also produce measurable changes in gene frequency. For this reason, scientists need information from as many consecutive generations as possible. Consistent changes in gene frequency in one direction would imply, though they would not prove, that the changes were due to selection. Because of the limits imposed by our own generation lengths, humans are not particularly good subjects for the study of natural selection. With rare exceptions, only three generations (a child, plus parents and grandparents) can be studied directly. Adding to the difficulty of study is the fact that relatively few populations as yet have accurate genetic records of past generations. If we had 10 or 20 generations of detailed records from some human population undergoing strong selection for, say, increased stature, we might see a permanent increase of several inches.

Vastly greater numbers of generations have been studied in some other organisms. For example, Woodworth, Leng, and Jugenheimer (1952) reported the results of selection in several lines of domestic corn. Their experiment had been going on for 50 years, with one generation raised per year. At the end of 50 generations, in one line the protein content had nearly doubled (from 10.9 percent to 19.4 percent), and the fat content had more than tripled (from 4.7 percent to 15.4 percent), with no sign that a limit was being reached. Over the course of about 10,000 years and perhaps half as many generations, over 200 breeds of domestic dogs have been derived from wolflike ancestors, producing such extremes as the tiny Chihuahua, which may weigh as little as 0.5 or 1 kg (1 or 2 lb) when fully grown, and the Irish wolfhound, which weighs over 68 kg (150 lb).

Numbers of Individuals.   Suppose that allele frequencies are studied for two generations in a large, closed population and it is found that the frequency measured in the second generation differs from that measured in the first. Can we be sure the difference is due to selection? Even if mutation, genetic drift, and gene flow can reasonably be excluded, the case for selection is not proved, chiefly because of the possibility for error that exists in all scientific work. An anthropologist or geneticist usually does not study every individual in a population. In most cases, complete sampling is impossible; every population includes some individuals who do not wish to be measured, to have their blood drawn, or to cooperate in any way. And even if everyone were willing to participate, usually it would be too time-consuming or expensive to study each person. Thus populations are usually represented only by samples of individuals chosen from them. It is always hoped, and often stated, that such samples are representative of the population from which they were drawn—but this is often not the case. A group chosen from almost any suburb in the United States, for example, would not include people from all economic levels or ethnic backgrounds, and it would probably not include those who were hospitalized or institution-alized.

As we saw earlier in this chapter, the larger the sample, the better the chance it will represent the population from which it was drawn. The need to measure gene frequency accurately is particularly great when only a slight shift from one generation to the next has taken place. The smaller the shift in allele frequency from one generation to the next, the greater the number of individuals who must be studied in order to discover the change.

Identification of Genotypes. Even if the entire population were studied or the sample from it were demonstrated to be representative, problems would remain. Data sheets may be labeled with the wrong name, careless handling may alter the chemical composition of a blood sample, or results may be recorded incorrectly. Any of these things can cause the gene frequency measured to differ by a few percentage points from its actual value in the population's gene pool. Thus errors may produce an apparent difference in gene frequency when none really exists, or they may mask a change that has taken place.

Other Problems. All these limitations taken together present a practical paradox. It seems difficult to catch selection at work unless the frequency differences between one generation and the next are very large. If we could rule out other forces of evolution so that only selection were left, such large-scale changes in gene frequency would require very strong selection pressures. But since most existing species have had long evolutionary histories in their present environments, we would expect that their gene pools would long ago have adjusted to the environmental factors that significantly affect survival. Stable or nearly stable gene frequencies should therefore be the rule at all or most loci. This is why most demonstrated cases of selection are normalizing or stabilizing and tend to keep gene and phenotype frequencies at a fixed point.

A very few loci might nevertheless show rapid changes in gene frequency. These changes could represent *transient polymorphisms,* cases where genetic variation is present while a new and beneficial mutant allele is replacing a gene that formerly predominated in the population, as apparently happened in the case of the gene that controls the ability to digest the lactose in milk. Or a rapid shift in gene frequency could take place because an earlier, balanced polymorphism had been disrupted. One or more of the selective coefficients could have changed, perhaps because of a shift in some critical environmental factor. Still, the chance of detecting any change in gene frequency due to selection should be slight. Opportunities would be limited to a few characteristics in a few organisms at a few points in their evolutionary history. One well-documented example in human evolution, the hemoglobin-S allele, is discussed in detail in Chapter 7. Because of all this, it is not surprising that selection is rarely detected; what is surprising is that evidence for it is ever found at all.

Avenues of Approach.   There are, however, some ways in which it might be possible to discover selection. First, we could look directly for a change in gene frequency in a situation where other forces of evolution could be ruled out as causes of the change. This approach would help substantiate cases of directional selection, or change in a consistent direction. A second alternative would be to survey populations for the existence of genetic polymorphisms which might be due to selection. Correlation of gene frequencies with factors in the biological or cultural environment could increase the likelihood of discovering whether such polymorphisms were due to selection. As a third approach we could measure the fitness values of different genotypes or of the same genotypes at different points in the life cycle. Consistent differences would suggest the operation of selection and might indicate areas for research on the precise causes of those differences.

Evidence for Selection in Natural Populations

Considering the problems described above, it would seem that we would have the best chance of observing selection in action in a population of short-lived, rapidly maturing organisms that produce numerous offspring with easily recognizable phenotypic characteristics that are known to be under simple genetic control. These conditions are met in many laboratory populations of organisms such as viruses and bacteria and in breeds of domestic plants and animals. But sometimes objections are raised against this evidence for selection because these populations are not "natural": the selective process is governed by human choice. Darwin himself answered this objection rather effectively, and since his time geneticists have used artificial populations to answer many questions about how selection operates. But selection in natural populations has also been observed. Two interesting examples are a color change in moths and head size in humans.

Industrial Melanism in Moths.   One of the best-documented cases of evolution in a natural population is an extraordinarily rapid color change in certain moth populations in England (Kettlewell, 1965). There are hundreds of species of moths in England. One of them, *Biston betularia,* was given its common name, "peppered moth," because of its coloring pattern, a sprinkling of black dots over the surface of a white body and wings. These insects fly about at night and spend their days clinging to the trunks of trees. For centuries these trees have been covered with lichens, minute plants that give the trunks a stippled appearance. The moths' coloration makes them virtually invisible against this background and protects them from birds.

The capture of a melanic moth—one predominantly dark in color— was recorded for the first time in 1849. Black moths remained rare for over a quarter of a century, but by 1886 the melanic form had become more common than the nonmelanic. Today in some parts of the British Isles, over 95 percent of the moths are black (see Figure 6-12). The dark color of the melanic moths is controlled by the presence of a single dominant gene, D. Its alternative allele is d. The dark moths would be either DD or Dd, and the light moths would be dd. If today over 95 percent of the moths are DD or Dd, then the remaining 5 percent or less are dd. Their frequency in the population would be equal to q^2. This information can be used to calculate the allele frequencies. If light moths constituted 4 percent of the population, then q^2 would equal .04, and q (the frequency of the d allele) would equal $\sqrt{.04}$, or .2. The corresponding frequency of the D allele, p, would be 1-q, or .8. If the population contained only 1 percent light moths, by the same reasoning q would be .1, and p would be .9. The extreme rarity of black moths just about a century ago suggests strongly that the D allele was maintained in the population by recurrent mutation alone. If moths have about the same mutation rates as humans, this would give a frequency of D around .00001. From there, the frequency of D has increased to approximately 1 or so. The D gene has therefore become about 10,000 times more frequent in 100 years.

FIGURE 6-12 This curve traces the increase in frequency of the dark (melanic) moths, from a level maintained by recurrent mutation in the middle of the nineteenth century to near-fixation at present.

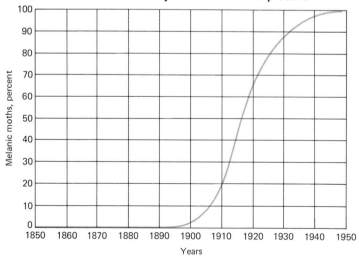

This is telling evidence for evolution in process in a natural population. What was the cause? The industrial revolution was well under way in England before 1800. At first, flowing water of the rivers was a sufficient source of power. But by the middle of the century, industrial expansion required additional energy. Coal was abundant, and vast quantities were used to generate steam for the mills as well as to heat the homes of an expanding population. Smokestacks and chimneys poured out enormous quantities of soot and smoke. In some parts of England, as much as 45 metric tons (50 tons) of particles per 2.6 km² (1 mi²) per month were deposited on the ground. This fallout affected the countryside as well as the industrial cities. Prevailing southwesterly winds carried the smoke over the counties of eastern England, as shown in Figure 6-13. In the affected areas, trees were covered with soot, the

FIGURE 6-13 Distribution of industrial centers (black circles) and melanic moths (shaded area) in the British Isles.

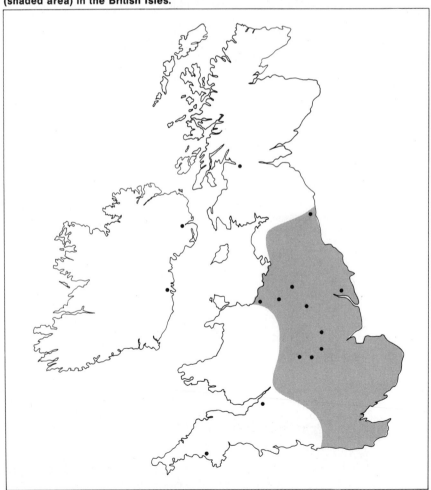

FIGURE 6-14 Color contrasts of moths and background in polluted and unpolluted areas. The moths shown here are *Lycia carbonaria*, the American counterpart of *Biston betularia*. Left, **melanic form**; right, **normal form**.
Source: Specimens from Frost Entomological Museum, Pennsylvania State University.

lichens were killed, and the trunks were blackened. Once this happened, the camouflage situation was reversed (Figure 6-14). The melanic moths were hidden, and the light-colored ones became easy prey for birds. The higher frequency of melanic moths in the soot-blackened areas was the first evidence of an evolutionary response to a selective agent in the environment. It was later tested directly by experimental studies in which paint-marked moths of different colors were released in polluted and unpolluted areas. In both cases, twice as many living moths with the appropriate camouflage were recovered as conspicuous ones. The results provide strong confirmation that color contributes to the probability of survival of an individual moth and the spread of its genes.

Brachycephalization. All the factors that make selection difficult to detect for any single-locus genetic trait also hamper the search for it in polygenic morphological traits. The study of selection in these more complex characteristics is limited by several additional factors:

1. Although most morphological traits are probably polygenic, their precise mode of inheritance (including the number of loci influencing a trait and the expression of the alleles at each loci) is often unknown.

2. The frequencies of alleles at each of the loci affecting a trait are not known.

3. The relative fitness of the various genotypes involved cannot be measured, since the genotypes themselves cannot be distinguished.

For most of the human evolutionary record, however, the only characteristics that can be studied directly are those preserved in fossil bones, such as size, shape, the extent of roughened areas where muscles were attached, anatomical details such as the number and placement of *foramina* (holes where blood vessels or nerves pass into or though the bones), and the proportions of various parts of the skeleton. The single most frequently calculated proportion is probably the *cephalic index,* which is (maximum head breadth × 100)/maximum head length. This index was devised in 1842 by the Swedish anthropologist Anders Retzius, who also coined several descriptive terms for heads of different porportions (Stanton, 1960). People with a cephalic index of 80 or above Retzius called *brachycephalic;* those with a cephalic index of below 75 he called *dolichocephalic;* and people whose heads were between these two ranges he called *mesocephalic*.

Retzius and the other physical anthropologists of his time considered the cephalic index to be a stable, highly heritable biological characteristic. Later work by Franz Boas (1910) showed that the cephalic index is subject to some environmental modification as well. Boas discovered that American-born children of broad-headed Jews from central Europe had cephalic index values lower (−2 points) than those of their parents. At the same time, children born here to long-headed immigrant Sicilians showed an increase in cephalic index values (+1.3 points). Regardless of the wide difference in their ancestral backgrounds, not only did the offspring of both groups resemble one another more closely than they did their parents, but both groups also converged on the appearance of Americans of European ancestry resident here for several generations. Similar results were obtained for other traits (width of face, stature, and so on) and were confirmed by several later studies. Humans thus show a degree of *developmental plasticity,* the ability to change bodily form (as reflected in anthropometric measurements and indices), within limits, in response to environmental rather than genetic influences.

Even when a substantial environmental component is known to affect a given polygenic characteristic such as stature, changes due to improving environmental conditions can reach a limit after a few generations (Damon, 1968). In contrast, progressive brachycephalization (shortening of the head from front to back) has been changing head shape in human populations over many parts of the world (Europe, India, North America, Polynesia) for several thousand years. Without denying environmental influence, Bielicki and Welon (1966) recently found some evidence for the operation of selection on cephalic index.

Although the selective mechanism has not been identified, patterns of fertility and mortality suggest that a phenotypic characteristic with a heritable component has been changing measurably over a number of generations.

To search for the operation of selection in recent human populations, Bielicki and Welon used anthropometric and demographic data collected during the late 1920s in the Military Anthropological Survey of Poland. These records included information on 6229 army recruits from the northeastern part of pre-World War II Poland (now part of the White Russian Republic in the Soviet Union), an area of 35,000 km² (about 13,500 mi²). At the time the area was studied, the population was overwhelmingly rural, poor, and illiterate—and had been for some time. Only 60 percent of the adults could read. Rye and potatoes were the basic crops, and yields of these staples were low. Fertility was high, with an estimated average of over six live births per family. Infant mortality was also relatively high, ranging from 1 to 25 percent during the first quarter of this century. Infectious diseases such as typhus, diphtheria, and scarlet fever were the main killers. Geographic and social mobility were low, with over 85 percent of the population, chiefly peasant stock, remaining in the same counties where they had been born.

All men included in the study had listed their occupation as farmer when inducted into the army. In addition to occupation, three other items of information collected were relevant to the study of selection in head shape: (1) the subject's cephalic index, (2) the total number of the subject's siblings born, and (3) the number of those siblings who were alive when the recruit was studied.

Table 6-6 shows the results of the study. The mean numbers of offspring born per family were higher for long-headed recruits. This finding might suggest that genotypes responsible for dolichocephaly are characterized by slightly greater average fertility than those pro-

---

TABLE 6-6   Cephalic index, fertility, and survival

Cephalic index category	Number of sibships	Mean number of offspring born per sibship	Mean number of surviving siblings*
Ultradolichocephalic (≤77.50)	544	6.316	4.16
Dolichocephalic (77.51–80.50)	1570	6.308	4.25
Mesocephalic (80.51–83.50)	2177	6.262	4.44
Brachycephalic (83.51–86.50)	1398	6.201	4.39
Ultrabrachycephalic (≤86.51)	540	6.192	4.31

*Transformed values; the reason and the formula for transformation are given in the source.
Source: T. Bielicki and Z. Welon. 1966. "The Operation of Natural Selection of Human Head Form in an East European Population," *Homo*, **15**(1):22–30.

ducing brachycephaly. Such an interpretation is possible but not conclusive because the fertility differences are not statistically significant. In contrast, the association between head form and mean number of surviving siblings was strong. Recruits with intermediate cephalic index values tended to have more living brothers and sisters than recruits with either extreme of head shape. In polygenically inherited shape traits, these intermediate phenotypes are most likely to be heterozygotes. When heterozygotes are favored, the result is stabilizing selection, which can lead to genetic equilibrium.

In the example here, however, some directional selection is evident as well, because although the mesocephalic recruits had the most surviving siblings, the recruits in the two short-headed categories had more surviving siblings than those in the long-headed categories. Bielicki and Welon concluded that in eastern Europe the process of brachycephalization has been due to selective pressures, acting chiefly through differential mortality and leading toward genetic equilibrium for a moderate degree of round-headedness. Additional studies would be desirable to show whether equilibrium has been attained or is still being approached, as well as to determine the precise mode of inheritance and the fertility of each genotype involved. It would also be desirable to know what advantage a shorter head confers on its possessor. Do longer-headed babies suffer a more severe birth trauma that lessens their chance of surviving? Skulls that are more spherical do have less surface area for each unit of enclosed volume than longer heads. This geometric relationship should make more spherical heads less susceptible to heat loss than heads of other shapes. Does the reduction in heat loss confer any benefit? It could, since 40 percent of the body's total heat loss is from the head and since severe heat loss can produce chilling and even death. Are longer-headed people more susceptible to the infectious diseases known to be prevalent in this population, and, if so, why? Even in the absence of this knowledge, though, Bielicki and Welon's study provides an interesting example of the possible effect of selection in a recent human population.

Indirect Evidence: ABO and Other Blood Group Polymorphisms

At the present time, there are some cases of possible selection in humans that have not yet been conclusively proved. Some of the most interesting of these concern the frequencies of various red blood cell antigens.

Rh Polymorphisms. Rh blood group polymorphisms are a particularly puzzling case. The Rh antigen differences are under direct genetic control, like the ABO blood group antigens. But the two systems differ with regard to antibody production. ABO antibodies are said to be *naturally occurring,* since human infants begin to produce the appropriate kinds at the age of 3 to 6 months. In contrast, Rh antibodies are not normally produced until an individual is exposed to an Rh antigen different from his or her own. This antibody formation can be triggered by transfusion of blood from individuals of a different Rh antigen type. Because there is a time lag between exposure to a foreign antigen and production of antibodies, difficulties due to Rh incompatibility are rare unless repeated transfusions are given. Exposure to foreign Rh antigens can also take place during pregnancy, leading to what is known as *hemolytic disease of the newborn.* This condition occurs only in matings between Rh-positive males and Rh-negative females. A man who is an RR homozygote will have offspring who are all of genotype Rr. A man who is Rr will have children half of whom are Rr and half of whom are rr. The circulatory systems of the mother and the embryo are separate, and so normally no direct mixing of blood occurs between them. But capillaries (the body's smallest blood vessels) from the maternal and embryonic circulation are in close contact in the placenta (Figure 6-15), and thus dissolved materials (gases and low-molecular-weight chemicals) can diffuse from one to the other.

Near the end of pregnancy, small tears often occur in the placenta, and a few blood cells from the embryo can enter the mother's system. When the mother is Rh-negative and the child is Rh-positive, this can stimulate the mother's antibody system to produce anti-Rh-positive

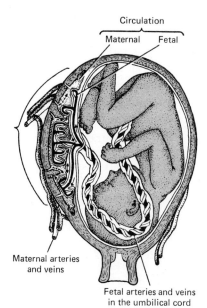

Circulation

Maternal   Fetal

Maternal arteries
and veins

Fetal arteries and veins
in the umbilical cord

FIGURE 6-15 The placenta.
Source: G. G. Simpson and W. S. Beck. 1965. *Life: An Introduction to Biology.* New York: Harcourt, Brace, and World. Fig. 9-26, p. 276. © 1965 by Harcourt, Brace, Jovanovich, Inc., and reproduced with their permission.

antibodies. The time lag in the production of these antibodies means that the first child is usually not harmed by them. But in later pregnancies, the antibody system has already been primed, and the mother can manufacture antibodies more quickly. These maternal antibodies enter the fetal system and destroy the embryo's red blood cells (Figure 6-16). The resulting deficiency of red blood cells can produce anemia, oxygen deprivation, brain damage, stillbirth, or early abortion and miscarriage. Hemolytic disease is not invariably fatal. If the blood of an affected newborn is replaced via a transfusion, the child can be saved. Furthermore, in the past few years passive immunization has become possible. With this technique, an Rh-negative mother who has just delivered an Rh-positive baby is given an injection of anti-Rh-positive antibodies. These will destroy any fetal cells in the mother's circulatory system before her own antibody-producing system is stimulated.

Rh incompatibility is not very common. In populations of European ancestry, the frequency of R is commonly .6, and that of r is about .4. Thus, 36 percent of males would be RR, and 48 percent would be Rr, for a total of about 84 percent Rh-positive. Rh-negative females would have a frequency of 16 percent. The frequency of incompatible matings would be .84 × .16, or about 13 percent. Because many males are heterozygotes, only 10 percent of pregnancies would involve incompatibility. And only about 3 percent of mothers in this situation produce significant amounts of antibodies, resulting in a further reduction in the number of infants affected. In all, only about 0.3 percent of all pregnancies in a population of European ancestry would lead to the symptoms described. Nonetheless, this is still an appreciable amount of selection, and all of it is directed against heterozygotes. Because of this, the frequency of the r allele should be declining. But it does not seem to be. This unexpected finding suggests that some other selective factor may be operating to produce a balanced polymorphism at this locus.

ABO Polymorphisms. Here, too, the suggestion that selection is producing polymorphism is intriguing, but proof remains elusive. Figure 6-17 shows the world distribution of ABO blood group alleles. If these were randomly distributed, points should be scattered all over the space enclosed by the sides of the triangle. Instead, they are clustered into a restricted region. This distribution is quite systematic. If a population lacks one allele, it is always B. If two alleles are absent, they are invariably B and A. In no human population is O ever absent. This is precisely what would be expected if there were selection against heterozygotes: the lowest-frequency alleles disappear first. But the fact that most populations have all three alleles in itself suggests the existence of balanced polymorphism. How the balanced polymorphism arose in the first place is quite another problem. The A and B antigens

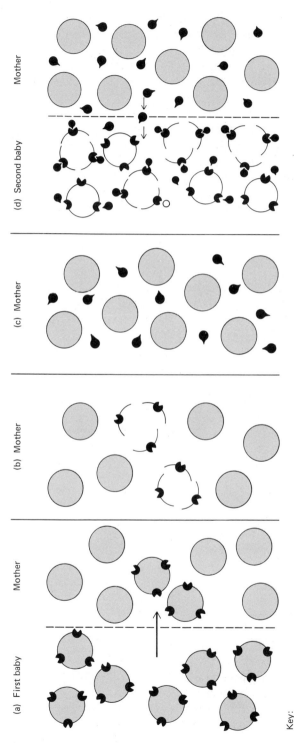

Key:

🦅 Rh antigen

🔻 Rh antibody

FIGURE 6-16 Rh incompatibility. (a) An Rh-negative mother has an Rh-positive baby, and some of the baby's red blood cells get into her circulation. (b) The baby's cells soon disappear naturally. (c) However, the mother may manufacture antibody to the Rh antigen. The first baby is not affected, because it has been born by the time the antibody appears. (d) But if the mother has a subsequent Rh-positive baby, the antibody may attack the baby's red blood cells, thereby giving rise to a possibly fatal anemia.

Source: C. A. Clarke. 1968. "The Prevention of 'Rhesus' Babies," *Scientific American,* **219**(5):46–52. Fig. on pp. 48–49.

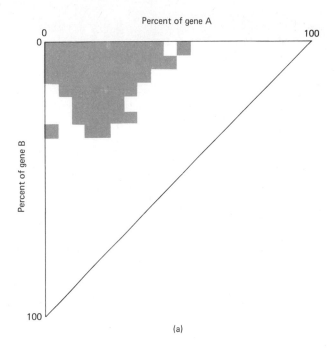

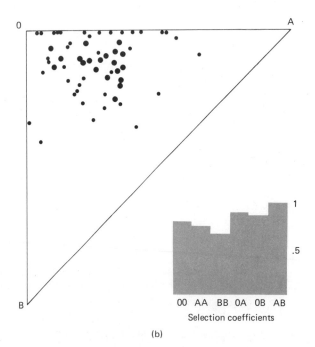

FIGURE 6-17 **World distribution of ABO blood group frequencies.** (a) **Limits of the world range in ABO blood group frequencies, in relation to the complete possible range.** (b) **Genetic drift for the ABO system. Compensation of incompatibility losses by a balanced system of favorable selection of heterozygotes, to simulate actual world distribution of gene frequencies.**

Sources: (a) Alice Brues. 1954. "Selection and Polymorphism in the ABO Blood Groups," *American Journal of Physical Anthropology*, **12**:559–597. (b) Alice Brues. 1963. "Stochastic Tests of the A-B-O Blood Groups," *American Journal of Physical Anthropology*, **21**: 287–299.

(as well as M-like and N-like substances and some of the Rh antigens), although not always on the surfaces of the red blood cells, are found in some tissues of the great apes. The work of Moor-Jankowski, Wiener, and Rogers (1964) has shown that the O phenotype, most common in humans, is absent in gorillas and not very common (about 15 percent) in chimpanzees. However, frequency variations aside, the simplest explanation for the existence of the ABO polymorphism in humans is that it was inherited from our nonhuman primate ancestors. Of course this explanation does not really solve the problem of how the polymorphism originated; it just displaces it by a few million generations.

Drift occurring within very small populations could increase the frequency of mutant alleles for new antigenic variants to appreciable frequencies in relatively few generations. Pleiotropism of the alleles that are responsible for ABO antigens is another possible answer (Brues, 1977, p. 218). Although heterozygotes seem to be selected against by maternal-fetal incompatibility, those who escape this hazard might exhibit *heterosis* (sometimes called *hybrid vigor*) and enjoy a physiological advantage during some other period in their lives. (Lerner, 1954, gives a good discussion of the various advantage that accrue to heterozygotes.) Brues (1963) has shown that by taking into account incompatibility effects but also assuming a selective advantage for all the ABO heterozygotes (AO, BO, AB) and a particular disadvantage for AA and BB homozygotes, a close approximation to the observed world variation at this locus can be simulated (see Figure 6-17).

There have been a number of attempts to correlate alleles of the ABO system with possible selective factors such as chronic diseases, infectious diseases, and diet. In the chronic disease category, it has been found that individuals of blood type A have a slightly greater risk of getting cancer of the stomach than those of other blood types. Type O individuals have a higher frequency of peptic ulcers. But these conditions normally occur late in life and may therefore have little impact on reproductive performance. Infectious diseases also seem to affect individuals of various blood types differently. Type A individuals contract smallpox more frequently and show more serious symptoms (Vogel, 1965; Vogel and Chakravartti, 1966); plague is more common in type O individuals; and so on.

Part of the attractiveness of the association between ABO antigens and infectious diseases lies in the hint of what one causal mechanism might be. Substances on the outer surface of the disease-producing bacteria are similar chemically to certain antigens. Thus, if the antigen on the surface of smallpox-causing microorganisms resembles the type A antigen, individuals of that blood group type couldn't very well produce antibodies against the disease, and thus their resistance would be lowered. Although this mechanism remains unproved, it has a strong logical appeal and merits further testing. Correlations between areas with high frequencies of certain alleles and diets high in some major component (for example, type B blood frequent in areas where the diet is high in fat, as among Eskimos) have been suggested (Kelso and Armelagos, 1963) but still remain largely unverified.

This general situation of uncertainty about the causes of ABO and other blood group polymorphisms is likely to continue for several reasons. One of the most serious is our inability to distinguish A and B homozygotes from heterozygotes other than by time-consuming family studies. But our inability to prove the operation of selection should not be taken as evidence that selection is not occurring. Selective changes in gene frequency of less than 1 percent per generation can bring about appreciable modification in a population's genome over the course of a few hundred generations and yet remain utterly masked by human errors in the attempts at measurement. For example, we know that humans have inherited differences in the ability to produce antibodies, which give a measure of resistance to infectious diseases. It is probable that because of the widespread use of antibiotics, some children now survive who would have died from infections in earlier times. The frequencies of genes that influence antibody production are probably changing even now, though at rates which remain unknown.

SUMMARY

1. Genetic drift is random change in gene frequency due to chance in a small population.

2. The smaller the population, the more likely it is that drift will occur, since gene frequencies in the sample of offspring born are less likely to be representative of gene frequencies in the population.

3. Drift is important in human evolution because for over 99 percent of that process, the basic breeding unit may have been very small, averaging fewer than 500 persons.

4. Drift can occur for a variety of reasons; two basic ones are generational drift (random generation-to-generation shifts in gene frequency in small groups of roughly constant size) and the founder effect, which can occur when a small and nonrepresentative sample is drawn from a larger population.

5. The probability of drift can be estimated by combining information on allele frequency and population size, using the formula

$$\sigma = \sqrt{\frac{pq}{2N}}$$

6. Drift can lead to the loss of all but one of the alleles at any locus, at which point fixation has occurred and there can be no further fluctuation in gene frequency.

7. Selection occurs when different genotypes produce unequal numbers of surviving offspring.

8. Fitness is the measure of this relative reproductive performance; the most fit genotype is the one that produces the most surviving offspring.

9. The theoretical operation of selection at the single-locus level is well understood; selection against various genotypes can cause an allele to increase, decrease, or remain at a constant frequency.

10. Whenever two or more alleles exist in a population above frequencies that would be produced by recurrent mutation alone, a polymorphism is said to exist; polymorphism may be balanced (giving stable allele frequencies) or transient (causing alleles to change gradually in frequency).

11. Because of various practical limitations, selection is far more difficult to demonstrate than to describe; however, there are several well-documented examples of selection in laboratory and natural populations of other animals, as well as in humans.

APPENDIX TO CHAPTER 6: SELECTION AND CHANGE IN GENE FREQUENCY

If we know the genotype frequencies and their fitness values in one generation, we can predict the outcome of one generation of selection. The result will depend completely on these numerical values. Hypothetical values are used in the cases below, but all are based on examples of real inherited conditions. In each case discussed below (Table 6-7), we want to know several things. First, does the gene frequency change? Second, if change in gene frequency occurs, what is its direction: which allele increases in frequency, and which decreases? Third, how fast does the gene frequency change?

TABLE 6-7   Effects of selection on genotypes at a single locus

Selection against	$\square$ = genotypes penalized		
1. None of the genotypes	AA	Aa	aa
2. One homozygote	AA	Aa	[aa]
3. One homozygote and one heterozygote	[AA]	[Aa]	aa
4. Heterozygote only	AA	[Aa]	aa
5. Both homozygotes	[AA]	Aa	[aa]

TABLE 6-8   No selection occurs when all genotypes have the same fitness value

Genotypes	AA	Aa	aa	Total
Generation 1: Gene frequencies: $p(A) = .5$				
$q(a) = .5$				
Genotype frequencies	.25	.50	.25	1.00
Fitness values	1.0	1.0	1.0	
Proportions left after selection = genotype frequency × fitness value	.25	.50	.25	1.00
Generation 2: Genotype frequencies = proportion left after selection	.25	.50	.25	
Total	1.00	1.00	1.00	
	.25	.50	.25	1.00
Gene frequencies: $p(A) = .5$				
$q(a) = .5$				

No selection against any of the genotypes is one of the conditions necessary for operation of the Hardy-Weinberg equilibrium. This establishes a good beginning point for comparison of other cases. The main features of this condition are presented in Table 6-8.

Selection against One Homozygote

The best-known cases here involve selection against the recessive homozygote. This is the same as saying that selection is operating against the recessive phenotype. Albinism is a good example because the most common form of the trait is caused by homozygosity for a recessive allele. Albinos are apparently normal in all respects except that they are unable to produce melanin and have a few resulting visual difficulties (for example, albinos are highly sensitive to sunlight because the retinas of their eyes lack pigment). Selection against albinos is probably due to a combination of factors. Albinos also suffer from skin cancer more often than those of the dominant phenotype. This probably produces some *differential mortality* (a lower probability of survival for some genotypes than for others). *Differential fertility* may stem from this if death occurs before the reproductive period has been completed. Fertility of albinos may also be reduced because their appearance makes it difficult for them to find mates. It is estimated that, probably because of all these factors, the fitness value of albinos is about 0.8 at the present time in the United States. The calculations in Table 6-9 are based on this figure. An initial gene frequency of $q(a) = .5$ has been used only to allow standardized comparison with other cases. The actual frequency of the a allele in European populations is .01 or less.

TABLE 6-9   Selection against one homozygote

Genotypes	AA	Aa	aa	Total
Generation 1: Gene frequencies: $p(A) = .5$				
$\qquad\qquad\qquad\qquad\quad q(a) = .5$				
$\qquad\qquad$ Genotype frequencies	.25	.50	.25	1.00
$\qquad\qquad$ Fitness values	1.0	1.0	0.8	
$\qquad\qquad$ Proportions* left after selection =	.25	.50	.20	.95
$\qquad\qquad\qquad$ genotype frequency × fitness value				
Generation 2: Genotype frequencies =				
$\qquad\qquad$ proportion left after selection	.25	.50	.20	
$\qquad\qquad$ Total	.95	.95	.95	
	.25	.53	.21	1.0
$\qquad\qquad$ Gene frequencies: $p(A) = .25 + \frac{1}{2}(.53) = .525$				
$\qquad\qquad\qquad\qquad\qquad\qquad q(a) = .21 + \frac{1}{2}(.53) = .475$				

*These are referred to as *proportions* rather than *percentages* because *percent* means "per hundred." Here the total is less than 100, making it necessary to divide by the total to find what the genotype frequencies are after one generation of selection.

After one generation of selection against the recessive homozygote, the frequency of the recessive gene has been reduced from .5 to .475. The frequency of its alternative allele, A, has increased from .5 to .525.

Selection against One Homozygote and One Heterozygote

This situation could just as accurately be called *selection against a dominant gene* or *selection against the phenotype that the dominant gene produces.* One inherited condition caused by an autosomal dominant gene is Marfan's syndrome. Individuals with Marfan's syndrome have extremely long bones in their fingers and toes and a "pigeon chest" that results because the rib cage grows in such a way as to produce a narrow, keel-like lump; other frequent symptoms include heart defects and a misplaced lens in the eye. Some have suggested that Abraham Lincoln had this syndrome; it has been reported among some of his relatives (Stern, 1973, p. 72). The fitness value of individuals with Marfan's syndrome is about 0.5. Selection may act through both differential fertility (due to appearance) and differential mortality (due to the heart defects). Table 6-10 shows the effect of one generation of selection, beginning with a gene frequency of A = .5.

Just one generation of selection has brought about a relatively large-scale change in gene frequency. This result is due partly to the high selective coefficient against the dominant phenotypes, but the sharp change in frequency is also attributable to the fact that since all the dominant alleles are expressed, selection can act against the entire

TABLE 6-10   Selection against one homozygote and one heterozygote

Genotypes		AA	Aa	aa	Total
Generation 1:	Gene frequencies: $p(A) = .5$ $q(a) = .5$				
	Genotype frequencies	.25	.50	.25	
	Fitness values	.5	.5	1.0	
	Proportions left after selection = genotype frequency × fitness value	.125	.250	.250	.625
Generation 2:	Genotype frequencies = proportion left after selection	.125	.250	.250	
	Total	.625	.625	.625	
		.2	.4	.4	1.0
	Gene frequencies: $p(A) = .2 + \frac{1}{2}(.4) = .4$ $q(a) = .4 + \frac{1}{2}(.4) = .6$				

sample of them in the population. This contrasts with selection against recessive genes, in which case only those genes present in homozygotes can be attacked.

Selection against the Heterozygote

At a single locus with two alleles (A and a) and three genotypes (AA, Aa, and aa) it might seem that selection against the heterozygote would have no effect. Elimination of one Aa individual would remove one allele of each type, but it would not seem to change things. This impression is accurate, but, as shown in Table 6-11, only in one special case: when $p(A) = q(a)$.

TABLE 6-11   Selection against the heterozygote when $p(A) = q(a)$

Genotypes		AA	Aa	aa	Total
Generation 1:	Gene frequencies: $p(A) = .5$ $q(a) = .5$				
	Genotype frequencies	.25	.50	.25	1.0
	Fitness values	1.0	0	1.0	
	Proportions left after selection = genotype frequency × fitness value	.25 0	.25	.50	
Generation 2:	Genotype frequencies = proportion left after selection	.25	0	.25	
	Total	.50	.50	.50	
		.5	0	.5	1.0
	Gene frequencies: $p(A) = .5 + \frac{1}{2}(0) = .5$ $q(a) = .5 + \frac{1}{2}(0) = .5$				

TABLE 6-12   Selection against the heterozygote when $p(A) \neq q(a)$

Genotypes	AA	Aa	aa	Total
Generation 1: Gene frequencies: $p(A) = .5$ $\qquad\qquad\qquad q(a) = .5$				
Genotype frequencies	.36	.48	.16	1.0
Fitness values	1.0	0	1.0	
Proportions left after selection = genotype frequency × fitness value	.36	0	.16	.52
Generation 2: Genotype frequencies = proportion left after selection Total	.36 .52	0 .52	.16 .52	
	.69	0	.31	1.0
Gene frequencies: $p(A) = .69 + \frac{1}{2}(0) = .69$ $\qquad\qquad\qquad q(a) = .31 + \frac{1}{2}(0) = .31$				

Regardless of whether selection against the heterozygote is strong or weak, as long as equal numbers of A and a alleles exist in the population to begin with, then equal numbers of them will be eliminated, and the balance will not shift in either direction. If the initial frequencies of these alleles are unequal, selection will widen the discrepancy. This is illustrated in Table 6-12.

Why does selection against the heterozygote cause a change in allele frequency when $p(A) \neq q(a)$? It is still true that as many A genes are removed as a genes. However, although the absolute numbers removed are the same, the proportions lost are different. This is not the case when p(A) = q(a), as shown in Figure 6-18.

To sum up, selection against the heterozygote can lead to an equilibrium only when the alleles at the locus are in equal frequencies. If either decreases even slightly in frequency for any reason (e.g., drift), further selection of the same type will act to reduce frequency still

FIGURE 6-18 Effects of selection against the heterozygote when $p(A) \neq q(a)$ and when $p(A) = q(a)$. When $p(A) \neq q(a)$, if the same absolute numbers of alleles are removed from each pile on the left, the A allele, which is lower in frequency, will disappear first, leaving only a in the population. When $p(A) = q(a)$, both alleles will be lost at the same rate, so that there is no net change in proportions of alleles left in the population.

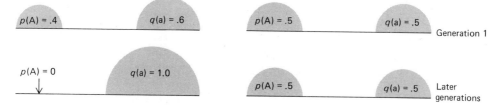

further. This would be expected eventually to eliminate the allele with the lower frequency from the population. Most known cases of selection against heterozygotes occur at human blood group loci.

Selection against Both Homozygotes

Here, by convention, the heterozygote would have a fitness value of 1.0, and each of the homozygotes would have a fitness value less than this. The fitness value of the two homozygous genotypes may be different. Unrelated factors can produce selection against each of the two genotypes, as in the case of sickle-cell anemia, discussed in Chapter 7. An idea of the direction and rate of change when selection favors the heterozygote can be gotten from Table 6-13. Here the alleles are equally frequent to begin with, and the aa homozygote has the lowest fitness value (W_3 = 0). This is the same as saying that all aa homozygotes are eliminated from the population as they are produced by segregation and fertilization each generation.

After just one generation of selection against both homozygotes, the frequency of the a allele has been reduced sharply, from .5 to .36. The frequency of the A allele has increased correspondingly, from .5 to .64. The information from these two generations alone suggests that if selection continued to operate in the same way, we might expect the change in gene frequency to continue, leading to eventual fixation of the A allele (and loss of the a allele), as shown in Figure 6-19.

It is an axiom of geometry that two points determine the orientation of a straight line and that such a line can in theory be extended beyond these points as far as one chooses. But the world of geometry is an idealized one; in the real world it is always risky to project a trend on

TABLE 6-13   Selection against both homozygotes when $p(A) = q(a) = .5$

Genotypes	AA	Aa	aa	Total
Generation 1: Gene frequencies: $p(A)$ = .5				
$q(a)$ = .5				
Genotype frequencies	.25	.50	.25	1.0
Fitness values	.75	1.0	0	
Proportions left after selection = genotype frequency × fitness value	.19	.50	0	.69
Generation 2: Genotype frequencies = proportion left after selection	.19	.50	0	1.0
Total	.69	.69	.69	
	.28	.72	0	

Gene frequencies: $p(A) = .28 + \frac{1}{2}(.72) = .64$
$q(a) = 0 + \frac{1}{2}(.72) = .36$

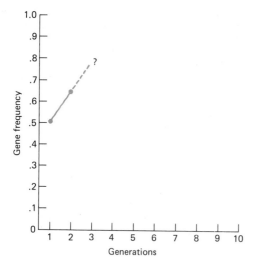

FIGURE 6-19 Selection against both homozygotes: increase in frequency of A from initial frequency of .5.

the basis of only two points. Specifically, the logical rules of geometry might not carry over into population biology. Thus we have no guarantee that the line which connects our two points (the gene frequency values in two successive generations) can be projected in a straight line beyond them. Plotting a few more points on the basis of different allele frequencies might help predict trends more accurately. This is done in Table 6-14.

The results in Table 6-14 indicate that when the initial frequency of the A allele is high (.9), one generation of selection against both homozygotes will reduce its frequency. This trend is in the opposite direction from the one based on Table 6-13, when the initial value of

---

TABLE 6-14 Selection against both homozygotes when $p(A)$ = .9 and $q(a)$ = .1

Genotypes	AA	Aa	aa	Total
Generation 1: Gene frequencies: $p(A)$ = .5				
$q(a)$ = .5				
Genotype frequencies	.81	.18	.01	1.0
Fitness values	.75	1.0	0	
Proportions left after selection =	.61	.18	0	.79
genotype frequency × fitness				
value				
Generation 2: Genotype frequencies =				
proportion left after selection	.61	.18	0	
Total	.79	.79	.79	
	.77	.23	0	1.0
Gene frequencies: $p(A)$ = .77 + ½ (.23) = .885				
$q(a)$ = 0 + ½ (.23) = .115				

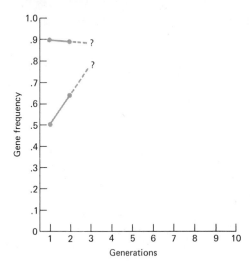

FIGURE 6-20   Selection against both homozygotes: convergence on intermediate allele frequency.

$P(A)$ was .5. Figure 6-20 contrasts these results. The situation in Figure 6-20 is intuitively reassuring. After all, simple logic suggests that as long as heterozygotes have the highest fitness value in the population, both alleles should be retained. But at what frequency? Figure 6-19 indicates that this should be somewhere between .5 and .9; projected changes in gene frequency seem to converge between these values.

The point of convergence of the lines marking the trends of gene frequency in the two sample calculations above marks an equilibrium position. Once it is reached, the allele frequencies should remain unchanged generation after generation. This equilibrium is independent of the initial gene frequencies (as Figure 6-19 suggests) and will be determined only by the selective coefficients of the homozygous genotypes. The situation occurs because selection against the AA homozygote removes only A alleles, while selection against the other homozygote removes only a alleles. This can be put in terms of an equation: $S_1 p = S_3 q$.

From this equation it is possible to derive a formula for the allele frequencies at equilibrium. If $S_1 p = S_3 q$ and if $q = 1-p$, then $S_1 p = S_3(1-p)$. Multiplying this out, we have $S_1 p = S_3 p - S_3 p$. Grouping terms gives us $p(S_1 + S_3) = S_3$. Solving for allele frequency, we find that $p = S_1/(S_1 + S_3)$.

An exact numerical value can easily be obtained now for the point which allele frequencies were converging in Figure 6-1. First, though, the selection coefficients must be found. If the fitness value of the AA homozygote $(W_1) = 0.75$, then $S_1 = 1 - .75$, or .25. And if $W_3 = 0$, $S_3 = 1 - 0$, or 1.0. Then, $p = S_3/(S_1 + S_3) = 1.0/(.25 + 1.0) = .8$, the equilibrium frequency of A. Likewise, $q = S_1/(S_1 = S_3) = .25/(.25 + 1.0) = .2$, the equilibrium frequency of a.

Selection against the two homozygotes will keep both A and a present in the population. When two or more alleles are retained at a locus in frequencies above those which are attributable to recurrent mutation alone, a genetic *polymorphism* exists. When these frequencies are stable, as when selection acts against both homozygotes, the situation is called a *balanced polymorphism.* The best-known example of a balanced polymorphism in humans involves the abnormal hemoglobin gene that produces hemoglobin S in homozygotes. In some parts of the world, stable frequencies of the gene that produce hemoglobins A and S have been maintained by selection against the homozygotes.

In the case of a balanced polymorphism, allele frequencies are sometimes said to be kept constant by *stabilizing selection* or *normalizing selection,* which favors the midrange while penalizing the extremes. More examples of normalizing selection are known from the study of natural populations than examples of any other type. But not all selection is stabilizing. As shown in the review of other examples of selection at a single locus, there can also be *directional selection,* which produces change in gene frequency in one consistent direction. But while one allele is being eliminated, there will be a considerable number of generations during which both genes will exist together in the population. This situation is known as a *transient polymorphism.* Selection, then, can lead to either stability or change (at differing rates) in gene frequency, depending on the particular circumstances.

SUGGESTIONS FOR ADDITIONAL READING

Calder, Nigel. 1974. "The Molecular Heresy," in *The Life Game.* New York: Viking. Chap. 3. (This chapter provides a nontechnical account—unfortunately, without a bibliography—of the use of evidence from molecular biology to reconstruct evolutionary relationships.)

Goodman, Morris, and Gabriel W. Lasker. 1975. "Molecular Evidence as to Man's Place in Nature," in Russell Tuttle (ed.), *Primate Functional Morphology and Evolution.* The Hague, Netherlands: Mouton. Pp. 71–101. (This is a recent and quite comprehensive account of the molecular evidence for evolutionary relationships among the primates, based partly on the assumption that gene replacement is slow and roughly constant. An extensive list of references is included.)

Johnson, Clifford. 1976. *Introduction to Natural Selection.* Baltimore: University Park Press. (This text, which can be read with profit by advanced undergraduate students, discusses the evidence for natural selection and the operation and measurement of selection in natural populations. The mode of presentation of this material combines descriptive and mathematical treatments.)

REFERENCES

Bielicki, T., and Z. Welon. 1966. "The Operation of Natural Selection on Human Head Form in an East European Population," *Homo,* **15**(1):22–30.

Bishop, J. A., and L. M. Cook. 1975. "Moths, Melanism, and Clean Air," *Scientific American,* **232**(1):90–99.

Boas, Franz. 1910. *Changes in Bodily Form of Descendants of Immigrants.* Senate Document 208, 61st Cong., 2d Sess.

Brues, Alice. 1963. "Stochastic Tests of Selection in the ABO Blood Groups," *American Journal of Physical Anthropology,* **21**(3):287–299.

Brues, Alice. 1977. *People and Races.* New York: Macmillan.

Cavalli-Sforza, L. L. 1969. "Genetic Drift in an Italian Population," *Scientific American,* **221**(2):30–37.

Crow, J. F., and M. Kimura. 1970. *An Introduction to Population Genetics Theory.* New York: Harper and Row. Chap. 8, "Stochastic Processes in the Change of Gene Frequencies."

Damon, A. 1968. "Secular Trends in Height and Weight within Old American Families at Harvard, 1870–1965. 1. Within Twelve Four-Generation Families," *American Journal of Physical Anthropology,* **29**(1):45–50.

Darwin, Charles. 1859. *The Origin of Species.* London: Murray.

Glass, B., M. S. Sacks, E. F. John, and C. Hess. 1952. "Genetic Drift in a Religious Isolate: An Analysis of the Causes of Variation in Blood Group and Other Gene Frequencies in a Small Population," *American Naturalist,* **86**(828):145–159.

Haldane, J. B. S. 1966. *The Causes of Evolution.* Ithaca, N.Y.: Cornell University Press.

Kelso, A. J., and G. Armelagos. 1963. "Nutritional Factors as Selective Agencies in the Determination of ABO Blood Group Frequencies," *Southwestern Lore,* **28**:44–48.

Kettlewell, H. B. D. 1959. "Darwin's Missing Evidence," *Scientific American,* **200**(3):48–53.

Kettlewell, H. B. D. 1965. "Insect Survival and Selection for Pattern," *Science,* **148**(3675):1290–1296.

Kimura, Motoo. 1968. "Evolutionary Rate at the Molecular Level," *Nature,* **217**(5129):624–626.

Lerner, I. M. 1954. *Genetic Homeostasis.* London: Oliver and Boyd.

Moor-Jankowski, J., A. S. Wiener, and C. M. Rogers. 1964. "Human Blood Group Factors in Non-Human Primates," *Nature,* **202**(4933):663–665.

Morris, L. N. 1971. "Natural Selection," in L. N. Morris (ed.), *Human Populations, Genetic Variation, and Evolution.* San Francisco: Chandler. Pp. 187–201.

Otten, C. M. 1967. "On Pestilence, Diet, Natural Selection and the Distribution of Microbial and Human Blood Group Antigens and Antibodies," *Current Anthropology,* **8**(3):209–226.

Shapiro, Harry L. 1962. *The Heritage of the Bounty.* Garden City, N.Y.: Anchor Books, Doubleday.

Simpson, G. G., A. Roe, and R. C. Lewontin. 1960. *Quantitative Zoology.* New York: Harcourt, Brace, and World.

Stanton, William. 1960. *The Leopard's Spots.* Chicago: University of Chicago Press.

Stern, Curt. 1973. *Principles of Human Genetics.* San Francisco: Freeman.

Tindale, N. 1953. "Tribal and Intertribal Marriages among the Australian Aborigines," *Human Biology,* **25**(3):169–190.

Vogel, F. 1965. "Blood Groups and Natural Selection," *Proceedings of the 10th Congress of the International Society of Blood Transfusion, Stockholm,* 268–279.

Vogel, F., and M. R. Chakravartti. 1966. "ABO Blood Groups and Smallpox in a Rural Population of West Bengal and Bihar (India)," *Humangenetik,* **3:** 166–180.

Woodworth, C. M., E. R. Leng, and R. W. Jugenheimer. 1952. "Fifty Generations of Selection for Protein and Oil in Corn," *Agronomy Journal,* **44**(2):60–66.

3

The Interpretation of Evolution

Chapter 7

Modern Human Variation and Its Interpretation

Evolutionary changes in populations can be brought about by mutation, gene flow, genetic drift, and selection. Each of these evolutionary forces can, on its own, cause changes in gene frequency. However, these forces do not act in isolation, but instead interact with one another to determine the frequencies of genes. In this chapter we'll begin by looking at some of the simpler models that have been devised to deal with interactions among the evolutionary forces that bring about human microevolution. From that basis we'll move on to a tougher task, that of examining the theories that have been put forth to explain the distribution of genetic and phenotypic characteristics among human populations in the world today—what is commonly referred to as *racial variation.*

MICROEVOLUTION: INTERACTIONS AMONG THE FORCES OF EVOLUTION

How much human variation is there at present? How is this variation distributed among the populations of the world? When and how did this variation originate? Will future patterns of variation differ from those of the past? We now know that the answers to these questions can come from an understanding of the forces of evolution discussed in preceding chapters. There is a difference, however, between understanding how gene frequency can change and finding the single correct explanation for a particular microevolutionary change within one population or for the difference in gene frequency between two populations. To see why this problem exists, consider the following question: Would it be possible for you to reconstruct in detail the plot of a movie you hadn't seen from just three consecutive frames of film? This is the problem that faces anthropologists who want to explain the causes of the variation we see in human populations today. Our knowledge of human variation at most loci is quite shallow, usually spanning three generations or less. Furthermore, there are many possible pathways to the patterns of variation we see. But the problem of finding the correct explanation is not a lack of possibilities; rather, it is an embarrassment of riches. For every change in gene or phenotype frequency, there are several different simple explanations, plus any number of plausible combinations of evolutionary forces acting together. The real challenge is to discover the correct answer.

In practice, such answers sometimes turn out to be surprisingly complex. Let us look, for example, at one search that was successful: the one that resulted in the discovery of the genetic control of sickle-cell anemia. In this case, anthropologists and geneticists were able to demonstrate the contributions of several forces of evolution to patterns of gene frequency observed in human populations distributed over the African continent.

Sickle-Cell Anemia: Culture and Microevolution

Sickle-cell anemia was first described early in the twentieth century (Herrick, 1910), but the molecular basis and genetic control of this disease were not worked out until the 1950s. Its clinical effects were well known: people who are homozygous for the S allele have a severe anemia which, without medical care, results in death before puberty. Despite this persistent drain of S alleles from the population because people suffering from the disease did not reproduce, in some areas of Africa the sickle-cell gene reaches frequencies of 10 to 20 percent. These contradictory observations led the eminent geneticist James V. Neel (1951) to point out that there must be some factor at work in these populations that tends to *increase* the number of sickle-cell genes.

Link with Malaria.   Explaining the high frequency of the S gene, which is deleterious when homozygous, in terms of forces of evolution was not easy. Drift was unlikely because the high frequencies of S did not always occur in small, isolated populations. There were systematic variations in the frequency of the gene, from quite high in some regions to sharply lower in others. But arguing that the high frequencies were due to gene flow did not really solve the problem either, since this implied the existence of populations in which the frequency of the S allele was higher still—which was also quite unlikely. Neel felt that the two most reasonable possibilities were frequent mutations to the S gene or selection favoring the AS heterozygote. Since the mutation rate from A to S did not seem to be high, this force of evolution could provide a dubious explanation at best (indeed, subsequent studies suggest that this mutation is exceedingly rare). Selection favoring carriers of the sickle-cell trait appeared to be the best possibility.

As early as 1949, J. B. S. Haldane had suggested that the frequencies of some hereditary anemias might be related to the distribution of malaria, which is common in tropical areas. This disease was known to be caused by protozoans (single-celled animals) such as *Plasmodium falciparum,* which lives as a parasite in human red blood cells. The organism is introduced into the bloodstream through the bite of a mosquito. It was suggested that people of certain genotypes were more resistant to the parasite's effects than others. The first direct evidence for this differential susceptibility was found in 1953, when the British geneticist Anthony C. Allison demonstrated that children with the sickle-cell trait (AS heterozygotes) were infected with malarial parasites less frequently than AA homozygotes (Allison, 1955). Furthermore, among adults inoculated with malaria, AS heterozygotes showed many fewer parasites than AA homozygotes. This suggested that sickle-cell heterozygotes had an advantage in terms of survival. Later studies indicated that heterozygotes might also have an advantage in terms of fertility. Malaria during pregnancy is a major cause of abortions, and so anything that would reduce the frequency or severity of malarial infection should result in more live births. In fact, studies indicate that AS females do have more surviving offspring.

Linguistic and Cultural Clues.   Although selection has been a major factor in determining the frequency of the sickle-cell gene in many populations, in some areas other evolutionary forces are also involved. The documentation of their role was the subject of a now classic anthropological study by Frank Livingstone (1958). Livingstone noted that most tribes in east Africa have frequencies of hemoglobin S that are in approximate equilibrium with the amount of malaria present but that in west Africa the pattern of distribution is extremely variable. Many tribes are even internally heterogeneous. The frequencies among the Mandingo in Gambia, for example, range from 6 to 28 percent, depending on the region sampled. But some general trends can be observed. As a rule, the higher frequencies in west Africa tend to occur toward the south, evidently following the north-south gradient in the distribution of malaria. All the populations with sickle-cell heterozygote frequencies greater than 15 percent inhabit areas where malaria is transmitted throughout most of the year. In these regions the average number of infective bites per person per year is always greater than five and sometimes ranges to 100 or more. As a practical matter, for the first 5 years of life every child struggles to survive against the parasite. Mortality is high, but those who do survive have become immune to the disease. Though they may contract it as adults, they rarely die from it. Among all these groups, malaria is securely established as the major cause of high frequencies of the gene for sickle-cell anemia.

In his detailed study of the distribution of the hemoglobin-S gene, however, Livingstone noted that some west African tribes living where malaria was prevalent had very low frequencies of the trait (Figure 7-1). His explanation for this anomaly has two parts. First, although the sickle-cell gene has been present in parts of west Africa for some time, tribes with low frequencies of this allele have until recently been comparatively isolated from those with high frequencies. Second, the environmental conditions responsible for high frequencies of the sickle-cell gene have been present among the low-frequency populations for a relatively short time, and so the spread of the sickle-cell gene is only now following the increase in selective advantage of the gene.

In support of his explanation, Livingstone used evidence from lin-

FIGURE 7-1   Distribution in Africa of (a) **malaria and** (b) **the hemoglobin S gene.**
Sources: Adapted from (1) F. B. Livingstone. 1967. *Abnormal Hemoglobins in Human Populations.* Chicago, Ill.: Aldine. (2) Arno G. Motulsky. 1960. "Metabolic Polymorphisms and the Role of Infectious Diseases in Human Evolution," *Human Biology,* **32**(1):28–62. Figs. 2, 3; pp. 43, 45.

(a) Distribution of falciparum malaria in Africa

(b) Areas in Africa with high frequencies of hemoglobin S

guistics, archeology, and the record of the cultural history of the African continent south of the Sahara. Most of the tribes in the area studied speak languages belonging to the Niger-Congo family, which is divided into seven subfamilies. All seven subfamilies are represented in west Africa. In central Africa, east Africa, and South Africa, in contrast, nearly all the Niger-Congo languages belong not only to one subfamily but also to a single subgroup of it (Bantu). The great linguistic diversity in west Africa suggests that the region has been inhabited for a long time by Niger-Congo speakers, whereas the linguistic uniformity in the Bantu-speaking areas argues for a relatively recent spread of those peoples (see also Phillipson, 1977). To draw a parallel to a more familiar situation, western Europe today is occupied by people speaking a great diversity of languages, whereas a single language, English, is spoken over most of North America because of very recent settlement beginning from Great Britain.

One key to the puzzle of sickle-cell gene frequency over sub-Saharan Africa came with the realization that the west African tribes with low hemoglobin-S gene frequencies are the tribes that have long been native to the area. Among them the sickle-cell gene seems still to be diffusing by gene flow rather than by large-scale migration and hybridization. Those areas with high sickle-cell gene frequencies are regions that have recently received substantial numbers of immigrants, chiefly Bantu. It looks, therefore, as if this abnormal hemoglobin gene was spread with the expansion of the Bantu peoples, in whom the frequencies were high.

Two questions remain to be answered: Why did the Bantu spread, and why did they have high sickle-cell gene frequencies? The spread of the Bantu seems to have depended on the adaptation of agriculture to the equatorial forest. Agriculture spread from Egypt to the rest of Africa around 3000 B.C. It had probably originated still earlier in Asia Minor, since the Egyptians cultivated crops such as wheat and barley, which are not native to Africa. Later the Egyptians domesticated African grains, such as millet and sorghum, and the spread of agriculture from Egypt to the equatorial forest region of central Africa was based on these crops.

Three factors prevented agriculturalists from penetrating the equatorial forests themselves: (1) the near impossibility of clearing the forests with stone tools, (2) the low yields of millet and sorghum, and (3) the poor quality of the soils, which become infertile after a few crops. The first problem was solved with the diffusion of ironworking, again from Asia Minor through Egypt, perhaps about 600 B.C. With iron tools, tropical vegetation could be cleared. The other limitations were overcome by the replacement of grains with cultivated root crops, chiefly yams and cassava, which yield two to four times as many calories per unit of land.

The conditions that made it possible for humans to thrive in tropical forests favored the multiplication of the mosquitoes that spread the malarial parasite. These mosquitoes cannot breed in water that is shaded or has a strong current. After forest trees are cut down, the soil rapidly loses the layer of humus that absorbs water; the soil now catches the water, which forms sunny pools kept filled by tropical rains. Human refuse and swamps within the cleared areas provide additional breeding places. When they are not breeding, the mosquitoes rest in the thatched roofs of African village huts. The hunters who occupied the forests before the agriculturalists arrived neither cleared the forests nor built permanent dwellings. They moved frequently, making it difficult for the mosquitoes to keep up with their food supply. Furthermore, the hunters' population densities were so low that they may have been below the critical size needed for the persistence of the parasites. In hunting groups, therefore, the sickle-cell gene should not offer the selective advantage it does to equatorial agriculturalists. The hunting groups studied so far meet this expectation: they have frequencies of the sickle-cell gene low enough to be explained in terms of very slight gene flow from neighboring agricultural groups.

It would appear that the cultural conditions associated with the agricultural revolution made possible both the rapid migration of Bantu peoples and the spread of their genes. The Bantu were subjected to ecological conditions that imposed a new hazard, malaria. This disease, in turn, selected for high frequencies of the sickle-cell gene. Through contact with other tribes, the sickle-cell gene is being spread by gene flow to still other areas. Besides providing an excellent example of the interactions among several forces of evolution, the case of sickle-cell anemia is one of the best-documented examples of the influence of culture on human evolution.

Population Size and Microevolution

The sickle-cell example demonstrates that although we can think of the forces of evolution in isolation from one another, they do not operate that way. Evolution in natural populations is the result of complex interactions among all of these forces, but it is greatly influenced by population size. The reasons for this have been worked out in considerable detail by Sewall Wright, an American who has contributed much to modern population genetics. Wright was one of the first to stress the importance of drift in evolution by showing that in the smallest populations this random process could outweigh the more systematic evolutionary forces (see Table 7-1). As population size increases, the other forces of evolution come into play and become relatively more

---

TABLE 7-1   Changes in gene frequency according to degree of determinacy

---

1. From systematic pressures (change in gene frequency can be determined, at least in theory)
   a. Recurrent mutation
   b. Selection (acting within a population)
   c. Gene flow
2. Changes which are unpredictable in direction but whose magnitude can be estimated
   a. Fluctuations in the systematic pressures above (**1**)
   b. Fluctuations due to accidents in sampling (i.e., genetic drift)
3. Unique events (completely unpredictable)
   a. A mutation favorable from the first
   b. A unique selective event
   c. Unique hybridization
   d. Swamping by mass immigration
   e. A unique reduction in numbers

Source: Adapted from S. Wright. 1949. "Population Structure in Evolution," *Proceedings of the American Philosophical Society,* **93**(6):471–478.

---

important. The more systematic forces of evolution are considered more significant than drift when

$$N_e \geq \frac{1}{4\mu} \quad \text{or} \quad \frac{1}{4s} \quad \text{or} \quad \frac{1}{4m}$$

where, as in earlier chapters, N_e is the effective population size, μ is the mutation rate, s is the selective coefficient, and m is the rate of migration.

It is possible to see what this means in terms of actual population size by substituting for μ, s, and m hypothetical values (but values comparable to those found in human populations) for some loci. For μ, a reasonable value would be about 1×10^{-5}; for s, perhaps 0.001 (a selective disadvantage of 1 part in 1000, so low as to readily escape notice); and for m, say, 0.01 (that is, about 1 percent gene flow per generation). If these values are substituted in the formula above, the following results are obtained:

$$N_e = \frac{1}{4\mu} = \frac{1}{4(1 \times 10^{-5})} = 25{,}000$$

$$N_e = \frac{1}{4s} = \frac{1}{4(0.001)} = 250$$

$$N_e = \frac{1}{4m} = \frac{1}{4(0.01)} = 25$$

Basically, the significance of these figures is that since they often exist in natural populations, the directional forces of evolution can be of very unequal strength. To put it another way, because they recur, mutation, selection, and gene flow can each have a directional effect on gene frequency, tending to push it in one direction. But they recur at vastly different rates.

Imagine a soccer ball labeled "gene frequency" being propelled across a field on which men wearing sweatshirts marked "drift" are stationed. Each time the ball reaches one of these men, he can kick it in any direction he pleases. The directions of the kicks really seem unpredictable, as if the kickers were absentminded or drunk. Several teams—gene flow, selection, and mutation—try in succession to move their balls across the field. The gene-flow group does this without too much difficulty; each man on the team of 25 pays attention to the ball's course and kicks it on ahead whenever it comes in his direction. The selection crew isn't nearly as good. The team of 25 just cannot make any headway against drift. Fortunately, in this game there are no rules governing team size, and so the selection group keeps adding more and more men. When their team size reaches 250, they are evenly matched against the drifters. Finally, the mutation team enters the field. If the selection players were individually ineffective, then these men are either nearly dead or stupid. Not one of the first 25 on the mutation squad is of much use, and even a group of 250 proves to be little better; none of them kicks the ball very often. Not until 25,000 of these players get involved is the effect of drift eventually overcome. Clearly, some teams must be much larger in size because their players are less effective individually than those on the smaller squads. The size of the group needed varies inversely with the effective strength of each force.

Explanations for Microevolution

Wright's calculations are useful in helping us predict the outcome of interactions among those forces of evolution that can determine gene frequency at a single locus. However, in interpreting human variation and evolution, anthropologists, geneticists, and other population biologists are interested in building on single-locus models in order to understand the processes affecting more complex polygenic characteristics and, indeed, the evolution of the entire human gene pool. At the present time, three major theories are offered as alternative explanations for the genetic variation now seen in humans and other sexually reproducing organisms. They are presented in simplified form here; for a more detailed discussion, see, for example, Lewontin (1974).

Classical Theory.   The classical view is that the typical individual in any population is homozygous at almost every locus for the "normal" allele (also called the *wild-type* allele by geneticists). New mutations are almost invariably harmful because they disrupt DNA sequences that have been finely tuned to optimal performance by thousands or millions of generations of evolution. In extremely rare cases, the mutant gene

```
+ + + + m + ... + + +        + + + + + + ... + m +
═══════════════════        ═══════════════════
+ + + + + + ... + + +        + + + + + + ... + + +
```

Key:
  + = Normal alleles
  m = Mutant alleles

FIGURE 7-2   The classical model of sections of genome in two randomly sampled individuals from the same population.

Source: R. C. Lewontin. 1974. *The Genetic Basis of Evolutionary Change.* New York: Columbia University Press. Fig. on p. 24.

would give an advantage and would be increased in frequency by selection until it replaced its alternative allele. While the process was going on, a polymorphism would exist at this locus, but only for a time. The new mutant would eventually become the normal allele, with a frequency at or near fixation (the point at which one allele reaches a frequency of 100 percent in the population). All individuals in the population would be heterozygous for rare, harmful genes at some loci—perhaps a few dozen to a hundred or so out of the tens of thousands of genes in the whole genome. This situation is shown hypothetically in Figure 7-2. Only a very small fraction of the population would be homozygous for one or another of these rare, harmful genes, and the affected homozygotes would be severely handicapped or die as a result. The probability of homozygosity at all loci is raised by inbreeding, which increases an individual's chances of getting two copies of the same gene from some ancestor common to both parents. Inbreeding can therefore result in a measurable increase in the frequency of those handicapped by diseases or malformations caused by homozygous recessive genes.

In sum, the classical model sees homozygosity as the normal condition, beneficial in all except an extremely small proportion of cases. Heterozygosity would be unusual and normally of no advantage; at best, it would be of no disadvantage.

Balance Theory.   This view of variation gets its name from the situation known to exist in cases of balanced polymorphism. Unlike classical theory, which sees heterozygosity due to balanced polymorphism as rare, this theory sees it as a common situation. Thus, if it were possible to microscopically examine the genomes of individuals in representative populations, they would look as shown in Figure 7-3. There would be no normal allele at any locus; most individuals would have two different members of a large series of multiple alleles. This array of alleles is thought to be maintained at most loci by stabilizing selection that favors the heterozygotes. Consequently, most polymorphisms detected in the population would be balanced rather than transient. Harmful and

$$\frac{A_3\, B_2\, C_2\, D\, E_5 \ldots Z_2}{A_1\, B_7\, C_2\, D\, E_2 \ldots Z_3} \qquad \frac{A_2\, B_4\, C_1\, D\, E_2 \ldots Z_1}{A_3\, B_5\, C_2\, D\, E_3 \ldots Z_1}$$

Key:

$A_1, A_2, A_3, \ldots A_n$ = Alternative alleles at one locus, with most individuals being heterozygous for different pairs from a multiple allelic series

D = Locus with only one allele in the population, an infrequent occurrence

FIGURE 7-3 The balance model of sections of genome in two randomly sampled individuals from the same population.
Source: R. C. Lewontin. 1974. *The Genetic Basis of Evolutionary Change*, New York: Columbia University Press. Fig. on p. 25.

approximately neutral mutations could still occur, but they would be a smaller proportion of new mutations than in the classical theory. With heterozygotes selected for, new harmful genes might not be gotten rid of so quickly since they would rarely become homozygous except among the offspring of close relatives, as a consequence of inbreeding.

Neoclassical Theory. As Lewontin has recently emphasized in *The Genetic Basis of Evolutionary Change*, the term *neoclassical theory* is a better label for a viewpoint that is today frequently called the *neutral mutation theory*. Natural selection plays a large role in this viewpoint, but it is a role limited almost exclusively to the removal of harmful mutations. In this sense, the neoclassical view resembles the classical theory. But it differs from the classical view in one important way: it attempts to account for a formerly unsuspected abundance of inherited biochemical variation uncovered by sophisticated new laboratory techniques such as electrophoresis and chromosome banding. The neoclassical explanation for the existence of these biochemical polymorphisms is based on the premise that physiological, chemical, and microstructural differences can be detected by the scientist, but not by the organisms in which they exist. The genes that produce them give rise to visible phenotypic characteristics that are apparently selectively neutral, and the frequencies of these genes are determined chiefly by random forces. Although harmful and neutral mutations may be the most common types, other possibilities are not absolutely excluded. Favorable mutants could arise, and balanced polymorphisms could be maintained at a few loci.

Relative Emphasis: Which Forces Predominate? We do not yet know enough to choose the one model that most closely corresponds to reality; to some extent, all fit the evidence for the pattern of variation at one point in time and of evolutionary change through time. But each can tell us something about the relative roles of different forces in evolution. In the classical model, an equilibrium is struck between mutation and selection; drift and gene flow occupy quite subsidiary positions. In the balance theory, mutation, gene flow, and drift can all influence gene frequencies, but the major emphasis is on selection: opposed selective agents are dynamically balanced against each other. In the neoclassical view, largely random factors (mutation and drift) have the upper hand; gene flow is given little attention, and selection is of lesser qualitative and quantitative importance.

We now know that the forces of evolution do not act independently of one another, but instead interact in complex ways, sometimes reinforcing one another and at other times acting in opposition. The several theories outlined above attempt to describe, in different ways, the nature of the balance among these evolutionary mechanisms. Before we can decide which of the models best describes the organization of genetic material in populations, much more research will be needed. Anthropologists are already at work on this task, searching for micro-evolutionary changes in human populations. In fact, the study of patterns of human variation began long before there was any formal science of anthropology or any systematic knowledge of the genetic basis of human characteristics. During the earlier period of study, a different set of beliefs and terms (such as *race*) came into use. In the remainder of this chapter we'll take a close look at this earlier theoretical framework which was devised to help understand human variation, and we'll see how it corresponds to the more detailed knowledge of the forces that shape human populations.

DISTRIBUTION OF HUMAN VARIATION

Human variation has traditionally been dealt with in terms of the concept of *race*, an idea that has existed far longer than the discipline of anthropology. Although the term was introduced into the formal tax-onomic hierarchy by Linnaeus and his immediate predecessors, the same word or its equivalent can be traced back to the ancient Greeks. And their free use of the concept suggests an even greater antiquity extending back beyond the reaches of recorded history.

Any word used over so long a period is bound to acquire a variety of different meanings. As a result, two people using the term in a discussion may not even realize they are talking about quite different things. Much of the confusion comes about because the concept of race overlaps two quite different realms: the social and the biological. Human societies are culturally defined groups whose members interact behaviorally in a number of ways. They are typically similar in terms of language, political beliefs, and dress, and they have many other habits and customs in common. Among the many aspects of life structured by these interactions is breeding behavior. As a result (though some would say this is a cause), members of many human societies also share some distinctive physical characteristics and may differ on the average from members of other societies.

Uses of Racial Classifications

In their characterizations of various human races (Figure 7-4), Linnaeus and many of his predecessors included language, dress, customs, and other purely cultural features along with physical appearance. Only recently have biological characteristics come to be accepted as the only valid criterion for the demarcation of human races. Thus, for example, Brace and Montagu (1977) defined a race as "a group of mankind, members of which can be identified by the possession of distinctive physical characteristics," (p. 389). Usually the members of a given race are thought to share certain biological or cultural features because they are residents of a particular geographic area—but not always. The Jews of Europe, for example, were (and sometimes still are) classified as a separate race, despite the fact that they had lived alongside, and intermarried with, non-Jews for centuries.

In country after country, Jews resemble other members of the surrounding population more closely than they resemble Jews from distant regions. For example, one group of eastern European Jews studied had a cephalic index of 86.3, being broad-headed like their European neighbors, and Yemenite Jews of Arabia were shown to have a cephalic index of 74.3, close to that of their long-headed Arab neighbors (Firth, 1958). Jews are not a race; they are a culturally defined subdivision of our species, members of a category defined by a common religion, traditions, and customs. Groups that share such cultural features may also constitute a population that shares biological characteristics, but one is not a necessary or even frequent consequence of the other.

It is now easy to see why the equation of nation with race is also an error, though a very common one. The citizens of a nation are all ruled by the same central government, participate in a common economic system, and so on. All the citizens of a country may also share a common language, but even this is not necessarily so. In the United States today English is the predominant language, but for a sizable minority of the population Spanish is the primary language. A similar situation exists in many other countries in Europe, Africa, and Asia. Nations are born through political processes; their boundaries often cut across populations that share biological features, and they may join together under the same national label peoples of very diverse ancestries.

Taxonomists of the eighteenth and nineteenth centuries did not pay particular attention to the very different mechanisms by which people come to share various cultural and biological features. As a result, racial categories were often defined in terms of language, dress, and other socially transmitted features as well as in terms of biologically inherited characteristics.

As part of their ordering of the natural world, the early taxonomists divided the human species into three or four or five subdivisions

MAMMALIA

Order 1.   Primates

Fore-teeth cutting; upper 4, parallel;
teats 2 pectoral.

1. HOMO

sapiens	Diurnal; varying by education and situation	
	Four-footed, mute, hairy.	Wild Man **(Ferus)**

Copper-colored, choleric, erect . . . . American
   Hair black, straight, thick; nostrils   **(Americanus)**
   wide, face harsh; beard scanty; ob-
   stinate, content free. Paints himself with fine red lines. Regulated
   by customs.

Fair, sanguine, brawny. European
   Hair yellow, brown, flowing; eyes blue;   **(Europaeus)**
   gentle, acute, inventive. Covered with
   cloth vestments. Governed by laws.

Sooty, melancholy, rigid. . . . . Asiatic
   Hair black; eyes dark; severe; haughty,   **(Asiaticus)**
   covetous, covered with loose garments.
   Governed by opinions.

Black, phlegmatic, relaxed. . . . . African
   Hair black, frizzled; skin silky; nose   **(Afer)**
   flat; lips tumid; crafty, indolent,
   negligent. Anoints himself with grease.
   Governed by caprice.

monstrous   Varying by climate or art.

Small, active, timid. . . . . Mountaineer
   (Alpini)

Large, indolent. . . . . Patagonian
   (Patagonici)

Less fertile. . . . . Hottentot
   (Monorchides)

Beardless. . . . . American
   (Imberbes)

Head Conic. . . . . Chinese
   (Macrocephali)

Head flattened. . . . . Canadian
   (Plagiocephali)

2. SIMIA

FIGURE 7-4 Human classification by Linnaeus.
   Source: Kenneth A. R. Kennedy. 1976. *Human Variation in Space and Time.* Dubuque, Iowa: Brown.
Fig. 3-1, p. 25. Based on a translation of the Latin 10th ed. (1758) and a facsimile ed. (English, 1806).

(Caucasians, Mongolians, Indians, etc.) that corresponded roughly to the world's major geographic areas: Europe, Asia, the Americas, Africa, and so on. But there were always some populations that did not fit very well into any of the major categories, which were believed to represent the original, or "pure," stocks of the human species. Where did Mestizos, the products of intermarriage between the Indian population of Mexico and their Spanish conquerors, belong? In which category should one place the Gypsies, who were thought to have come from India to Europe, where they supplemented their gene pool with occasional additions from surrounding populations? What about the Lapps, thought by many to represent a mixture between northern Europeans and Asians? Or the South African Bushmen, the southeast Asian Negritos, and all the other populations whose origins were unknown? When these and other problem populations were included in racial taxonomies at all, they were usually treated as hybrids of major, or pure, races and given special names of their own. The assumption was that the human species had been composed of just a few races in the not-too-distant past. These earlier races were thought to have been the original subdivisions of mankind, kept separate and pure by isolation. This is now known to be absolutely wrong; even small local populations, whether of humans, baboons, mice, moths, or fruit flies, are genetically diverse. As Table 7-2 shows, every individual human is heterozygous at thousands of loci, and there is good evidence to indicate that such stored variation is and always has been necessary for the survival of populations. But until 25 years ago, the idea of pure races was accepted as a given by anthropologists. This premise led, in turn, to a further hope: that by analyzing the biological characteristics of the hybrid groups, anthropologists would be able to reconstruct the percentages of the major racial stocks that had made up the ancestry of the problem populations. The objective here was to work out a classification of human populations based entirely on nonfunctional phenotypic variation.

This set of assumptions, hopes, and goals gave rise to a whole program of research that occupied anthropologists for the better part of two centuries, from the time of Johann Friedrich Blumenbach's *On the Natural Variety of Mankind* (1775) to Carleton Coon's *Origin of Races* (1962). Researchers set out to discover traits that differed from one major racial stock to the next and were stable (in modern terms, traits that would be unaffected in expression by the environment or any force of evolution other than gene flow). For such a program to work, stable, nonfunctional traits were thought to be an absolute necessity. One trait after another was proposed to meet this need: skin color, hair form, body size and form, nasal index, cephalic index, and so on (Figures 7-5 to 7-9 show the distributions of these physical features). Sometimes groups of traits were used. Thus, in 1946 E. A. Hooton, a major figure in American physical anthropology, put forth the following

TABLE 7-2   Genetic differences between individuals

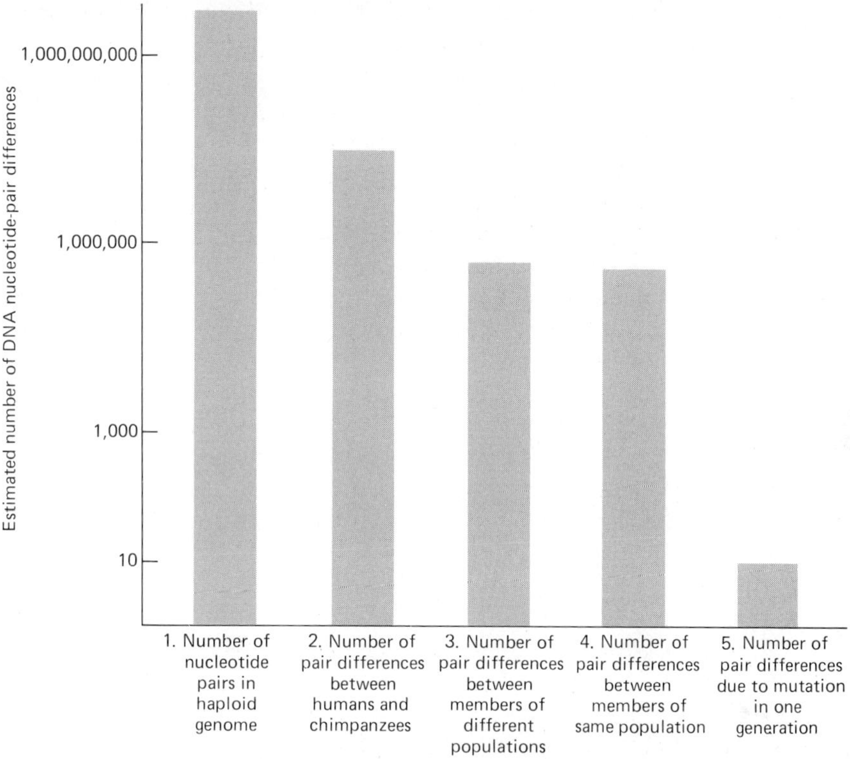

Note: This logarithmic scale shows the numbers of genetic differences estimated to exist between individuals of varying degrees of relatedness.

Source: Modified from W. F. Bodmer and L. L. Cavalli-Sforza. 1976. *Genetics, Evolution, and Man.* San Francisco, Calif.: Freeman. Fig. 19-16, p. 590.

definition: "A race is a great division of mankind, the members of which, though individually varying, are characterized as a group by a certain combination of morphological and metrical features, *principally nonadaptive,* which have been derived from their common descent" (p. 397; italics added).

Evidence against Pure Races

Despite the continuation of research based on nonadaptive traits, their very existence had been questioned long before. As early as 1910, Franz Boas had pointed out that children born in America differed from their European immigrant parents in the form of the head. Moreover, the differences tended to be greater in proportion to length of residence in the United States. His conclusions were reaffirmed in even more rigorous

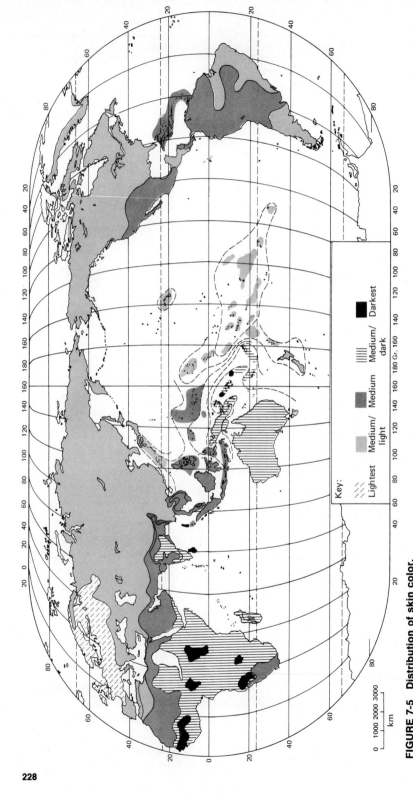

FIGURE 7-5 Distribution of skin color.
Source for Figures 7-5, 7-6, 7-7, 7-8, and 7-9: Renato Biasutti. 1967. *Razza e popoli della terra.* Vol. 1. Turin, Italy: Unione Tipografico—Editrice Torinese. Tables VI, VII, VIII, IX, X.

Key:

Lightest

Medium/ light

Medium

Medium/ dark

Darkest/ dark

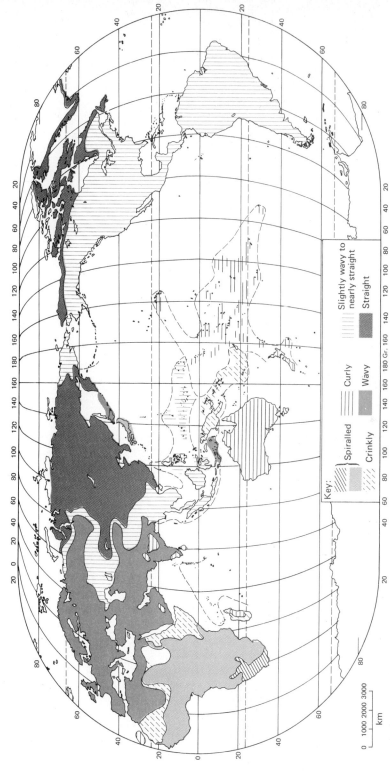

FIGURE 7-6 Distribution of hair form.

Key:

- Spiralled
- Curly
- Slightly wavy to nearly straight
- Crinkly
- Wavy
- Straight

km
0 1000 2000 3000

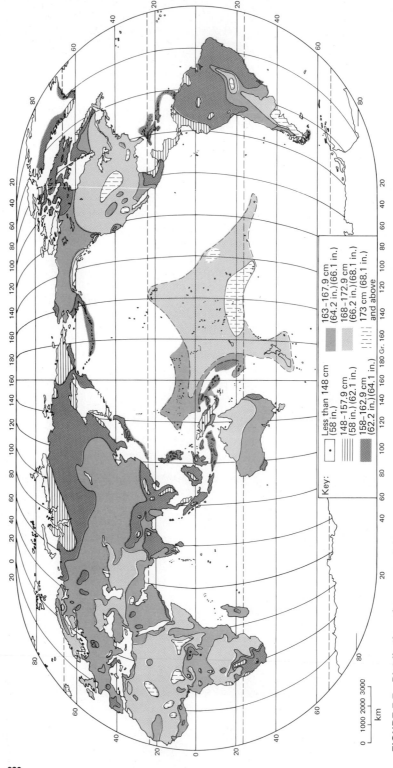

Key:

Less than 148 cm
(58 in.)

148–157.9 cm
(58 in.) (62.1 in.)

158–162.9 cm
(62.2 in.)(64.1 in.)

163–167.9 cm
(64.2 in.)(66.1 in.)

168–172.9 cm
(66.2 in.)(68.1 in.)

173 cm (68.1 in.)
and above

0 1000 2000 3000
km

FIGURE 7-7 Distribution of average stature.

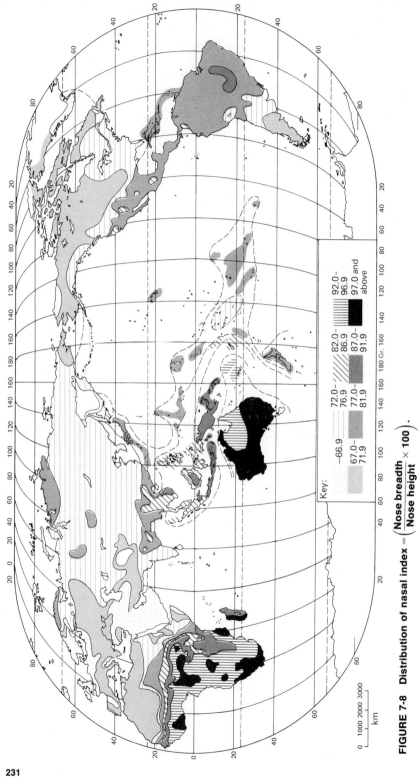

Key:

—66.9	72.0–76.9	82.0–86.9
67.0–71.9	77.0–81.9	87.0–91.9
	92.0–96.9	97.0 and above

FIGURE 7-8 Distribution of nasal index $= \left(\dfrac{\text{Nose breadth}}{\text{Nose height}} \times 100\right)$.

0 1000 2000 3000
km

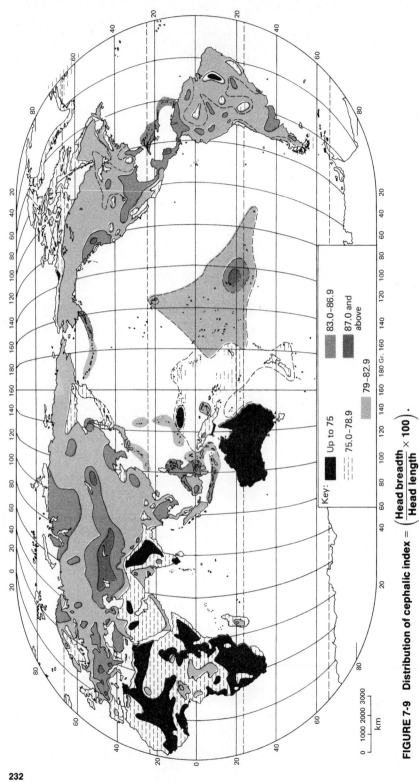

FIGURE 7-9 Distribution of cephalic index $= \left(\dfrac{\text{Head breadth}}{\text{Head length}} \times 100 \right)$.

Key:

Up to 75	83.0–86.9
75.0–78.9	87.0 and above
79–82.9	

form by the work of Harry L. Shapiro. In *Migration and Environment*, published in 1939, Shapiro described his comparisons of three groups: Japanese who had emigrated to Hawaii, their relatives who stayed in Japan, and the emigrants' children raised in Hawaii. Like Boas, Shapiro found that the measurements and proportions of the head differed between Japanese-born parents and their Hawaiian-born children. Males born in Hawaii had heads that were shorter from front to back and broader from side to side than those of immigrant males. As a result, in males the cephalic index changed by 2.6 points–a difference that was over six times the standard deviation of the measurement. Other physical features thought to be stable and hence racially based had changed as well. For example, both immigrants and their Hawaiian-born descendants developed longer and narrower noses (and hence lower nasal indices) than those found in Japanese who stayed in their home country. Sitting height, a measure of trunk length, also changed, increasing significantly in the Hawaiian-born over their immigrant parents or relatives in Japan. The magnitude and rapidity of these changes argued that the environment of Hawaii could directly modify the expression of genes that determine major visible physical features.

Once it was demonstrated that physical features could be altered so easily by the environment, they were of no value as racial markers. The result, however, was not the abandonment of the idea that stable, nonfunctional traits could be used to reconstruct relationships among human populations. Instead, the findings spurred a search for new nonfunctional traits. In the 1940s and 1950s it was widely believed— perhaps *hoped* would be a better word—that blood groups would fill this void. The argument that blood groups were nonfunctional was made soon after they were discovered, early in this century. Blood groups were invisible to the eye, were apparently unmodified by any environmental setting still compatible with life, served no known purpose, and differed in frequency from population to population. Surely such traits would make ideal racial markers. The definition of race was modified to encompass simply these inherited genetic variants as well as polygenically controlled morphological variations. For example, Dobzhansky (1944) said: "Races are defined as populations differing in the incidence of certain genes but actually exchanging or potentially able to exchange genes across whatever boundaries (usually geographic) separate them" (p. 252).

The most influential exponent of the use of blood groups as a basis for racial classification was W. C. Boyd. In the introductory chapter of *Genetics and the Races of Man* (1950), Boyd stated his position clearly:

The sort of character we shall be led to choose as being relatively non-adaptive will probably be the characters for which we cannot imagine any survival value. (Of course the fact that we cannot imagine any usefulness in evolution of a character does not prove that such usefulness does not exist, but such characters are at any rate to be preferred to those which obviously have high survival value.) The bony structures obviously have high survival value, and we shall hardly select the more important features of them. Among the racial characters which we would be tempted to pick out at the present time as non-adaptive, there are certain serological features of the blood, such as the genes O, A, B, M, N, etc.; many other characters, such as the direction of hair whorls, general body hairiness, (probably) tooth cusp patterns, fingerprint patterns, etc., might be considered (pp. 26–27).

The maps in Figures 7-10 to 7-15 show the distributions of some of these serologic traits.

But neither Boyd nor any of those who sought to devise racial classifications ever objectively demonstrated the nonfunctional nature of these traits. The usual argument (or assumption) was that human survival must generally depend on such qualities as intelligence, keen vision, muscular strength, and resistance to disease. By comparison, a difference in type of blood group antigen was felt to be too trivial to have any influence. Therefore, as Boyd suggested, frequency of a given feature would be stable from one generation to the next.

Theoretical Arguments against Nonfunctional Value

The idea that an inherited trait which does not seem important to humans cannot influence survival or reproduction is probably just a deceptively simple fallacy. It is easy to show this by means of an example. Recall that virtually the only well-documented case for the operation of selection in humans is that of the hemoglobin-S gene, where the AS heterozygote has a fitness value of 1.0, the AA homozygote has a fitness value in the vicinity of 0.75, and the SS homozygote has a fitness value near 0.0. As shown in Figure 7-16, under such conditions the S gene would go from an initial frequency of .000001 to its equilibrium value of .2 in just 50 generations. This evolution, rapid by any standards, is the result of quite stringent selection.

It is worth considering in this instance how readily such an extreme case of selection would be detected. The substantial reduction in fitness of the AA homozygotes results from reductions in fertility and increases in mortality that are spread over their lifetimes. Its most readily quantifiable component seems to be high mortality among young children. But even in tropical Africa, where malaria is endemic, this disease kills only 15 to 20 out of every 1000 young children, while the total mortality over the same time span is between 200 and 500 per 1000. Given these statistics, it is not surprising that it was difficult to obtain evidence for malaria as a selective factor and for the AS genotype as a genetic

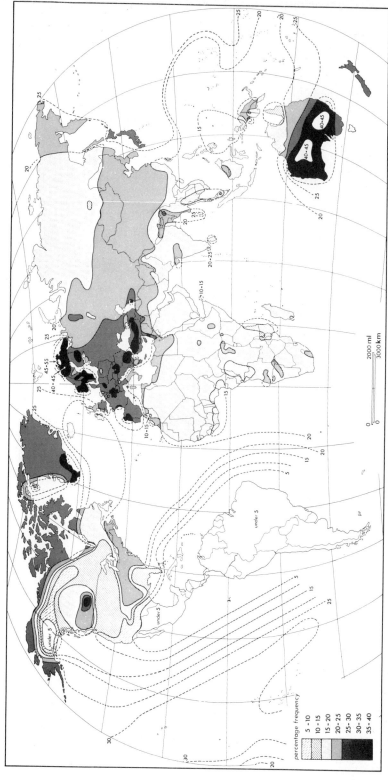

FIGURE 7-10 ABO blood group system: distribution of the A gene in the indigenous populations of the world.

Source for Figures 7-10, 7-11, 7-12, 7-13, 7-14, and 7-15: A. E. Mourant, A. C. Kopec, and K. Domaniewska-Sobsczak, 1976. *The Distribution of the Human Blood Group Polymorphisms* (2d ed.). London and New York: Oxford University Press. Maps 1, 2, 3, 17, 23, 24.

percentage frequency
5 - 10
10 - 15
15 - 20
20 - 25
25 - 30
30 - 35
35 - 40

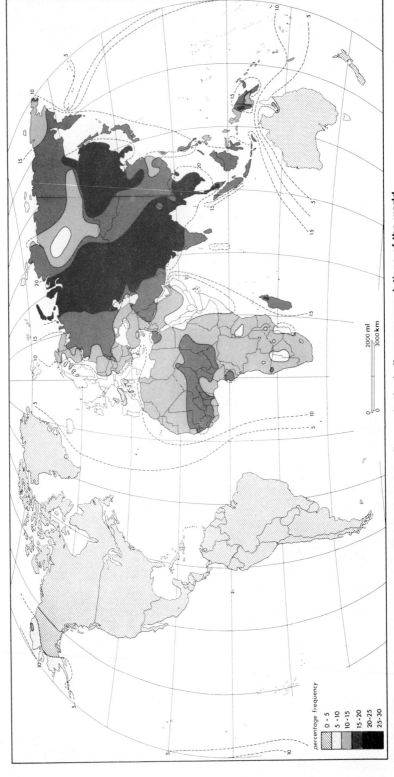

FIGURE 7-11 ABO blood group system: distribution of the B gene in the indigenous populations of the world.

percentage frequency

	0 - 5
	5 -10
	10 -15
	15 -20
	20 -25
	25 -30

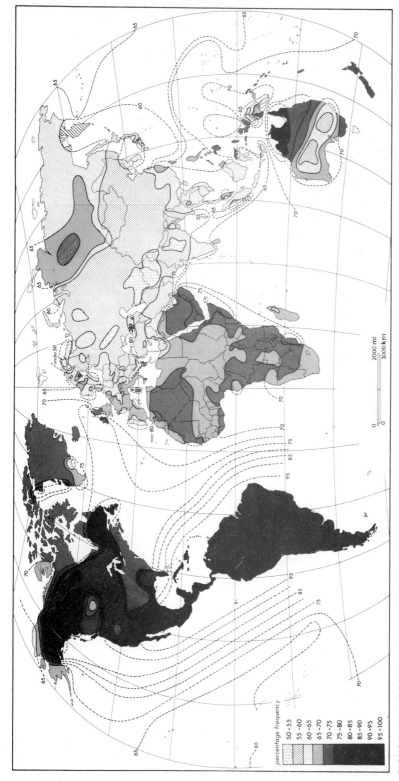

percentage frequency

	50 - 55
	55 - 60
	60 - 65
	65 - 70
	70 - 75
	75 - 80
	80 - 85
	85 - 90
	90 - 95
	95 - 100

FIGURE 7-12 ABO blood group system: distribution of the O gene in the indigenous populations of the world.

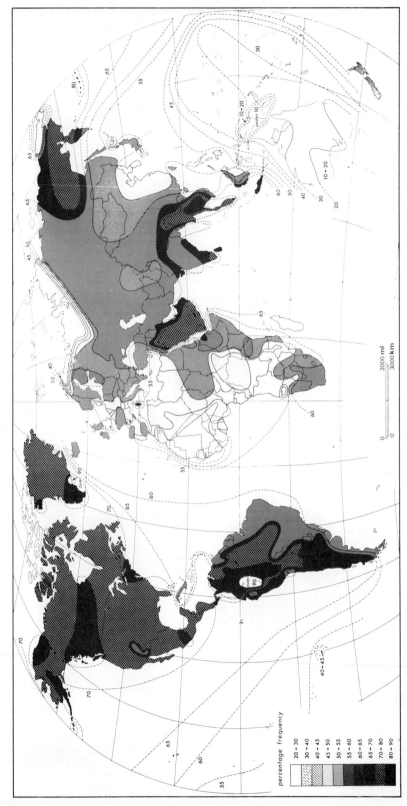

FIGURE 7-13 MN blood group system: distribution of the M gene in the indigenous populations of the world.

percentage frequency

20 – 30
30 – 40
40 – 45
45 – 50
50 – 55
55 – 60
60 – 65
65 – 70
70 – 80
80 – 90

0   2000 ml

0   3000 km

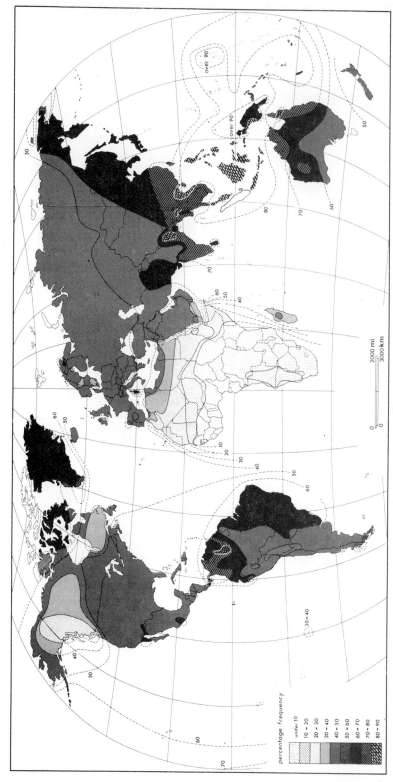

FIGURE 7-14 Rh blood group system: distribution of the C gene in the indigenous populations of the world.

percentage frequency

under 10
10 - 20
20 - 30
30 - 40
40 - 50
50 - 60
60 - 70
70 - 80
80 - 90

2000 ml
3000 km

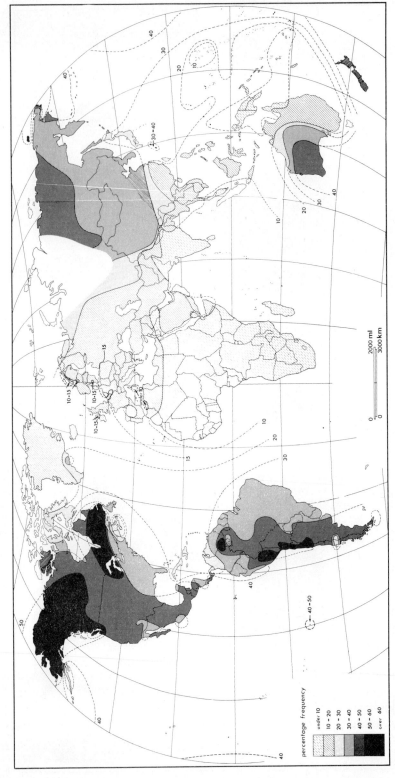

FIGURE 7-15 Rh blood group system: distribution of the E gene in the indigenous populations of the world.

percentage frequency

under 10
10 – 20
20 – 30
30 – 40
40 – 50
50 – 60
over 60

2000 ml
3000 km

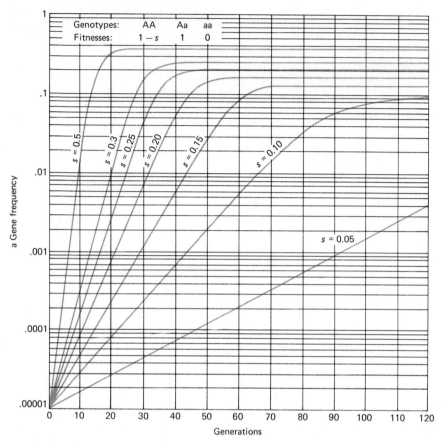

FIGURE 7-16 **Change in gene frequency under different fitness values. This graph shows the approach to equilibrium for the frequency of a gene *a*, a recessive lethal, for different fitness values of the dominant homozygote. In each case, the heterozygote is assumed to have a selective advantage (*s* > O), and the initial frequency of *a* in the population is assumed to be 1 in 1 million. Note that smaller values of *s* (small advantages for the heterozygote) lead to slower approaches toward equilibrium and to smaller equilibrium values for the gene frequency of *a*.**

Source: W. F. Bodmer and L. L. Cavalli-Sforza. 1976. *Genetics, Evolution, and Man.* San Francisco, Calif.: Freeman. Fig. 9.5, p. 317.

feature conferring resistance. In contrast, look at a hypothetical example in which there are again two codominant genes at one locus but in which the population exists in circumstances such that one homozygote has a fitness value only 0.01 less than that of the heterozygote. In such a case, if one allele started out at a frequency of 1 in 1 million, it would reach a frequency of .2 only after 2486 generations and would be near fixation (at a frequency of .99) after another 1196 generations. With a human generation length of 20 years, these periods would be about 50,000 years and about 24,000 years, respectively (see Table 7-3 for

TABLE 7-3   Number of generations required to change gene frequency when genotype selected against has a fitness value of 0.99*

Mode of expression of autosomal gene favored by selection ($W = 1.0$)	Change in gene frequency			
	From .001 to 1%	From 1 to 50%	From 50 to 99%	From 99 to 99.999%
1. Codominant	1384	918	918	1384
2. Dominant	1256	560	10,260	9,989,692
3. Recessive	9,989,692	10,260	560	1256

* Formulas used in calculations:

1. No dominance (codominance): $t = \dfrac{2}{s} \ln \dfrac{p_t(1 - p_0)}{p_0(1 - p_t)}$

2. Dominant favored: $t = \dfrac{1}{s} \left[\ln \dfrac{p_t(1 - p_0)}{p_0(1 - p_t)} + \dfrac{1}{1 - p_t} - \dfrac{1}{1 - p_0} \right]$

3. Recessive favored: $t = \dfrac{1}{s} \left[\ln \dfrac{p_t(1 - p_0)}{p_0(1 - p_t)} - \dfrac{1}{p_t} + \dfrac{1}{p_0} \right]$

where p_t is the frequency of the favored gene at time t and p_0 is the initial frequency.

Source: Modified from J. F. Crow, and M. Kimura. 1970. *An Introduction to Population Genetics Theory*. New York: Harper and Row. Table 5.3.1, p. 194.

some other examples and the formulas by which such calculations are made).

When the fitness values of two genotypes differ by very little, gene frequency changes very slowly—probably too slowly to be detected by humans, even if enormously large samples could be observed under controlled conditions. Nevertheless, even a period of about 74,000 years is small in relation to the total time span of hominid evolution of several million years. Other implications follow from these calculations. Even if two genotypes have no apparent (detectable) influence on human survival or reproduction, fitness differentials that we are unable to observe could have significant influences on the genetic constitution of populations or the phenotypic appearances of their members.

Human Biological Differences

Most human biological characteristics probably are functional, and their distribution reflects the operation of several of the forces of evolution, not just gene flow. The first widely read systematic discussion of these ideas came in 1950 with the publication of *Races: A Study of the Problem of Race Formation in Man,* by Coon, Garn, and Birdsell (Figure 7-17 shows the book's table of contents). Races were still considered to be the major units of study, but these three anthropologists realized that the characteristics of these major groups could be approached on a trait-by-trait basis. Following the introduction of this approach, it was no longer sufficient to say that a particular phenotypic trait was a racial characteristic; such a designation did not explain either the trait's origin

Contents

xi

xii RACES

CONTENTS xiii

xiv RACES

FIGURE 7-17 Table of contents from *Races*.

Source: Carleton S. Coon, Stanley M. Garn, and Joseph B. Birdsell. 1950. *Races: A Study of the Problems of Race Formation in Man.* Springfield, Ill.: Charles C Thomas.

or its function. However, if the population frequencies of each trait or allele were plotted on a map of the world, correlations might be found with various environmental factors that shaped the forces of evolution acting on the loci influencing the trait. Now the study of human variation could move beyond mere description and classification to a fuller understanding of the causes of our similarities and differences, which is the goal of most recent anthropological studies of living populations. The next section discusses some of the factors that shape the genetic differentiation of human populations and the varying degrees of distinctiveness that can result.

CONCEPT OF RACE: RECENT APPLICATIONS

In the opinion of many anthropologists, three major factors affect differentiation: (1) the strength of barriers to gene flow, (2) the size of the population that becomes isolated as a result, and (3) the length of time that such isolation has existed. A few specific examples show how these factors operate to pattern the gene pools of specific human populations and result in the differences that have been called *racial*.

Isolate Populations

The Dunkers, discussed in Chapter 6, provide a familiar starting point. Their history is known in detail. The group is small, consisting of a few hundred individuals who have been isolated for six or seven generations. The Dunkers have been kept apart from the surrounding populations largely by self-imposed cultural barriers—chiefly religion. Differences in gene frequency probably have been brought about by drift but maintained by low gene flow. These differences in gene frequency distinguish the contemporary Dunker population from both its ancestors and its present neighbors.

The Crow Indians of the western United States have existed as an entity distinct to a moderate degree from other American Indian populations longer than the Dunkers have been isolated from their neighbors. The most reliable current estimates place the entry of human populations over the Bering Straits into the New World at between 10,000 and 50,000 years ago. Within a few thousands years, the descendants of these Old World hunters had spread over the entire continent, subdividing in the process into a number of groups that differed in language and other aspects of culture. As Americans speak English, a member of the Indo-European language family, the Crow speak one of the languages of the Siouan family. Groups speaking these Siouan tongues are spread widely over the western plains. Several centuries ago, the Crow diverged from other plains groups, giving up agriculture and abandoning settled villages to resume a nomadic life. Their divergence in way of life has been linked to the introduction of the horse to the plains area early in the eighteenth century. The history of the Crow populations is not as detailed as that of the Dunkers, but it is known that these Indians have constituted a separate group much longer than the Dunkers have and that this separation has been marked by cultural and probably genetic distinctions.

Africa south of the Sahara contains a great number of populations separated from North Africa by a vast desert and from Europe and Asia by a combination of mountains (the Atlas range) and large bodies of water (the Mediterranean and Red Seas). Yet their isolation is not complete; the Sahara is a filter rather than an impassable barrier. The desert has been crossed regularly by caravans and occasionally by armies. In addition, the Sinai peninsula forms a bridge between Africa and Eurasia, and Africa and Europe nearly touch at Gibraltar. But for tens of thousands or hundreds of thousands of years, gene flow between sub-Saharan Africa and other parts of the world was relatively restricted. This limited access is reflected in the similarities between populations over most of the continent.

On the basis of genetic and phenotypic differences of the sort shown for just one locus in Figure 7-18, many anthropologists would call the characteristics that distinguish sub-Saharan Africans from non-African populations *racial differences*. Fewer would apply this term to traits setting off the Crow from other American Indian groups, though they might apply it to the distinction of Amerindians from either Europeans or Africans. Fewer still—perhaps none—would speak of the Dunkers as a separate race. Nonetheless, the identifiable pattern of gene frequencies that distinguishes the Dunkers from the surrounding populations is the result of the operation of forces of evolution through time—as are the differences between Africans and populations that occupy other major geographic areas. The differences between one case and the next are matters of degree, not of kind. Logically, therefore, if the differences between Africans and non-Africans are racial differences, then the differences between the Crow and other Indians, or between the Dunkers and non-Dunkers, are racial differences. In these terms, then, races can be seen as populations that differ from one another in the frequencies of genes at one or more loci, with the differences being caused by the operation of evolutionary forces. At first glance this redefinition of race in genetic terms seems to successfully blend traditional beliefs with new scientific knowledge. Examples such as the Dunkers do pose problems, since few anthropologists would feel comfortable calling them a separate race. But perhaps such difficulties could be dealt with by setting up a standard of some minimum amount of genetic distinctiveness that would have to be reached for a population to qualify as a race.

Adjustments of this sort cannot so readily solve a different kind of problem having to do with the concept of race, however—one that is raised by modern genetic knowledge. Once it is realized that similarities and differences between human populations reflect not only the extent of gene flow but also the operation of other forces of evolution such as selection and drift, the interpretation of human microevolution becomes more complex, as we will see in the next section.

Does Similarity Imply Relationship?

Physical resemblances, which were studied long before gene frequencies, are still often used to recognize racial identity. Similarity is thought to indicate a relationship; difference, a lack of relationship. The choice of physical characteristics for these purposes is quite arbitrary, and

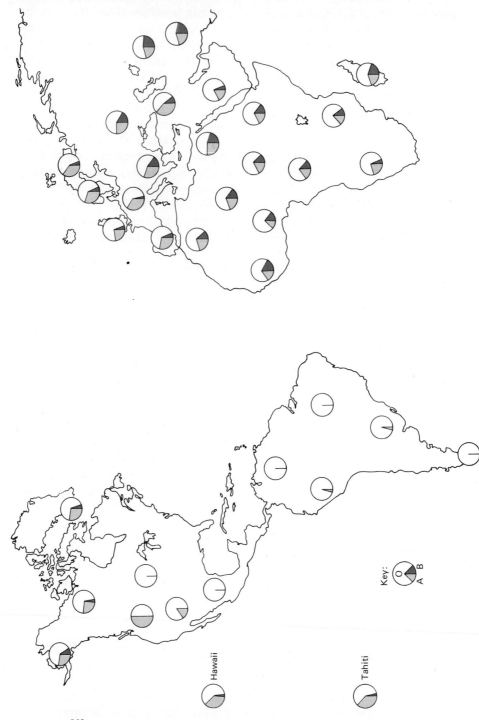

FIGURE 7-18 Variation in ABO gene frequency within and between populations.
Source: Alice Brues. 1977. *People and Races.* New York: Macmillan. Figs. 40 and 41, pp. 208–210.

Hawaii

Tahiti

Key:

O
A   B

usually quite obvious traits are used. Egyptian wall paintings of about 3500 B.C. show members of different tribes with different skin colors. Although first used by ordinary people, skin color and other visible traits such as hair and eye color, hair form, facial features, and body size were adopted by early anthropologists as criteria for categorizing different human groups. Their use made possible a convenient although rather mechanical subdivision of the human species into several "races." This type of classification led to misconceptions because the assumption was that two groups which looked alike were related by descent from common ancestors. Populations all over the world were grouped together under a common name and presumed to be derived from a single ancestral group. One example—the presumed common descent of African Pygmies and Oceanian Negritos—was discussed in Chapter 5. However, such a view ignores a well-documented fact: when quite unrelated groups are subjected to a similar set of evoutionary forces for a long enough period of time, they can come to resemble one another more closely than they did to begin with. Small stature, for example, could have evolved independently among dark-skinned populations in several widely separated parts of the world. Phenotypic similarity does not necessarily imply close genetic relationship.

In another direction, when populations can be shown to be genetically distinct, they are sometimes recognized as separate races, no matter how they are related historically to other groups or what their physical resemblances are. On this basis, the Dunker isolate in Pennsylvania would, at least by one commonly used definition, be racially distinct from its German ancestors. One paradox raised by these contrasting phenotypic and genetic criteria of racial affinity is shown in Figure 7-19. According to anthropometric traits (external body measurements such as stature, chest girth, and head length), American Indians are more similar to people of European ancestry, but data on gene frequency group them with Australian Aborigines. Since these answers are different, both cannot be correct as direct measures of population relation-

FIGURE 7-19   Trees of descent. (a) **Partial tree based on differentiation in anthropometric traits.** (b) **Corresponding tree based on differentiation in genetic markers.**
Source: W. F. Bodmer and L. L. Cavalli-Sforza. 1976. *Genetics, Evolution, and Man*. San Francisco, Calif.: Freeman. Fig. 19-14, p. 587.

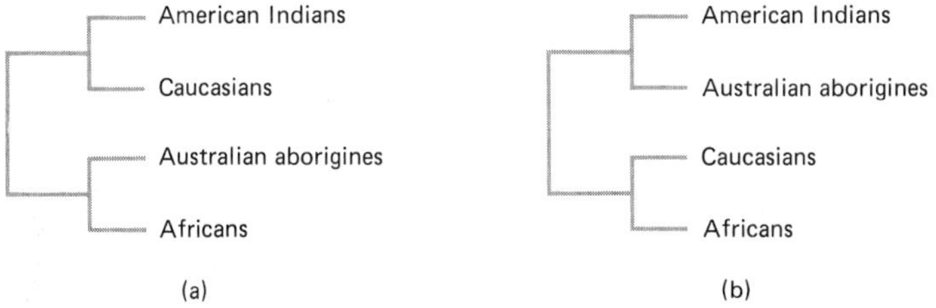

(a)   (b)

ships. More traits must be studied before we will have a reliable picture of the ancestral links between present human populations. Meanwhile, it is useful to realize that each of the traits studied does tell us something useful about the action of the environment on a particular trait and on the loci that influence it.

Causes of Confusion, and Some Solutions

Confusion results when people use the word *race* without defining what they mean. At present, the term may be used to imply three different— and somewhat contradictory—things: physical similarities, historical relationships, and genetic isolation. Whether the concept of race is useful enough to be retained in discussions of human evolution is a debatable point, and those who are interested in human variation and its causes have devoted much time and effort to resolving the debate.

One solution is to use different terminology. Perhaps the most enduring and widely used approach has followed Garn's (1961) replacement of the single term *race* with three hierarchical subdivisions: geographic race, local race, and microrace.

Geographic races are spatially delimited collections of similar populations. Human geographic races correspond broadly with the major continents, and so they are sometimes called *continental races*. Some geographic races, however, are spread over noncontinuous land areas such as the Pacific Islands. The existence of geographic races is usually attributed to physical barriers—mountains, deserts, oceans—that existed in earlier times. These populations were subjected to broadly similar environmental conditions, and gene flow with other areas was restricted. Africa south of the Sahara is one such region. A listing of human geographic races is given in Table 7-4, and their distribution is shown in Figure 7-20.

Local races are thought to correspond more closely to actual breeding populations. They are isolated by distance, geographic features, or

---

TABLE 7-4   Human geographic races

1. Amerindian
2. Polynesian
3. Micronesian
4. Melanesian-Papuan
5. Australian
6. Asiatic
7. Indian
8. European
9. African

Source: Stanley M. Garn. 1961. *Human Races.* Springfield, Ill.: Charles C Thomas. Pp. 117–124.

FIGURE 7-20 Human geographical races. Polar-projection map showing the limits of the nine geographical races listed in Table 7-4.

Source: Modified from Stanley M. Garn. 1961. *Human Races*. Springfield, Ill.: Charles C Thomas. Fig. 21, p. 118.

cultural barriers and are largely endogamous. Each local race could become adapted to local selection pressures, and any gene flow would normally occur with other local subdivisions of the same geographic race. There are many local races—the Pacific Negritos and the Pygmies of Africa's Ituri forest and elsewhere in sub-Saharan Africa, for example. Among the groups in North America, the Crow would be considered to be one local race. In *Human Races*, Garn (1961) briefly describes some 32 local races from all over the world in what is only a partial listing.

Microraces are statistically distinct populations that cannot be defined as breeding populations. Isolation by distance is thought to play a major role in maintaining the differences in gene frequency of these groups, whatever their original causes. These differences can exist between one large city and the next within the area of one local race. When large enough samples can be studied, statistically significant differences are often found in the frequencies of alleles at one or more loci between any two human populations. Because of this common variation in gene frequency, there is almost no limit to the numbers of microraces that could be defined. Among examples we have discussed, the Dunkers of Pennsylvania could be considered a separate microrace.

Even with these terminological refinements, there are still problems with the biological concept of race. Not all geographic races are the same size. The Micronesian geographic race occupies a series of islands so small that they do not even appear on the map in Figure 7-18. On the other hand, just before the fifteenth century the Amerindian geographic race occupied the combined area of North and South America, an extensive region. Nor do all geographic races differ from one another genetically to the same extent. What has been defined as a separate geographic race on the Indian subcontinent differs less in many characteristics from the European geographic race than either does from the African geographic race. The same problem exists at lower levels in the racial hierarchy. Local races on the same continent diverge by different degrees, and the same patterns of variation appear among different microraces.

There are also more fundamental problems. One is that populations differ not just from place to place but also over time. Traditional racial classifications, with their static categories, are not designed to deal with temporal change. Another problem is that different traits can show contrasting patterns of distribution. Initially, races were defined on the basis of visible phenotypic traits such as skin color, hair color, and hair form. For various reasons, these traits showed broadly similar patterns of occurrence over wide areas. The distribution of each of these traits can be described in terms of a *cline,* which is a regular variation in frequency from population to population over space. Clines for any two traits can be either concordant or discordant. Concordant clines show a similar pattern of spatial variation, while discordant clines reflect different patterns, as when one of two traits increases in frequency from north to south, while the other increases from east to west. Many externally visible traits such as skin and hair color show clines that are concordant.

Shortly after the blood groups were discovered, their genes were found to differ in frequency from population to population. This led to the redefinition of races as populations that differ in the frequency of genes at one or more loci, rather than merely as groups of people who differ on the average in appearance. The two views could be reconciled for a time, since the visible phenotypic traits were also known to have an inherited basis. This basis was more complex genetically and more subject to environmental influence, but these were differences of degree, not of kind. In recent years, however, a significant conceptual problem

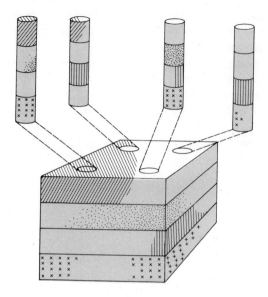

FIGURE 7-21 Discordant variation in four characters.
Source: Paul R. Ehrlich and Richard W. Holm. 1964. "A Biological View of Race," in Ashley Montagu (ed.), *The Concept of Race.* New York: Free Press. Fig. 3, p. 170.

has begun to emerge. As more and more genetic traits have been discovered and as their distributions have been mapped, many have been found to have patterns of distribution that are discordant among themselves, particularly the more traditional traits such as skin color. This situation was introduced in Figure 7-19 and is represented schematically in Figure 7-21. As a result, we can now see that when enough different traits are studied, each population will show a unique pattern of genetic and phenotypic characteristics.

Discarding the Concept

The problem is not that our species cannot be subdivided. Rather, it is that many different traits can be used as the basis for such partitioning, and the different classifications simply do not all coincide. Should we then discard traditional racial categories? Those who take this approach (for example, Livingstone, 1964) have been said by their critics to believe in "the nonexistence of human races." It would be more accurate to say that they believe races exist, but more as mental concepts than as objects in the external world. One very sensible comment on this situation was made by the French anthropologist Jean Hiernaux (1964): "Classification is not a goal in itself, but a tool, a very useful one indeed when it works. When it does not, discarding it will not withdraw any scrap of knowledge, but on the contrary force us to face the facts as they are in their full complexity" (pp. 198–199).

Human genetic and phenotypic complexity is extensive, but its causes are no longer a puzzle. Gene frequencies differ from population to population in response to the sum total of evolutionary forces that have acted on each group's ancestors in the past. Not all forces influence genes at all loci in the same way, and therefore different loci will show different patterns of geographic variation. At the present time anthropologists know the specific causes of some of this variation; the distribution of abnormal hemoglobin genes, for example, coincides rather nicely with the distribution of malaria. In other cases a good guess about the cause of genetic variation can be made, but evidence does not yet exist; the relationship between skin pigmentation and intensity of ultraviolet radiation is one such example. There are many other cases in which a pattern of geographic variation is known in considerable detail, although its cause (or causes) remains in doubt, as with the ABO blood groups. New variants are being discovered almost daily, and their geographic distributions still need to be studied in human populations all over the world.

There is no dispute about the existence of the variation or about its patterns of distribution. The important disagreement concerns only whether the standard racial categories that have been used to subdivide the human species are adequate to summarize the patterns of variation and understand their causes. An end of this debate probably will not come for some time, but major conceptual gains have already been made. Few anthropologists feel any longer that labeling a gene a *racial marker* explains anything. Most would agree that the causes for each trait must be investigated separately. Lack of agreement about the adequacy of the traditional racial categories does not mean that we are not continuing to learn more; it means simply that we do not agree on how to describe what we know. Asking whether races exist is like asking whether $\sqrt{-1}$ exists, or whether the waves discussed by twentieth-century physicists exist, or whether the ether believed in by nineteenth-century physicists existed. None of these terms refers to one simple thing that can be held or touched. All are abstractions, reflecting views of how properties of the world are organized. The wave proved to be a useful concept in dealing with some problems in physics; ether did not. The concept of race is still being tested.

SUMMARY

1. Patterns of modern human genetic variation are produced by interactions among the forces of evolution; one well-documented example of the complex determinants of gene frequencies is the distribution of the allele for hemoglobin S in west Africa.

2. The balance between systematic or recurrent forces of evolution (mutation, selection, and gene flow) and genetic drift is influenced to a considerable extent by population size.

3. At present there are three major alternative models that attempt to account for the extent and distribution of genetic variation in genomes of humans and other sexually reproducing organisms: the classical theory, the balance theory, and the neoclassical theory.

4. At the phenotypic level, the concept of race has been used to summarize and explain human variation.

5. Before there was scientific knowledge about the nature, extent, and distribution of genetic variation, the concept of race came into use to summarize and explain patterns of human phenotypic variation.

6. Some of the earliest characterizations of human races were based on language, dress, and other purely cultural features as well as biological characteristics.

7. When biological characteristics alone were used, these features were at first believed to be stable (unaffected by environmental influences or evolutionary change) and nonfunctional; consequently, all races could be seen as belonging to a few original pure stocks or relatively recent mixtures of these.

8. In recent years races have been redefined in genetic terms as populations that differ in the frequencies of one or more genes; groups that show such differences have been further subdivided into geographic races, local races, and microraces.

9. When large numbers of genetic and phenotypic characteristics are studied, each population is found to be unique in some ways, and over broad areas of the world some frequency clines show different patterns of distribution that cut across traditional racial groupings. These facts have led some anthropologists to suggest that the concept of race may have decreasing usefulness in helping scientists understand the patterns of human variation in the world today.

SUGGESTIONS FOR ADDITIONAL READING

Coon, C. S. 1966. *The Living Races of Man.* New York, Knopf. (This companion volume to Coon's *Origin of Races,* cited in this chapter, is the most recent comprehensive study of human races from the traditional point of view.)

Montagu, M. F. A. (ed.). 1964. *The Concept of Race.* New York: Macmillan. (Montagu's book is a collection of 11 essays critical of the race concept and its usefulness in understanding the causes and distribution of human variation.)

Montagu, M. F. A. (ed.). 1975. *Race and IQ.* New York: Oxford University Press. (In this collection of 15 essays a number of anthropologists and biologists examine what is known and what is still unknown about the relationship between the two singly problematic ideas of race and intelligence.)

REFERENCES

Allison, A. C. 1955. "Aspects of Polymorphism in Man," *Cold Spring Harbor Symposia in Quantitative Biology,* **20**:137–150.

Blumenbach, J. F. 1975. *De generis humani varietate nativa (On the Natural Variety of Mankind).* Doctoral dissertation, University of Göttingen.

Boas, F. 1910. *Changes in Bodily Form of Descendants of Immigrants,* Senate Document 208, 61st Cong., 2d Sess.

Boyd, W. C. 1950. *Genetics and the Races of Man.* Boston: Little, Brown.

Brace, C. L., and M. F. A. Montagu. 1977. *Human Evolution.* New York: Macmillan.

Coon, C. S., S. M. Garn, and J. B. Birdsell. 1950. *Races: A Study of the Problems of Race Formation in Man.* Springfield, Ill.: Charles C Thomas.

Coon, C. S. 1962. *The Origin of Races.* New York: Knopf.

Dobzhansky, T. 1944. "On the Species and Races of Living and Fossil Man." *American Journal of Physical Anthropology,* **2**(3):251–265.

Firth, R. 1958. *Human Types.* New York: Mentor Books.

Garn, S. M. 1961. *Human Races.* Springfield, Ill.: Charles C Thomas.

Haldane, J. B. S. 1949. "Disease and Evolution," *La Ricerca Scientifica,* **19,** Suppl. 1:3–10.

Herrick, J. B. 1910. "Peculiar, Elongated and Sickle-shaped Red Corpuscles in a Case of Severe Anemia," *Archives of Internal Medicine,* **6**(5):517–521.

Hiernaux, J. 1964. Discussion of "Geographic and Microgeographic Races," by M. T. Newman, *Current Anthropology,* **4**(2):198–199.

Hooton, E. A. 1946. *Up from the Ape.* New York: Macmillan.

Lewontin, R. C. 1974. *The Genetic Basis of Evolutionary Change.* New York: Columbia University Press.

Livingstone, F. B. 1958. "Anthropological Implications of the Sickle Cell Gene Distribution in Africa," *American Anthropologist,* **60**(3):533–557.

Livingstone, F. B. 1964. "On the Nonexistence of Human Races," in M. F. A. Montagu (ed.), *The Concept of Race.* New York: Macmillan. Pp. 46–60.

Livingstone, F. B. 1967. *Abnormal Hemoglobins in Human Populations.* Chicago: Aldine.

Neel, J. V. 1951. "The Inheritance of the Sickling Phenomenon with Particular Reference to Sickle Cell Disease," *Blood,* **6**(5):389–412.

Phillipson, D. W. 1977. "The Spread of the Bantu Language," *Scientific American,* **236**(4):106–114.

Shapiro, H. L. 1939. *Migration and Environment.* New York: Oxford University Press.

Chapter 8

Human Adaptability

Human populations occupy extremely varied environments, from scorching deserts to frigid arctic wastes and from humid jungles at sea level to mountains so high that the air is noticeably thin and deficient in oxygen. In all these settings people survive, reproduce, and carry on characteristic cultural activities. The capacity of our species to adjust to and surmount these diverse environmental challenges is the measure of *human adaptability.* Genetic changes play an important role in the adaptive process. If a population's gene frequency is not optimum for its environment, the forces of evolution can continue to operate until it is. For example, high frequencies of the hemoglobin S gene were built up in many African populations living in malarial areas. Now where the malaria has been eliminated by the use of DDT and other chemical sprays, the AS heterozygotes no longer have any biological advantage— but SS homozygotes still die from a severe anemia. Although selection has been reducing the frequency of this gene as a result, a transient polymorphism will exist at its locus for many generations to come. In the absence of equilibrium, the forces of evolution will act on the existing pool of genes in a population, selecting those alleles which increase the population's fitness. Over the long run, the population should approach equilibrium. But over the long run, we are all dead.

That inescapable restraint on human affairs was pointed out by John Maynard Keynes, a practitioner of the dismal science of economics. His observation is just as essential a truth for evolutionists: the human life span is finite, and individuals do not evolve—populations do. Fortunately, other biological and behavioral mechanisms make it possible for most people to survive and function adequately while evolution repatterns the gene pools of the populations to which they belong. In some cases these adjustments provide enough shielding to lessen the amount of genetic change necessary for equilibrium.

This chapter will discuss the types of adaptive mechanisms that are available to human populations in all environments and the ways in which anthropologists can determine the array of mechanisms used in any particular case. Specific examples of human adaptation to heat, cold, and high altitude will be presented as detailed case studies of human adaptation.

ADAPTIVE MECHANISMS

A systematic treatment of the adaptive mechanisms available to human populations was provided by Baker (1966), whose analytical framework is followed here with a slight modification in terminology. Baker discussed four major means by which a population can achieve a better fit with its environment: (1) behavioral adaptation, (2) physiological acclimatization, (3) developmental acclimatization, and (4) genetic adaptation. There is a fundamental difference between any of the first three categories and the fourth. Only the last, genetic adaptation, results in permanent change in the gene pool from generation to generation, as a result of the operation of the forces of evolution. The first three adaptive mechanisms are distinguished chiefly by the relative lengths of time over which they operate and also in part by the nature and methods of the disciplines that study them. The behavioral, physiological, and developmental processes of adaptation are grouped together and set off from genetic adaptation in one more way: genetic adaptation underlies all the others. The capacities of populations for psychological, physiological, and developmental adjustments have been shaped over the course of evolution by selection for increased behavioral, functional, and morphological flexibility. Nonhuman primates such as chimpanzees, for example, show a range of behavior in response to different environments greater than that shown by most mammals, but they have less behavioral flexibility than humans. Although any given environmental stress is likely to bring several of these adaptive mechanisms into play more or less simultaneously, it is useful to analyze each component of the adaptive process separately.

Behavioral Adaptation

Included in this category are all the psychological adjustments people can make in response to environmental variables, as well as other behavioral mechanisms that operate at the group level. The complex

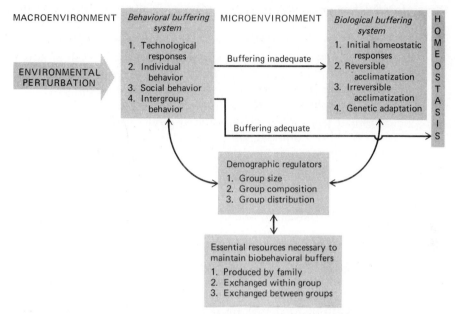

FIGURE 8-1   Factors influencing the maintenance of homeostasis.
Source: R. Brooke Thomas. 1975. "The Ecology of Work," in Albert Damon (ed.), *Physiological Anthropology*. New York: Oxford University Press. Chap. 4, fig. 4-2, p. 70. Copyright © 1975 by Oxford University Press, Inc. Reprinted with permission.

has been referred to as a *behavioral buffering system* (see Figure 8-1). When the buffering is complete, the organism attains *homeostasis:* the normal state of a relatively unvarying internal environment.

Some of the response individuals make to environmental challenges may be largely internal accommodations. For example, in temperate regions with seasonally changing climates, some degree of discomfort during summer and winter is expected and accepted. When the discomfort exceeds a person's range of tolerance, however, he or she may modify habitual activity patterns. In summer, for example, physical activity may be reduced, since exercise raises the level of metabolic activity which produces heat as a by-product. Ultimately, technological devices are brought into play, if available: someone who is uncomfortably warm might switch on a fan or an air conditioner, thus creating a different *microenvironment* (the organisms's immediate surroundings, regardless of climatic region). The organism-environment interaction is affected at all these stages, but with increasing emphasis on modification of the external environment. As will be discussed in Chapters 13 to 15, the ability of humans to modify the environment by cultural means has increased dramatically over the course of the evolution of our species.

Physiological Acclimatization

In physiological acclimatization, accommodation to the environment is attained chiefly by adjustment of the internal physical and chemical state of the organism. This adjustment is a dynamic process that takes place as individuals face constant minor changes in day-to-day conditions. Sometimes environmental conditions exceed the normal limits of tolerance, producing physiological strain. Unless this burden can be reduced and homeostasis reestablished, dysfunction and ultimately even death can result.

Again, heat provides a good example of a commonly encountered type of stress that can produce physiological strain. None of the physiological mechanisms humans use for heat control are unique to our species. Like other animals, as long as the surrounding environment is not warmer than our bodies, we lose heat by radiation, convection, and conduction. *Radiation* is the loss of energy (as photons, or packets of energy) from a warm surface to other, initially cooler objects; any intervening air is also warmed, though radiation can occur even in a vacuum. *Convection* can occur only if some fluid, such as air, is present; in convection, heat lost from warm objects is picked up by passing currents. In *conduction,* heat is lost directly to molecules of other materials. Many solids, particularly metals, are good conductors, as are some liquids, such as water; air, particularly when still, is by contrast a good insulator.

When temperature in the surrounding environment becomes high relative to that of the human body or when heat is produced by exertion, various physiological mechanisms are triggered. One of the first is peripheral vasodilation, an increase in diameter of the blood vessels near the surface of the body. Warm blood is shunted from the body core to the skin, whose temperature increases. The skin can then give up some of this heat via radiation, convection, or conduction. With a further increase in heat stress, sweating begins. Not all sweat glands begin to function at the same time; more become involved as body temperature increases. Secretion of sweat can range from very little to about 2 l (a bit under 2 qt) per hour. In sweating, heat is lost from the body to the moisture that evaporates. If conditions are favorable—a moderate level of activity, a radiant heat load that is not too high, and low humidity—sweating can stabilize the skin temperature at about 35 to 36°C (95 to 97°F).

Through internally controlled physiological processes such as peripheral vasodilation and sweating, normal humans have the capacity to regulate body temperature in response to a wide range of heat stresses. However, people do not respond identically to the same stress. Among the factors that cause different responses to heat, prior exposure or acclimatization seems to be quite important. Physiological acclimatization can be demonstrated experimentally by exposing to heat stress healthy individuals who are used to the lower temperature levels of a cooler climate. The maximum adjustment to the new, higher levels of heat and activity occurs after just 5 to 10 days.

After acclimatization has occurred, internal body temperature does not show as much of an initial increase as before in response to a given stress. Instead, it either remains level or tends to rise very slowly during exposure. Much the same is true for heart rate, which does not increase as much in response to heat and exercise after acclimatization to those combined stresses. Production of sweat, on the other hand, begins sooner, peaks more rapidly, and holds at a relatively high level for a longer period. In areas where the climate is hot but dry, this increased sweating helps lower skin temperature through evaporation. If exposure to heat stress is stopped, the physiological adjustments described above are lost just about as quickly as they were gained. This ready reversibility gives members of human populations a good degree of functional flexibility and allows our species to live in wide range of environments.

Developmental Acclimatization

Developmental acclimatization, like the short-term psychological and physiological adjustments just described, is a nonevolutionary type of change. It occurs within each person's lifetime and requires no shift in gene frequency. Various environmental factors can have direct influences on developmental processes. For example, food shortages during critical phases of the growth period can permanently reduce stature and overall body size. The ability of the genotype to produce a phenotype thus shaped by, and suited to, a given environment is often referred to as *developmental plasticity.* After the environmental molding process has been completed, however, the design is fixed, much as when an initially plastic statue fashioned from a pliable lump of clay is fired in a kiln. Just one of the many potential phenotypes for any genotype at conception has developed. Most of the better-known examples of developmental plasticity concern quantitative characteristics of shape and form resulting from complex growth processes that are still incompletely understood.

Some type of developmental acclimatization probably occurs in all environments. In populations living in areas inhabited by their ancestors for many generations, it is difficult, if not impossible, to separate effects of developmental acclimatization from effects of genetic adaptation. For this reason, most documented cases of developmental acclimatization have come from studies of migrants, such as those by Boas, Shapiro, and Hulse mentioned in Chapter 7.

Scientific knowledge about the developmental changes conditioned by particular environmental agents is still spotty and uncertain, but some generalizations can be made. For example, variations in adult body size (measured by height and weight) have often been interpreted as responses to climate, chiefly to temperature. One generalization about this relationship is *Bergmann's rule,* which states that in a widely distributed species of warm-blooded animals, body size varies inversely with environmental temperature. The average body size in populations inhabiting cold areas tends to be larger than the average body size in populations of the same species living in warmer areas. Increased body size results in conservation of body heat because a larger body has less surface area in proportion to volume. This idea can be demonstrated by using as examples the two cubes shown in Figure 8-2. The cube on the right is larger than the cube on the left in all respects: linear measurements, area of outer surface, and enclosed volume. However, these three measures of the size of the two cubes do not differ in the same ratio. The length of a side of the larger cube is twice that of the smaller cube. But the surface area of the larger cube is four times as great as that of the smaller cube, and the volume of the larger cube is eight times that of the smaller cube.

Body weights in human populations all over the world show a reasonably good fit with predictions based on Bergmann's rule. Northern Europeans are large; Congo Pygmies and South African Bushmen are considerably shorter. In a systematic survey of 116 human populations, D. F. Roberts (1953) demonstrated that the inverse correlation between male body weight and average annual temperature is a relatively persuasive −.60. More recently, however, M. T. Newman (1975) has suggested that the correlation between mean annual temperature and adult body weight is due much less to needs for heat regulation than to amount and kinds of food consumed. Most of the developed countries of the world are located in temperate zones, where soils are relatively rich and where moisture is distributed in the amounts needed for the

FIGURE 8-2   Ratio of surface area to volume and Bergmann's rule. As any body becomes larger, its volume increases more rapidly than its area. For animals, this means that the heat produced by cells is conserved more readily by larger bodies.

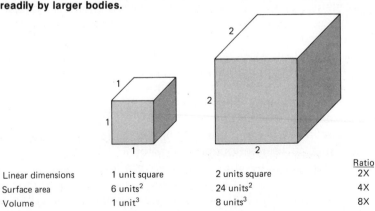

			Ratio
Linear dimensions	1 unit square	2 units square	2X
Surface area	6 units^2	24 units^2	4X
Volume	1 unit^3	8 units^3	8X

production of crops and the raising of domestic animals. As a result, more and better-quality food is produced—and consumed—per capita in these countries. In tropical and subtropical zones, on the other hand, soils are generally poor, and rainfall is often extremely low or extremely high. As a result, crop and livestock production are difficult and unpredictable. In addition to these basic problems of production, insects and diseases flourish, reducing yields still further. What food is available must be divided among the members of populations that are steadily increasing in numbers. It is little wonder that, on the average, members of these populations are small in size. Climate in the tropical or subtropical areas they inhabit does influence development, but perhaps indirectly through diet as well as directly in terms of temperature.

Allen's rule is another generalization about the relationship of body form to temperature. Allen noted that warm-blooded animals living in colder regions generally have shorter extremities than members of the same species inhabiting warmer areas, where limbs (and, in some species, ears) tend to be longer in proportion to body size and weight. The functional explanation seems to be that long, slender extremities have a larger surface area from which they can dissipate heat; shorter, stubbier projections minimize the amount of exposed surface and consequently conserve heat. Human populations living where temperatures are extreme conform relatively well to Allen's rule. For example, Nilotic blacks have greatly elongated limbs in proportion to body size, whereas Eskimos have much shorter and thicker arms and legs (Figure 8-3). Differences in shape are harder to account for in purely nutritional terms than differences in body size. Anthropologists still have much to learn about the relative roles of genetic and direct environmental influences on the development of size and shape features in humans.

Genetic Adaptation

Since this adaptive process has been discussed extensively in earlier chapters, there is no need to review the forces of evolution here. Instead, we will look at what fraction of human variation is due to genetic adaptation rather than shorter-term adaptive mechanisms that act within the life span of any individual. Table 7-2 (page 227) gives one indirect measure of the genetic fraction. It is based on recently developed techniques of molecular genetics, introduced in Chapter 3, that make it possible to estimate the degree of genetic difference between the genomes of any two organisms. A sample of the proteins of each is taken, and their amino acid sequences are analyzed. Any differences between the amino acid sequences of the two organisms are translated into differences in DNA nucleotides. The percentage of nucleotides differing in this sample is then taken to be representative of the extent of variation in the entire genome, the size of which is known from

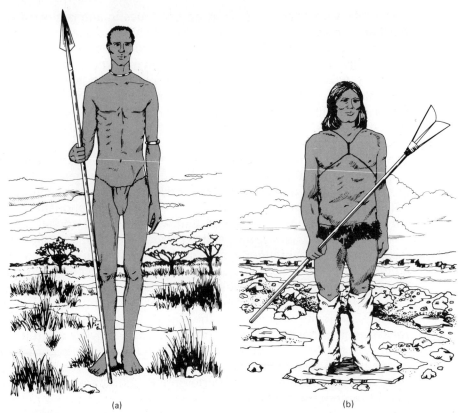

(a)   (b)

FIGURE 8-3   (a) **Nilotic black;** (b) **Eskimo. The proportionately greater body surface of the Negro enhances dissipation of body heat; the proportionately greater bulk of the Eskimo has the opposite effect.**

independent measurements of the DNA content of an entire cell's nucleus.

As can be seen from Table 7-2 (page 227), the haploid genomes of any two randomly chosen humans from the same or different populations are likely to differ by about half a million nucleotides. This number sounds very great, but it is actually a small fraction (about 0.00017) of the haploid genome. This difference of half a million nucleotides can also be compared with another standard: the amount by which the genome of any randomly chosen human differs from that of our close relative, the chimpanzee. The haploid genome of a chimp probably differs from that of a human of today at about 1 percent of its nucleotide positions, or 30 million nucleotides. To state this in other terms, any two humans are likely to differ, at the nucleotide level, only about one-sixtieth as much from each other as either human differs from any typical chimp. By any of these objective measures, rough as the calculations may be, we humans are a very homogeneous lot.

The inference that one human doesn't differ much genetically from another comes as a surprise to scientists and ordinary people alike. Everyone knows that people come in a great variety of sizes, shapes, and colors. A look at the population of any large city confirms the accuracy of this observation. However, there are nearly as many DNA nucleotide pair differences within members of any major geographic group (American Indians, Africans, Orientals, Caucasians, and so on) as there are on the average between any two of these groups. This finding of a relatively low amount of intergroup genetic difference implies that the strong differences in visible features such as skin color, hair form, body proportions, and similar phenotypic traits may cause us to overestimate the extent of inherited differences between human populations.

There are several possible explanations for the disparity between differences at the phenotypic and genotypic levels. Perhaps visible features respond more rapidly to selection than less readily detectable traits such as biochemical variants. Another possibility is that visible traits may have a higher component of environmentally caused variation than genetic polymorphisms such as blood groups. Human blood group antigens, at least, develop according to the underlying genotype under virtually any environmental conditions compatible with survival. Yet another possibility is that the loci used to estimate human genetic diversity make up a nonrandom and biased sample that for some reason underestimates the amount of diversity in the entire genome. These alternatives are, of course, not mutually exclusive. All could provide part of the explanation for the human diversity we see around us.

DETERMINING THE TYPE OF ADAPTATION

Determining the relative roles of the different adaptive mechanisms used by a human population living in a given environment is of considerable interest, as well as practical value. Evidently, not all human populations are suited equally well to all environments, and so knowledge of these limitations can help suggest cultural and technological innovations to protect populations that move to new environments. The study of adaptive mechanisms takes many different forms. Psychological and physiological adjustments, for example, can be studied fruitfully under controlled experimental conditions in the laboratory. Similar approaches are just not possible for either developmental or genetic mechanisms. For one thing, ethical considerations rule out the possibility of subjecting humans to stresses for the years it would take to demonstrate how some environmental factor irreversibly shaped a particular developmental trait. Similarly, it would not be morally justifiable to move a human population from the region long occupied by its ancestors into a new region just to see whether offspring born in a new environment differed significantly from their parents. For reasons of their own, however, many human populations do live in environments extreme enough—in terms of altitude, temperature, or other factors—to cause notable stresses. Occasionally, too, a human group will move from an extreme environment to one quite different. Anthropologists have quite often been able to utilize these natural, voluntary experiments to measure the relative contributions of different mechanisms to the survival and functioning of human populations. The accuracy with which this can be done, however, depends on a number of factors in the situation studied.

Research on our origins (discussed in detail in Chapters 12 to 15) has securely established that the earliest humans were tropical mammals. Over the course of several million years, our ancestors increased in numbers and gradually expanded over the surface of the earth. As a result, humans have been exposed to more different environments than any other species. But understanding how humans adapt to these environments is a complex task. Most environments impose several different stresses on the populations that inhabit them, and these populations have numerous alternative and overlapping mechanisms for responding to the challenges. The situation is further complicated by the fact that human populations have occupied the areas in which they now live for different periods of time, and thus it is often difficult to prove that a particular characteristic is an adaptation to some factor in the present environment rather than the heritage of genetic adaptation to a stress encountered in the past but no longer important.

In some populations of Burma described by Hulse (1957), for example, virtually all people have epicanthic eye folds (Figure 8-4); in others this feature is completely absent, and in still other groups intermediate frequencies are seen. Language, culture, and tradition identify those with the highest frequency of eye folds as recent immigrants from China, where their ancestors had been subject to a given array of selective factors for tens of thousands of years. Among the Burmese

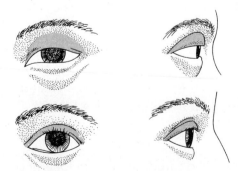

FIGURE 8-4 Mongoloid eye fold. *Above,* **Chinese man.** *Below,* **north European woman.**
Source: Alice Brues. 1977. *People and Races.* New York: Macmillan. Fig. 27, p. 116. By permission of Macmillan Publishing Co., Inc.

populations that lack eye folds are the Tamils, who had lived for thousands of years in south India, where the population also lacks eye folds. Native Burmese, those who have resided longest in their present habitat, are among those with intermediate frequencies (about 72 percent) of epicanthic eye folds. None of this information explains why a particular frequency exists in a given group or what its selective value might be. But information on the original area in which each population evolved should help in establishing a correlation between the trait and the environmental factors that may have determined its distribution. Much the same pattern appears for the skin color of populations that are now all found in Burma. The original Burmese are rice farmers who have resided longest in the hot coastal lowland areas. They are lighter than the Tamils from sun-scorched south India, but darker than the Chinese from the higher latitudes in north Burma.

Our knowledge of the operation of adaptive mechanisms, then, comes mostly from the study of populations that have occupied extreme environments (high altitudes, deserts, the arctic) for relatively long periods of time, combined with more controlled research on populations exposed to new stresses for shorter periods of time. In these studies it is necessary to remember that we are usually looking at the influence of one major stress, such as heat or cold, confounded with numerous other influences.

Adaptation to Heat

Humans are sweatier, thirstier, and more naked (at least in terms of a natural coat of hair or fur) than other primates. R. W. Newman (1970) speculated that this unique trio of characteristics evolved in the relatively well-watered grasslands of the tropics before our dispersal to other climates. Despite these identifying marks, humans exchange heat with the surrounding environment by the same four means used by all other mammals: conduction, convection, radiation, and evaporation. We rely particularly heavily on the last for cooling. In assessing the nature of adaptive responses to heat stress, it is necessary to divide hot environments into two types: those which are hot and dry, such as deserts; and those which are hot and moist, such as the humid tropics.

Moist Heat in the Tropics.   Most nonhuman primates, such as our nearest living relatives, the chimpanzees and gorillas, are largely tropical forest animals. Conditions where they live are relatively stable, with daytime ground-level temperatures within highs of about 28 to 32°C (82 to 90°F), very high humidity, and little radiant heat from the sun (Read, 1968; Richards, 1952). With the temperature of the surrounding moist air just a bit below body temperature, there is little stress on an animal's cooling mechanisms as long as activity levels are not high (R. W. Newman,

1970). Increased movement can step up metabolic rates to double or treble resting levels. Adaptation to such activity is possible in humans through physiological acclimatization. Blood vessels near the body surface dilate (enlarge), allowing more blood to be shunted to the skin, where some heat can be lost from increased radiation. This mechanism is more effective in humans, who lack a coat of insulating hair. Sweating also increases. However, because the surrounding air is already saturated with moisture, little increase in the level of evaporation is possible. Humans have the same total number of sweat glands (about 2 million) as other primates of comparable size. One functional difference, though, is that we have fewer of the *apocrine* sweat glands, located at the bases of hairs, and many more *eccrine* sweat glands, which produce a watery secretion that cools as it evaporates. The eccrine glands are spread over exposed skin areas. The difference between humans and apes is one of degree; we retain some apocrine glands, chiefly in the armpits and genital area; apes have a sprinkling of eccrine glands on their palms and soles (Weiner, 1973). Under conditions of humid heat, when little sweat can evaporate, humans adapt by lowering activity levels or engaging in strenuous exercise during the cooler periods of the day, such as at dusk and dawn.

Dry Heat in a Desert Environment.   In his study of human adaptation in the Sahara desert, Briggs (1975) provided a rather comprehensive overview of the environmental factors with which desert dwellers must cope. Cloudless skies and 8 to 10 hours of sunlight per day combine to produce a high level of solar radiation. Much of the sunlight that strikes the land surface is reflected back, producing significant ground radiation as well. Temperatures may be very high during the day (to a maximum of 72°C, or 162°F), but they can plunge at night (to as low as −15°C, or 5°F); a normal daily range may be as much as 90°F (50°C), with less variation in the shade. The high daytime temperatures intensify the effects of low moisture levels, producing extremely low relative humidity. Water supplies are generally scarce, and most of those which do exist have a very high mineral content because of evaporation. Because of all these factors, vegetation is sparse, and so there is little to interfere with moving air currents. Sandstorms may strike as often as once a day every other week. Hot, dry winds greatly speed up water loss from the skin surface.

Within our species there is little firm evidence for clear-cut inherited differences between populations in the adaptive response to heat stress in such environments. There are, of course, inherited differences in body build and skin color, but these characteristics are also influenced by developmental and shorter-term physiological acclimatizations. Desert dwellers do tend to have a high ratio of skin surface area to body volume, as would be predicted by Bergmann's and Allen's rules. There is also some developmental contribution to the body build of Saharan populations. Among the Tuareg and some other Saharan nomads, for example, there is a positive correlation between body weight and social status (richer people work less and eat better). The South African Bushmen, most of whom are native to the Kalahari Desert, have the small stature (149 to 161 cm, or about 58 to 64 in) and low body mass that conform to climatic rules. Bushmen also possess a number of other notable physical features, among them a scalp covered with tightly curled ("peppercorn") hair and isolated fat deposits over the buttocks (a condition called *steatopygia*) and thighs (*steatomeria*). Although these visible traits are sometimes said to be specific adaptations to desert life, Tobias (1964) has shown that people of this physical type were formerly much more widely distributed and occupied nondesert areas. In short, although a number of the Bushmen's physical features may be highly heritable, we do not know for sure that they are adaptations to desert life.

The frequency of darker skin colors in equatorial regions suggests that higher levels of melanin pigmentation are adaptive—but to what? Baker (1958) demonstrated that American blacks are less tolerant of desert conditions because their dark skins absorb more radiant energy than white skins, although there were no significant differences in sweating between blacks and whites matched for body weight, surface area, or degree of acclimatization. This discovery was followed by the work of Weiner and his colleagues, who showed that the amount of light absorbed by African blacks was twice that absorbed by light-skinned Europeans. In areas with large amounts of sunlight, exposed areas of dark skin add to the heat load that must be gotten rid of to maintain a constant body temperature. On the other hand, high levels of skin pigmentation can screen out harmful ultraviolet rays that cause mutations leading to skin cancer. Furthermore, fair-skinned people are more susceptible to sunburn, which can produce symptoms from mild localized reddening to widespread blistering accompanied by the release of a chemical (histamine) that causes a sharp drop in blood pressure, leading to dizziness and other disturbances. On balance, the most adaptive skin color for most areas of high solar radiation is probably that found in many populations in the Mediterranean area: a light olive with the capacity to darken considerably by tanning rather than burning.

But humans in desert areas rarely expose extensive areas of skin to the sun; clothing, an important cultural adaptation, is used to provide shielding and insulation. The ideal garments for dry heat are light in color and weight, loose-fitting, and made with an open weave to allow evaporation of sweat from the skin surface. The dress of native Saharans such as the Tuareg conforms rather closely to these specifications. Baggy trousers and one or two loose shirts cover the body, while the head is shielded with a loosely wound turban. Clothing of this sort has been shown to reduce the rate of heat absorption by as much as 100 cal per hour (Adolph, 1947). An inventory of the material culture, practices, traditions, and beliefs of desert peoples, of course, would require far more space than we have here, but many such factors are geared toward fostering survival in a hostile environment (see Briggs, 1975).

Adaptation to Cold

Though humans began to evolve in the tropics, thousands of generations ago some populations expanded into cooler areas of the world, eventually colonizing northern Europe, northern Asia, the Arctic, and the southernmost tip of South America near the Antarctic. Humans have now been in these environments long enough for the forces of evolution to have brought about genetic adaptations to cold.

Steegman (1975) has provided one of the more recent comprehensive reviews of the evidence on the responses of human populations to cold. To begin with, he found that, as is true of groups living in warm climates, many human populations living in cold areas have adapted in conformance with Bergmann's and Allen's rules; they have higher body weights, larger chests, and shorter extremities. In addition, noses are narrower and longer in cold climates, possibly a more direct response to the need to moisten as well as to warm breathed-in air (see also Wolpoff, 1968). Heads tend to be rounder in cold, dry climates; this shape has relatively less surface area in proportion to volume. How genetic and developmental influences combine to produce these features, however, is still mostly unknown.

Populations from major geographic areas of the world can also be ranked on the basis of physiological responses to cooling of the extremities. Hands and feet, far removed from the body core, have relatively little heat-generating muscle mass of their own in proportion to surface area. In experimental tests to detect the presence of physiological mechanisms that resist cold, a hand or foot of an otherwise dry and warm person is immersed in a container of water kept at a constant low temperature. Response to the cold stress is measured by how warm the skin stays or by how much heat the water picks up over a set period of time. Under the conditions of this test the finger temperature of adult male blacks simply drops to that of the coolant as their peripheral blood vessels contract. In the hands of adult male Europeans, the same blood vessels also contract, but they open up periodically to allow some brief rewarming. Under natural conditions this circulatory fluctuation may prevent serious tissue damage. Steegman inferred that the African ancestors of American blacks simply retained a tropical pattern of heat conservation adequate to guard against moderate cooling in an area where the temperature never dropped to freezing. Europeans, in contrast, evolved in areas of severe cold, and adaptions to such conditions were developed and retained. Steegman thinks this difference between Africans and Europeans has a genetic basis. Adult natives of Asiatic origin, such as Eskimos and American Indians, have shown a cold response superior to that of Europeans, although the evidence for an inherited basis of this superiority is not as clear-cut.

Some acclimatization to cold does occur, but the evidence so far suggests it is less important than what are evidently inherited differences. For example, Baker (1966) showed that Quechua Indians from warmer lowland areas of Peru gave responses similar to those of closely related Quechua who had grown up in the colder highlands. Both groups maintained warmer hands than Peruvian university students of European ancestry when all three groups were exposed to the same standardized cold stress discussed above.

Cultural mechanisms to lessen the effects of cold are extremely diverse. Australian Aborigines often sleep huddled around campfires with skin cloaks drawn around them, but on especially cold nights many sleep with their dogs for extra warmth. Eskimos, who are exposed to more extreme cold for longer periods each year than most other human groups, have perfected insulated boots, parkas, and houses from an extremely restricted supply of raw materials, chiefly animal skins and furs. They also subsist on a diet that is very high in animal fat, which produces twice as much heat per unit of weight than either carbohydrates or protein. And, of course, many human groups use drugs or alcohol to promote warmth. Most investigations of the effects of these substances show that they do less to preserve warmth than to create the impression of well-being under uncomfortable conditions—as most readers already know.

Adaptation to High Altitude

Another well-studied example of human adjustment to an extreme environment is adaptation to high altitude. About 12 percent of the earth's population lives in mountainous areas, all of which share certain common features that produce identifiable stresses on human populations. For example, mountainous terrain is almost by definition difficult to traverse, requiring higher levels of muscular activity to get about than would be needed in more level plains areas. Moreover, the same sloping, rocky land that increases exercise levels and consequently requires greater caloric intake also tends to reduce the nutritional base available for the resident population. Temperature and moisture usually decrease with altitude, and winds increase in duration and velocity. As a result, the range of crops that can be grown at high altitudes is restricted, and above certain elevations farming must give way to herding. The previously mentioned stresses, including rough terrain, cold, aridity, high winds, and a limited nutritional base, can be found, singly or in combination, in other environments. The same is true of high levels of solar radiation, which are found not only at high altitude, where the thinner atmosphere provides less of a screen, but also in equatorial areas, where the sun's rays strike the earth nearly vertically rather than obliquely.

One stress, however, is unique to populations living at high altitude: reduced atmospheric pressure. Atmospheric pressure measures the weight of a blanket of air that covers the earth's surface to a height of about 805 km (500 mi). The oxygen in the air makes up about 21 percent of the atmosphere by volume. At sea level, atmospheric pressure ensures that enough oxygen diffuses across the membranes of the lung and into the bloodstream at a rate sufficient for all needs (Harrison, Weiner, Tanner, and Barnicot, 1977). Since the thickness of the atmospheric blanket decreases with altitude, the corresponding pressure also drops, and the air at high altitude is less dense as a result. As oxygen pressure falls with increasing altitude, the rate of transfer across the lung membranes also decreases. For unacclimatized visitors, at levels of about 3000 m (about 10,000 f) the lower amount of oxygen available makes strenuous physical activity difficult; above about 4000 m (about 13,000 f) breathing becomes difficult even when the person is at rest. The resulting oxygen deficits at these altitudes are called *hypoxia.*

High-altitude areas are found on every major continent: in the Rocky Mountains of North America, the Andes of South America, the Alps of Europe, the Ethiopian high plateau of Africa, and the Himalayas of Asia. In all, over 10 million people live permanently at high altitude, and some of these areas—parts of the highlands of South America, for example— have been inhabited for nearly 10,000 years. Severe hypoxic stress can reduce fertility and increase mortality. According to Baker (1969), as early as 1500 the Spanish complained about the "thinness of the air" in the Peruvian highlands and moved their capital to the coast. On the basis of his study of the historical records, Monge (1948) reported that the production of a live child by Spanish parents in Peru was an extremely rare event. Several congenital abnormalities are markedly more common among children born at high altitude, as are a number of diseases that are serious in adulthood. Some of these are mentioned below; others are discussed in more detail by Mazess (1975). The opportunity for genetic adaptation has existed in high-altitude areas occupied for a long time, although none has yet been proved to exist. Shorter-term adjustments, however, are certainly evident. (The discussion here of the responses to high altitude follows the systematic coverage of Harrison and his colleagues, 1977.)

On ascent to high altitude, the breathing of a sea-level native becomes deeper and more rapid. This adjustment brings about a limited increase in the oxygen supply, but not enough to prevent the symptoms of hypoxia: fatigue, headache, loss of attention, and an overconfident feeling like that produced by drinking a bit too much alcohol. The change in breathing pattern also leads to a pronounced reduction in the levels of carbon dioxide in the lungs and blood. Loss of carbon dioxide changes the acid-base balance of the blood to a more alkaline level, resulting in *alkalosis,* the symptoms of which may include dizziness, nausea, and vomiting. The combination of hypoxia and alkalosis produces what is called *mountian sickness.* Some form of this disease afflicts nearly all visitors to high altitude for the first few days or weeks, and in long-term residents it may become severe and chronic.

With continued residence at high altitude, some degree of acclimatization does result within about 10 days. The rate of breathing falls (though it does not become as low as in natives), and this in turn decreases the degree of alkalosis. Reduction in alkalosis is aided by the kidneys, which excrete a more alkaline urine. Other physiological adjustments include increases in total blood volume and in the number of red blood cells and hence in the total amount of hemoglobin in circulation; these changes enable the blood to carry more oxygen. In high-altitude Peru, this increase in red cell count may be 30 percent above sea-level norms. Other physiological acclimatizations to high altitude include a more ready release of oxygen from the blood to the tissues, increased heart output, and even chemical changes at the cellular level that enable the body to extract energy from sugar anaerobically (without using oxygen).

There are also developmental responses to high altitude. The increase in blood flow through capillaries may be due to growth of more of these tiny blood vessels. Girls who grow up at high altitudes in Peru have the *menarche* (first menstruation) later than those who grow up at low altitudes. In highland Peru, height and body weight are reduced during the growing period. Some, but apparently not all, of these growth changes may be due to nutritional deprivation (Frisancho and Baker, 1970). In any case, height increases within one generation when high-altitude populations move to low altitudes. In contrast, development of the characteristic large, barrel-shaped chest of high-altitude Peruvian Quechua Indians is not limited by the factors that depress stature at high altitude. Moreover, chest size does not decrease among children of high-altitude Quechua who have migrated to low altitudes. On the basis of these observations, Beall, Baker, Baker, and Haas (1977) have suggested that chest proportions may be under genetic rather than developmental control.

Heat, cold, and high altitude are but a few of the stresses to which human populations are exposed; others include poor nutrition and high levels of infectious disease. But, although the stresses may be different, the basic adaptive mechanisms are much the same. Human biology and human culture have responded well to most environmental challenges, no matter what their form.

SUMMARY

1. Many characteristics are now known to help members of human populations function, survive, and reproduce better in the environments in which they live; such characteristics are said to be *adaptive.*

2. There are four major forms of human adaptation: cultural adaptation, physiological acclimatization, developmental acclimatization, and genetic adaptation. The first three adjustments can occur within the life span of any individual; the last requires genetic change between generations.

3. It is probable that much human variation is due to short-term adjustments that take place within the life span of the individual, since the number of nucleotide pair differences between humans from different major geographic subgroups is about equaled by the differences within any one of these groups.

4. Adaptation to moist heat involves physiological acclimatizations (such as increased sweating and the shunting of more blood to the skin) and behavioral adaptations (such as decreasing activity levels and engaging in strenuous activity only during cooler periods); the genetic changes leading to a reduction in body hair and an increase in the number of eccrine sweat glands probably began in tropical areas.

5. Genetic adaptations to dry heat may include some changes affecting skin color and perhaps part of the relative increase in body surface area, though developmental processes also influence body form, physiological acclimatization increases tolerance of dry heat, and cultural elements such as clothing and shelter help decrease heat stress.

6. Adaptations to cold include decreases in surface area, as well as changes in head shape and nose form. These derive from what is probably a blend of genetic and developmental influences. Physiological acclimatization increases tolerance to chilling, and cultural practices are extremely important.

7. Adaptation to high altitude occurs partly via physiological adjustments that increase blood volume and red cell counts, facilitate release of oxygen to the tissues, and accelerate heart rate. Developmental changes may increase the number of capillaries and produce larger chests and shorter stature, though there is some evidence that the alterations in body proportions may also be under some genetic control.

SUGGESTIONS FOR ADDITIONAL READING

Damon, A. 1975. *Physiological Anthropology.* London: Oxford University Press. (This recently published volume contains chapters that discuss human adaptation to a variety of environmental factors such as light, heat, cold, high altitude, nutritional levels, work, noise, and disease. The introduction discusses the place of physiological studies in anthropology.)

REFERENCES

Adolph, E. F. 1947. *Physiology of Man in the Desert.* New York: Interscience.

Baker, P. T. 1958. "The Biological Adaptation of Man to Hot Deserts," *American Naturalist,* **92**(867):337–357.

Baker, P. T. 1966. "Human Biological Variation as an Adaptive Response to the Environment," *Eugenics Quarterly,* **13**(2):81–91.

Baker, P. T. 1969. "Human Adaptation to High Altitude," *Science,* **163**(3872): 1149–1156.

Beall, C., P. T. Baker, T. S. Baker, and J. D. Haas. 1977. "The Effects of High Altitude on Adolescent Growth in Southern Peruvian Amerindians," *Human Biology,* **49**(2):109–124.

Briggs, L. C. 1975. "Environment and Human Adaptation in the Sahara," in A. Damon (ed.), *Physiological Anthropology.* New York: Oxford University Press. Pp. 93–129.

Harrison, G. A., J. S. Weiner, J. M. Tanner, and N. A. Barnicot. 1977. *Human Biology* (2d ed.). London: Oxford University Press.

Hulse, F. T. 1957. "Some Factors Influencing the Relative Proportions of Human Racial Stocks," *Cold Spring Harbor Symposia on Quantitative Biology,* **22:** 33–45.

Mazess, R. B. 1975. "Human Adaptation to High Altitude," in A. Damon (ed.), *Physiological Anthropology.* New York: Oxford University Press. Pp. 167–209.

Monge, C. 1948. *Acclimatization in the Andes.* Baltimore: Johns Hopkins.

Newman, M. T. 1975. "Nutritional Adaptation in Man," in A. Damon (ed.), *Physiological Anthropology.* New York: Oxford University Press. Pp. 210–259.

Newman, R. W. 1970. "Why Man Is Such a Sweaty and Thirsty Naked Animal: A Speculative Review," *Human Biology,* **42**(1):12–27.

Newman, R. W. 1975. "Human Adaptation to Heat," in A. Damon (ed.), *Physiological Anthropology.* New York: Oxford University Press. Pp. 80–92.

Read, R. G. 1968. "Evaporative Power in the Tropical Forest of the Panama Canal Zone," *Journal of Applied Meteorology,* **7**(1):417–424.

Richards, P. W. 1952. *The Tropical Rain Forest: An Ecological Study.* London: Cambridge University Press.

Steegman, A. T. 1975. "Human Adaptation to Cold," in A. Damon (ed.), *Physiological Anthropology.* New York: Oxford University Press. Pp. 130–166.

Shapiro, H. L. 1939. *Migration and Environment.* New York: Oxford University Press.

Thomas, R. B. 1975. "The Ecology of Work," in A. Damon (ed.), *Physiological Anthropology.* New York: Oxford University Press. Pp. 59–79.

Tobias, P. V. 1964. "Bushman Hunter-Gatherers: A Study in Human Ecology," in D. H. S. Davies (ed.), *Ecological Studies in South Africa.* The Hague, Netherlands: Junk. Pp. 67–86.

Weiner, J. S. 1973. "The Tropical Origins of Man," *Addison-Wesley Module in Anthropology,* no. 44. Reading, Mass.: Addison-Wesley.

Wolpoff, M. H. 1968. "Climatic Influence on Skeletal Nasal Aperture," *American Journal of Physical Anthropology,* **29**(3):405–424.

Chapter 9

The Place of Humans among the Primates

Humans are *primates,* members of a taxonomic order within a wider group, the *mammals.* But what are mammals, and why are primates classed as a subdivision among them? The majority of large animals familiar to us are mammals: dogs, cats, mice, horses, rabbits, and so on. Like us, all these living mammals have hair or fur, maintain a very constant body temperature, nurse their babies on milk, and nourish them before birth through a placenta. Possession of this last feature of the circulatory system, described in Chapter 6, is the reason why these animals are all called *placental* mammals. A few less familiar animals such as the American opossum and native Australian kangaroos and koalas are *marsupial* mammals. Marsupial mammals have no placenta: the embryo developing inside the mother receives little or no food from her and is very immature when born; after birth, the young animal crawls into a pouch on the mother's belly, attaches to a nipple, and finishes developing there. Mammals originated about 200 million years ago, and placentals and marsupials did not diverge until tens of millions of years later.

Primates are placental mammals. The word *primate* itself can mean either "highest" or "first," and both meanings fit our order. In terms of anatomy, biochemistry, and physiology humans are as complex as any other animal, and our behavior is by far the most highly elaborated that has ever evolved on the planet. Yet primates are among the most ancient groups of mammals that have survived, with some members living at the time of the first burst of placental mammalian evolution, about 100 million years ago. Over this vast span of time so many diverse primate species have evolved that it is difficult to find many characteristics that all share in common, though the earliest and latest to appear can be linked together by a series of intermediate forms. Thomas Huxley (1863), staunch defender of Darwin's theory of evolution and a noted scientist in his own right, remarked: "Perhaps no order of mammals presents us with so extraordinary a series of gradations as this—leading us insensibly from the crown and summit of the animal creation down to creatures, from which there is but a step, as it seems, to the lowest, smallest, and least intelligent of the placental Mammalia" (p. 98). In the century since Huxley made this comment, his conclusion has gained in force. Not only has human culture allowed the members of our species to create within every climatic zone the subtropical microenvironment we need, but it has also permitted us to escape the confines of the earth itself and to briefly inhabit the moon, about 385,000 km (239,000 mi) away. At the other extreme, some scientists now suggest that the boundaries of the primate order should be stretched beyond the lowest limits known to Huxley to include the tree shrews, which are tiny, small-brained animals that scurry about in the brush during the twilight hours hunting for the insects that form a large part of their diet.

Between the two extremes marked by humans and tree shrews stands a whole array of other animals, ranging from the mouse lemur, which weighs a few grams, to the gorilla, which may exceed 227 kg ($1/4$ ton). Humans and gorillas, along with the other apes and the monkeys, are often referred to as *higher* or *anthropoid* (humanlike) *primates.* Mouse lemurs are among the less familiar *prosimians* ("premonkeys"), which are sometimes termed *lower primates.* Some prosimians did evolve before the earliest anthropoids; however, as we will see later, the dividing line between the two groups is quite arbitrary, and the terms *higher* and *lower* are really not much help in understanding evolutionary relationships among the primates. All primate species differ in morphological characteristics, physiology, biochemistry, and behavior. They even differ greatly in their geographic distribution. The orangutan is now confined to the forested areas of Borneo and Sumatra; macaque monkeys range all the way from Japan to north Africa. Knowing about these other animals is not just interesting; by studying our relatives, we are likely to gain a better understanding of ourselves and some direct practical benefits as well. For example, the results of medical research on the nonhuman primates most similar to us can be applied to humans with far less risk than the results of experiments with mice. When fed a diet high in cholesterol and fats like that eaten by many Americans, baboons develop the same types of fatty plaques that narrow arteries and bring on heart attacks or strokes. To the extent that these large

African monkeys are more like us biologically than rats or other more common experimental animals, their reactions can tell us more accurately how humans might respond to diets, drugs, or other treatments.

Which living primates are closest to humans biologically? Where do we stand in relation to all the others? These are the questions with which we'll be concerned in this chapter. One reason for these concerns has already been given: knowledge of this sort can lead to some practical benefits from medical research. Systematic study of living primates can also help us understand our own origins. How can we learn about the characteristics of ancestors long dead from species now alive? Consider the parallel situation within human families. Two distant cousins are less likely to resemble each other than two sisters because the sisters have closer common ancestors (their parents) and share a larger proportion of genes by descent from them. Furthermore, by analyzing the features shared by the sisters, it should be possible to reconstruct some details of the parents' appearance. Similarly, by comparing species of living primates, we should be able to tell which of them shared recent common ancestors and which are linked only by more distant relatives. And just as in the case of human families, we should be able to build up a list of the features to be expected in common ancestors at various stages.

Following the discussion of primate classification is a section in which some features of several living primates are described in detail. Primary emphasis is given to skeletal and dental features, since these are the ones most likely to turn up as fossils. The chapter closes with a discussion of the type and extent of variability to be found within species of living primates. To extend the image used earlier, it is necessary to know how different two children of the same parents can be, lest we mistake two brothers for distant cousins. Now let's start by looking at what types of relatives we have.

CLASSIFYING THE PRIMATES

What makes it possible for us to group baboons and gorillas, mouse lemurs and humans, into a single order? The basic idea, as noted above, is that all species classified as primates share a set of characteristics produced by genes inherited from common ancestors. This grouping by degree of similarity and inferred closeness of common ancestry is the basis for the classification of all organisms.

The historical background of classification was introduced briefly in Chapter 2. As noted there, the basic form followed in modern taxonomy was originated by Linnaeus, though the idea that the classifications devised reflect evolutionary relationships is a much more recent one. George Gaylord Simpson, whose major work on classification was published in 1961, is a modern scientist who has done much to shape present ideas in this area. In a more recent book Simpson and his colleague William Beck (1965) pointed out that classification involves three steps: (1) recognizing and describing groups of organisms, (2) fitting these groups into a formal *hierarchy* (or series of grades), and (3) giving names to the groups in the hierarchy. The hierarchy referred to here is a nested set of categories, each higher in rank and more inclusive than the one below. The same set of categories is used for classifying all organisms; those most commonly used are shown in Table 9-1, which gives the taxonomic placement of the human species.

The basic unit of classification is the *species,* which was introduced in Chapter 4 as the unit of evolution that includes all the populations that are capable of interbreeding and exchanging genes. This property gives the species its unique standing as a taxonomic category. Units below the species, such as local populations, are less distinctive in appearance because of continual gene exchange with groups similar to themselves.

In Chapter 4 breeding populations of one species were likened to several teabags in the same pot of water. Even if each bag contained tea of a different flavor to begin with, after a bit of soaking the distinctions would be lost, and any flavor description would be of the blend, not its components.

For categories above the species the opposite problem exists. A higher taxonomic category such as the *genus* lacks unity because it includes several species with gene pools that have been separated for a considerable period of time, allowing differences to develop between them. To carry our earlier image a bit further, a genus with its component species could be represented by a collection of pots, one filled with tea, a second with coffee, a third with broth, and a fourth with water.

If we are able to sample two pots, we can tell the coffee from the tea quite objectively; the same is true for most species, though their members must be analyzed, measured, and described rather than just tasted. But in turning to the question of whether all four of our beverage "species" belong in the same genus, we face quite a different type of decision. Many people might feel that all share the important generic property of being liquids, particularly liquids served at meals. Yet some people might feel that two genera are needed here: one for broth, which has some food value, and another for the other three drinks, which do not. Others might feel that there should be one genus for tea and coffee, which are usually served at the end of a meal; another genus for soup, which is traditionally an early course; and yet another genus for water, which is commonly drunk throughout a meal. Still other classifications could be based on the temperatures at which the liquids are served, how they are consumed, and so on. It is evident that, although in higher taxonomic categories observation still plays a role in the decisions

TABLE 9-1 Taxonomic placement of the human species

Taxonomic category	Specific taxon*	Common characteristics
Kingdom	Animalia	Cells surrounded by sea water or a constant internal environment that resembles sea water by being rich in salt; food derived from ingestion and metabolism of complex chemicals rather than captured from the sun via photosynthesis; highly mobile and reactive; etc.
Phylum	Chordata	Animals with a stiffening rod of cartilage called a *notochord* along the back at some stage in their life cycles; most are vertebrates, having a true spine or backbone.
Class	Mammalia	Vertebrates in which females possess milk glands, or *mammae*; mammals are *homeothermic*, maintaining a highly constant body temperature, a process aided by a skin that is usually covered with hair or fur for insulation and a variety of circulatory characteristics such as an efficient four-chambered heart.
Order	Primates	Characteristics of this order and its subdivisions are discussed in the text.
Suborder	Anthropoidea	
Infraorder	Catarrhini	
Superfamily	Hominoidea	
Family	Hominidae	
Genus	*Homo*	
Species	*sapiens*	

* A taxon is any group of organisms that is recognized as a formal unit at any level in a hierarchical system of classification.

made, opinions are also important. This subjective element characterizes classification at all higher taxonomic levels, and it explains why several different classification schemes have been developed to fit the primates. One commonly used classification of the primates is given in Table 9-2, pages 282–283. All we can ask is that each classification used fit what we know about the evolutionary relationships of the species involved.

Each high taxonomic category is defined in terms of the common or shared characteristics of the subunits it contains—a genus in terms of its component species and a family in terms of its genera. For this reason, no formal definitions of these higher categories will be given here. However, discussions of the appearances and ways of life of many primate groups are given in Chapter 10.

Comparative Approach

Classification on the basis of similarities and differences is not just a matter of observation. The marsupial Tasmanian "wolf" looks like the true wolf, even though the two are only distantly related as mammals. Birds and bats are similar in being able to fly, though they are not closely related; their wings are *analogous* (similar in function), but not *homologous* (similar in evolutionary origin). However, the idea that degree of genetic or phenotypic similarity implies closeness of relationship has gained support as we have accumulated more and more evidence. That evidence has come from several sources and from several disciplines—anatomy, embryology, physiology, ethology. Data from all these disciplines are thought to reflect correspondences in the DNA sequences that make up the genomes of the organisms being compared, although to differing degrees. Primate behavior probably reflects a relatively smaller fraction of genetically coded information than primate anatomy. The most direct measure of genetic similarity would come from comparison of DNA sequences from different species. In fact, it is now actually possible to do this, although, unfortunately, few primate genomes have been compared as yet. But some recent biochemical studies go a long way toward filling this gap. Proteins are but two steps removed from the primary genetic material, each being coded for by a messenger RNA molecule whose nucleotide sequence mirrors that of the DNA in the nucleus. Amino acid sequences of molecules such as hemoglobin and haptoglobin have been directly compared, but this time-consuming process has not yet been applied to many other proteins. Some of these newer techniques hold much promise of increasing our understanding of primate relationships and evolution. However, let's begin our study of these patterns of relationship with a review of the anatomical evidence that has traditionally served as the mainstay of classification.

Anatomical Evidence. Before molecular evidence was available, anatomical features served as the major basis for diagnosing relationships among various groups. Scientists compared the bones, skin, muscles, and organs of an animal of one kind with the corresponding parts of another in order to see how similar or different they were. From these resemblances or contrasts, degrees of relationships were inferred. The process begins with a detailed study of one species and then progresses to a search for other relatives. For example, because gorillas and chimpanzees have the same numbers and types of teeth and very similar skeletons (large overall, with broad chests, long arms, short legs, no tail, and so on), they are believed to be close relatives and descendants of a relatively recent common ancestor. Some features, fewer in number, show both apes to be similar to Asian and African monkeys; all have 32 teeth when adult, for instance.

As anatomists start with one species and search for its more and more distant relatives, the characteristics that link the whole group together usually become fewer and less clear—like ripples at a distance from their source. This makes it difficult to find many features that are common to all the primates, though anatomists have long tried to do this. Thus, in 1873 the anatomist St. George Mivart described a primate as "an unigiculate, claviculate, placental mammal with orbits encircled by bone; three kinds of teeth at least at one time of life; brain always with a posterior lobe and a calcarine fissure; the innermost digits of at least one pair of extremities opposable; hallux with a flat nail or none; a well-marked caecum; penis pendulous; testes scrotal; always two pectoral mammae" (p. 507). The meanings of these forbidding-sounding anatomical terms are not as important here as a major shortcoming of Mivart's approach. Later scientists who studied primates have pointed out that not a single one of these traits is found only in primates; many (e.g., the presence of clavicles, bones which brace the shoulders) are shared with a large number of other mammals, and some (opposable digits, for instance) are present in primates to very different degrees.

Evolutionary Approach

As an alternative to Mivart's approach, Wilfred LeGros Clark (1959) suggested that the primate order be defined as one whose members are united less by common anatomical characteristics than by several major trends:

1. Preservation of a pattern of limb structure similar to that found in more primitive mammals, with retention of skeletal elements (all five digits on the appendages and the presence of clavicles, for example) which are reduced or lost in other mammals.

2. Increased mobility and manipulative ability of the digits.

TABLE 9-2 Living members of the order primates

Order	Grade	Suborder	Infraorder	Superfamily	Family	Genus	Common name
Primates	Prosimian	Prosimii	Lemuriformes	Tupaioidea	Tupaidae	Tupaia	Common tree shrew
						Dendrogale	Smooth-tailed tree shrew
						Urogale	Philippine tree shrew
						Anathana	Madras tree shrew
						Ptilocercus	Pen-tailed tree shrew
				Lemuroidea	Lemuridae	Lemur	Common lemur
						Hapalemur	Gentle lemur
						Lepilemur	Sportive lemur
						Phaner	Fork-marked dwarf lemur
						Cheirogaleus	Mouse lemur
						Microcebus	Dwarf lemur
					Indridae	Indri	Indris
						Avahi	Avahi
						Propithecus	Sifaka
					Daubentonidae	Daubentonia	Aye-aye
			Lorisiformes	Lorisoidea	Lorisidae	Loris	Slender loris
						Nycticebus	Slow loris
						Arctocebus	Angwantibo
						Perodicticus	Potto
						Galago	Bushbaby
			Tarsiiformes	Tarsioidea	Tarsiidae	Tarsius	Tarsier
	Monkey	Anthropoidea	Platyrrhini	Ceboidea	Callithricidae	Callithrix	Marmoset
						Cebuella	Pygmy marmoset
						Saguinus	Pinche
						Leontideus	Golden lion marmoset
						Callimico	Goeldi's marmoset

282

Infraorder	Superfamily	Family	Genus	Common name	
Catarrhini		Cebidae	*Cebus*	Capuchin	
			Saimiri	Squirrel monkey	
			Aotus	Douroucouli	
			Callicebus	Titi	
			Pithecia	Saki	
			Chiroptes	Bearded saki	
			Cacajao	Uakari	
			Alouatta	Howler monkey	
			Ateles	Spider monkey	
			Lagothrix	Wooly monkey	
			Brachyteles	Wooly spider monkey	
	Cercopithecoidea	Cercopithecidae	*Cercopithecus*	Guenon	
			Erythrocebus	Patas monkey	
			Cercocebus	Mangabey	
			Papio	Baboon	
			Mandrillus	Mandrill	
			Theropithecus	Gelada	
			Macaca	Macaque	
			Cynopithecus	Celebes black ape	
			Colobus	Guereza	
			Nasalis	Proboscis monkey	
			Presbytis	Langur	
			Simias	Island langur	
			Rhinopithecus	Snub-nosed langur	
			Pygathrix	Douc langur	
	Hominoidea	Hylobatidae	*Hylobates*	Gibbon	Ape
			Symphalangus	Siamang	
		Pongidae	*Pongo*	Orangutan	
			Pan	Chimpanzee	
			Gorilla	Gorilla	
		Hominidae	*Homo*	Human being	Human

3. The replacement of short, laterally compressed claws by wide, flattened nails and the development of touch-sensitive pads on the digits.

4. Increased importance of the visual sense, marked by greater protection of the eye by surrounding bone and greater overlap of the visual fields, giving better depth perception.

5. Decreased functional importance of the sense of smell and diminution of structures associated with it.

6. Reduction in the number of teeth.

7. Progressive shortening of the snout.

8. Increased size of the brain, particularly of the cerebral cortex, which also has become highly convoluted.

9. Increased duration of all phases of the life cycle, including gestation (associated with modifications of the placenta and related structures to ensure improved nourishment and protection of the fetus), childhood (related to the growing complexity of social structures and the increasing importance of individually learned information), and even adulthood (when learned information may enable individuals to contribute to the survival of the group even after their own reproductive period is over).

The existence of these general trends was first inferred from comparative study of the many living representatives of the order listed in Table 9-2. For example, in the century-old study by Thomas Huxley cited at the beginning of this chapter, the number of teeth was shown to range from higher numbers in lemurs to lower numbers in apes and humans. The trends exist because each species has adapted to a set of evolutionary forces whose action has been structured by the *ecological niche* that its members have occupied through time. The characteristics of this niche are determined by all the factors in the physical environment that influence the animals and by the entire way of life developed by the species, including everything from form to behavior (see Chapter 10 for a further discussion of this concept). Since primates are united by these common trends, we can assume that their niches shared some common elements.

Many of the trends can be seen as progressive responses to the conditions of an arboreal way of life. Most members of the primate order have been adapting to life in the trees for tens of millions of years. As long as a particular set of ecological conditions continued to exist, each successive generation faced a similar array of selective forces, and any slight adaptive changes in the gene pool of one generation modified the accumulated genetic heritage of thousands of past generations. Obviously, the ability to make a wide range of movements in a shifting, three-dimensional network of branches high above the ground was of paramount importance. The need to judge distances accurately and make decisions rapidly made binocular vision and a large brain of greater value than a good sense of smell. Monkeys and apes, for example, frequently make incredibly long leaps without hesitation. The rewards for these dangerous acrobatics include safety from predators and access to the rich supply of food provided by the fruits of deciduous trees. Following a probable dietary shift from insects to fruit, the numerous sharp-cusped teeth that had been present in the ancestors of primates were no longer needed, and so decreases in number and changes in shape could occur. With fewer teeth and a smaller olfactory region, reduction of the snout could occur. The decrease in number of offspring at each birth and the increased care given to those remaining can also be seen as requirements for a highly mobile life in precarious surroundings. Life in the trees demanded increased maternal surveillance and attention, as well as a capacity for cooperation and rapid learning on the part of the infant. Thus there were—and still are—strong selective pressures operating to shape the trends observed in these characteristics.

Are the Trends Real? Recently several biologists (Martin, 1968; Schwartz, Tattersall, and Eldredge, 1978) have reacted against the use of evolutionary trends as a basis for classifying animals as primates. Their objection is that such trends, which have been constructed by the self-defined "highest" primates, may be artificial and deceptive. Primates have not evolved in a straight line, from the earliest prosimians to present humans; populations have divided and differentiated, giving rise to many different species with rather diverse ways of life and characteristics suited to them. In order to answer questions like the one posed in the heading of this section, we must have some knowledge of the diversity of anatomical traits that can be found in living primates. The discussion that follows in the next section ("Representative Living Primates") concentrates on features of the skeleton, since these will be of the most direct use when we begin, in Chapter 12, to search the geological record for the fossilized remains of our relatives who lived in earlier periods.

REPRESENTATIVE LIVING PRIMATES

This section presents an overview of the characteristics of a number of living primates. The skeleton of each primate, though obviously a functioning whole in the living animal, will be discussed as if it consisted of four major subsystems: the skull, forelimbs, trunk, and hindlimbs. Each of these regions, in turn, has its own subunits, and these parts all have one or more functions.

Functional Units of the Skeleton

Primate skeletons contain about 200 bones. Professional physical anthropologists and anatomists have to know the names of all these and of many detailed features on them. For our purpose here, however, it will suffice to look at just the major parts of several regions of the body that serve a few important functions. All the parts discussed are illustrated in Figure 9-1, page 288.

Skull. The skull is subdivided into the face and the braincase. The face includes a concentration of sense organs—eyes, nose, tongue, and, in some prosimian primates, large whiskers (*vibrissae*) that have a very keen sense of touch. Other important facial components are the jaws and teeth, used chiefly in processing food. All primates have two sets of teeth (the *deciduous* teeth, or baby teeth, and the *permanent* teeth of adults); this succession is a piece of structural evidence for the existence of a life cycle with discrete segments. The teeth are structurally differentiated (into incisors, canines, premolars, and molars) according to function (for catching, holding, and grinding food). The number of teeth of each type can be summarized by a *dental formula,* which shows how many teeth there are of each type in one half of the upper and lower jaws. For example, dogs have a dental formula of $\frac{3142}{3143}$; this is quite similar to that of the earliest placental mammals, which had dental formulas of $\frac{3143}{3143}$. Behind the face is the dome-shaped braincase, which houses the animal's information-processing center. Nerves leading to and from other parts of the body are joined into the spinal cord, which enters the base of the skull through the *foramen magnum* (from the Latin *magnum,* "big," and *foramen,* "hole").

Forelimbs. The forelimbs are used to move about and to handle objects. The hands of most primates are *pentadactyl* (having five digits), as were the hands of the earliest vertebrates to colonize the land hundreds of millions of years ago. The full ancestral set of fingers is still very useful in grasping beetles, bananas, branches, and babies. Each hand joins at the wrist to a forearm, which contains two bones, the *radius* and the *ulna*. These rotate quite freely around each other; thus, for example, when the arm is extended, the hand can be turned from a palm-up to a palm-down position. In the upper arm a single bone, the *humerus,* goes from the elbow to the shoulder, where it joins two others: the *scapula* and the *clavicle*. The scapula is a flat bone shaped like a right triangle that lies along the back of the rib cage. In a quadruped, one straight edge of the scapula is parallel to the spine, the second edge juts out perpendicularly to a shallow socket for the head of the humerus, and the third edge slopes from this socket back toward the spine. The clavicle starts at the front of the *sternum,* or breastbone, in the middle of the chest, and extends to the point where the scapula and humerus meet; the S-shaped clavicle makes it possible to extend the forelimb perpendicularly to the main axis of the body as well as parallel to it. In animals which lack a clavicle, such as the dog, the forelimb can be moved only parallel to the body.

Trunk.   The trunk is the main part of the body, the section between the shoulders and the hips. It has two main regions: the chest, which is made up of several pairs of ribs that arch from the spine to the sternum, and the abdomen, which is the area behind the chest in a quadruped. The chest protects the heart and lungs, and movements of the ribs aid in breathing. In the abdominal region the remaining internal organs concerned with digestion and other vital functions are supported beneath the spine's arch by strong, flexible walls of muscle and skin.

Hindlimbs.   The hindlimbs begin with the *pelvis,* which is formed from three bones (the *ilium, ischium,* and *pubis*) that grow together. This composite structure is firmly attached to the spine and is thus well-suited to resist the stresses produced by running, jumping, and climbing. The hip joint consists of a socket in the side of the pelvis that receives the rounded head of the *femur,* or upper leg bone. At the other end this bone meets the *tibia* and *fibula* of the lower leg. The feet of most primates are also pentadactyl, but they are usually structured more for support than for manipulation.

   Having had this brief introduction to the major regions of the skeleton, let's go on to see how the basic parts have been modified to fit the ways of life of several representative primates.

A Prosimian: The Mouse Lemur

The mouse lemur (*Microcebus murinus;* Figure 9-1) is a tiny prosimian that weighs approximately 60 g (a bit over 2 oz) and is covered with soft brownish or grayish fur. Mouse lemurs are found only on Madagascar, where they live in forested areas along the coasts. They build nests of dry leaves in bushes or holes in trees, where they sleep during the day. Mouse lemurs are nocturnal, coming out at night to feed, mostly on insects but also on some fruit and possibly small animals.

   The skull of the mouse lemur is very similar to that of the tree shrew, *Tupaia. Microcebus* has a moderate-sized snout tipped with a *rhinarium.* This is a furless area of granular skin that stays moist to help capture odors, an indication of the importance of the sense of smell to this tiny predator, which would also be easy prey for larger hunting animals. As befits a nocturnal animal, the ears are large. The eyes are large too,

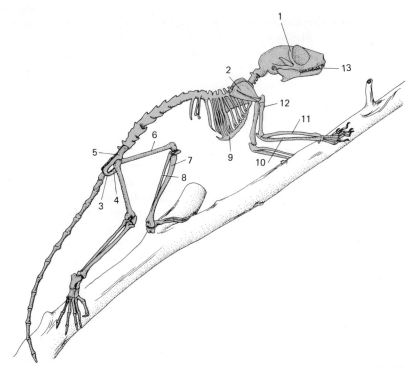

**FIGURE 9-1   Skeleton of mouse lemur, *Microcebus murinus:* 1, postorbital bar;
2, scapula; 3, ischium; 4, pubis; 5, ilium; 6, femur; 7, tibia; 8, fibula; 9, sternum;
10, ulna; 11, radius; 12, humerus; 13, tooth comb.**

enabling the mouse lemur to make the most of what dim light is available
to see by. The base of the snout separates the eyes and directs their
gaze partly to the side. This gives a wide range of vision that is useful
in watching for predators and prey, but it reduces by a bit the overlap
of the visual fields needed for depth perception. A *postorbital bar* arches
behind each eye; this flat strip of bone provides protection for the eye
and separates it from adjacent muscles.

The dental formula of *Microcebus* is $\dfrac{2133}{2133}$. The upper incisors are
short pegs that are planted vertically in the jaw, and the canine tooth
that follows them in the upper jaw is a simple cone that rises from a
rounded base to a single point, or *cusp,* that projects about twice as far
as the crown of any other tooth. In the lower jaw the canine is long and
flattened like the incisors, with all three teeth jutting outward to form
a *tooth comb* that is used for grooming the fur. In both the upper and
lower jaws the premolars rise from bulbous bases to roughly conical
crowns. The upper and lower molars differ a bit from each other. In the
upper molars the cap, or crown, of enamel has three main cusps and

several smaller ones, with the outline forming a rough triangle. The first two lower molars each have four major cusps; the last molar has five, the first three cusps again forming a triangle, which is separated by a deep trough from the last pair of cusps. All the molar cusps are quite long and sharp and thus are useful for piercing the bodies of insects to extract their juices.

A mouse lemur's braincase has a relatively flat base that continues in a line from the roof of the mouth. The result is that the base of the entire skull is relatively straight, and the braincase is more behind the face than above it. The brain itself is one of the smallest and simplest found in primates.

Mouse lemurs have prehensile hands. They are capable of seizing and grasping small objects, such as insects, with one hand; thus they do not need to use both hands, as a squirrel does, or to use the mouth for grabbing and holding, which a dog must do. The rest of the forelimb skeleton fits the general pattern described earlier.

These small lemurs have a chest that is narrow, flattened from side to side, and relatively deep. This produces a center of gravity that is low and centered over any branch on which the animal stands. The spinal column forms a simple arch that supports the internal organs; the arch can be bowed and straightened rapidly in making leaps. The narrow, rod-shaped pelvis joins this flexible spine to the hindlimbs. In addition to occasional leaps in which its long tail is needed for balance, this tiny lemur gets about by scurrying rapidly like a mouse.

A Monkey: The Guenon

The guenon (genus *Cercopithecus;* Figure 9-2) is an African forest monkey of medium to large size. Some species of guenons spend a considerable amount of time on the ground, but most are arboreal, staying in trees to sleep, play, mate, and eat. Their diet consists of a variety of vegetable foods such as leaves, shoots, and fruit. They are highly social animals, and some species form groups with as many as 50 members.

The monkeys have small faces in proportion to the size of their skulls. The snout is small, and no rhinarium is present, indicating that smell is a sense of relatively little importance. In contrast, the orbits of the eyes are large and directed more forward than to the sides. This produces good binocular vision and depth perception, which are valuable in arboreal life. The eye is sheltered in a complete bony socket.

The same structurally differentiated categories of teeth are present as in prosimians, but there are further numerical reductions in some categories. Guenons, like all Old World monkeys, have a dental formula $\frac{2123}{2123}$ $\Big($ South American monkeys have a dental formula of $\frac{2133}{2133}$, while

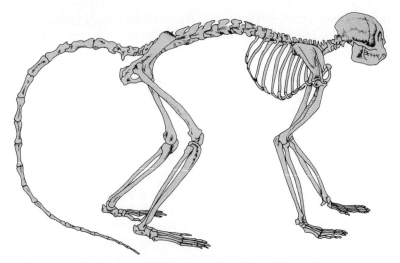

FIGURE 9-2   Skeleton of Old World monkey (Cercopithecus).
Source: W. E. Le Gros Clark. 1959. *The Antecedents of Man.* Edinburgh: Edinburgh University Press. (By permission of the publisher.)

the smaller marmosets have a dental formula of $\frac{2132}{2132}$). Monkeys usu-
ally have wide incisors, with the upper central pair often being quite broad. The canines project beyond the plane formed by the tips of the other teeth and are especially large in males. Each of the two upper premolars are *bicuspid;* that is, they have two cusps that are used in chewing. In the lower pair the second premolar also has two cusps, but the one right behind the canine has only one major cusp. This point is large and forms one blade of a pair of shears; the other blade is the upper canine, which slices against the first lower premolar. Because of its role in this shearing mechanism the first lower premolar is said to be *sectorial* (scissorlike). The large lower canine fits into a short gap, or *diastema,* between the upper canine and the lateral incisor, as a knife fits into a sheath. Like a knife, this tooth can inflict a nasty wound in a fight. The molar teeth of Old World monkeys have four principal cusps (sometimes five, especially on the third molars), divided into an anterior and posterior pair. The two cusps of each pair are connected by a ridge of enamel, and thus the surface of this tooth has two sharp blades that are used to chop up leaves, which are then eaten.

The monkey's smaller facial skeleton is bent partially beneath the vault of the skull, which houses a brain that is relatively large and has a surface (the *cerebral cortex*) covered with numerous folds that increase the area.

Guenons, like most monkeys, are pentadactyl (in a few monkey species one finger is reduced in size for specialized types of branch-swinging locomotion). The thumb is usually functionally opposable to the other fingers, making it possible to manipulate even small objects easily. In many species the forelimbs are nearly equal in length to the hindlimbs, which facilitates quadrupedal running along branches and on the ground. The scapula is basically triangular in shape, as it is in prosimians, but in the more arboreal monkeys the edge parallel to the spine is more elongated, which gives greater stability to the shoulder and thus allows it to absorb the considerable shocks that result from swinging and jumping.

The spinal column has a simple arch and serves the same purpose as in *Microcebus* and other quadrupedal prosimians. In the larger tree-living monkeys the ilium is slightly wider than in prosimians, giving additional area for the attachment of the powerful muscles of the hind leg.

Of course, not all monkeys fit exactly the pattern described here, but none show the extent of difference from the others that any of the apes do from the monkeys.

An Ape: The Gorilla

The African ape (*Gorilla gorilla;* Figure 9-3) is the largest living primate. Despite the ferocious image they have acquired in popular literature, gorillas are usually shy, slow-moving vegetarians that live in small, stable social groups. Although genetic evidence marks them as close relatives of ours, their anatomy contrasts sharply in some features.

Gorillas have a broad muzzle which projects ahead of the skull to a considerable degree. This is due less to highly developed olfactory structures (which are really quite small for the size of the animal) than to the necessity of accommodating the large teeth used in biting and chewing the coarse, bulky diet of plants. The considerable stress this places on the jaws is countered by the *simian shelf,* a bony buttress that grows where the two halves of the lower jaw join together.

The dental formula in the gorilla and other apes, $\frac{2123}{2123}$, is the same as in all living Old World monkeys. Behind the row formed by the incisors, the remaining teeth are usually arrayed in two rows that are parallel or arch convergently toward the rear. The incisors are large, with the upper central ones being markedly wider than those usually seen in monkeys. The canines are large tusks, especially in males. Diastemas are frequently present. As in the monkeys, both upper premolars are bicuspid, and the lower ones differ, sectorial fore and bicuspid aft. The molar teeth of apes typically have five large cusps, separated from one another by grooves. Three of these, which are especially deep, meet to form a crude Y; hence the term *Y-5 pattern,* which is used to describe the crowns of these teeth.

Like those of monkeys, the eyes of apes give good depth perception

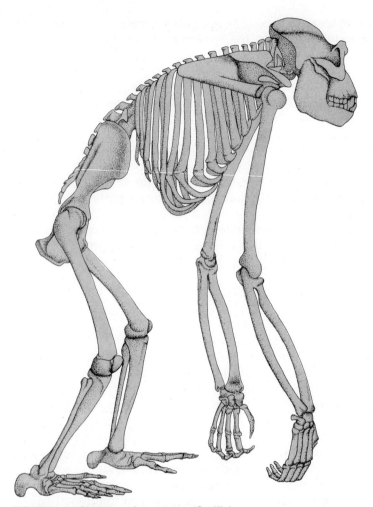

FIGURE 9-3 Skeleton of great ape (Gorilla).
  Source; Redrawn from Bernard G. Campbell. 1966. *Human Evolution.*
Chicago, Ill.: Aldine. (By permission of the author and publisher.)

and are nestled within sockets of bone in the front of the skull. Adding
to the protection of the eyes is a shelf of bone that projects from the
forehead to form a *supraorbital* torus, or brow ridge. The cranial base
is flexed sharply downward, and the foramen magnum is moved even
farther toward the center of the cranial base than in monkeys. This shift
is not sufficient to counteract the projection of the heavy face, however,
and so the skull is poorly balanced on the vertebral column and must
therefore be supported by neck muscles attached to a heavy *nuchal
crest.* In most males and a few females, the skull is surmounted by a
central bony ridge, the *sagittal crest,* which provides an additional area
of attachment for the muscles used in chewing.

The elongated, prehensile (grasping) hands are equipped with fingers that are shorter than those of monkeys, and the thumb is also relatively shorter. When the gorilla is moving by means of the arms, the four flexed fingers form a hook that is used to support the body; the thumb is not used to any appreciable extent in this activity. The great toe in the foot is also opposable to the other toes, allowing the foot as well as the hand to be used in grasping branches. Since they possess feet that resemble hands in structure and function, apes are sometimes referred to as being *quadrumanous* (four-handed). The greatly elongated forearm is another structural modification. When the animal is suspended by the hands, the elbow acts as a fulcrum about which the humerus may be rotated by contraction of the biceps and associated arm muscles to lift the whole body in climbing. In the apes the vertebral border of the scapula is even more elongated than in the arboreal monkeys. This elongation provides more leverage for the muscles that rotate the scapula, and it allows further extension of the arm upward from the trunk. This is another adaptation for moving by means of the arms, as is the placement of the scapula toward the back of the chest. Principally as a result of the elongation of the forearm mentioned above, the arms are much longer than the legs, and so the animal is semierect when standing and walking on the ground.

The gorilla's chest is broadened from side to side and narrowed from front to back, increasing the distance between the shoulder and hence indirectly widening the span between the hands when climbing or moving through the trees. The section of spine below the chest is proportionally shorter in gorillas and other apes than in monkeys. This helps to make the trunk less flexible (and perhaps decreases the chances of injury to the lower back from swinging or climbing). Adding to the increased stability of the trunk, the upper crest of the ilium is expanded, providing a greater area for the anchorage of muscles associated with movement at the hip and support of the internal organs. Although the external tail has disappeared, four vertebrae have been retained below the pelvis; they curve underneath the pelvic opening, providing additional support for the internal organs. Some similar adaptations to upright posture of a different sort are discussed below.

The Human: *Homo sapiens*

Our species (*Homo sapiens;* Figure 9-4) is the only living primate that habitually walks upright on two legs. Since no further introductions are necessary, let's move on to see what major structural features accompany our behavioral uniqueness.

One notable feature of the human face is the very slight degree to which it projects from beneath the skull. Neither the dental nor the olfactory elements are greatly developed. The lower jaw is thin and has

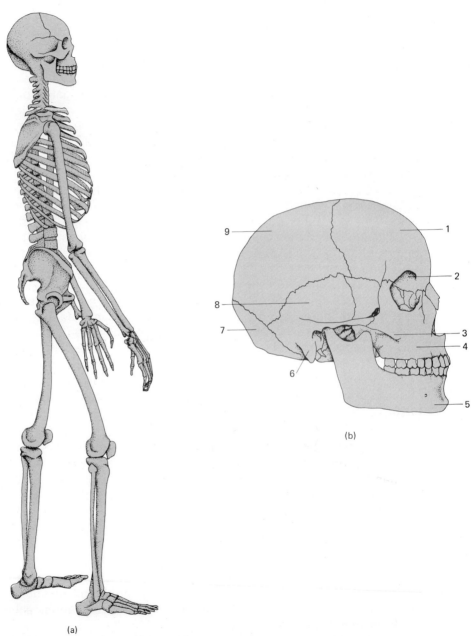

(a)

(b)

FIGURE 9-4   (a) **Skeleton of human** *(Homo).* (b) **Human skull: 1, frontal; 2, orbit; 3, zygomatic; 4, maxilla; 5, mandible; 6, mastoid process; 7, occipital; 8, temporal; 9, parietal.**
Source for *a:* Redrawn from Bernard G. Campbell. 1966. *Human Evolution.* Chicago, Ill.: Aldine.

no simian shelf. A functionally similar reinforcement is provided by the chin, which bridges the midline on the outer surface of the jaw.

The dental formula is the same in humans as it is in all living Old World monkeys and apes: $\frac{2123}{2123}$. In the upper and lower jaws the rows of teeth are arranged in smooth semicircles or ellipses. Diastemas are only very rarely found. The incisors are relatively narrower than those of the apes, although the upper central incisors are typically wider than the others in both jaws. The canines are small flattened ovals, sometimes with a small point that projects just slightly above the tips of the other teeth. All the premolars are bicuspid, but the anterior lower ones have an inner cusp smaller than the outer one. The smaller molars bear rounded cusps which are somewhat closer together than those of the apes. Some molars retain a Y-5 cusp arrangement, but in others one cusp is often very small or absent, leading to a frequent alteration of the basic Y-5 hominoid pattern to one designated as +4.

The orbits of the eyes are roughly flattened ovals which lie beneath separate ridges of bone that vary from barely perceptible to moderately projecting, depending on the individual and the population. The cranial base is flexed to a great degree, and so the face is tucked beneath the rounded forehead. The foramen magnum is located nearly at the center of the cranial base; thus the whole skull is neatly balanced on the vertebral column. Because of this, there is no need for any great development of the neck muscles to support the skull. The area of attachment of these muscles on the skull is thus reduced from a heavy crest to the barely evident nuchal line. Similarly, the braincase itself is so large—usually ranging from 1100 to 1300 cc, though from as little as 1000 to as much as 2000 cc in extreme cases—and the jaws are so small that the uppermost point of attachment on the skull of the muscles that move the lower jaw is marked by the slight temporal line, rather than a sagittal crest. The hands and feet are five-digited, but they have diverged as greatly from each other in form as they have in function. The hand, which has a shorter palm and a longer thumb than are found in the apes, is capable of precise manipulation as well as a strong grasp. Changes in the foot include the complete loss of opposability of the great toe, a general shortening of all the toes, and a complex of other changes that have transformed the grasping foot of the ape into a stable platform for support and bipedal walking and running. The forelimbs are essentially the same as those of the great apes, but the arms are shorter than the legs, reversing the condition found in the gorilla. The form and placement of the scapula, as well as the articulation of the humerus with it, closely resemble those in the apes.

As in the apes, the chest is wide from side to side and shallow from front to back. The entire erect trunk is supported by a spinal column with a pronounced S curvature. The arch points toward the rear in the region of the chest and curves forward in the lower back. The spinal column terminates in several vertebrae fused to form the coccyx, which curves under to form part of the floor of the pelvic basin.

The pelvis shows an exaggeration of the condition that separates apes from monkeys. The ilium is reduced in height and extended backward, while the iliac crest itself is greatly expanded. The first of these changes realigns the muscles to make bipedal walking easier; the second stabilizes the trunk in an erect posture. The articulation of the pelvis with the sacral portion of the spine is brought closer to the socket for the femur, so that weight is transmitted through it directly from the spine to the leg. The ischium is shortened, allowing full extension of the hip.

These four examples provide an introduction to some of the structural and functional features that help us understand our place among the living primates. Even if we made other comparisons and looked at more characteristics of more species, we would still have to conclude, as Huxley (1863) did, that: "Whatever system of organs be studied, the comparison of their modifications in the ape series leads to one and the same result—that the structural differences which separate Man from the Gorilla and Chimpanzee are not so great as those which separate the Gorilla from the lower [primates]" (p. 96).

Comparisons among living primates can tell us who our closest relatives are, but to discover the physical features of our actual ancestors, we must deal directly with the fossil record. In order to make reasonable sense of this record, in turn, anthropologists and paleontologists have to deal with the idea of change through time and all its consequences. The study of norms and averages is not enough; scientists must go beyond the concept of "typical" characteristics and the assumption that every specimen is representative of a whole population. Each species consists of individuals that differ anatomically, and the frequencies of these characteristics change through time, along with the frequencies of the genes that influence them. It follows that some specimens which differ in appearance may have belonged to the same population, and others which resemble each other may have been members of different populations. To interpret the record of past primate evolution correctly, therefore, it is necessary to know something about the extent of the variation that occurs in skeletal characteristics.

VARIATION IN THE HOMINOIDS: SOURCES AND SIGNIFICANCE

Most primate taxonomists of the nineteenth and early twentieth centuries worked with only a few specimens and had faulty or nonexistent data on the locations where these specimens had been found. Because there was no way they could know the degree of variability possible within local subpopulations of a single species, let alone among the representatives of an entire species' gene pool, they frequently designated each new individual as a specimen of a different species or genus. This spirit carried over into the interpretation of fossil remains, and thus nearly every new primate fossil was also given its own Latin name. For several centuries, the whole idea of variation within a species as well as from species to species was a sort of nuisance to taxonomists and other biologists interested in cataloging and studying the natural world. Each species had only one ideal *type* specimen; it simply did not occur to anyone to look at these specimens as just one of a range within a larger pool.

Variation is much more than a classification problem; it is a crucial dimension of biological reality, and one with important implications. The first of these is philosophical: the size and complexity of the human genome are such that, given probable levels of heterozygosity, every individual who has ever lived (except members of a pair of monozygotic twins) is likely to have had a unique genotype as well as a unique record of experiences. This individuality, which shows up as variability at the phenototypic level, lends scientific support to the belief that each person, from no matter what subpopulation, is and ought to be treated as unique. The second implication is practical: without variation, evolution—and even survival—would simply be impossible. A uniform population might be perfectly adapted to a given set of conditions, but should those conditions change, it would be uniformly ill adapted. The fact that populations vary means that at least some stable, heritable modifications may be suited to the new environmental conditions and thus allow the group to survive. Variation is the raw material that supplies the potential for evolution. It is on the variation stored within the gene pool of a population that selection can act, successively changing rare genetic variants to significant minorities, on occasion elevating these minorities to majorities, and eventually possibly even resulting in their fixation at a given locus. Third, awareness of systematic patterns of variation, including genetic polymorphisms, can help prevent us from formulating misleading hypotheses about the biological relationships between living human populations. One example of such an idea appeared in Boyd's classic *Genetics and the Races of Man* (1950). When studies were first made of the blood groups of North American Indians, it was found that group O was by far the most common, group B was extremely rare, and group A occurred chiefly in tribes in which there was considerable likelihood of European admixture. It was therefore suggested that the ancestors of the American Indians had left Asia before the mutations which gave rise to the genes producing groups A and B. Instead, studies of blood groups in nonhuman primates show that A, B, and O antigens are found in the apes and thus are a basic part of our primate heritage. They are not the result of very recent evolution affecting only some human subpopulations.

Unless we know the extent of morphological variation in living populations, we cannot interpret fossil finds correctly. And the mistakes we may make will skew our interpretation of the whole course of human evolution. Fossil remains merely exhibit certain morphological and metrical characteristics; it is we who classify them by names that are interpretations of these characteristics. Assigning two specimens different Latin names indicates a belief that the two were so different that they could not have been members of the same species. But without knowing what the range of differences can be, we are just making guesses on the basis of whatever surviving evidence happens to come to our attention.

Although there is a large and growing body of literature on variation in nonhuman primates of all kinds, our discussion will be limited to that shown by the hominoids. Especially useful for comparative purposes are the two semiterrestrial African great apes, which bracket humans in overall body size: the chimpanzee is on the average slightly smaller, and the gorilla is considerably larger. Among members of any given species population, including the hominoids, there are several major sources of variation in addition to the normal range that would be found in a group whose members were of the same sex and of the same approximate age: differences in age, sex, and geographic area occupied, as well as the effects of abnormal or pathological processes such as disease and injury.

Individual Variation

The extent of individual variation is summarized in Tables 9-3 to 9-5. Two cautions are needed to interpret and use this information: (1) Up to a point, the range of variation in any sample depends on the number of specimens studied, since larger samples have a greater probability of including more and greater deviations from the average (the number of specimens on which the range is based is given in each table), and (2) members of larger animal species appear at first glance to be more variable than those of smaller species. Thus, the difference in size between the largest and smallest adult male chimpanzees is 62.0 kg (28.2 lb), whereas the largest and smallest gibbons differ by only 3.6 kg (7.9 lb). Do chimpanzees really vary more in body size than gibbons? If we express the range as a percentage of the average for each species, we discover that the range in males of both species is the same—61 percent of the average body weight.

The tables tell us not only that there is a wide range of individual variation among the members of any single species but also that there is considerable overlap between the ranges of different species. Thus,

TABLE 9-3   Variations in body weight in living hominoids

Common name of genus	N	$\bar{X}$	Range, males		N	$\bar{X}$	Range, females	
			kg	As % of average			kg	As % of average
Gibbon	35	5.9	4.3– 7.9	61	23	5.8	4.6– 6.8	38
Orangutan	10	78.0	58.3– 90.8	42	11	38.0	32.7– 45.4	33
Chimpanzee	10	45.9	31.8– 60.0	61	10	39.9	31.1– 55.0	60
Gorilla	21	151.8	140 –180	26	?	92.5*	75 –110	38*
Human	6677	71.7	45.2–128.7	116	1905	57.7	37.5– 91.9	94

* Average not available. Figures given are based on midpoint of range.

Sources: *For gibbons,* Adolph Schultz. 1933. "Observations on the Growth, Classification, and Evolutionary Specialization of Gibbons and Siamangs," *Human Biology,* 5(2):212–255, 385–428. *For orangutans,* (1) Marcus W. Lyon. 1908. "Mammals Collected in Eastern Sumatra during 1903, 1906, and 1907 with Descriptions and New Species and Sub-species," *Proceedings of the United States National Museum,* 34(1626):619–679. (2) Marcus W. Lyon. 1911. "Mammals Collected by Dr. W. L. Abbott on Borneo and Some of the Small Adjacent Islands," *Proceedings of the United States National Museum,* 40(1809):53–146 (with corrections by the author). (3) Adolph Schultz. 1941. "Growth and Development of the Orangutan," *Carnegie Institute Contributions to Embryology,* 29(182):57–110. *For chimpanzees,* Adolph Schultz. 1954. "Bemerkungen zur Variabilität und Systematik der Schimpansen," *Säugetierkundliche Mitteilungen,* 2(1):159–163. *For gorillas,* J. R. Napier and P. H. Napier. 1967. *A Handbook of Living Primates.* New York: Academic. *For humans,* (1) Robert M. White et al. 1971. *The Body Size of Soldiers. U.S. Army Anthropometry—1966.* (72-51-GE.) Washington, D.C.: GPO. (2) Charles E. Clauser et al. 1972. *Anthropometry of Air Force Women.* (AMRL-TR-70-5.) Washington, D.C.: GPO.

although male gorillas are on the average much larger than men, some of the men in the sample were larger than some of the gorillas. When we compare men and male chimpanzees, we find that the similarity in size is even greater; there is extensive overlap in the body-weight ranges of both. Moreover, females of these species are even more similar in size than males.

Of course, the fossil record never provides whole bodies for comparison, and the fact that only bones and teeth are found makes the job of classification far more difficult. If we had complete bodies, a number

TABLE 9-4   Variations in cranial capacity in living hominoids

Common name of genus	N	$\bar{X}$	Range, males		N	$\bar{X}$	Range, females	
			cc	As % of average			cc	As % of average
Gibbon	95	104	89– 125	35	85	101	82– 116	34
Orangutan	96	424	320– 540	52	111	366	276– 494	60
Chimpanzee	70	396	322– 500	45	63	355	275– 455	51
Gorilla	400	535	420– 752	62	173	458	340– 595	56
Human	Hun-dreds	1400	1100–1700	43	Hun-dreds	1300	1000–1600	46

Source: Adolph Schultz. 1968. "The Recent Hominoid Primates," in S. W. Washburn and P. Jay (eds.), *Perspectives on Human Evolution.* New York: Holt, Rinehart and Winston. By permission of the author and publisher.

TABLE 9-5   Variations in length of molar (right M₁)

Common name of genus	N	X̄	Range, males		N	X̄	Range, females	
			mm	As % of average			mm	As % of average
Gibbon	10	6.01	5.5– 6.5	17	10	5.80	5.2– 6.5	22
Orangutan	20	13.63	12.0–15.0	22	20	12.53	11.0–13.5	20
Chimpanzee	64	11.36	10.4–13.2	25	64	11.12	9.9–12.2	21
Gorilla	20	15.40	13.0–17.5	29	20	14.88	13.6–16.5	20
Human	20	10.60	9.5–12.5	28	20	10.78	10.0–11.5	23

Sources: *For gibbons, orangutans, and humans,* Aleš Hrdlička. 1923. "Variation in the Dimensions of Lower Molars in Man and Anthropoid Apes," *American Journal of Physical Anthropolgy,* **6**(4):423–438. *For chimpanzees,* E. L. Schuman and C. L. Brace. 1954. "Metric and Morphologic Variations in the Dentition of the Liberian Chimpanzee: Comparisons with Anthropoid and Human Dentitions," *Human Biology,* **26**(3):239–268.

of anatomical features would enable us to differentiate clearly between, say, male apes and men. And as we go further back in time toward the common ancestor of living apes and men, we would expect the differences between members of the two lines to decrease, thus making classification even more difficult and less certain. For example, as Table 9-5 shows, the degree of overlap in the size of some ape and human teeth is even greater than the degree of overlap in their body weights. Moreover, this dental similarity goes beyond size alone; detailed resemblances are also found in shape and cusp patterns. The Y-5 cusp pattern, for example, is typically found on the molar teeth of many apes, and the +4 pattern is found on most human molars—but there are many exceptions to this generalization, as Figure 9-5 shows. Because of these resemblances, those who have studied the teeth of primates have stressed how difficult it is to distinguish between the unworn molars of some apes and those of men. When even a moderate amount of wear has taken place, certain identification of such teeth verges on the impossible.

But although we are all familiar with continuous, quantitative variations in height, weight, and so on, among the people and animals we know, we tend to think of the various parts of the body as being fixed and unvarying in type and number, at least among the members of any single species. This impression is reinforced by statements in anatomy texts such as, "The adult human skeleton consists of 206 named bones" or "There are 7 vertebrae in the neck, 12 in the thorax, and 5 in the lower back, and 5 fused vertebrae form the sacrum." As Table 9-6 shows, however, these too are only averages; the different types of spinal vertebrae found in hominoid species also vary in frequency. *Most* humans have a total of 17 thoracic and lumbar vertebrae, as is true of *most* gorillas and chimpanzees. But 7 out of every 100 humans examined have one less lumbar vertebra, the same number found in the average orangutan. Similarly, a typical human has five sacral vertebrae—but so

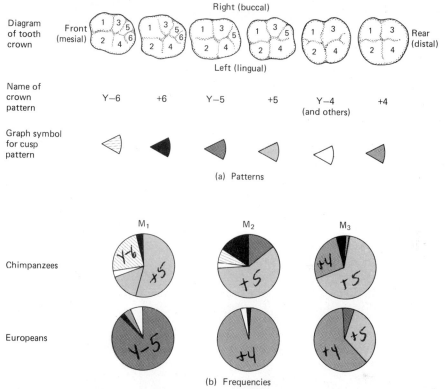

FIGURE 9-5 **Mandibular molar patterns and their frequencies in chimpanzees and humans.** (a) **Patterns; in all cases, a right lower molar is shown. Numbers designate cusps, which are called: 1, protoconid; 2, metaconid; 3, hypoconid; 4, entoconid; 5, hypoconid; 6, accessory cusp.** (b) **Frequencies.**

Source: E. L. Schuman and C. L. Brace. 1954. "Metric and Morphologic Variations in the Dentition of the Liberian Chimpanzee: Comparisons with Anthropoid and Human Dentitions," *Human Biology,* **26:**239–268.

does the average orangutan. The atypical one person in four who has six sacral vertebrae shares this distinction with a majority of the African apes.

These few examples show that the extent of variability in even the most easily measured characteristics can be quite significant. More complex morphological characteristics that are more difficult to measure also show interesting variations. For our purpose, which is simply to show that great variations can be found in even small samples of individuals of each species, photographic evidence is sufficient. Figures 9-6 and 9-7 show the skulls of a number of adult chimpanzees and gorillas. All the animals have complete sets of permanent teeth and are thus technically adult, although some are older than others, as indicated by greater degrees of tooth wear. Intraspecific differences may be seen in the degrees of facial protrusion, the size of the braincase, the degree

TABLE 9-6 Variations in the number of vertebrae of different types in living hominoids (expressed as percentages)

Common name of genus	N	Number of thoracic and lumbar vertebrae						Number of sacral vertebrae						
		15	16	17	18	19	Average	3	4	5	6	7	8	Average
Gibbon	319	—	—	5	72	23	18.2	3	42	51	4	—	—	4.6
Orangutan	127	19	74	7	—	—	15.9	—	3	59	36	2	—	5.4
Chimpanzee	162	—	29	68	3	—	16.8	—	1	36	55	8	1	5.7
Gorilla	81	—	43	56	1	—	16.6	—	1	36	56	6	1	5.7
Human	125	—	7	91	2	—	17.0	—	3	72	24	1	—	5.2

Source: Adolph Schultz. 1968. "The Recent Hominoid Primates," in S. W. Washburn and P. Jay (eds.), *Perspectives on Human Evolution.* New York: Holt, Rinehart and Winston. By permission of the author and publisher.

FIGURE 9-6 Comparison of skulls of adult male chimpanzees: (a) front view; (b) side view. Note differences in shape of skull vault, brow ridges, nasal opening, and degree of facial projection. Scale: ¼ natural size.

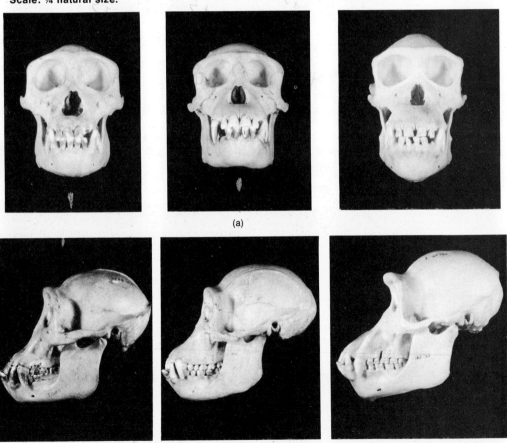

(a)

(b)

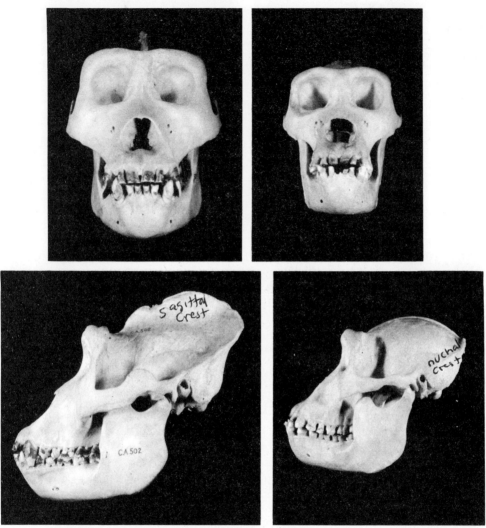

FIGURE 9-7 Comparison of skulls of (left) **adult male** and (right) **adult female go-rillas. Note differences in overall size and, in particular, development of sagittal and nuchal crests. Scale: ¼ natural size.**

of development of the sagittal crest and brow ridge, the projection of the cheekbones, and the size and shape of the nasal opening and the orbits of the eyes.

Variation Due to Sex

Some idea of the effects attributable to secondary sexual differences can be gotten from Table 9-7, which shows size differences in several characteristics. These measures are based on averages for males and

TABLE 9-7   Sexual dimorphism in several metric characters*

Common name of genus	Female average expressed as percent of male average		
	Body weight	Cranial capacity	Length of M_1
Gibbon	98	97	97
Orangutan	49	86	92
Chimpanzee	87	90	98
Gorilla	61	86	97
Human	81	93	102

* Based on Tables 9-3 to 9-5.

females of each species. The most marked differences between males and females are in body weight. Sagittal crests are more frequently found in male hominoids than in females, and when present they are usually larger (see Figure 9-7). Mature male gorillas have a saddle of white or silvery hairs across the lumbar region not found in most females of comparable age. Adult male orangutans usually have cheek pads of fatty and fibrous tissue that project as wide flanges, enlarging the face; they also have large, inflatable throat sacs and a more conspicuous beard of orange hair.

Why is it important that we be aware of the amount and patterns of sexual dimorphism among the hominoids? Any decision on taxonomic status and phylogenetic position is a matter of opinion on the part of the observer. When making such judgments, we should be aware that differences due to sex in contemporary primate species are sometimes very great. An illustration is provided in Table 9-7. As we will see in succeeding chapters, the differences it shows are greater than many others that distinguish fossil specimens generally thought to belong to different species, genera, and even families.

Variation Due to Age

The effect of age on morphology is most easily seen at the extreme, as shown in Figure 9-8. This picture also points up another factor to be borne in mind as we search for human ancestors. Immature apes very often look much more humanlike than adults of the same species, as can be seen more clearly from the series of hominoid skulls in Figure 9-9. The human skull retains throughout adult life a general appearance similar to that found only in very young hominoids of other species. For this reason, our species has been called a "fetalized ape." It is also said that humans are "feminized apes." These descriptive analogies lead us to the conclusion that the most characteristic features of apes are displayed by adult males. The ultimate sizes of the nuchal and sagittal crests, for example, are determined by the need for areas of attachment

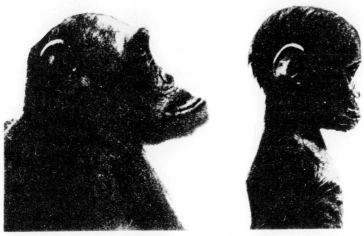

FIGURE 9-8 Contrast between (left) **an adult and** (right) **a juvenile chimpanzee.**
Source: Gustav Schenk. 1961. *The History of Man.* Stuttgart: Chr. Belser Verlag.
(By permission of the author and publisher.)

by muscles that continue to enlarge throughout much of adult life. Thus these crests usually have not completed their development even by the time the individual has a full set of permanent teeth. Full crest development is found only in individuals whose permanent teeth have become quite worn, which suggests advanced age. Similar changes in strength that come with age affect bones in many other areas of the body.

When dealing with unidentified skeletal or fossil material, there is commonly a problem in distinguishing the effects of sex from those of age. The situation becomes especially difficult when a specimen is very incomplete or fragmentary, as many fossils are. Young apes, and particularly immature female apes, will appear more like humans than adult males that belonged to the same breeding population. For this reason, we should be cautious in assuming that a fossil which looks like a human *was* one in life. Unless there is sufficient independent evidence to indicate that it was ecologically, behaviorally, and reproductively different from contemporaries whose remains are more apelike in appearance, we should be reluctant to assume that it was not also an ape.

Geographic Variation

No other hominoid (ape or human) species has at the present time as wide a distribution over the earth as ours does; most inhabit relatively restricted ranges. Even when taken together, the chimpanzee and gorilla occupy an area of central Africa smaller than the United States. The primary reason for these variations in distribution is climate. Although

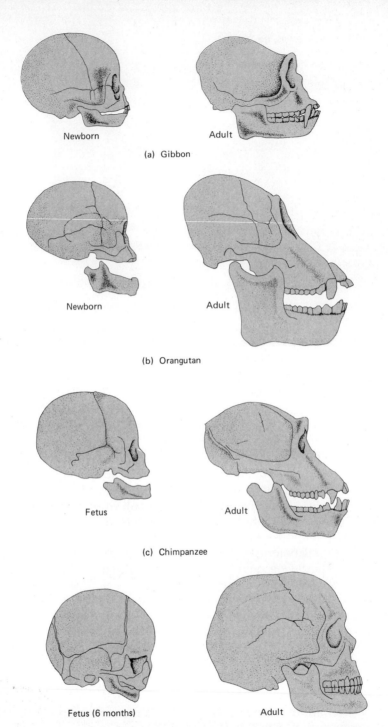

Newborn   Adult

(a) Gibbon

Newborn   Adult

(b) Orangutan

Fetus   Adult

(c) Chimpanzee

Fetus (6 months)   Adult

(d) Human

FIGURE 9-9 Changes with age in hominoid skulls.
Source (for apes): Adolph Schultz. 1956. *Primatologia*. Vol. 1. Basel: S. Karger AG.
(By permission of the author and publisher.)

there are some exceptions, apes are basically inhabitants of tropical rain forests and woodlands. Yet apes do not seem to be confined to such regions by direct climatic effects, such as a need by the animals themselves for warm temperatures. Some gorilla populations are found on mountain slopes where the temperature drops as low as 10°C (50°F). Instead, climate appears to act indirectly by governing the kinds of food and shelter available. In all likelihood, apes could not live very long where a prolonged period of cold like that which characterizes winter in temperate areas causes trees to shed their leaves.

A second important factor limiting their distribution is the very success of our own species in expanding its range, a process that has been continuing for a very long time. Forest trees are felled with chain saws, as they were earlier with axes. Where once gorillas fed alone, cattle are now permitted to pasture. Paths, roads, and highways divide what were formerly wide expanses into ever-smaller blocks. People passing along these routes sometimes spread infectious diseases, including polio, to their nonhuman primate relatives. The shrinking domain of the apes is a predictable consequence of our alteration of nature. Fossil evidence indicates that the area inhabited by apes during some eras in the past was many times greater than it is at the present, and even without that evidence we can see that this is possible. The map in Figure 9-10, for example, shows the present distribution of the gorilla. Since the areas in which it is found are not connected, they must be remnants of a formerly larger territory.

In spite of their limited range, however, the remaining groups of gorillas still show considerable variation from area to area. What is more, these morphological variations are correlated with differences in significant features of the habitat, including altitude, temperature, and rainfall. Distinctive regional patterns have evolved in response to selection pressures caused by these environmental differences. The effect of these differences is intensified by an absence of or reduction in gene flow between regional populations which are now separated by the Congo Basin. The two groups separated from each other most in distance, ecology, and appearance are the western lowland gorillas and the eastern mountain gorillas, and they differ in a whole series of characteristics.

The mountain gorilla is about 5.1 cm (2 in) taller and 16 kg (35 lb) heavier. Its chest is broader, and it has a longer trunk and neck and shorter limbs. The hands, too, are shorter and broader, and the toes are more frequently joined to one another by webs of skin. Both skin and hair are usually jet black in color (as opposed to the grayish or brownish tint of the western lowland form), and the hair is longer. Examination of a mountain gorilla's skeleton usually reveals a longer and narrower pelvis, a spine with a slightly lower average number of vertebrae in the thoracic and lumbar regions, and a scapula with a more sinuously

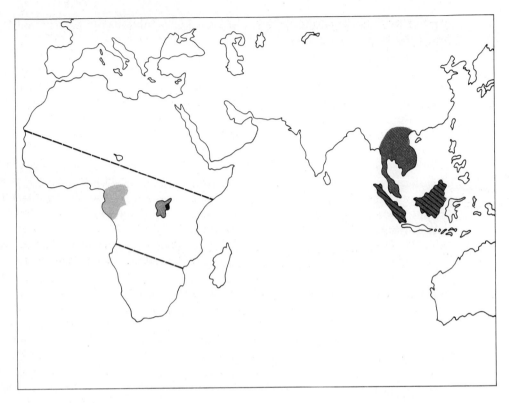

Key:

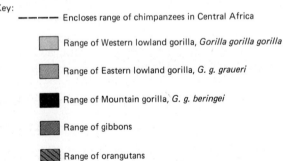

— — — — — Encloses range of chimpanzees in Central Africa

Range of Western lowland gorilla, *Gorilla gorilla gorilla*

Range of Eastern lowland gorilla, *G. g. graueri*

Range of Mountain gorilla, *G. g. beringei*

Range of gibbons

Range of orangutans

FIGURE 9-10 Present distribution of hominoids.
Sources: *Gorilla*, Colin P. Groves. 1970. "Population Systematics of the Gorilla," *Journal of Zoology*, **161**:287–300. *All others*, J. R. Napier and P. H. Napier. 1967. *A Handbook of Living Primates*. London: Academic Press. (By permission of the author and publisher.)

curved border on the edge adjacent to the spine. In addition, its skull is narrower, and it has a higher face and a longer palate. The lower face is pierced near the chin by an opening that admits a nerve serving some of the teeth. This inlet is known as a *mental foramen* (from the Latin *mentum*, "chin," and *foramen*, "hole"). In the mountain gorilla there are frequently several of these openings, rather than just the one found in most lowland animals. Some of these differences are illustrated in Figure 9-11.

In addition to the western lowland and eastern mountain gorillas, there is also an eastern lowland group (Groves, 1967). Members of this third group are somewhat intermediate in appearance, showing on the whole a mixture of the features found in the other two. They resemble the mountain gorillas more in size, coloration, and nose form, but their foot and scapula are much like those of the western lowland population.

All the differences just discussed are of relatively small degrees, and there is significant variation within each local population in most features. Because the populations differ only on the average, an animal from any one region should not be expected to show the whole array of distinctive characteristics. In the past, when fewer specimens were available for study and ranges and patterns of variation were only incompletely known, newly discovered specimens were often classified as members of different species or even different genera. As we now know, morphological distinctiveness does not in itself imply genetic isolation, nor can it be used as a sure guide that such isolation existed.

FIGURE 9-11 Some characteristics differentiating western lowland gorillas from eastern mountain gorillas.

Sources: (1) Colin P. Groves. 1970. "Population Systematics of the Gorilla," *Journal of Zoology,* **161**:287–300. (2) Adolph Schultz. 1934. "Some Distinguishing Characters of the Mountain Gorilla," *Journal of Mammology,* **15**:51–61. (3) C. Vogel. 1961. "Zur systematischen Untergliederungen der Gattung *Gorilla* anhand von Untersuchungen der Mandibel," *Zeitschrift für Säugetierkunde,* **26**:1–12.

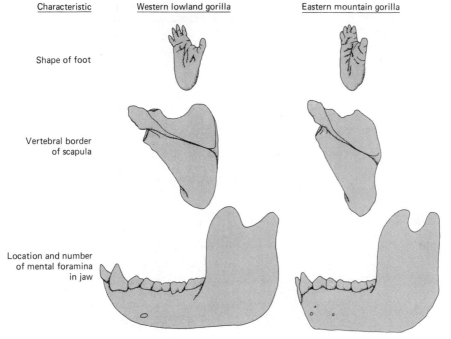

| Characteristic | Western lowland gorilla | Eastern mountain gorilla |

Shape of foot

Vertebral border of scapula

Location and number of mental foramina in jaw

Atypical, Abnormal, and Pathological Variations

It is somewhat difficult to say what constitutes abnormal variation, chiefly because it is not always easy to define the ranges of normality. Nature scarcely ever uses the same mold twice, and this is particularly true in the case of the primates. Extreme cases, like the chimpanzee suffering from a bone disease (probably rickets) in Figure 9-12, present little problem. The same cannot be said for the gorilla whose lower jaw is shown below in Figure 9-13; here the variation is very small. We usually accept as normal those traits which are found in a majority of the individuals examined. A good example of this is the dental formulas given earlier, which provide basic characteristics for classification. Yet when large enough samples are examined, even these formulas show some variation, as Table 9-8 shows.

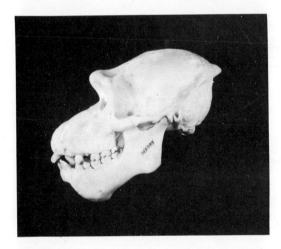

FIGURE 9-12 An extreme case of abnormal variation: an abnormality of bone growth has substantially changed the shape and proportion of the skull. Scale: ¼ natural size.

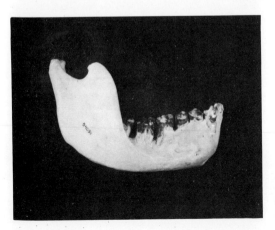

FIGURE 9-13 Slight variation (reduction of canine teeth so that they do not project) in the lower jaw of a female gorilla. Scale: ¼ natural size.

TABLE 9-8 Variations in numbers of teeth in hominoid primates

Common name of genus	Number of specimens	Extra teeth						Absence of teeth					Total number of specimens varying	Percentage of total number of specimens varying
		Class of tooth varying				Number of specimens varying	Percentage of specimens varying	Class of tooth varying			Number of specimens varying	Percentage of specimens varying		
		I	C	P	M			I	P	M				
Gibbon														
Skulls with mandibles	306	1	—	—	1	2	0.7	3	1	7	11	3.6	13	4.2
Orangutan														
Skulls with mandibles	229	—	1	—	16	17	6.8	1	1	5	7	2.4	25	8.5
Skulls only	38	—	—	—	—	—		—	—	—				
Mandibles only	28	—	1	—	2	3		—	—	—				
Chimpanzee														
Skulls with mandibles	467	3	—	1	13	15	2.9	—	2	8	9	1.6	24	4.3
Skulls only	93	—	—	—	1	1		—	—	—				
Gorilla														
Skulls with mandibles	546	6	—	—	22	27	4.4	2	1	1	4	0.6	34	5.0
Skulls only	129	—	—	—	3	3		—	—	—				

Source: F. Colyer. 1936. *Variations and Diseases of the Teeth of Animals.* London: © John Bale, Sons, and Danielsson. By permission of the author and publisher.

Change through Time

Abnormal variation overlaps and merges into the other categories previously discussed. This should not be surprising; if we believe that evolutionary change can and does occur through time, then at some point a trait that began as a deviation from the norm in a single individual must become more frequent, and possibly more pronounced, as it is acted on by evolutionary forces. If we could have observed the earlier course of primate and human evolution, we might have seen some individual aberrations become local geographic variants and then eventually turn into characteristics typical of a whole species.

How rapidly can such changes occur? Our information on this point comes largely from studies on populations of domestic and laboratory animals. When such populations have been subjected to artificial selection, in some cases the mean of the characteristic selected for has been changed by as much as a full phenotypic standard deviation from the original condition in as few as 10 to 20 generations. In the next chapter we will look at the ways in which living primates adapt to their environments—at how change through time can fit them for their ways of life.

SUMMARY

1. Populations of living primates can be classified into various nested categories on the basis of their similarities and differences.

2. The species is the only one of these taxonomic categories that can be defined and measured objectively; assignment of a species to all the other high taxonomic categories, from the genus up, involves more subjective judgments.

3. Living primates are extremely diverse, ranging from some insectivore-like prosimians to humans with complex behavior patterns and an elaborate material culture.

4. Despite their diversity, members of the primate order are united by a number of common trends: preservation of a primitive limb structure, increased manipulative ability, replacement of claws by nails, increased importance of the visual sense, decreased importance of the sense of smell, reduction in the number of teeth, shortening of the snout, increased brain size, and increased duration of the gestation period and other phases of the life cycle.

5. Within each species of primate there is considerable morphological variation among individuals of a single generation due to differences in sex, age, and geographic origin, as well as the influence of various diseases and accidents; in addition, evolutionary change through time can dramatically extend the range of variation within a sample composed of several successive generations.

6. Although primate species and other taxonomic categories are defined on the basis of certain common characteristics, many of these characteristics vary within species as well.

SUGGESTIONS FOR ADDITIONAL READING

Clark, W. E. LeGros. 1971. *The Antecedents of Man.* (Rev. ed.) Chicago, Ill.: Quadrangle. (This is the revised edition of one of the most informative basic handbooks of primate biology.)

Eimerl, S., and I. DeVore. 1966. *The Primates.* New York: Time-Life. (This is an elementary-level text; it is, however, well written, and the illustrations are superb.)

Napier, J. R., and P. H. Napier. 1967. *A Handbook of Living Primates.* New York: Academic. (This is the most complete and reliable modern handbook of the primates now available; each order is illustrated and discussed in detail.)

REFERENCES

Boyd, W. C. 1950. *Genetics and the Races of Man.* Boston: Heath.

Clark, W. E. LeGros. 1959. *The Antecedents of Man.* Edinburgh: Edinburgh University Press.

Groves, C. P. 1967. "Ecology and Taxonomy of the Gorilla," *Nature,* **213**(5079):890–893.

Huxley, T. H. 1863. *Evidence as to Man's Place in Nature.* London: Williams Norgate.

Martin, R. D. (1968). "Towards a New Definition of Primates," *Man,* **3**(3):377–401.

Mivart, St. George. 1863. "On *Lepilemur* and *Cheirogaleus* and on the Zoological Rank of the Lemuroidea," *Proceedings of the Zoological Society of London,* 484–510.

Napier, J. R., and P. H. Napier. 1967. *A Handbook of Living Primates.* New York: Academic.

Schultz, A. 1968. "The Recent Hominoid Primates," in S. L. Washburn and P. Jay (eds.), *Perspectives on Human Evolution.* Vol. 1. New York: Holt, Rinehart and Winston. Pp. 122–195.

Schwartz, J. H., I. Tattersall, and N. Eldredge. In press. "Phylogeny and Classification of the Primates Revisited."

Simpson, G. G. 1961. *Principles of Animal Taxonomy.* New York: Columbia University Press.

Simpson, G. G., and W. S. Beck. 1965. *Life: An Introduction to Biology.* New York: Harcourt, Brace, and World.

Chapter 10

![banner]

Primate Ecology and Behavior

In Chapter 9 we looked at some of the major anatomical features of living primates, including humans, and at the possible range of variations within species. If we look now at a greater variety of living primates against the background of the environments in which they live, we can more easily see and understand the relationships among structure, behavior, and environment. This is much the same type of approach we used in looking at human adaptability in Chapter 8.

Each species of nonhuman primate occupies a much more restricted geographic area than we do, in large part because humans rely much more on culture and other nongenetic adaptive mechanisms. However, this difference is only one of degree; not all the characteristics of nonhuman primates are strictly genetically determined. Our knowledge of genetic variation and adaptation in nonhuman primates does not yet compare with the wealth of data we have for humans, but an increasing number of studies in the field and in the laboratory are beginning to document the range of strategies used by various species. Perhaps the most important finding so far is that behavior provides a flexible interface between a population and its environment in all species of our order.

Although scientists usually study nonhuman primates under controlled conditions in the laboratory or more naturalistic conditions in the field, most people see these animals only in zoos. But even under these less-than-ideal conditions, some useful observations can be made. One can feel a kinship with distantly related lemurs while watching several juveniles chasing one another about in play or while observing a few adults huddled together resting or grooming one another's fur. The locomotor function of a gibbon's long arms becomes strikingly evident from the daring drops and soaring leaps the animal can make from bar to bar or tree to tree; few other creatures move with such grace and speed. The intense flush of a male mandrill's brightly colored face mirrors the strong emotion evident in his stiff-legged stance and intense stare at the humans just in front of his cage. The bluffing charge of an adolescent gorilla bears a close resemblance to the jostling and banter of the teenagers watching him.

Wherever and however it is gathered, information about the behavior of nonhuman primates is not only of interest in itself but also of direct use in the study of human evolution. Modern humans are the surviving tip of a branch of evolution that goes back at least several million years. The behavior of our human ancestors can never be observed directly; it is gone, and only traces remain in the bone and stone of the fossil record. Nor does any living primate species preserve completely unchanged the behavior characteristic of an actual stage in our evolution. But the study of primates now living in areas like those in which the evolutionary divergence of the earliest humans may have occurred (tropical *savannas,* which are dry grassland areas with a few scattered trees, or their woodland fringes) may give us additional clues to past human behavior. Let us begin by looking at some of the major environmental features of the regions inhabited by nonhuman primates and at a few of the adaptive responses of these species.

There are many different species of living primates, each genetically unique and nearly all phenotypically distinctive as well. The characteristics that we see have been shaped directly and indirectly by the environments which the living primates and their ancestors have inhabited for many generations. Only by studying the interaction between present primate populations and their environments can we come to better understand the complex influences that have shaped past evolution in our order. In this chapter we'll look at some of the major features of the surroundings in which primates live so that we can grasp the behavioral implications of the structural features that were presented in Chapter 9.

DETERMINANTS OF PRIMATE DIVERSITY

Where They Live: Geographic Distribution

Primates, with few exceptions, are animals of the tropics; most live in or near the region around the equator bounded by the tropic of Cancer and the tropic of Capricorn. In Africa and South America, some range

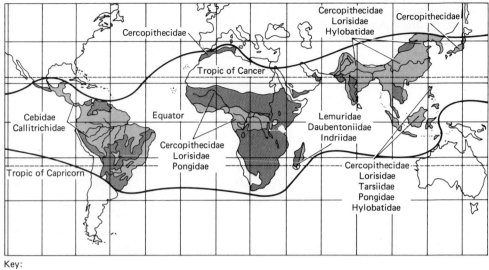

Key:

─── Primates are not found outside these limits except in small communities introduced by humans (e.g., Gibraltar, Mauritius)

[shaded] Forest, including tropical rainforest, swamp forest, mangrove, secondary forest, deciduous forest (temperate and tropical), monsoon forest, and montane forest

[dark] Grassland (savanna and steppe) including forest outliers, gallery forest, wooded steppe, thorn forest, Mediterranean scrub, montane meadow

[white] Desert, dry upland, or permanent snow

FIGURE 10-1   Distribution and habitats of living nonhuman primates.

Sources: (1) A. Jolly. 1972. *The Evolution of Primate Behavior.* New York: Macmillan. Fig. 3, p. 24. Modified from (2) J. R. Napier and P. Napier. 1967. *A Handbook of Living Primates.* New York: Academic Press. Figs. 4 and 5, following p. 378.

to 30° south latitude, and macaque monkeys in Japan are found up to 41° north latitude (see Figure 10-1). In higher latitudes there are fewer hours of sunlight per day, and the growing season of plants is shorter. These factors combine to reduce the availability of food, which is probably a more immediate limiting factor on the distribution of these animals than low temperature itself.

Within the broad belt inhabited by nonhuman primates, three rather distinct major regions can be recognized: (1) South America, (2) southeast Asia, and (3) Africa. The families Cebidae (New World monkeys) and Callithricidae (marmosets) are found only in South America, and within the African region Madagascar is an isolated subdivision. Aside from humans, the only primates found there now are members of the prosimian families Lemuridae (lemurs), Indridae (indris, sifakas, and avahis), and Daubentoniidae (aye-ayes); these live free nowhere else in the world. The families Lorisidae (lorises, pottos, and galagos), Cercopithecidae (Old World monkeys), and Pongidae (great apes) are found in both mainland Africa and Asia. However, Asia boasts a greater diversity; it has in addition the families Tupaiidae (tree shrews), Tarsiidae (tarsiers), and Hylobatidae (gibbons).

Conditions of Life: Primate Habitats

The geographic area that a group occupies is simply a map location. In contrast, a *habitat* is its physical environment—warm or cool, wet or dry, hilly or level. The habitat of a species may be quite uniform over its entire range, or it may differ somewhat from area to area. Each habitat contains a natural *community* of plant and animal species, and the whole set of functional relationships of any species to its habitat and community defines its *ecological niche:* food sources, predators, diseases, and other interactions (Whittaker, 1970).

The tropics as a region are no more uniform than the temperate zone familiar to most of us; variations in topography, temperature, rainfall, soil types, and many other factors combine to produce differences in plant communities. Although these communities grade into each other at their boundaries, four major types can be recognized: tropical rain forest, savanna, steppe, and desert. These are shown schematically in Figure 10-2, which represents a 2100-km (1300-mi) strip in west central Africa from the equator to the tropic of Cancer. Table 10-1 lists the main characteristics of these zones and the primates typically found in them.

FIGURE 10-2   Vegetational zones in tropical Africa.
Source: J. R. Napier and P. Napier. 1967. *A Handbook of Living Primates.* New York: Academic Press. Fig. 6, p. 379.

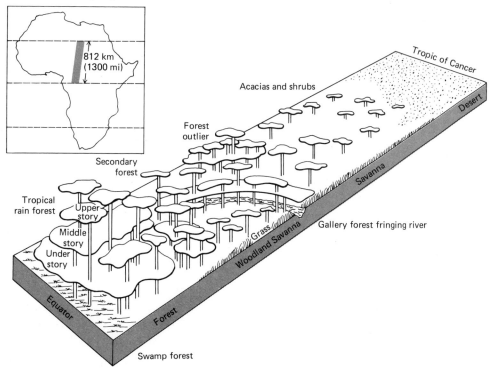

TABLE 10-1 Vegetation zones and primate faunas of tropical Africa

Vegetation zone	Subtype	Vegetation and climate	Primates
Type I: Tropical rain forest Alternative terms: Moist forest Lowland rain forest Tropical high forest Forêt dense		Three strata constituting an open and closed canopy with emergents Temperature—steady with narrow range Rainfall—high Relative humidity—high	Guenons except vervets Colobus monkeys Chimpanzees Gorillas Mangabeys Mandrills Pottos, golden pottos Bushbabies: Allen's, Demidoff's, needle-clawed Talapoins
	Mangrove	Specialized mangroves lining estuaries and creeks to tidal limits	De Brazza's guenons Allen's swamp guenons White-collared mangabeys
	Secondary forest	Tropical rain forest that has been cultivated and subsequently abandoned	
	Swamp forest	Similar but more open and irregular in structure	
	Montane rain forest Alternative terms: Highland forest Cloud forest	900–2400 m (3000–8000 ft)(depending on climatic conditions); varies from evergreen forest to woodland with tree fern and bamboo thickets.Lianas	Chimpanzees Gorillas Colobus monkeys Savanna baboons
	Bamboo forest	2100–3000 m (7000–10,000 ft); stands of bamboo from 6–11 m (20–35 ft); ground cover sparse	

Vegetation zone	Subtype	Vegetation and climate	Primates
Type II: Savanna Alternative terms: Sour veldt (South Africa) High grass (east Africa)	Woodland	Trees 6–15 m (20–50 ft) high Grass 2–5 m (6–15 ft) high	Vervets Patas monkeys Savanna baboons Bushbabies: Senegal and thick-tailed
	Open savanna	Trees widely spaced Grass 2–5 m (6–15 ft) high	
	Forest outliers Alternative terms: Bowl forest Kurmi Copses	Islands of tropical rain forest; occur in hollows and ravines where soil conditions are favorable	Vervets Colobus monkeys Savanna baboons Mangabeys Bushbabies Pottos
	Gallery forest Alternative terms: Riverine forest Fringing forest	Tropical rain forest on river banks	
Type III: Steppe Alternative terms: Thornland Sweet veldt (South Africa) Short grass (east Africa) Desert grass Orchard steppe	Wooded steppe	Open and closed woodlands or thickets Short grasses	Vervets Patas monkeys Savanna baboons Hamadryas baboons Geladas
Type IV: Desert		Sandy soil with rocky sublayer; sparse shrubs bloom during short, irregular rainy season	Savanna baboons (occasionally)

Sources: Modified from (1) A. Jolly. 1972. *The Evolution of Primate Behavior.* New York: Macmillan. Table 2, pp. 26, 27. After (2) J. R. Napier and P. Napier. 1967. *A Handbook of Living Primates.* New York: Academic Press. Table I, p. 381.

Each major vegetation zone can be divided into subtypes, as shown in Table 10-1. Some of this diversity is two-dimensional, and some is three-dimensional. The rivers that flow through open savanna, for example, may be lined by tall trees and other plants. A vegetation map of the area would show these gallery forests along the rivers, then woodlands, then open grassland. Different animals will be found in each of these habitats: savanna baboons in the dry, open areas; patas monkeys and thick-tailed bushbabies in the woodlands; and vervets, mangabeys, and pottos in the riverine forest. In a region of this sort, we could use a simple two-dimensional vegetation map to locate various nonhuman primates and to study the ecological factors that govern their distribution.

In a tropical forest, however, we need to consider and map the vertical dimension as well. The canopy of vegetation in tropical rain forests, which may extend as high as 61 m (200 ft) above the ground, can itself be divided into several "stories," or strata. These include an *understory,* with crowns of trees from 8 to 15 m (25 to 50 ft) above the ground; a *middle story,* extending from 8 to 37 m (50 to 120 ft); and an *upper story,* extending from 37 to 46 m (120 to 150 ft) or more. These levels differ in more than just height above the ground. In the understory, branches of adjacent trees interlace and are further bound together by vines, lianas, and other woody creepers into a *closed canopy.* In the middle story, crowns of trees often join each other, but they form a less continuous layer. In trees of the upper story, branches from different heights within the crown sweep upward to form wide, almost flat islands of greenery. And above them tower occasional *emergents,* taller trees up to 61 m (200 ft) in height.

Each story provides a different three-dimensional matrix for different activities—feeding, traveling, mating, fighting, resting, and sleeping. Most primates use more than one story because each differs in type, amount, and distribution of food. A species may feed in one story, travel through another, and rest or sleep in yet a third. But because primates are intelligent animals with flexible patterns of behavior, their populations are not scattered unpredictably through different forest layers.

For example, in some areas chimpanzees build sleeping nests in trees, may also be arboreal for one-half to three-quarters of the daylight hours, and yet come to the ground to feed on insects and even to flee from humans. The highly arboreal red colobus monkeys sleep in the upper story, travel from place to place through the upper and middle stories, and have been seen feeding at all three levels. Of course, less vertical variety is possible in savanna and steppe zones, but even there trees, when available, are used by ground-feeding baboons for sleep and refuge.

What They Live On: Primate Diets

Most primates are omnivorous; the bulk of their food comes from plants, supplemented by small but nutritionally significant amounts of animal protein in the form of insects or occasional birds, eggs, or small mammals. The actual range of diet, however, is considerable. At one extreme lie the aye-ayes, the mouse lemurs, the tarsiers, the lorises, the galagos, and the New World night monkeys and pygmy marmosets. All live mostly on insects, and all but the marmosets are nocturnal. Their feeding habits, however, are very different. Lorises can maneuver slowly through a tangle of flexible branches toward an insect, making scarcely perceptible movements until within snatching range. Their hip and ankle joints are structured to allow an extraordinarily great range of mobility, and a circulatory trick (called *retia mirabelia*—literally, "wonderful nets"—consisting of tangled bundles of veins and arteries) feeds a constant supply of fresh blood to the feet, which must hold firm while the rest of the body is carefully hauled into position. Tarsiers, in contrast, are able to pounce on their prey from several feet away. Their leaps are made possible by a special modification of the foot skeleton in which the tarsal segment (the instep, or arch, of the foot) is greatly elongated. Aye-ayes can also jump with agility from branch to branch like other lemurs, but while feeding they crawl slowly along a branch, tapping with the thin middle finger to locate insect burrows in the wood. Promising holes are probed with this wirelike finger, sometimes after the animal has enlarged them by gnawing busily with its front teeth, which are shaped like sharp, sturdy chisels. Continued growth of these teeth throughout life compensates for the wear on the cutting edges.

At the opposite extreme are the various leaf-eating primates. One or more species of these are found in each major geographic area: langurs in Asia, colobus monkeys and gorillas in Africa, indris on Madagascar, and howlers in South America. Adaptations among leaf eaters are not as diverse as those found among insect-eating primates. Most have an enlarged digestive tract with a multicompartmented stomach or longer intestines to accommodate the greater amounts of food that must be eaten (leaves have a lower concentration of nutrients than fruits, seeds, insects, or meat). And since leaves must be chopped or ground to break down their high content of fibrous material, dental modifications include pairs of sharp ridges or crests that run across the chewing teeth of many Old World monkeys, and the increased surface area of these same teeth in gorillas.

In feeding as well as other aspects of primate behavior, however, the key features are adaptability and flexibility. Tarsiers and other insect eaters do not eat only insects; they capture lizards and other small animals using much the same techniques. Grass-eating baboons sometimes catch and eat termites, as do the more omnivorous chimpanzees, who gather these choice morsels more efficiently than baboons do. Instead of grabbing at the individual insects as they emerge, chimps insert a stripped twig or grass stem into a hole in the termite nest. The termites attack the intruding object and hold onto it with their mandibles even as it is withdrawn, only to be stripped off by the mobile lips of the waiting chimp. (See "Chimpanzee Tool Use" on pages 324–325 for other examples.)

Much the same behavioral flexibility is shown when primates eat meat. Although meat is not a major component of the primate diet, it is eaten with apparent relish when found. A monkey feeding on fruit may discover and pillage a bird's nest concealed among the leaves. A grazing baboon may find, kill, and eat a hare or young antelope lying hidden in the grass. Baboons have also been observed to hunt antelope, sheep, and other animals. Most often, young or half-grown animals are taken by an adult male or female baboon that ambles into a herd without causing panic. The victim is seized by a hind leg, flipped onto its back, disemboweled, and partially eaten while still alive (Dart, 1963). This hunting may occur because of nutritional needs, scarcity of more traditional foods, or other factors. A similar kind of hunting also occurs among chimpanzees. ("Chimpanzee Hunting," pages 326–327, discusses the significance of this hunting behavior for human evolution.)

How They Move About: Primate Locomotion

All primates are capable of a wide range of limb movements, but members of a given species commonly use a limited subset of the total possible range. These dominant patterns enable us to categorize primates by method of locomotion (see Table 10-2). The primates in any one category typically execute the given range of activities with greater ease than members of most other species.

Locomotor categories are not just styles of movement used in getting from place to place; they can also be correlated with habitats and the food sources utilized in them. *Vertical clinging and leaping* is characteristic of a number of prosimians. Some of them feed on insects, and others on leaves, but all share small body size and a tendency not to venture often to the ground, where predators lurk. Their locomotor style could make it possible for them to cross gaps from one arboreal feeding area to the next in safety. To choose another example, *quadrupedal climbing* (using all four legs while the body is more or less horizontal) is good for reaching fruits, leaves, and nests located along sturdy branches in the forest canopy. But slender branch ends would be off limits to all but the smallest arboreal quadrupeds. Larger animals reach flowers and tender buds that grow at the tips of branches by using the arm-hanging posture of *brachiation*.

TABLE 10-2 Locomotor classification of the primates

Category	Subtype	Activity	Primates
1. Vertical clinging and leaping		Leaping in trees and hopping on the ground	Bushbabies, avahl, hapalemurs, lepilemurs, sifakas, indris, tarsiers
2. Quadrupedalism	(i) Slow-climbing type	Cautious climbing—no leaping or branch running	Pottos, goldon pottos, slow lorises, slender lorises
	(ii) Branch-running and branch-walking type	Climbing, springing, branch running, and jumping	Mouse lemurs, dwarf lemurs, fork-marked lemurs, lemurs, marmosets, tarmarins, titis, sakis, uakaris, night monkeys, cebus monkeys, squirrel monkeys, guenons
	(iii) Ground-running and ground-walking type	Climbing, ground running	Macaques, baboons, mandrills, geladas, patas monkeys
	(iv) New World semibrachiation type	Arm-swinging with use of prehensile tail; little leaping	Howler monkeys, spider monkeys, woolly spider monkeys, woolly monkeys
	(v) Old World semibrachiation type	Arm-swinging and leaping	Colobus monkeys, proboscis monkeys, all langurs
3. Brachiation	(i) True brachiation	Gibbon type of brachiation	Gibbons, siamangs
	(ii) Modified brachiation	Occasional brachiation, climbing, knuckle-walking	Chimpanzees, gorillas, orangutans
4. Bipedalism		Striding	Humans

Sources: Modified from (1) J. R. Napier and P. Napier. 1967. *A Handbook of Living Primates.* New York: Academic Press. Table I, p. 385. (2) A. Jolly. 1972. *The Evolution of Primate Behavior.* New York: Macmillan. Table 4, p. 35.

CHIMPANZEE TOOL USE

Many species of monkeys and apes have been observed using tools in captivity or in the wild. It is evident that the use of natural objects to perform certain tasks, sometimes after modification to suit them better to the job required, is a behavior pattern distributed widely among nonhuman primates. Because chimpanzees are believed to be especially closely related to humans, tool use by these great apes has been studied with particular attention.

Since the classic laboratory studies begun by Wolfgang Köhler in the 1920s, chimpanzees have been known to use available materials for a variety of purposes: sticks have been used to pry open box lids or to grub in the ground for roots; straw, cloth, or paper has been piled up in sleeping places; and feces or clay has been smeared onto cage walls or onto the animals themselves. Although some of these activities might have represented attempts by bored animals to occupy their time, others could have been explained in functional terms from even the little that was known at that time about apes living in their natural habitats. Free-living chimpanzees were reported to make nests in trees. Furthermore, in their arboreal locomotion these large apes would have to gauge the strength of limbs, branches, and vines and to manipulate these as weight was shifted from one point to another. In effect, chimps spend a good part of their time moving about in a three-dimensional maze of ropes and levers.

The tool-using capacities of chimps were tested quite systematically by Köhler and other scientists. Many of the animals proved to be quite able to use sticks to work switches or to draw in food that was otherwise beyond their reach. Beyond this, some tests uncovered actual toolmaking abilities. Chimps were able to join several short sections of tube together to make a long pole, and they could uncoil a length of wire to get a piece long enough to hook in desired objects on the other side of a fence. When food was concealed within a narrow pipe, one female proceeded to chew a piece of wood into splinters thin enough to tease out the prize. Solutions to some of these problems were reached only after repeated trials and errors, but others clearly involved an insightful grasp of novel approaches that had not been tried before. Faced with bananas hung out of reach over his head, one chimp led the investigator under the fruit and then tried to climb up him to reach it!

In recent years field studies have provided numerous well-documented instances of chimpanzees modifying and using natural objects in the wild. Some of these simple devices are really multipurpose tools. Thus twigs are used not only to capture termites but also to extract honey from the underground nests of bees and to probe into crevices in tree limbs to search for insects. Larger sticks may also be used to poke at objects of interest (such as dead snakes) and to pry open narrow

(a)   (b)

(a) This chimpanzee is inserting a grass stem into the opening of a nest of termites; the termites will seize the stem and allow themselves to be drawn into the chimp's waiting lips. (b) Here the chimp is using a sponge made of chewed leaves to extract water from a small hollow in the tree trunk.

Source: J. van Lawick-Goodall and H. van Lawick. 1965. "New Discoveries among Africa's Chimpanzees," *National Geographic*, **128**(6):802–831. Pp. 827, 829. Photographs, Hugo van Lawick; © National Geographic Society.

crevices, and on occasion they are brandished in threat displays. Leaves, too, are used in a variety of ways, for example, to wipe off dirt, excrement, food residue, or blood. Chimps have also been observed to briefly chew a mouthful of leaves and then dip the crumpled mass into a small water-containing hollow in a tree trunk. The crude sponge was found to pick up eight times as much water on a single dip as the animal's bare fingers.

As a result of these observations, tool use and toolmaking can no longer be considered to be exclusively human endeavors requiring a brain the size of ours to accomplish. Instead, these activities are part of a behavioral complex that we share with other advanced nonhuman primates—and quite possibly with prehumans as well.

Sources: (1) J. van Lawick–Goodall and H. van Lawick. 1965. "New Discoveries among Africa's Chimpanzees," *National Geographic*, **128**(6):802–831. (2) J. van Lawick–Goodall. 1971. *In the Shadow of Man*. New York: Dell.

CHIMPANZEE HUNTING

Like tool use, hunting is an activity formerly believed to be limited to humans among the primates but now known to occur in other primate species as well.

In terms of nutritional impact it would probably be artificial to make a distinction between hunting and the catching of insects by some prosimians, the collecting of nestlings by many monkeys, or the taking of small mammals (hares, young antelope) by baboons as these ground-feeding primates chance on them while foraging for dietary staples such as grass stems. In all these cases small amounts of animal protein are added to a basically vegetarian diet. But when animal protein is obtained by some larger primate species (baboons, chimps) through hunting activities, different behaviors are involved, and these have different evolutionary implications from those of collecting activities.

Regular observations over the course of a decade at the Gombe National Park in Tanzania have documented nearly a hundred cases in which chimpanzees killed and ate other mammals, plus over one-third

The male chimpanzee in this drawing has just captured a young monkey, which it is killing by smashing it against the ground. Consumption of the prey will follow swiftly.

Source: Redrawn from J. van Lawick-Goodall. 1971. *In the Shadow of Man.* New York: Dell.

as many other instances in which capture of prey was unsuccessfully attempted by chimps. A number of features were common to all these occurrences. To begin with, the prey was limited to relatively small (about 9 kg, or 20 lb) animals such as newborn or very young bush pigs, infant and juvenile baboons, and adult monkeys of smaller species (such as *Colobus*). These were hunted by adult chimpanzees of both sexes, with the hunters working alone or in small groups of up to five animals.

In group hunts coordination of activities was evident (for example, one chimp would pursue the prey while others blocked likely escape routes), but the means of communication were not clear; there was no evidence of vocal signals or hand gestures. Whether a single animal or a group was involved, hunts generally followed a set sequence with three phases: *pursuit* (which included quick seizure of the prey, sometimes preceded by a longer period of chasing or stalking), *capture* (which ended with the tearing apart of the prey, a common method of simultaneously killing and dividing the prey among the hunters), and *consumption* (first by the successful hunter or hunters, and then by others who begged for scraps of meat).

Several features of the consumption phase are of particular interest. As a rule, the entire carcass of the captured animal is eaten, including viscera, bones, skin, and hair; quite often wads of leaves are chewed along with the meat and other parts of the kill, perhaps to prolong the meal. Rudimentary tool use may also be involved here; occasionally, wads of leaves are used to mop out the last vestiges of brain from an opened skull. In any case, it is evident that the meat is not only actively sought but also savored. Another indication of the value placed on meat is the degree of sharing that goes on. An animal more dominant than the hunter may snatch the prey, but this is rare. It is far more common for other chimpanzees, even those far higher in status, to search on the ground for dropped scraps or to beg for pieces from the possessor.

In many cases hunting temporarily alters the usual social order in the group, so that the successful hunter acquires what might be called *situational dominance* (temporarily higher status in a particular context). It is at least possible that repeated success in hunting could lead to generally higher status and the enhanced reproductive potential that this often confers. Hunting is probably too infrequent an event among present chimpanzees for this to have much effect in selecting for whatever physical or behavioral traits enhance success in hunting. However, if circumstances arose in which chimpanzees or animals like them would be increasingly dependent on meat for survival, it is not difficult to see how characteristics related to hunting behavior could be selected for.

Sources: (1) J. van Lawick–Goodall. 1971. *In the Shadow of Man*. New York: Dell. (2) G. Teleki. 1973. "The Omnivorous Chimpanzee," *Scientific American*, **228**(1):32–42.

Many locomotor categories can be correlated with anatomical modifications visible even in the skeleton. The elongated middle section of the tarsier's foot, mentioned earlier, marks this animal as an extreme vertical clinger and leaper, and as a rule vertical clingers and leapers have hindlimbs that are longer than their forelimbs. The spine of the slow-climbing loris lacks an *anticlinal vertebra,* which normally forms the keystone in the arched spine of more active quadrupedal springers. Brachiators have longer front than hind limbs; the reverse is true of vertical clingers and leapers. The longer forelimbs give them a longer span between hands, permitting a wider gap to be covered in each hand-over-hand swing. This span is further expanded by the shape of the chest, which in brachiators is widened from side to side (as in humans) rather than deep from front to back (as in most other mammals, such as domestic cats). Even the hands of some part-time brachiators are modified. Thumbs are often short in relation to the other fingers, which form a branch-grasping hook when the arm is extended. Gorillas and chimpanzees are called *knuckle-walkers* because when on all fours on the ground, they support the front of the body on these turned-under fingers. Structure, ecology, and behavior, then, seem to be strongly interrelated. Because of such relationships, it is possible to reconstruct many aspects of a primate's behavior from selected portions of its skeleton.

MAJOR GROUPS OF LIVING PRIMATES

In the preceding sections we surveyed some of the diverse ecological and behavioral dimensions of primate adaptation. From this basis we'll move on to an overview of the ways of life of a number of representative groups of living primates. The aim is to see what problems are posed by these life-styles and how the animals meet these challenges.

Taxonomic names, although they are sometimes relics of earlier mistakes in diagnosis, often convey useful information in a shorthand form. The names of the two primate suborders—Anthropoidea and Prosimii—are cases in point. The monkeys and apes of the Anthropoidea are rather like humans in their biological characteristics. The Prosimii (premonkeys) represent a sort of halfway point between higher primates and other mammals such as the insectivores; as the name suggests, prosimians preceded monkeys in time (see Chapter 12).

The reason for looking at the ways of life of a few of the prosimians here is to get some idea of the characteristics that might have been found in the ancestors from which higher primates arose. As you will see, generalizations will not come easily because the prosimians are divided into three infraorders (Lemuriformes, Lorisiformes, and Tarsiiformes), whose members show as great diversity in ecology and behavior as they do in physical form. However, in addition to adaptive diversity to fit a number of different life-styles, living prosimians do show some differences in evolutionary grade, ranging from the debatably primate tree shrews to the nearly anthropoid tarsiers.

Prosimii

Prosimians existed for millions of years before any members of Anthropoidea evolved, and their long history is one reason why the animals of this suborder are adaptively so diverse. Many species are also highly specialized to fit very narrow environmental niches. In some cases, there is an evident trade-off between anatomical specialization and behavioral flexibility. Prosimian primates typically have brains that are smaller and simpler than those of monkeys, apes, or humans, and their behavior patterns are more stereotyped and less open to modification through individual experience. Perhaps this is why prosimians have continued to survive in settings effectively isolated from the competition of higher primates.

Lemuriformes. The lemurs (superfamily Lemuroidea), for example, are found only on Madagascar and the immediately adjacent Comoro Islands, which separated from the continent of Africa long before primates branched off from other mammals. Perhaps a few early primates accidentally crossed the Mozambique Channel on rafts of floating vegetation. These early colonizers might have resembled the present-day mouse lemur (*Microcebus*), which is chiefly insectivorous but has a skull strikingly similar to that of the tree shrew, which is either a very primitive primate or quite a close relative. Once the colonizers were on Madagascar, repeated speciation (splitting into two or more closed gene pools—see Chapter 11) produced the great variety of known lemurs, which were able to flourish in the absence of competition from any higher primates until the arrival of humans several thousand years ago. In addition to the mouse lemur and the aye-aye, mentioned earlier, the range of living species includes quadrupedal leaf and fruit eaters of the genera *Hapalemur* and *Lemur* and some large, long-legged vertical clingers and leapers of the genera *Indri* and *Propithecus*. Some extinct forms (discussed in Chapter 12) extended the range of diversity even more. Social behavior within this superfamily ranges from that of the nearly solitary nocturnal dwarf and mouse lemurs, through the small family groups among the indris, to the partly terrestrial diurnal lemurs, which live in groups of up to several dozen members of mixed sexes and ages. Most lemurs are territorial; solitary individuals or social groups scent-mark their territory with urine or with oily secretions from special glands on the arm, chest, or neck. These territories may be vigorously defended against outsiders by bluff or fighting.

Lorisiformes.   The *Lorisiformes* are small, relatively simple-brained prosimians of mainland Africa and Asia. Most are slow-moving quadrupeds that clamber about the trees in search of insects and small animals, a diet they supplement with fruit and leaves. The African galago (bushbaby), however, is a much more active vertical clinger and leaper. The social behavior of most of these animals tends to be solitary. Although not geographically isolated from higher primates, the lorises, galagos, and pottos are protected from competition with them by a number of factors. They feed at night, whereas monkeys and apes feed during the day. In addition, they are smaller in size and more insectivorous, and they have different locomotor habits.

Tarsiiformes.   The Tarsiiformes (tarsiers) of southeast Asia are also active nocturnal animals. Their anatomical specialization for hopping locomotion, mentioned earlier, exceeds that of the galagos and enables them to pounce quickly on the insects and small animals that make up the bulk of their diet. In a number of characteristics—relatively large brain, forward-directed eyes protected in nearly complete sockets of bone, and smaller snout—tarsiers are the most advanced of living prosimians. The distance from tarsiers to some of the more primitive members of the suborder Anthropoidea is not a very great gap to have been bridged by past evolution.

Anthropoidea

The suborder Anthropoidea is divided taxonomically into the infraorders Catarrhini and Platyrrhini. This mirrors a geographic split between the higher primates of the Old World (chiefly Africa and Eurasia) and those of the New World (the living nonhuman primates of Central and South America). Differences in a few anatomical characteristics correspond to this geographic subdivision. One of these is the shape of the nose, which has rounded, divergent nostrils in New World monkeys (*platyrrhine,* "wide-nosed"); in Old World higher primates the nostrils are narrower and closer together (*catarrhine,* "narrow-nosed"). Another distinction, less readily visible in living animals but preserved in skeletons or fossils, is the number of teeth. The monkeys of Central and South America all have one extra premolar tooth in each quadrant of the jaw, for a total of four more than are found in any living Old World monkeys, apes, or humans. There are still other differences between the two, but they are not as universal. Asian and African monkeys, for example, lack the prehensile (grasping) tail found in some (but not all) New World primates. In fact, to the casual observer, there seems to be more of a similarity between Old and New World monkeys, which have been evolving separately for tens of millions of years, than among the monkeys, apes, and humans of the Old World, which may share closer common ancestors. Perhaps the colloquial term "monkey" in fact refers to a *grade* or stage attained independently more than once by the primates as they evolved from different prosimian ancestors.

Platyrrhini: The Howler Monkey.   Platyrrhine monkeys, as a group, tend to be more uniformly adapted to life in the trees than their Old World relatives, and for this reason one example will illustrate the general patterns and characteristics of this group of primates.

Howlers are monkeys of the genus *Alouatta* and were the first free-living primates to be studied by Clarence Ray Carpenter, a pioneer in modern methods of field investigation. The social groups he observed lived on the island of Barro Colorado in the Panama Canal Zone, a tiny and isolated part of the howler's total range, which extends from the coastal forests of Mexico through parts of Brazil. Carpenter's classic study (1934) remains a useful scientific document on the way of life of highly arboreal social primates in a prime tropical forest environment.

The forests provide ample food for these relatively large monkeys, which have an average body weight of nearly 7 kg (15 lb). They eat some buds, flowers, and fruit, but leaves make up most of their diet. Bulky, low-calorie vegetation is stored and digested in a rather large stomach. Howlers gather much of their food near the ends of branches, a potentially precarious task made somewhat safer by their ability to hold onto several branches at once by using their prehensile tails as well as their hands and feet. Howlers rarely venture to the ground. They even mate in the trees and give birth there, typically to one offspring, which the mother may nurse for up to 2 years. The young grow up in the supportive and instructive setting of a social group; adults of both sexes are known, for example, to help the young across wide gaps in the forest canopy.

Although social groups can range from 2 to 45 animals, the average group numbers 18; a typical composition would be 3 adult males, 8 mature females, 3 infants, and 4 juveniles. Lone adult males are sometimes seen and are apparently able to survive on their own. Each social group travels over a familiar area, which is known as the *home range*. Within this is a *territory* that is defended against intrusion by other groups of the same species. Territories are fluid and shifting; they are not fixed locations that can be neatly circled on a map. Howlers defend the spot where they are, and they tend to be in certain places more frequently than others. Defense rarely involves actual fighting. Two groups may engage in vigorous *display behavior:* shaking branches and breaking off twigs and leaves, all to the accompaniment of loud vocalizations, the characteristic howls for which these primates are known. Howlers give their calls regularly each dawn; these whoops and roars can carry for miles, and evidently they serve as a spacing mechanism that lets each group learn where other groups are so that they can be avoided.

Within any given social group, interactions tend to be low-key; conflict and fighting are rather rare when the animals are not crowded or short of food or other resources. Some group members show *dominance* over others; *subordinate* animals may be threatened or briefly attacked by higher-ranking individuals. There are separate *hierarchies* of social status or rank among males and females, with female ranks being less sharply defined. Overall, social cohesiveness prevails, probably because behavioral roles are learned and tend to remain fairly stable over long periods of time.

Catarrhini. The catarrhine primates include the superfamilies Hominoidea (humans and our closest relatives, the apes) and Cercopithecoidea (which contains many species of monkeys). Before moving on to a closeup view of our nearer kin, let's take a look at some of the groups of monkeys. While not as similar to us in structure as the apes are, Old World monkeys do share with humans some important behavioral characteristics such as living in organized social groups. In these groups behavioral interactions regulate access to food, sleeping places, mates, and other desirable features of life that are in short supply. Social behavior even leaves its mark on some physical characteristics, such as the relative body sizes of males and females. With this as an introduction to our review, let's look briefly at what the study of modern monkeys can tell us about our ancestors, and possibly about ourselves.

Cercopithecoidea. The superfamily Cercopithecoidea contains but a single extant family, the Cercopithecidae. The family in turn can be subdivided into two subfamilies: the Cercopithecinae and the Colobinae. Genetic evidence allows us to recognize two major groups among the cercopithecines: the genus *Cercopithecus* (with its close relative *Erythrocebus*) and all the others. The genus *Cercopithecus* is itself subdivided into nearly two dozen species. Most of these are quite arboreal, though the grivets (*C. aethiops*) are partially terrestrial in some areas, and the patas monkeys (usually placed in the genus *Erythrocebus*) are terrestrial animals with a light body and long legs for swift running. The members of different species are not very diverse anatomically, but they show a great range of diploid chromosome numbers, from 48 to a high of 72 (Chiarelli, 1975). These chromosomal differences may have been selected for in order to keep different species from interbreeding in areas where their ranges overlapped.

The remaining cercopithecine monkeys are split taxonomically into several genera and still more species. With the exception of the African mangabeys (genus *Cercocebus*), which include mostly arboreal species, they are all substantially terrestrial and highly social. The most widely distributed African genus is the common baboon, *Papio,* which ranges over nearly all sub-Saharan Africa. Drills and mandrills (genus *Mandrillus*) are found in the vicinity of Nigeria and Cameroun, and geladas (genus *Theropithecus*) in Ethiopia. Drills, mandrills, and geladas are also called *baboons* because of some characteristics they share with members of *Papio;* among these are a doglike projecting muzzle and front and back legs of about equal length. All baboons are relatively large primates, with adult weights in the range of about 10 to 40 kg (22 to 88 lb) and with most of the variation due to differences in population and sex. In some groups, males are twice as heavy as females and commonly have much larger, daggerlike canine teeth. The macaques (genus *Macaca*) extend the range of semiterrestrial and terrestrial monkeys still further. They range from north Africa, through much of Asia, to northern Japan; the Celebes black "ape" (*Cynopithecus*) is a tailless close relative of the macaque with a limited distribution in southeast Asia. Macaques are smaller than baboons, with a body-weight range of about 3 to 18 kg (approximately 6 to 40 lb). Like baboons, they have large canine teeth, particularly the males.

The wide geographic areas inhabited by mangabeys, macaques, and baboons, as well as differences in physical appearance and social behavior, provide some justification for the variety of taxonomic names assigned to them. But this diversity is much less evident at the genetic level: all have a common diploid chromosome number of 42, with further shared structural details in individual chromosomes. Pairs of many of the commonly recognized species and even genera have been mated in zoos and have produced viable and fertile offspring (Chiarelli, 1975). Because of their terrestrial existence, complexity of social behavior, and wide distribution, these populations of higher primates are about the closest nonhuman parallels we have to our own species. But their genetic interrelationships, although substantial, are still less close than those of geographically separated populations of humans; we are genetically far more similar than they.

Common baboons live in large troops that average from 30 to 50 members, although some have been observed to contain as many as 200 animals, including individuals of both sexes and of all ages. Within the troop, social relationships are structured by a dominance hierarchy that is much more pronounced than the one seen in arboreal monkeys such as howlers. The top ranks are usually held by adult males in their prime, with females and younger males occupying subordinate positions. In most areas these baboons take to the trees to seek safety and to sleep. During the day they range over the savanna, where they feed on grass stems, seeds, roots, and other vegetation, supplemented by occasional bits of animal protein ranging from insects to small mammals and birds. In open country, troop size and structure provide some protection against predators. On the move or at rest, the dominant males are concentrated in the middle of the troop near the more vulnerable females and the young. When challenged by a predator, the dominant males put themselves between the threatening animal and more vulnerable troop members (DeVore and Washburn, 1963). In all, the troop's social structure is protective, supportive, and tightly ordered. It must be. In this open terrestrial setting populated by ground-living predators such as the big cats, a lone baboon is a dead baboon.

In other habitats troops of the same species show variations on this aspect of social life. Harding (1977) recently described the behavior of olive baboons that range over a large cattle ranch near Gilgil, Kenya, on the floor of the Great Rift Valley. Ranchers in this arid region had greatly reduced populations of predators in order to protect the cattle. Nearly all the lions had been shot, and as many of the leopards as possible had been live-trapped and removed to the national parks. In the pattern of progression across open territory that Harding saw, the leading section of the baboon troop tended to be composed of adult males, adult females, and subadult males; juveniles and infants clustered toward the middle, and adult females were in the last segment of the troop. The same pattern has been seen by primatologists studying baboons elsewhere in Africa. The grouping behavior of baboons while moving is evidently not a rigidly fixed, species-specific characteristic. Certain patterns exist as possibilities and are used in some settings but not in others. The determining factors in a given situation probably include the experience of a troop with its habitat and the temperaments of different individuals, which are likely to be conditioned by the ages and sexes of the animals. Adult males, with their large canine teeth and body size about twice that of females, are more likely to take risks in a threatening situation. The behaviors that are important in allowing baboons to survive seem to be as open and variable as the habitats they occupy.

Hamadryas baboons (Kummer, 1968) live in the even drier environments found at the easternmost tip of central Africa and across the Gulf of Aden into the southern part of the Arabian peninsula. In this habitat food is not abundant, and there are no high, strong trees. Since they have no suitable sleeping trees, troops consisting of several hundred animals congregate to sleep each night on ledges along steep cliffs. During the day, however, the troops split up into foraging units, each consisting of just one male (or rarely two) plus a few females. In addition to larger body size, males are distinguished by their large manes and side whiskers (which further enhance their bulk) and by their belligerent behavior. The females are kept within a few feet of the male by his constant attention, threats, and bites on the neck of those who stray. The resulting tightly knit daytime units are able to exploit scarce and scattered resources that could not support the entire troop together.

The geladas of the Ethiopian highlands, in a habitat like that of the hamadryas baboons, have a similar social structure; large herds at night on sleeping cliffs and one-male daytime foraging units (Crook, 1966). The gelada's diet is high in small, tough morsels such as seeds, stems, and bulbs. To process these items, the canines have become reduced in size, and the chewing surfaces of the cheek teeth (premolars and molars) have expanded; C. Jolly (1970) has called this set of features the *T-complex,* which he views as a specialization for grinding abrasive foods.

Over the great expanse of Africa and Asia where they are found, macaques and baboons are divided by many taxonomists into about 18 different species. Many of these species interbreed in zoos, and a few are known to do so in nature, producing viable, fertile offspring. However, even if the number of distinct, closed gene pools of these ground-living Old World monkeys is fewer than the number of species recognized by taxonomists, it is still clear that the macaque and baboon populations are much more diverse genetically and phenotypically than the human populations that are distributed over the same areas of Africa and Asia. Why? Spears, guns, fire, and fences protect us from potential predators, and so we do not need to sleep sitting up on tree branches rough enough or rocks hard enough to require the evolution of special ischial callosities on our buttocks. Agriculture, irrigation, and transport systems of roads and railroads make the diets of humans far more similar than those of the monkeys dependent on gathering what grows in a territory consisting of a few square miles; in response, our teeth and skulls show much less variation. These Old World monkeys give us an indirect measure, then, of the extent to which culture blunts the impact of the natural environment on the human gene pool.

The Colobinae are relatively large, predominantly arboreal monkeys. Most of them, commonly known as *langurs,* live in the forests of Asia, although the genus *Colobus,* which gives its name to the subfamily, is native to Africa. In the African genus the thumb is reduced to a short stump, and the rest of the fingers are used together as a hook for swinging beneath branches. The diets of all these related Asian and African forms have a high component of leaves, which are digested in a stomach that is large and divided into several somewhat separated sacks. It is a bit like the stomach of a cow or other ruminant, and it serves the same function: to aid in the digestion of a bulky diet with a low concentration of nutrients.

Colobines live in social units that vary in size from species to species and within each species. Groups may have just one adult male or as many as 10; typically there are several adult females for each adult male and a variable number of offspring. One adult male serves as leader of the social group, selecting trees in which to feed and sleep and choosing the routes traveled over from place to place. The dominant position of the lead male is clear, and overall the social structure of groups appears to be stable, with little actual fighting. This social structure is similar to that of other arboreal monkeys such as howlers; in contrast, tense social relationships exist among terrestrial monkeys.

Hominoidea. This is the superfamily to which we belong, along with the true apes, all of which lack tails and are equipped with long arms, short legs, and broad chests. Of the Hominoidea, the most distantly related to us are the gibbons and siamangs of the family Hylobatidae. These small arboreal apes of southeast Asia live in highly territorial family groups consisting of a monogamous pair of adults and their immature offspring.

The family Pongidae includes the orangutans of Borneo and Sumatra (genus *Pongo*) and the chimpanzees and gorillas (respectively, *Pan* and *Gorilla,* though increasingly these last two are being lumped into a single genus, *Pan,* in recognition of their close relationship). The major structural features of gorillas were discussed in Chapter 9. Here we are interested in exploring the interrelationships between some pongid physical features and the behavior of the animals in their natural environments.

Sexual dimorphism is particularly pronounced in the great apes. Why does such sexual dimorphism exist? We can get some help in answering this question by looking at the functions of various physical characteristics. The ridge of bone that forms the sagittal crest provides an additional area of attachment for the temporal muscles, which supply most of the power for chewing. More food, and hence more chewing (which wears down the teeth, making large ones useful), would be required to reach and maintain the larger body size of males. But this tells us only that there is an association between two dimorphic characteristics; it does not account for the existence of dimorphism in the first place.

One approach to the problem is to examine the distribution of dimorphism among various species and correlate this pattern with factors that influence the selective forces acting on primate populations. Sexual dimorphism is usually most pronounced in species which have a regular system of male-male competition. Larger males enjoy an advantage over smaller ones in contests that involve either physical strength or even the threat of its use. Yet this is not always true: gibbons show by far the least sexual dimorphism of any hominoid, but field observations indicate that male gibbons spend about 6 percent of their waking hours in threat displays and conflict with other males. So male-male competition, although an important factor in determining the degree of sexual dimorphism, is not a complete explanation. We also need to know the context in which conflict occurs.

Body-size differences are greatest in terrestrial and semiterrestrial primates. Ground-foraging baboons and macaques are much more dimorphic than the colobines and guenons, which feed mainly in the trees. Among apes, the gibbons are not an exception; they are the most arboreal and least dimorphic. But the correlation between type of habitat and degree of dimorphism is not perfect. Differences in size between male and female orangutans, for example, are very much like those between semiterrestrial male and female African apes, which spend much less time in the trees. Orangutans, however, may be more arboreal because of their present habitat, part of which is tropical forest with a swamp floor. These Asian great apes are still far less confined to the trees than gibbons; recent field studies suggest that most long-distance travel by orangs is done on the ground. And the orangutan's large size makes it far less tempting prey for predators of the region, which include the snow leopard in Borneo and the tiger in Sumatra. This may be one explanation for the larger size of male orangutans. They spend more time on the ground than females and would therefore be exposed to predators more often. It is also possible that in the relatively recent past the orangutan enjoyed not only a wider distribution but also a habitat much more like that of the other great apes. Now it is found only in isolated areas of Sumatra and Borneo, but once it ranged over China and much of southeast Asia.

One cause of sexual dimorphism in primates that live on the ground, then, is defense. Males are larger than females because defense is logically a male function. Since one male can fertilize the ova of a number of females, males are, up to a point, more expendable than females. Females would also be at a disadvantage in physical combat when they are carrying young. In addition, the loss of a pregnant female would be a double danger to the troop; if the mother were killed or severely injured, her unborn infant would also die. Sexual dimorphism in size correlates with differences in other biological factors and with the differentiated social roles of males and females. The existence of these roles, in turn, is dictated by ecological constraints. Defense of the troop by males is less necessary in arboreal species because independent flight from predators is far easier in the trees than on the ground.

To explain the marked degree of sexual dimorphism in terrestrial primates as being due to the advantage of having males larger for purposes of defense is to tell only half the story, however. In open grassland the supply of vegetable food is often less abundant and regular than it is in the forest. Limited resources therefore mean a compromise between the maximum size of individuals (for defense) and the maximum population number and density (for long-term survival). Sexual dimorphism is one possible answer: with a few large males well suited for defense, females can remain (or become) smaller and more numerous. In the tropical forests where the arboreal gibbons live, in contrast, suitable food plants are abundant but highly dispersed. Sexual dimorphism is not so marked, but neither can the large bands characteristic of ground-living primates exist. The primary and enduring gibbon social group is a permanently mated adult pair and their immature offspring. The frequent conflicts between males of adjacent groups suggest that larger males might enjoy a competitive advantage, enabling their group to range over a wider area, but the potential advantages of larger size would be balanced by the greater amount of food these males would consume. Among more dimorphic primates, males typically make up considerably less than half of the social groups. Larger size could confer the right to exclusive use of a larger territory, but among the higher socially-oriented and group-oriented primates, it usually means a dominant position in a group that holds a certain territory in common.

High status and the right to range freely are intangible benefits, much like membership in an exclusive country club or social fraternity. Nonhuman primates, like their human counterparts, might seek dominant positions for the feeling of satisfaction they bring and for the access they allow to more tangible rewards—in feeding, mating, and a number of other activities. To the extent that morphological and behavioral traits which enhance an animal's chances of success are heritable, there can be genetic consequences as well.

Apparently, competitive social interactions serve to identify those animals in the group which are stronger, more aggressive, and more persistent. Although there are exceptions, often animals with these qualities are males at the height of their physical prime. These are the animals best suited to succeed in competition not only with other members of the same species but also with predators. Thus, a high-status position confers privileges and also entails obligations. The larger, more dominant males protect the smaller and weaker members of the troop. Even the largely solitary male orangutans perform this function. They are not often found with females and young, but still they may offer a shield like the presence of a patrolling police officer. The lone male orang cannot be physically present everywhere at once, and yet the very uncertainty of his location and the possibility that he will appear with little or no warning may serve to discourage predators.

What about other aspects of hominoid social behavior? An increasing body of evidence points toward the African great apes as our closest living relatives. Both gorillas and chimpanzees have now been studied scientifically in their natural habitats, but so far primatologists have spent many more hours of field time watching chimpanzees than they have spent observing gorillas. In part because of this more extensive knowledge, chimps are very often used as models for some aspects of the behavior of our immediate prehuman ancestors.

Chimpanzees (*Pan troglodytes*) have been studied in their native habitats by scientists since the end of the nineteenth century. For the most part these earlier studies—for example, that of Nissen (1931)—yielded relatively little information on such important topics as diet, daily range, group size, and social interactions within and between groups. Systematic work on chimpanzees in the wild began in the early

1960s (Bramblett, 1976). Among these investigators was Jane Goodall, an Englishwoman who worked with chimpanzees in what was then the Gombe Stream Reserve, near Lake Tanganyika, an area of mixed gallery forest, open woodland, and grassland. Two other Britons, Vernon and Frances Reynolds, concentrated on chimpanzees living in the Budongo forest of Uganda, and other forest-living groups were studied by Adriaan Kortlandt of Holland and by Junichiro Itani of Japan.

Though some of the best-known populations are found on savannas, most chimpanzees are forest dwellers, and they feed mainly on fruits, supplemented by other vegetable foods and some animal protein. As is usually the case for nonhuman primates, most of the day is spent in feeding, which alternates with rest, bouts of play, and other social activity. In contrast to groups of baboons, chimpanzee groups are much more open to exchange of members. In fact, some scientists who have studied chimpanzee behavior in the wild question whether there is any permanent association beyond that between a mother and her offspring, since the composition of most groups seems to be so fluid. Goodall sees chimpanzees as grouped into large communities of 50 or more members, with all animals seldom together but capable of reorganizing and interacting socially with one another on contact. Recently, however, two Japanese investigators (Kawanaka and Nishida, 1974) have found chimpanzee groups that consist of a permanent core of adult males and a more fluctuating membership of adult females. In addition to their greater cohesiveness, groups of males also seem more mobile and exploratory than groups of females with young. Reynolds (1976) sees this as the basis for a division of labor along sexual lines, with males discovering food and with females caring for the young, and he points out the similarity of this pattern of social organization to that seen in human hunter-gatherers.

As noted earlier, chimpanzees in certain areas make a variety of tools, and on occasion they hunt and kill other animals for food. Thus, there seems to be a considerable basis for the belief, held by many anthropologists, that chimpanzees possess biological and behavioral capabilities which could have served as the starting point for a primitive human way of life. Whether these characteristics were ever actually present in our direct ancestors, and when, will be considered in Chapter 12.

SUMMARY

1. Although we can never directly observe the behavior of our nonhuman primate ancestors, the study of living nonhuman primates enables us to see the relationships among structure, behavior, and environment in a variety of precultural species related to our own.

2. Most nonhuman primates are native to the tropics, probably being limited to these areas mainly by the availability of food.

3. Nonhuman primates occupy a variety of habitats, or environmental settings, and a given habitat may have a number of ecological niches occupied by different primate species.

4. Most primates are omnivorous, eating a mixture of different plant and animal foods; however, some species eat principally insects, while others live almost exclusively on leaves.

5. All primates are capable of a wide range of limb movements, but each species has a dominant pattern of locomotion related to its anatomical characteristics.

6. Among the taxonomic groups of living primates there are a number of prosimian species; many of these have highly restricted geographic ranges, and some prosimians are highly specialized to fit very narrow ecological niches.

7. Among the higher primates of the suborder Anthropoidea there are many species of monkeys and apes. Nearly all belong to moderate-sized or large social groups in which the flexibility of organization correlates broadly to habitat differences such as tropical forest versus dry savanna.

SUGGESTIONS FOR ADDITIONAL READING

Jolly, A. 1972. *The Evolution of Primate Behavior.* New York: Macmillan. (This readily available paperback gives a survey of recent findings about primate behavior and its relevance to the understanding of human behavior. Particular stress is placed on the ecological settings in which the animals live and the extent to which these shape ways of life.)

De Vore, I. (ed.) 1955. *Primate Behavior.* New York: Holt, Rinehart and Winston. (Subtitled "Field Studies of Monkeys and Apes," this volume is a collection of articles that describe the behaviors of higher primates living in the wild.)

Jay, P. C. (ed.) 1968. *Primates.* New York: Holt, Rinehart and Winston. (This book followed the one edited by De Vore. As its subtitle, "Studies in Adaptation and Variability," suggests, there is a shift in emphasis to the comparison of populations of the same species living in different habitats, with the objective of understanding the causes of behavioral adaptations.)

REFERENCES

Bramblett, C. A. 1976. *Patterns of Primate Behavior.* Palo Alto, Calif.: Mayfield.

Carpenter, C. R. 1934. "A Field Study of the Behavior and Social Relations of Howling Monkeys (*Alouatta palliata*)," *Comparative Psychology Monographs,* **10**(48):1–168.

Chiarelli, B. 1975. "The Study of Primate Chromosomes," in R. Tuttle (ed.), *Primate Functional Morphology and Evolution.* The Hague, Netherlands: Mouton. Pp. 103–127.

Crook, J. H. 1966. "Gelada Baboon Herd Structure and Movement: A Comparative Report," *Symposia of the Zoological Society of London,* **18**:237–258.

Dart, R. A. 1963. "The Carnivorous Propensity of Baboons," *Symposia of the Zoological Society of London,* **10**:49–56.

DeVore, I., and S. L. Washburn. 1963. "Baboon Ecology and Human Evolution," in F. C. Howell and F. Bourlière (eds.), *African Ecology and Human Evolution.* New York: Viking Fund Publication no. 36. Pp. 335–367.

Harding, R. S. O. 1977. "Patterns of Movement in Open Country Baboons," *American Journal of Physical Anthropology,* **47**(2):349–353.

Jolly, A. 1972. *The Evolution of Primate Behavior.* New York: Macmillan.

Jolly, C. 1970. "The Seed-Eaters: A New Model of Hominid Differentiation Based on a Baboon Analogy," *Man,* **5**(1):5–26.

Kawanaka, K., and T. Nishida. 1974. "Recent Advances in the Study of Inter-unitgroup Relations and Social Structure of Wild Chimpanzees of the Mahali Mountains," *Symposia of the 5th International Primatological Society.* Tokyo: Japan Science Press. Pp. 173–185.

Kummer, H. 1971. *Primate Societies.* Chicago: Aldine.

Napier, J. R., and P. H. Napier. 1967. *A Handbook of Living Primates.* New York: Academic Press.

Nissen, H. W. 1931. "A Field Study of the Chimpanzee," *Comparative Psychology Monographs,* **8**(1):1–122.

Reynolds, V. 1976. *The Biology of Human Action.* San Francisco: Freeman.

Whittaker, R. 1970. *Communities and Ecosystems.* New York: Macmillan.

Chapter 11

From Microevolution to Macroevolution

Species are genetically closed populations that must therefore pursue their own independent courses of evolution. The survey of the primates in Chapters 9 and 10 showed the degree to which species, even within the same order of mammals, can differ. Let us look now at how species are recognized, how they can come into existence, and what evolutionary fates they can have after they have attained reproductive isolation. This new information will enable us to see how variations within populations can be converted into differences between species—that is, how the small-scale results of microevolutionary processes can be converted into changes major enough to be recognized as macroevolutionary transformations that cross the boundaries of species and higher taxa. Later in the chapter we will see how the past evolutionary record is reconstructed, and we will look at the types of problems that can arise in the process of such work. An appendix at the end of the chapter discusses how fossils and the rocks that contain them can be dated so that the course of macroevolution can be traced over vast spans of geologic time.

RECOGNIZING SPECIES

Ernst Mayr, in *Animal Species and Evolution* (1965), sees definitions of species arising from three separate theoretical concepts: (1) the typological species concept, which is based on appearance; (2) the nondimensional species concept, which is based on whether populations interbreed where they overlap; and (3) the interbreeding-population (or multidimensional species) concept, which rests on a decision about the possibility that two populations separated by time or space could potentially interbreed if they did come together. Mayr's discussion of these alternative ideas is based on half a century of field research and thinking about the problems involved in defining species and recognizing them in nature. His ideas are worth considering in some detail before we move on to see how species arise through evolution.

Those who use a typological species concept define members of two separate species by saying that they look different. Members of a single species look more alike. They are not identical, but they are similar enough to seem to be variations on a basic or ideal type. In this view there are practical limits to the range of variation that can exist within any one species. When an individual falls outside these limits, it must belong to a different species. The typological species concept, then, leads to species' being recognized primarily on the basis of morphological differences.

The nondimensional species concept is derived from a study of the relationships between two populations living in the same place (*sympatric*) at the same time (*synchronic*). If two sympatric and synchronic populations do not interbreed, they are different species; if they do interbreed, they must belong to the same species. This test has one great advantage: it is based on events that can actually be observed in nature. It has another advantage as well, in that morphological differences are not always reliable guides to reproductive behavior. Some populations that appear to be extremely similar or even identical nevertheless do not interbreed. (These phenotypically similar but reproductively isolated populations are called *sibling species*.) On the other hand, some reproductive communities include highly polymorphic and polytypic populations (Figure 11-1). But what do we do about populations that do not overlap in time and space?

The multidimensional species concept overcomes this difficulty, at least in theory, by viewing species as groups of populations that actually do or potentially can interbreed. With this approach in mind, we can ask questions such as: Can primates as different in appearance and as widely separated as the gelada baboons (*Theropithecus gelada*) of Ethiopia and the rhesus monkeys (*Macaca mulatta*) of India interbreed? The answer to this question is known; when brought together in zoos, they have interbred and produced living offspring. Populations that are *allochronic* (living at different times) and *allopatric* (living in different areas) can now be identified, as well as those which are sympatric and synchronic. But the multidimensional species concept cannot always be applied objectively because we cannot measure potential to interbreed directly; we must infer it from degree of similarity or difference.

FIGURE 11-1 Similarity is not always a reliable guide to closeness of relationship. The butterfly at the left is *Danaus plexippus* (the monarch butterfly), which has a foul taste and is therefore avoided by insect-eating birds. In the center is *Limenitis archippus,* which is not closely related to the monarch but avoids being preyed upon by mimicking its appearance. At the right is *Limenitis arthemis,* which differs sharply from its close relative *Limenitis archippus.*
Source: Specimens from Frost Entomological Museum, Pennsylvania State University, courtesy of K. C. Kim.

This seems to bring us back to appearances, but with an important distinction. According to the multidimensional species concept, phenotypic differences are not important in themselves; they are only clues as to whether reproductive isolation would exist if the populations were synchronic and sympatric.

NEW SPECIES: THE PROCESS OF SPECIATION

As long as populations can exchange genes, the differences between them will not become too extreme. But if gene flow is restricted over a long period of time, the gene pools of the populations may become so different that exchange is no longer possible. At this point, one species has already diverged into two. This brief account makes *speciation* sound like a single event, but really it is a process that extends over many generations and has several separate stages: subdivision of the gene pool, independent evolution of the separate pools, and overlap of ranges with reproductive isolation (Figure 11-2).

Stages in Speciation

Subdivision of the gene pool is the starting point. There are various views of how this splitting of one gene pool into two or more isolated ones can occur. The idea now favored is that interruption of gene flow is most commonly due to geographic isolation. Such isolation can begin when a river changes its course and divides what had formerly been the continuous range of a species population. Something akin to this took place in 1914, when the Chagres River in Panama was dammed to form Gatun Lake, which is part of the Panama Canal. As the water of the artificial lake rose, a mountain ridge was turned into an island. The howler monkeys on the island were then cut off from others of their species on the surrounding shores of the lake. There has not yet been any study of possible genetic effects of this isolation. However, we do know that populations of African green monkeys (*Cercopithecus sabeus*) introduced into the island of Saint Kitts in the West Indies during the seventeenth century have developed some small but recognizable anatomical differences from their parent populations in the Old World.

Belief in *geographic speciation* tends to go hand in hand with acceptance of the balance theory of evolution, according to which populations are seen as always having abundant genetic variation. Quite

FIGURE 11-2   Stages in speciation.

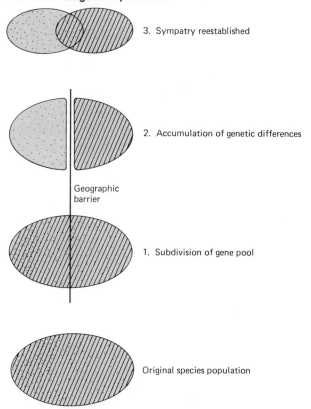

3. Sympatry reestablished

2. Accumulation of genetic differences

Geographic barrier

1. Subdivision of gene pool

Original species population

a different picture would follow logically from the classical model of evolution, in which populations consist of individuals that are homozygous at nearly all loci. The classical model fits comfortably with a typological species concept; speciation would have to involve the appearance of a new "type," depending on the origin of at least one new favorable mutation. More probably, several would have to take place at the same time and to survive in the face of heavy odds favoring chance loss. Commonly scientists who favor the classical view also tend to accept models that involve the possibility of *sympatric speciation*, the origin of two reproductively isolated groups within a population. However, sympatric speciation has not been demonstrated convincingly in natural populations, and experimental attempts to bring it about under laboratory conditions have also given conflicting results.

Independent evolution of the separate gene pools follows subdivision of the original population. Once gene flow is eliminated, other forces of evolution can act independently on the two groups. If the populations are very large, the pattern of recurrent mutations may not be very different from one population to the next, but the sequence of rare mutations may be very different. Subdivisions of one larger population into two or more smaller ones should automatically increase the magnitude of changes attributable to genetic drift in each. Overall, these two forces of evolution should tend to increase the divergence between the two populations at a rate roughly proportionate to the length of time they remain separated.

The effect of selection is somewhat less predictable, though probably quite important in most cases. Each isolated population is almost certain to experience an environment different in some ways from any other, and so the set of selective forces acting on each is also likely to be different. On balance, selection should also lead to divergence in most cases, but not in proportion to time alone; divergence is more likely to be a product of time and degree of environmental difference. If separation lasts long enough, speciation should result. A minor paradox here is that by a strict application of a nontypological species concept, as long as the two populations remained separated, it is not possible to tell for certain whether speciation has occurred. Only when the ranges of the two populations overlap can it be determined with certainty whether speciation has taken place. The test is simple enough in theory. If interbreeding does not take place, two species exist. If interbreeding occurs freely, there is only a single species, regardless of how polytypic its populations may be. Our own very diverse human species is a good example of the extent to which members of populations can look different and yet interbreed with no problem. But in the real world not all cases fit one of these two extremes; divergence of two gene pools sometimes reaches the stage where reproductive isolation is only partial. For example, a smaller proportion of hybrid offspring might survive than those from matings between parents from the same population.

At this point a number of *isolating mechanisms* come into play; these are the factors that keep one species reproductively isolated from others. A list of isolating mechanisms is given in Table 11-1. If reproductive isolation is not complete at the time the two populations become sympatric, divergence might be increased by *character displacement*, owing to selection that increases phenotypic differences between two closely related species to prevent waste of gametes in matings of genetically incompatible animals. For example, two species of the small seed-eating birds known as rock nuthatches are found in Eurasia—one in the east, and the other in the west. Over this very wide range the two species populations are quite similar overall, but where they overlap in Iran the bill of one species is greatly enlarged, while that of the other is much reduced. This probably lets them utilize different food sources, which may help reduce the frequency of contact and competition. The isolating mechanisms would ensure that matings between members of the different populations would result in fewer offspring than matings within either population. If hybrid matings produced fewer offspring, then fewer copies of genes would be transmitted to subsequent generations. Increasing reproductive isolation is often accompanied by other morphological, physiological, and behavioral changes that further differentiate the ecological niches exploited by the two populations.

TABLE 11-1   Isolating mechanisms, in order of decreasing efficiency

Mechanisms that prevent interspecific crosses (these act prior to mating)

1. Potential mates do not meet.
   a. Seasonal isolation (members of two different populations utilize the same area at different times).
   b. Habitat isolation (members of two different populations utilize different parts of the same region at the same time).
2. Potential mates meet but do not mate (ethological or behavioral isolation).
3. Mating is attempted, but no transfer of sperm results (mechanical isolation).

Mechanisms that reduce success of interspecific crosses (these act after mating takes place)

4. Sperm transfer takes place, but egg is not fertilized (gametic mortality).
5. Egg is fertilized, but zygote dies (hybrid mortality).
6. Zygote produces first-generation hybrid of reduced viability (hybrid inviability).
7. First-generation zygote is fully viable but partly or completely sterile, or it produces fewer second-generation offspring (hybrid sterility).

Source: Modified from Ernest Mayr. 1965. *Animal Species and Evolution.* Cambridge, Mass.: Harvard University Press. Table 5-1, p. 92.

Examples

Complete reproductive isolation of related populations and free inter-breeding between them are the two extremes. More interesting are the borderline cases because they illustrate the process of speciation itself. Let's look at two—the leopard frog and the herring gull.

Leopard Frogs.   The scientific name of these small spotted frogs is *Rana pipiens.* They are found in swamps, ponds, and streams over most of the United States and into Canada and Mexico. As is true of many species of small animals, few of these frogs stray far from their places of birth, and populations are quite localized. Nevertheless, matings do take place between members of adjacent populations, and thus gene flow joins even widely separated groups like links in a chain.

If frogs from Vermont are mated with frogs from New Jersey, about 480 km (300 mi) to the south, their fertility is as high as when there is a cross between frogs from two adjacent ponds in either place. The same results are obtained when the parents are drawn from Vermont and Wisconsin, 1300 km (800 mi) apart: fertility is normal. However, when frogs from Vermont are paired with frogs from Louisiana, 2090 km (1300 mi) to the south, the embryos develop more slowly and show some morphological abnormalities, though none great enough to affect survival. When individuals from Vermont and Mexico are bred, nearly all the embryos die in the earliest stages of development (Figure 11-3). In this species, for two populations anywhere in the range up to about 1600 km (1000 mi) apart, fertility is normal; beyond that point, repro-

FIGURE 11-3   Leopard frogs: one species or many?
Source of data: J. A. Moore. 1950. "Further Studies on *Rana pipiens* Racial Hybrids," *American Naturalist,* **84:**247–254.

Results of matings over long distances

Vermont to New Jersey: normal fertility and viability

Vermont to Florida: many embryos die

Mexico

Vermont to Mexico: nearly all embryos die in earliest stages of development

ductive success goes down as distance goes up. Genes *can* be passed from one end of the range to the other, but only by being filtered through the intervening populations in the web of relationships. We can only speculate on what is happening, but it is likely that the gradual passage gives the rest of the genome a chance to make compensating adjustments to the new genes.

Herring Gulls.   Populations of herring gulls are found in western Europe, eastern Asia, and North America, but not all are identical in appearance. Populations differ in visible traits such as the color of the feet (yellow versus pink) and the orbital rings around the eyes (yellow or purplish red). Distinguishable populations are enclosed by single dark lines on Figure 11-4. Interbreeding occurs among all at the point of contact, as

FIGURE 11-4   Herring gulls: one species or many? The map shows circular overlap of gulls of the *Larus argentatus* group. Subspecies A, B, and C evolved in Pleistocene refuges; D evolved in North America into a separate species (*L. glaucoides*). When A expanded, after the Pleistocene, probably from a north Pacific refuge (Yukon? Alaska? Kamchatka?), it spread across all of North America and into western Europe (*argentatus*). Here it became sympatric with *fuscus* (B3, B4), the westernmost of a chain of Eurasian populations.

Source: Modified from Ernst Mayr. 1965. *Animal Species and Evolution.* Cambridge, Mass.: Belknap (Harvard University Press). Fig. 16-7, p. 509. Copyright © 1963 by the President and Fellows of Harvard University.

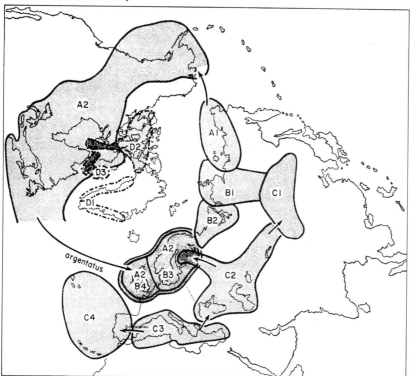

indicated by the broad continuous arrow in the figure, with one notable exception. Relatively recently, gulls from North America have crossed the Atlantic and now overlap the range of some European populations (the area enclosed by double lines). In this instance, gene exchange does *not* occur.

The birds themselves have conducted the same type of experiment that scientists performed with leopard frogs, with much the same results. Populations at the extremes of the distribution do not exchange genes. They act like separate species and are designated as such. (*Larus argentatus* from North America overlaps *Larus fuscus* in Europe.) However, the intermediate populations interbreed as if they were members of the same species. This behavior can be taken as an indication that all the interconnected populations belong to the same species. Which view is correct? Neither or both, really—all these gull populations exchange genes even though not all can do so directly. The problem is one of terminology, not of understanding. Forcing cases of this sort into one pigeonhole (several species exist) or another (only one species exists) simply sacrifices information for a comparatively trivial gain in neatness. This realization should be reassuring rather than disturbing. After all, if we could devise a definition of species that left no room for transitional situations, how then could one species ever arise from another?

AFTER SPECIATION: EVOLUTIONARY FATES

The evidence presented in Chapter 10 demonstrated that speciation has taken place many times during the course of primate evolution. It is difficult to say precisely how often primate gene pools have been subdivided by the speciation process, but it is safe to say that primate populations have speciated hundreds of times, and this estimate may be too low by an order of magnitude. All the resulting species differ from one another genetically, and in most cases there are obvious differences at the phenotypic level. But although speciation can occur again and again, it is not an inevitable event that takes place at regular intervals. Various other responses (adaptive radiation, no change, extinction, phyletic evolution) can lead to the diverse patterns of evolution we see in the fossil record of primate and human evolution. Let us look at these alternatives (Figure 11-5).

Adaptive Radiation

An *adaptive radiation* occurs when one ancestral population gives rise to multiple descendant species. In the original sense in which the term was introduced by Henry Fairfield Osborn in *Age of Mammals* (1910),

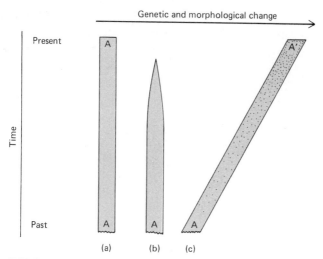

FIGURE 11-5   Patterns of evolution:
(a) **no change;** (b) **extinction;** (c) **phyletic evolution.**

a primitive species living in a large, isolated area that is diverse in climate, topography, and vegetation can speciate repeatedly, producing very diverse descendants. From the starting point of a ground-living, four-footed mammal, evolutionary diversification can produce four major lines (later expanded to five by other theorists): cursorial, or terrestrial running; scansorial, or arboreal climbing; fossorial, or burrowing; aquatic, or swimming; and volant, or flying (Figure 11-6). More recently the concept of adaptive radiation has been broadened, and now it is often used to describe a great variety of cases in which one species gives rise, usually over a relatively short period of geologic time, to several other species that occupy well-differentiated ecological niches.

No Change

It is theoretically possible that a population might not change at all through time. In a strict sense this would seem unlikely, given the number of different ways in which change in gene frequency can take place. But there are various ways in which equilibrium states can be maintained. Recurrent mutations can be balanced by selection, polymorphisms can be maintained by opposed selective forces, and so on. Situations like this are determined by the interaction of the population and its environment. In time, each species population comes to occupy a particular niche in its habitat; as a result, the population interacts with its physical environment and other species in it, and it can be defined by a unique set of morphological, physiological, and behavioral

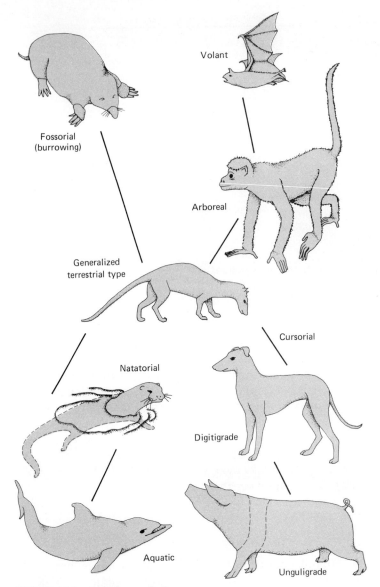

FIGURE 11-6   Adaptive radiation.
   Source: Maurice Burton. 1956. *Living Fossils.* London: Readers Union. Used with
permission.

characteristics. The array of features that define a niche has dynamic
stability; thus if some significant element is altered, other elements will
adjust in response to this change. For example, the one-male daytime
feeding groups of hamadryas baboons probably represent an adjustment
to a habitat where food is too sparsely distributed to support larger
troops. Given all the potential ways in which change can occur, it is

perhaps a bit surprising to find that some populations do remain unchanged for quite long periods of time (this is shown schematically in Figure 11-5). So-called "living fossils" such as the horseshoe crab, the cockroach, and the opossum (see Figure 11-7 and "Conservatism and Change: The Coelacanth and Its Relatives") have persisted virtually unchanged over tens or hundreds of millions of years.

FIGURE 11-7 Living fossils. These animals represent species that have remained unchanged over very long periods of time. (a) The opossum originated over 100 million years ago; (b) the cockroach, about 300 million years ago; (c) the shark, about 400 million years ago; and (d) the horseshoe crab, about 600 million years ago.

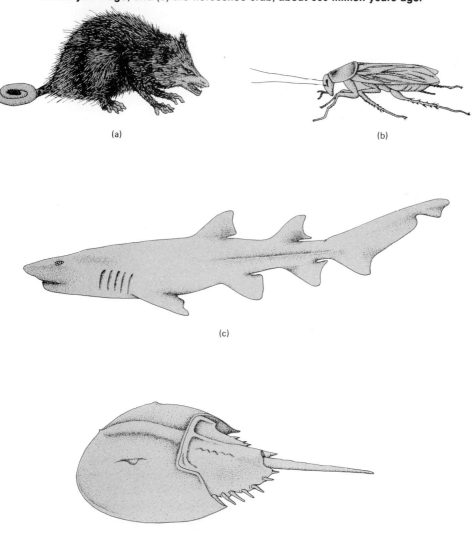

(a)

(b)

(c)

(d)

CONSERVATISM AND CHANGE: THE COELACANTH AND ITS RELATIVES

Perhaps the coelacanth is the most dramatic example of a living fossil. This is a member of a very ancient order of fish which was formerly believed to have become extinct about 70 million years ago. However, in 1938 a living specimen of one species, *Latimeria chalumnae,* was caught off the east coast of South Africa.

These fish have continued to live on, apparently unchanged, over a span of time which includes all primate evolution and during which humans evolved from a form originally small and ratlike in appearance.

There is no objective basis for the assertion, sometimes made, that some groups of animals are "inherently progressive," while others, like the coelacanths, are "inherently conservative." It is known, for example, that other crossopterygian fish (the rhipidistians) closely related to the coelacanths were anything but conservative. The rhipidistians are thought to have given rise directly to the amphibians, and through them to reptiles, birds, and mammals as well.

It seems rather more likely that the crucial difference between the two closely related crossopterygian orders lies not in any inherent difference in genetic plasticity but in the nature of the environments they inhabited. Coelacanths represent a group adapted to living in the unchanging ocean depths. The rhipidistians, on the other hand, inhabited relatively shallow swampy pools containing water which was frequently deficient in dissolved oxygen; in these conditions, survival depended on the evolution of lungs and external nostrils which could use atmospheric oxygen. Also, because these shallow ponds not infrequently dried up, rhipidistian limbs gradually evolved from bony stalks which originally supported fins into stubby and weakly supported legs which nonetheless allowed a clumsy form of locomotion across stretches of land to pools which still contained water.

The coelacanths adapted to a constant milieu and remained unchanged; the rhipidistians adapted to a changing environment—and consequently became transformed themselves. Their changing niche required the rhipidistians to gradually evolve the components of an

Extinction

Extinction can take place for a number of different reasons—shifts in climate, a reduced supply of accustomed foods or the disappearance of these foods, the appearance of new diseases or predators, or competition from another species. However, the immediate cause must always be the same: the gene pool of the population cannot accom-

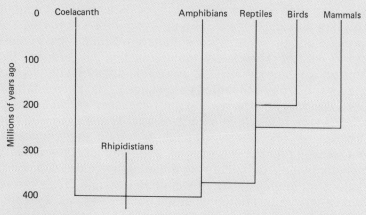

The coelacanth.

adaptive complex which in retrospect we recognize as that which characterized the earliest terrestrial vertebrates. This pattern of stress and response, which has emerged many times in the study of evolution, is now generally referred to as *Romer's law* in honor of the paleontologist S. A. Romer, who recognized that many evolutionary innovations such as this can best be explained as having originated because at the time of their appearance they represented a rather minimal—and hence conservative—response to changed or changing circumstances.

modate with sufficient rapidity to the stresses placed on it. As a result, the species declines in numbers until none are left. Probably the most dramatic familiar example of recent extinction is that of the passenger pigeon, formerly found in eastern North America. Two centuries ago these birds were so abundant that their flocks actually darkened the skies as they passed. At roosting sites, tree branches were broken by the combined weight of their numbers. Each year hundreds of thousands were killed and sold in markets by the barrel. Their destruction was relatively easy because the birds were not especially intelligent; a man walking by swinging a pole could sweep dozens from the branches on which they sat. Their disappearance was probably further hastened by the fact that only one egg was produced at each sitting. But whatever the particular combination of causes, the last known survivor of the species died in captivity in 1914. Well-documented cases of extinction of nonhuman primates are discussed in Chapter 12.

Phyletic Evolution

Phyletic evolution occurs when a species changes in appearance through time but when there is no splitting of one population into two or more reproductively isolated units, even though one or more characteristics may be completely transformed in the process. Transformation through time by phyletic evolution is very much like the divergence over space that accompanies isolation by distance in the examples of the leopard frog and the herring gull, discussed earlier in this chapter. Whether genetic differentiation occurs over space or through time, the end points may be clearly distinct, but they are connected by a continuously intergrading series of populations. This series can be divided only by arbitrarily drawing a line between two slightly different populations. In the fossil record it is often possible to label the ends of such a distribution as different species, but in many cases we are spared the embarrassment of drawing the species-demarcating line between parents and their offspring only by the very incompleteness of the evidence. Phyletic evolution is also intriguing because it provides a mechanism through which populations can disappear without becoming extinct. Some animal populations now living may be the direct descendants of earlier species that have changed their phenotypes gradually through time in response to forces of evolution, without splitting of the gene pool anywhere along the line. By the definition given above, disappearance of the diagnostic phenotypic characteristics of the earlier populations does not mean extinction of the population or its gene pool.

DEBATE ABOUT HUMAN EVOLUTION

The record of evolution contains examples of all the events just described (Figure 11-8). This is true even when we focus on a particular group of animals such as the primates. But there is a difference between understanding how these events can occur in general and deciding which of them most accurately represents what actually occurred in the ancestry of any given living species. Generalizations about the course

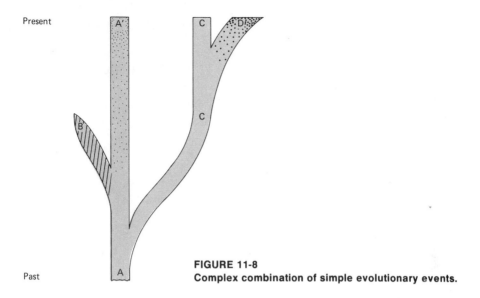

Present

Past

FIGURE 11-8
Complex combination of simple evolutionary events.

of human evolution rest on a body of evidence derived from observations of living or recently dead animals and from the study of fossil remains and a set of assumptions used to interpret this evidence. Put this way, the whole thing seems simple and straightforward—but obvious points are sometimes the easiest ones to overlook. For example, fossil remains are sometimes regarded as facts—which they are not. A fact is a proposition about the material world; a fossil bone is merely a piece of that world. The *existence* of the bone is a fact, but the *significance* of that bone is simply a matter of interpretation.

Evidence

Jaws and teeth are among the most commonly found fossil remains of primates because they are particularly dense, though other bones are sometimes found. The durability of bones and teeth is a side effect of their chemical composition: they consist of a protein matrix impregnated with minerals, chiefly calcium and phosphorus. These fossils are used to reconstruct lines of relationship among past populations. The procedure is based on the reasonable assumption that, like other components of the body, bones indirectly reflect the linear sequence of nucleotide bases in one or more segments of DNA in the chromosomes. But bones mirror more than this. The sizes and shapes of bones reflect levels of nutrition, and the areas of muscle attachment on them can reflect levels and types of exercise and other environmental effects during the course of growth and development. As a result, skeletal and dental remains show genetic relationships, but neither perfectly nor exclusively.

The fossil record of primate evolution is incomplete, both qualitatively and quantitatively. Not all parts of any individual are equally likely to be found, and the remains of very few individuals are preserved. There are many reasons for this shortage of primate fossils. Most primates have been relatively small mammals, and so their bones could easily be crunched to bits by hunting and scavenging carnivores. Then, too, most primates have lived in warm, moist, tropical forests, where decomposition is usually rapid and complete. Realizing this limitation on the nature of the evidence helps us keep the paleontological record in its proper perspective. Like the shells of fossil mollusks, the jaws, teeth, and other bones are studied not because they are more important than other parts of the organisms but because they are all that is left. Of course, jaws and teeth do reflect the size of the animal and give important clues about its diet, but the fossil record can supply nowhere near as abundant or detailed evidence as can be gotten from the study of living forms. And studies of living primates, although extremely detailed, have their limits too. Comparative studies of living primates may suggest that the various morphological types point to the broad outlines of an evolutionary sequence, but this can never be more than a logical supposition without the historical information from the fossil record.

The study of the living primates is therefore not a substitute for the study of fossil forms, nor can paleontology be pursued as an isolated discipline. If we combine the two, however, the study of human evolution can approach the methods of experimental science. If that seems like a contradiction, it is only because we have the stereotyped idea that an experiment consists of tinkering with a process to see whether its outcome can be influenced. In the broadest conception, an experiment is any procedure for testing a hypothesis. Comparative studies of living primates can suggest ideas that can be "tested" when appropriate fossil material is discovered.

Interpretations

As Tattersall and Eldredge (1977) emphasize, hypotheses about the course of evolution can be formulated at three levels of complexity, each one further removed from the available evidence than the last. The simplest level of analysis is called a *cladogram*. This is a diagram that shows the distribution of characteristics within a group of related organisms. These characteristics (or *character states*) may be *primitive* (present in the common ancestor of the group of related organisms) or *derived* (as in the case of an evolutionary novelty found in just one descendant). Figure 11-9 shows a simple cladogram in which A, B, and C each have one or more unique derived character states that set them apart from the others and in which B and C share at least one common

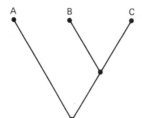

FIGURE 11-9   A simple cladogram.
Source: I. Tattersall and N. Eldredge. 1977. "Fact, Theory, and Fantasy in Human Paleontology," *American Scientist,* **65**:204–211. Fig. 1, p. 204.

character state that is not present in A. A *phylogenetic tree* (so named because it begins with an ancestral trunk in the past and has branches to represent lines that evolved later) is the next more complex type of analysis. In addition to the information contained in the cladogram, a tree embodies a hypothesis about the nature of the evolutionary relationships between the organisms. For example, in the hypothetical phylogenetic tree in Figure 11-8, A′ is represented as having descended from A by phyletic evolution. A phylogenetic tree can also include information on the temporal relationships among the populations. Even more complex than the phylogenetic tree is the *scenario,* which is Tattersall and Eldredge's term for a verbal explanation of a phylogenetic tree. The scenario usually explains why the postulated evolutionary events took place, usually in terms of adaptation and ecology. The text of "Conservatism and Change: The Coelacanth and Its Relatives," pages 354–355, particularly the fourth paragraph, gives a brief scenario for the evolution of a living fossil and some related species.

Tattersall and Eldredge are of the opinion that much of the disagreement concerning the interpretation of evolution, particularly human evolution, happens because hypotheses are most often stated in the form of scenarios—the level at which they are the most complex and difficult to test. Let us look at some of the difficulties encountered in interpreting past evolution in detail.

PROBLEMS IN INTERPRETING THE EVIDENCE

The Present as a Guide to the Past

Only a fraction of all the animal species that have ever existed are alive now. On the grounds of probability alone, therefore, it would seem that not all the primate adaptive patterns that have ever existed can still be found today. Some traits may no longer be seen in any living species, and there may have been species that combined features of two or more different grades or levels of evolution such as monkey or ape (see "*Megaladapis* and *Archaeolemur:* Two Primates of the Past That Do Not Fit Present Patterns"). As a result, our view of the characteristics

MEGALADAPIS AND ARCHAEOLEMUR:
TWO PRIMATES OF THE PAST THAT DO NOT FIT PRESENT PATTERNS

Isolated since at least the Eocene, the lemurs of Madagascar diversified into a number of distinctive adaptive patterns. In all, 16 genera evolved there, but only 10 of these survive. There is substantial evidence that the extinct forms were destroyed by humans relatively recently, sometime within the last 10 or 15 centuries. Indeed, some of the species now gone may have survived until the time of European contact, although by then they were much depleted by aboriginal hunting and habitat destruction.

Megaladapis

Megaladapis is an atypical primate. It certainly does not fit the usual prosimian stereotype of a small, arboreal, and nocturnal animal. A mouse lemur could wriggle through the ring of bone that surrounded the socket of Megaladapis' eye, and it might be able to nest within its braincase. Some Megaladapis skulls are nearly 51 cm (20 in) long, as large as those of yearling cows.

The resemblance does not stop with size; there are other superficial similarities to cattle. In both, the skull is long, low, and robust; the foramen magnum opens to the rear, as is typical for quadrupeds; the upper incisors have been lost; and the size of the eye orbits indicates that activity was carried on chiefly during daylight hours.

It is most likely that Megaladapis was an herbivore, probably browsing on tree leaves and twigs. But because no living individuals have been systematically studied, this reconstruction of its way of life is not a certainty. Other conjectures abound; some scholars have suggested that

of earlier primate populations may be distorted by being limited to what we happen to know. Our most direct evidence for past primate and human evolution comes from the fossil record, which is quite incomplete. Millions of years of evolution over areas as large as a continent may be represented by only a few fragmentary fossilized bones, or there may be no such evidence at all. Under such circumstances our reconstructions of the past are based as much on theories as on evidence, or more. These theories, in turn, are only as good as the assumptions on which they are based. In the next section of this chapter ("Generalizing from Individuals to Populations") we will look at some of these assumptions and at the problems that can arise when these assumptions are used to reconstruct our own origins. Are prosimians, monkeys,

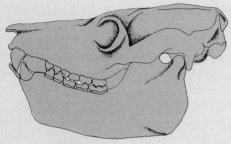

Megalapidis.

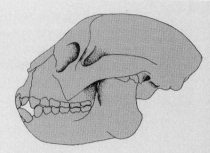

Archaeolemur.

it was aquatic, perhaps like a hippopotamus, while others have likened it to a gorilla or even a cave bear.

Archaeolemur

Archaeolemur has analogies within the primate order, but not necessarily with other prosimians. Its braincase is rather large and globular; its face projects relatively little; its eye orbits are directed forward, with the result that the visual fields of the eyes overlap considerably; its upper incisors are vertically implanted; and its molars have two ridges of enamel like those of *Cercopithecus* monkeys.

The list of these traits brings to mind a monkeylike image, which is reflected in the drawing of the skull above. A reconstruction of the way of life of *Archaeolemur* as being similar to that of one of the living cercopithecoid monkeys should not be far wrong.

Source: W. E. Le Gros Clark. 1963. *The Antecedents of Man.* New York: Harper and Row. Fig. 58, 59, p. 145.

apes, and humans natural categories that adequately represent an evolutionary series of stable adaptive plateaus which existed in the past? Or are they useful only in terms of the present? To answer these questions, we need to deal with two problems: one has to do with language, and the other with logic.

Let us look first at the linguistic problem. Words like *monkey, ape,* and *human,* although sometimes used by scientists, are really colloquial terms, part of everyday language. They are also open to interpretation. Take the term *human,* for instance. This label evokes a picture of a creature set off from other animals not only by certain anatomical features (a large brain, a much reduced upper face tucked beneath it, a prominent chin, an S-curved spine, forelimbs shorter than hindlimbs, an expanded ilium, and so on) but also by a particular way of life characterized by a set of behavior patterns, including the social transmission of abstract ideas, dependence on complex material culture, and spoken language. These features are common to all living men and women and are not found in even our closest primate relatives. They are, in every commonsense way, human characteristics—but it is unlikely that all of them evolved at the same time. Terms like *human origins* and *early man* are thus less obvious and less clear than they seem at first. They do refer to past events and populations that preceded our own, but these populations would show differing degrees of "human" traits. Thus, the term *human* will mean different things to different anthropologists, depending on what criteria they feel are most important; unless the criteria are clearly stated, the result will be much unnecessary and unenlightening disagreement and no discussion of real issues. Scientific terms like *Hominidae* and *hominid* are therefore preferable because they are not so ambiguous. The term *Hominidae,* for example, means the family that includes all the populations ancestral to our present species but to no other species of living primate.

The second problem is a logical issue: When fossil evidence is absent or not very abundant, *horizontal classification* is often used to express the relationships among groups. Platyrrhines and catarrhines have traditionally been included together in the suborder Anthropoidea, even though they may have had different prosimian ancestors. The idea that both Old and New World monkeys represent a similar stage of evolution also reflects a horizontal system of classification. As fossil evidence becomes more abundant, a *vertical classification* may be substituted for the horizontal one. Regardless of their phenotypic appearance, fossils of a different grade may be classified as hominid if it can be demonstrated that they belonged to populations ancestral to living members of our species and not to other species of primates. That is, they are hominids if it can be shown that they belong exclusively to a hominid lineage and do not represent ancestors common to other lineages as well.

Generalizing from Individuals to Populations

The various current theories of the genetic structure of populations can lead to different ideas about the range of morphological variation within species populations. For example, the balance theory usually leads to an expectation of greater genetically based variation than the classical theory. But all the theories presented in Chapter 7 have some features in common with one another and with the earlier and more limited views of the evolutionary process described in Chapter 2. In all cases, individual specimens (living or fossil) are taken as the point of departure for comparative study. Newly discovered specimens are compared with known material. In addition, all share the assumption that similarity between two forms indicates relationship and that dissimilarity implies a more distant relationship. There, however, the theoretical resemblance ends.

The major point of departure of modern evolutionary theory is the emphasis it places on the population rather than the individual. This change derives from a shift in our view of what the individual member of a species represents. As long as species were thought to be relatively uniform, any randomly chosen individual would of course be representative, or typical, of the rest. With the greater knowledge gained by many studies of living populations in the field and in the laboratory, however, came the realization that populations vary over space and through time. No single individual or small sample of specimens can be considered representative of the genetic and morphological variability found in the entire population. Even if the fossil record were satisfactory from a qualitative standpoint, as it would be if we found occasional complete skeletons of earlier primates, the fact that few specimens are found would still make it difficult to determine the relationships among the populations that these fragments represent. The degree of difference between two fossils can be determined simply by examining the specimens, but a decision about the significance of this difference— whether the two individuals belonged to the same species population or to different species—can be made only against a background of knowledge of the range of variation. To understand primates of the past, it is necessary to know a considerable amount about primates of the present. It is important to know about the diversity of existing species, the differences among them, and the ranges of variation within each.

Different views of the variation to be expected within a species population at any one point in time, or in an evolving phyletic line through time, will lead to different positions on the number of separate species represented in any section of the fossil record. Two emotionally loaded colloquial terms are sometimes used to describe scientists who reach different conclusions on such matters. *Splitters* are taxonomists who tend to divide populations represented by a given body of fossil remains into two or more species. *Lumpers* tend to see fewer species in the same body of evidence. It is not an invariable rule, but those whose background and training have fostered a population viewpoint seem most frequently to be lumpers. A typological frame of reference is very often recognizable in the work of the splitters. From the information base built up in previous chapters, it is easy to see why different conclusions are sometimes reached. A typological approach to defining species would almost inevitably lead to oversplitting, since nearly every individual in a sexually reproducing population can be found to differ in some way from every other and since every difference can serve as the criterion for a new type.

As a practical matter, the remains of a single individual must be used with caution in reconstructing the characteristics of a population. With a single skeleton, no matter how complete, the possibility remains that the individual might not be representative of the population it belonged to when alive. If a single specimen differs from previously known material, an important question must be asked: What are the relative odds that the new specimen is (1) typical of a species not previously known or (2) an unusual member of a population already familiar (see "One Tooth: A New Species or an Atypical Member of a Known Species?")?

ONE TOOTH: A NEW SPECIES OR AN ATYPICAL MEMBER OF A KNOWN SPECIES?

In 1932 the American paleontologist G. E. Lewis found an isolated tooth near the village of Chinji in the Punjab region of India in geological deposits more than 5 million years old. He later established this tooth as the type specimen of a new primate genus and species called *Adaentotherium incognitum*. There have been many widely diverging opinions as to what this tooth represents, but no other fossil resembling it has been found since.

Recently, however, a clue regarding its identity has been discovered. In the American Museum of Natural History the lower jaw of a very large adult male mountain gorilla was found to contain a fourth molar—that is, an extra tooth. This is unusual in structure, being smaller than M_3 and rounded in outline, bearing only three cusps. But it shows many general and detailed resemblances to the fossil tooth of *Adaentotherium*.

Such extra molar teeth are infrequent in modern apes, but not unheard of. The recorded frequency of fourth molars ranges from 3.4 percent in one sample of gorillas to 14 percent in a large collection of orangutan skulls.

It has not been established beyond doubt that the single odd tooth found in India is a fourth molar of an early pongid. However, it seems more likely that a tooth which occurs only infrequently in modern apes might be found only once in the fossil record than that an entire genus and species might be represented by just one tooth in a region that has been searched often by paleontologists.

Using Single Traits to Define Taxonomic Relationships

In living pongids, the canines of both the upper and lower jaws project beyond the levels of adjacent teeth. There is also a *diastema*, or gap, between the lateral incisor and the canine in the upper jaw, and the premolar usually has a single cusp with a diagonally sloping outer surface. Some traits in this pongid list are occasionally found in humans, though, and some of the traits typical of the human dental apparatus (short, incisorlike canine; no diastema; and a bicuspid lower premolar behind the canine) turn up once in a while in living apes. Does this mean that such features are useless for differentiating between apes and humans? No, but it does mean that it is risky to assign a specimen to a species population, or to decide taxonomic affinity, on the basis of one trait or a few traits that are treated as if they were independent.

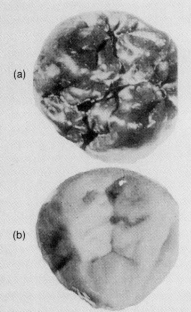

(a) **Tooth of "*Adaentotherium incognitum.*"**
(b) **Fourth molar of mountain gorilla.**

Source: Modified from I. M. Tattersall and E. L. Simons. 1969. "Notes on Some Little-Known Primate Fossils from India," *Folia Primatologia*, **10**:146–153. Fig. 3, p. 150.

A SYNTHETIC MORPHOLOGICAL PATTERN: THE PILTDOWN HOAX

On December 5, 1912, a brief note in the British journal *Nature* announced the discovery of a human skull and mandible in a gravel pit near Piltdown Common in Sussex, England. With the human bones were found primitive stone tools, two fragments of the molar tooth of an ancient elephant, a worn piece of the molar of a mastodon, and teeth from a hippopotamus, as well as the remains of other animals that no longer live in Britain. The entire collection certainly seemed quite ancient, and the human fossils were taken to be much older than any previously discovered in the British Isles.

The cranium was incomplete but could be restored as shown above. It differs little from that of living humans except in its great thickness, which is about twice that normally found in modern humans. The piece of lower jaw was far more primitive in appearance, closely resembling that of an ape like a chimpanzee or orangutan, although the molars were worn flat—more a hominid pattern than a pongid one. The morphological discrepancy between cranium and mandible led some authorities to hold that two species were represented rather than one. However, one of the more eminent anatomists of the time, Grafton Eliot Smith, held that the association of an apelike jaw with a human braincase should not be at all surprising to anyone who was familiar with recent research on human evolution. After all, since the area of the cerebral cortex is three times larger in humans than in apes, it was reasonable to assume that the brain had led the way in human evolution. And the total pattern presented by the Piltdown skull matched this expectation.

It was not until 1953 that a more lasting explanation was provided for this very excellent fit between the Piltdown evidence and the theory that existed at the time it was found: the evidence had been manufactured to fit the theory.

It is far more informative to analyze a complex of characteristics in a given functional subsystem (such as the dentition) to see whether they are related to one another and, if so, how. The anterior teeth of living pongids are not separate structures independently inherited. As the mouth closes, the lower canine slides into the diastema ahead of the upper one, and as the upper canine passes between the lower canine and the premolar, it shears against them. The efficient cutting mechanism that results is part of an integrated structural complex. Complexes such as this are said to show a *total morphological pattern* (a term invented by the British anatomist Le Gros Clark; see Le Gros Clark, 1964). Such complex patterns cannot be recognized with any

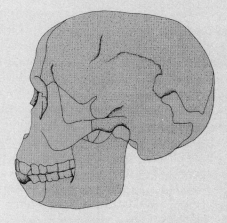

Restoration of Piltdown cranium.
Source: E. A. Hooton. 1931. *Up from Ape.* New York: Macmillan.

Piltdown was a forgery. The lower jaw looked like an ape's because it was an ape's. The humanlike pattern of wear had been produced by filing the crowns down. The bone in the jaw was still fresh, not mineralized like the skull, but both had been stained with the same chemical solution to make them appear more alike. Numerous other chemical tests confirmed that the evidence had been faked and that the scientific community as well as the public had been victimized by a fantastic hoax. Further details about the background of the event and the personalities involved are given in *The Piltdown Men,* by Ronald Millar.

The Piltdown hoax shows that well-qualified scientists can be misled by thinking that bones found together in the ground must have belonged together in life. Although purposeful frauds are in all probability quite rare, it should be remembered that misleading associations of the parts of different animals can occur by accident as well as design.

Source: R. Milar. 1972. *The Piltdown Men.* New York: St. Martin's.

degree of certainty in fragmentary, isolated specimens. Pooling several fragmentary pieces of individuals found in different places at different times does not improve the situation. On the contrary, it may result in the creation of a character complex that never existed in nature (see "A Synthetic Morphological Pattern: The Piltdown Hoax").

Reasoning from a Part to the Whole Animal

The problems we have just discussed deal with how representative a single individual is of the population from which it was drawn. The difficulty is compounded if the fossil material consists of isolated teeth or small fragments of jaws. Our objective in the study of human evolution is not to find fossils and describe the details of the bones themselves but rather to build from these pieces of evidence an image of how the living animal looked and acted. The animals themselves are complex mechanisms, although to facilitate our study, different subsystems (locomotion, feeding, reproduction) can be looked at separately. However, any single structure can serve a variety of functions. Forelimbs might be used for running on the ground, swinging from branches, picking fruit, slapping as part of an aggressive display, and grooming. Similarly, teeth in the front part of the jaw could be used for tearing the tough outer covering from a piece of fruit, scraping the soft flesh from the rind, threatening another member of the troop, and smoothing ruffled fur. For each of these functions there is probably some optimum degree of expression, but the actual structure that developed in an average individual drawn from the population would be a compromise based on all these independent needs. This is another reason why it is misleading to study a specimen in terms of single traits and to seek single causes for what we see only as isolated parts.

But although an organism is a complex whole, not all its subsystems are interrelated to the same extent. Different subsystems can evolve at different rates, producing what is called *mosaic evolution*. For example, adult members of all living species of Old World higher primates (monkeys, apes, and humans) typically have the same number of teeth (32). However, the locomotor system of humans differs as radically from those seen in apes as those of apes differ from what is seen in monkeys. The fact of mosaic evolution therefore affects our ability to reconstruct an organism and its way of life. If we have several relatively complete skeletons, all similar to one another, we are on fairly firm ground. But if part of only one subsystem—say, the dentition—is present, we are not. This is particularly true if the specimen comes from a segment of the fossil record where lineages are in dispute or if the parts show new or unfamiliar characteristics (see "On Reasoning from One Part to the Whole Animal").

Is Evolution Reversible?

Organisms are complex wholes consisting of a number of structurally and functionally integrated subsystems. Any individual displays just part of the gene pool of its species population. Further, the populations of the many species that live together in a given area interact to form an ecological community, which is itself a complex system. Some variations in individuals do not noticeably influence the gene frequency equilibrium of a species' gene pool, and some species-wide changes can have little impact on the rest of the community. But over the long run in evolution

ON REASONING FROM ONE PART TO THE WHOLE ANIMAL

Paleontologists commonly have attributed to them the ability to reconstruct the appearance of an entire animal from one tooth or bone. This apparent magic can be worked fairly easily for many recent animals. For example, the molar teeth of orangutans can be distinguished from those of chimpanzees and gorillas by the secondary crenulations, or wrinkling, of the enamel on the teeth of the Asian apes. Given one such tooth, an experienced student of primate biology could rattle off approximate body weight, arm length, details of foot and hand structure, facial appearance, and even hair color. But this would be no more magical than the process by which detectives, given a thumbprint of a known criminal, can easily state sex, approximate height and weight, eye color, and natural hair color. The process in both cases involves merely the recall of an already known pattern from one highly characteristic part.

The problem is a very different one when detectives find a print that cannot be matched from their files. They cannot reconstruct an appearance from the print alone, because neither height, weight, nor any other detail of appearance correlates very closely with fingerprint pattern. In such instances a picture must be built up laboriously from many other clues and must await confirmation until the suspect is caught. This is the same type of problem faced by paleontologists who come across part of a skeleton of a previously unknown animal. They must build up a picture of what the rest of the animal looked like on the basis of experienced conjecture. But they cannot be certain until an entire skeleton is found. Sometimes sobering surprises are in store.

In 1850 in the Great Plains region of the United States, part of a skull was found by Joseph Leidy. On the basis of its teeth he described it as a ruminant (a group that includes sheep, goats, deer, cows, and related animals) and named it *Agriochoerus* ("wild hog").

In the early part of this century Gerritt S. Miller, on an expedition from the American Museum of Natural History, found in the same region a complete hind foot, with all bones in place, of an entirely unknown type of animal. It presented quite a puzzle; the ankle bones were like those of ruminants, but the toe bones at the end of the foot were thin and must have ended in sharp claws like those of many predatory carnivores. Because of this odd set of traits it was named *Artionyx* ("clawed, even-toed" animal).

Shortly afterward, W. B. Scott of Princeton University noted that the skull of *Agriochoerus* was known, but its feet were not, while the feet of *Artionyx* were known, although no skull of it had yet been described—

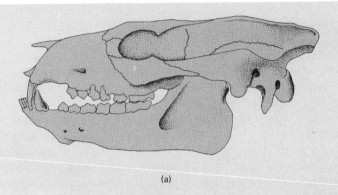

(a)

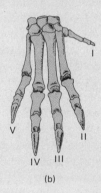

(b)

(a) **Skull of *Agriochoerus latifrons*.** (b) **Right front foot of *Agriochoerus latifrons*.**
Source: W. B. Scott. 1937. *A History of Land Mammals in the Western Hemisphere.* New York: Macmillan, Figs. 233, 234, p. 373.

and yet both animals were about the same size and had lived at the same time in the same places. The next summer two specimens were found with teeth and feet unquestionably associated, clearly confirming Scott's theory that not one but two species were represented.

This is not the only case of its kind known. Nearly exactly the same sequence of events took place with the fossil remains of chalicotheres—bulky, claw-toed near relatives of the horse. In this case as well, the puzzle was finally solved by the discovery of skeletons with the disputed bones naturally connected.

Source: G. S. Miller and J. W. Gidley. 1931. "Part II: Mammals and How They Are Studied," in *Warm-Blooded Vertebrates.* Washington, D.C.: Smithsonian Scientific Series, vol. 9. Pp. 196–198.

most changes, however slight, have ripple effects. An important consequence of the interaction of individuals and species is that once a number of related, sequential changes have taken place, the weight of accumulated evidence shows that a return to an earlier state is unlikely. Evolution, in other words, is irreversible.

The principle that evolution cannot be reversed is also applied to levels below that of an entire ecological community. Within single phyletic lines it should remain a valid generalization for a change that has a complex genetic basis. Once a series of stepwise changes has occurred, it is unlikely that the sequence could ever be reversed in precisely the same order. But sometimes this principle, often referred to as *Dollo's law*, is applied overenthusiastically. For example, it may be asserted that a phyletic line whose members had increased in size could not later have descendants of a smaller average size. Or it may be said that a change of shape or structure could not ever be "undone." As shown in "Irreversibility of Evolution: Evidence from Paleontology and Genetics," this is not always true. In reality, all the statements about the reversibility or irreversibility of evolution should be given as probabilities. The more complex the change, for example, the smaller the likelihood that it could be reversed.

IRREVERSIBILITY OF EVOLUTION: EVIDENCE FROM PALEONTOLOGY AND GENETICS

That evolution is irreversible is a principle stated in its most familiar form by Louis Dollo (1857–1931), a prominent Belgian paleontologist. Like many principles, it is generally true, but exceptions do exist. Two are discussed below.

Paleontology: Return of a Lost Structure

Cats come close to being pure carnivores. As such they have little need for the grinding surfaces that molars normally supply. Typically, in felines only one molar tooth is left in each jaw, giving a dental formula of $\frac{3131}{3121}$. This condition has existed for tens of millions of years. Yet in

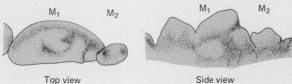

Top view   Side view   **Dentition of northern lynx.**

one living species of cat, the northern lynx, M_2 is found in appreciable frequencies. Out of a total of 53 skulls from Finland and Sweden, 5 have this tooth present—a frequency of nearly 10 percent. But the second molar has not been found in any fossil lynx. This may therefore be regarded as evidence of a probable evolutionary reversal, return of a structure that had been lost, possibly through suppression or switching off of the genes that produced its expression.

Genetics: How a Lost Structure Can Return

Guinea pigs normally have four toes on each front foot and three on each hind foot (below left). This is a reduction from the original pentadactyl (five-toed) structure of the foot in the ancestors of all land-living vertebrates. Additional toes that occur sporadically among guinea

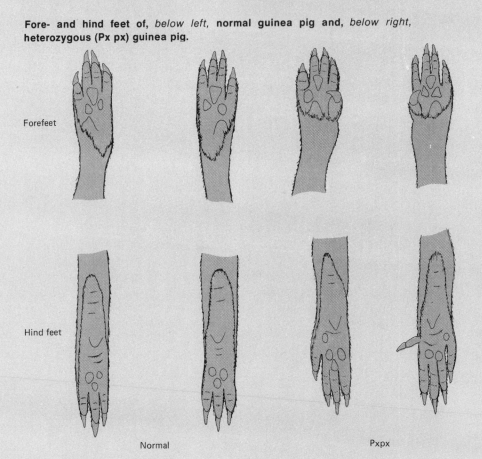

Fore- and hind feet of, *below left,* **normal guinea pig and,** *below right,* **heterozygous (Px px) guinea pig.**

Forefeet

Hind feet

Normal

Pxpx

pigs are said to be *atavistic* structures (representing the recurrence of structures typical of ancestors more remote than the parents).

Genetic crosses indicate that when little toes recur, they are typically under polygenic control. If selected for, they can be fixed in stocks (designated D) which then breed true for the trait.

In one case a normal pair of guinea pigs became the parents of an offspring bearing not only little toes but also a thumb on each front foot and a big toe on one hind foot. This trait complex proved to be under the control of a gene (Px) which was semidominant; over half of the heterozygotes had both thumb and little toe present on one or both sides, and other animals had lesser degrees of expression (thumbs but not little toes, etc.). Crosses between the stock with polygenically controlled little toes (D) and Px heterozygotes produced nearly 100 percent presence of little toes and thumbs in the heterozygotes.

At first glance it would seem that these traits might be controlled by genes which had been reduced to very low frequencies in the course of evolution, but not yet completely lost, and that careful breeding from the rare individuals in which they occur restores the ancestral condition. But there is strong evidence against this idea. Individuals who are homozygous PxPx are all monsters and are aborted early or are dead at birth. They have bulging foreheads, sometimes with the brain protruding; microphthalmia (tiny eyes); and nostrils connected with the mouth by symmetrical clefts. Their hind legs are rotated so that they face the belly, and the feet on these limbs are approximately twice the normal width. Each foot has a large number of small, clawed toes (as many as 12 on one foot, and as many as 44 in total on all four feet).

It is extremely improbable that the Px allele, which produces a gross monstrosity when homozygous, can be an ancestral gene. More probably, during the course of evolution the genes controlling the development of the ancestral pentadactyl foot have been kept, but are now overlaid by newly evolved hereditary factors. These new genes act to suppress certain parts of the foot as the embryo grows. The Px gene could thus represent a mutant which modifies these suppressor genes.

In this case evolution has not really been reversed, since the genetic basis of the new pentadactyl foot is not the same as that found in ancestral populations. But in the fossil record only structures are seen, not genotypes. And it is possible that in some cases the phenotypic appearance of more recent populations could quite closely match that of earlier ancestors.

Sources: (1) B. Kurtén. "Return of a Lost Structure in the Evolution of the Felid Dentition," *Societas Scientiarum Fennica, Commentationes Biological,* **26**(4):1–12. Fig. 2, p. 10. (2) S. Wright. 1935. "A Mutation of the Guinea Pig, Tending to Restore the Pentadactyl Foot When Heterozygous, Producing a Monstrosity When Homozygous," *Genetics,* **20**:84–107. Figs. 1, 2, p. 90.

Convergent and Parallel Evolution

This assumption operates in the opposite direction of one mentioned earlier—that degree of similarity can be equated with closeness of genetic relationship. *Convergence* refers to the evolution of similarity between members of separate phyletic lines whose ancestors were less alike. There are numerous good examples of convergence. The most impressive is found in the resemblance of species of marsupial mammals in Australia (wolflike carnivores, catlike carnivores, anteaters, mouse-sized seed eaters, etc.) to the corresponding species that fill similar ecological niches on other continents. A closer look, however, shows that the functional similarities follow from different inherited structural bases (Figure 11-10). *Parallel evolution* is the term usually applied to lineages which were rather alike to begin with and which both went through a similar series of evolutionary changes at about the same rates. Old World and New World monkeys, which have evolved independently from prosimian ancestors, are often cited as an example of parallel evolution. The probable closeness of starting points is what distinguishes parallelism from convergence.

To detect either one, a fossil record of some time depth and completeness is necessary. If this is not available, all we can see is the similarity of end points. This similarity could be due to parallelism, to convergence, or even to divergence from the same ancestors (see Figure 11-11). Unless fossil remains are accurately dated, ancestors and descendants (which resemble one another in some features and yet differ in others) might be mistaken for contemporaneous populations whose resemblances are due to convergent or parallel evolution.

Parallelism and convergence do occur in the course of evolution, and they have probably operated at various times in the long history of primate evolution. But they are often overworked as explanations for resemblances between fossil specimens or the populations they are presumed to represent. Parallelism, in particular, proceeds from similar genetic and phenotypic potentials, and the degree of parallelism is usually in proportion to the degree of relationship. Whenever parallelism seems to be an especially striking coincidence, the alternative possibility of similarity due to ancestral-descendant relationship should also be carefully examined.

Significance of Geological Age

Good evidence of the geological age of a specimen is extremely important for correct placement in an evolutionary sequence. Without it, for example, a fossil may be treated as if it represented a population as old chronologically as it appears to be. But we know from many examples that not all populations of a given lineage evolve at the same

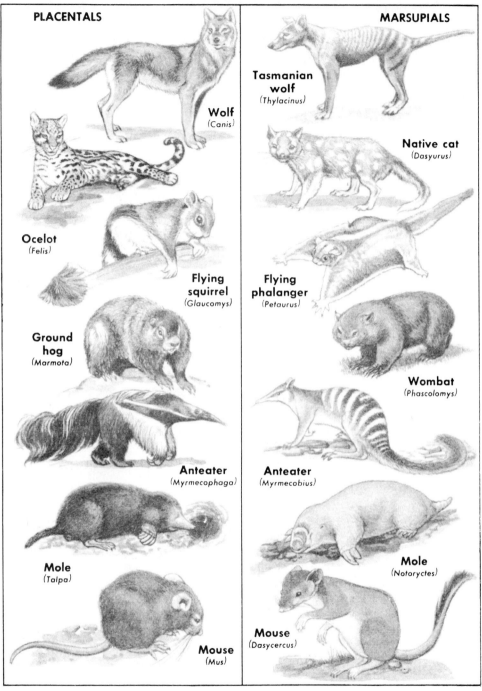

PLACENTALS

Wolf
(Canis)

Ocelot
(Felis)

Flying
squirrel
(Glaucomys)

Ground
hog
(Marmota)

Anteater
(Myrmecophaga)

Mole
(Talpa)

Mouse
(Mus)

MARSUPIALS

Tasmanian
wolf
(Thylacinus)

Native cat
(Dasyurus)

Flying
phalanger
(Petaurus)

Wombat
(Phascolomys)

Anteater
(Myrmecobius)

Mole
(Notoryctes)

Mouse
(Dasycercus)

FIGURE 11-10 Convergent evolution of marsupial and placental mammals.

Source: G. G. Simpson and W. S. Beck. 1965. *Life: An Introduction to Biology* (2d ed.). New York: Harcourt, Brace, and World. Fig. 8-7, p. 500. © 1965 by Harcourt Brace Jovanovich, Inc., and used with their permission.

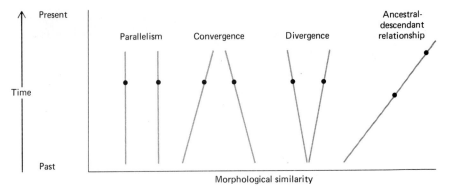

FIGURE 11-11 Similar degrees of resemblance produced by different courses of evolution. The dark dots indicate populations sampled; all pairs are the same distance apart on the morphological scale.

rate. In the opposite direction, specimens that look phenotypically modern may appear in older contexts. There is no infallible general rule for resolving paradoxes of this sort. Sometimes geological and morphological sources of evidence, while apparently contradictory, are both upheld. The coelacanth, discussed earlier, is a case in point. Sometimes either geological or morphological data will prove faulty on further analysis. A specimen may appear modern because it is in fact recent and was mixed with deposits of earlier ages by an accident such as burial. Alternatively, a bone that is superficially modern may be genuinely old. It is essential in such cases not to confuse possibility with probability or certainty.

Because knowledge about the geological age of fossils is so important in reconstructing the course of human evolution, many different methods have been developed to assign dates to the specimens themselves or to the geological deposits in which they are found. *Relative dating* methods give only the sequential positions of two specimens, allowing us to determine which preceded the other. *Absolute* or *chronometric dating* techniques make it possible to estimate the dates at which populations lived. Examples of these various techniques are outlined in the appendix at the end of this chapter.

Once the sequence of populations has been established and the amount of time that separates them has been estimated, it becomes possible not only to assess how much one specimen differs from another but also to determine whether the differences are too great to have been produced by phyletic evolution in the time that separates the populations from which they were drawn. Of course, not all dating techniques are equally useful in the study of human evolution. Each dates a different segment of the geologic time scale (see Table 11-2). The potassium-argon technique spans the longest period of time and can be used whenever rock samples of the appropriate chemical

TABLE 11-2   The standard geologic column and time scale (as recognized by the United States Geological Survey)*

Eras	Systems (of rocks) Periods (of time)	Series (of rocks) Epochs (of time)
Cenozoic (recent life)	Quaternary (an addition to the old tripartite 18th-century classification)	Recent Pleistocene (*most recent*)
	Tertiary (third, from the 18th-century classification)	Pliocene (*very recent*) Miocene (*moderately recent*) Oligocene (*slightly recent*) Eocene (*dawn of the recent*) Paleocene (*early dawn of the recent*)
Mesozoic (middle life)	Cretaceous (chalk) Jurassic (Jura Mts., Europe) Triassic (from tripartite division in Germany)	

Paleozoic (ancient life)	Permian (Perm, a province in Russia)
	Carboniferous systems (from abundance of coal in these rocks)
	    Pennsylvanian
	    Mississippian
	Devonian (Devonshire, England)
	Silurian (an ancient British tribe, the Silures)
	Ordovician (an ancient British tribe, the Ordovices)
	Cambrian (Roman name for Wales)	

Precambrian	Many local systems and series are recognized but no well-established worldwide classification has yet been attained.	

* The same terms are used, as indicated, both for the rock sequences and for the time intervals during which these strata accumulated. Definitions in italic are from the Greek.

Note: Although local names are frequently used for series and epochs within Mesozoic and older strata, the generally recognized subdivisions are simply *lower, middle,* and *upper* series (*early, middle,* and *late* epochs) of the respective systems. Thus, where it is convenient to talk of smaller units, we may speak of the upper Cretaceous series (of strata) which accumulated during the late Cretaceous epoch (of time).

Source: John S. Shelton. 1966. *Geology Illustrated.* San Francisco: Freeman. Fig. 291, p. 296. Modified from Gilluly, Waters, and Woodford. 1959. *Principles of Geology.* San Francisco: Freeman.

composition are available. The uranium-series and fission-track techniques are also useful for much of the record, and so they provide alternatives to the potassium-argon method or independent checks on it. Unfortunately, despite the many dating methods available, the vast majority of hominid fossils found so far remain undated. Many were discovered before sophisticated dating techniques had been invented, and in many cases the sites from which the specimens came have been destroyed. And not all recently found hominid remains are accompanied by materials suitable for dating. For this reason, relative dating techniques based on correlation of geological strata or fossils contained within them are still useful despite their lack of precision.

There are other less evident but nevertheless important limitations on the dating of hominid specimens. Except for radiocarbon dating, the various chronometric techniques do not directly date the fossil remains themselves, but only the mineral formations somehow associated with them. The confidence with which the ages of the actual fossils can be specified depends to a great degree on knowledge of the connections between them and the datable materials. In the great majority of cases, chronometric techniques are used to date rock formations that lie above or below sediments containing the fossils. Therefore, the dates can do no more than provide upper or lower limits on the age of the bones. A more precise determination would require knowledge of their location in the sedimentary strata and of the rates at which these sediments were deposited. Even if the mineral sample dated is immediately adjacent to the fossil, some uncertainty about the age of both will remain. Dating techniques are the outcome of laboratory procedures, all of which have some associated error factor. This is shown by the form in which dates are correctly given. A mineral sample might be given an age of 2.61 ± 0.62 million years old, which means that there are about 2 chances out of 3 that the mineral, and perhaps the associated fossil, is somewhere between 1.99 (that is, $2.61 - 0.62$) and 3.23 ($2.61 + 0.62$) million years old. What this unavoidable inaccuracy means can be seen by imagining that we have not one fossil but two, each from a different site. If both were coincidentally associated with rocks, which, when dated, gave the same age of 2.61 ± 0.62 million years, it is possible that the two individuals represented by the fossils lived over a million years apart. This is rather a wide range considering that the bulk of well-documented hominid evolution spans only about 3 million years.

FROM EVIDENCE TO CONCLUSIONS

Finding the earliest members of any taxonomic category requires the interpretation of observed similarities and differences as well as the study of fossil remains. The major decision is whether any two individual specimens are from the same lineage or from two different lineages. In the course of phyletic evolution, slight genetic changes can accumulate generation by generation even in one continuous, unbranching lineage. Members of time-successive populations will therefore usually differ from one another in appearance. When two such populations are separated by a very long gap in the fossil record, they might be as distinct from each other as two different species of the same genus living now. The magnitude of this difference in appearance between ancestors and their direct descendants is sometimes expressed by even giving them different formal taxonomic names. But the distinctions that can sometimes be made between two evolutionary stages may disappear as new specimens are found to fill the previous gap.

Speciation probably begins most often by geographic isolation. A new species develops when a group first becomes isolated geographically from its parent population. During the isolation, each group evolves characteristics suited to its environment. Fossil remains are attributed to hominids on the basis of their morphological characteristics, but it is unlikely that the speciation process begins with large-scale morphological changes. After a gene pool is subdivided, there are gradual shifts in the frequencies of alleles at various loci in the resulting subdivisions. Only after the passage of time do the phenotypic characteristics of individuals reflect these underlying genetic changes to any great extent. As was stressed earlier, even in living populations for which we have many data, phenotypic characteristics are imperfect guides to the presence or absence of isolating mechanisms that could restrict gene flow. With living organisms, the existence of population boundaries can be checked against observations of actual breeding behavior. This is not possible, of course, when we are dealing with fossil evidence, but the problem is not an insuperable one. By the time populations have reached the stage where their members can be recognized as hominids on the basis of one or a few fragmentary fossils, their gene pools should have been distinct for a considerable period of time. Evidence for the existence of a distinct population might take the form, for example, of diagnostic portions of several individuals together in an appropriate ecological setting. When such evidence is not available, the belief that a single fragmentary specimen is hominid must remain as uncertain as the evidence on which it is based.

The most distinctive physical characteristics of hominids evolved in conjunction with behavioral attributes and in a sense are secondary reflections of them. William King Gregory, a distinguished early student of primate evolution, referred to the unique features of each species as *characters of habitus* because (like derived character states) they evolve as the result of selective pressures activated by the particular habits or way of life of the animal. He thus differentiated them from *characters of heritage,* which are primitive character states inherited from ancestors in an earlier stage of evolution (before the assumption of present life habits) and which have been retained either because they are still useful or because they haven't yet been eliminated by subsequent adaptations (see Gregory, 1922).

Unique hominid traits can be seen merely by comparing humans with other living primates. To discover when these characters first appeared, the geological record must be searched for the remains of forms that appear to show their first faint stamps. To discover why hominid populations evolved at all, it is necessary to examine collateral evidence—the associated remains of other animals and of plants, and even the mineral composition of the sediments in which all are buried. Only then is it possible to attempt to reconstruct the appearance and way of life of the animal in the context of the conditions that selected for its newly evolving characteristics. Thus, although the fossil remains of earlier primates provide us with important cornerstones for reconstructing the sequence of human evolution, any interpretation based on them must be combined with what we know about the nature, extent, and causes of variation in living populations and about the processes of evolutionary change. If living primates are seen as forming graded series of steps from humans downward to insectivores, then the primate fossil record consists of the chips that fell when stairways were carved step by step from sloping inclines. To find the earliest hominids, we must search for and interpret whatever of these bits still remain.

SUMMARY

1. Species have been defined on the basis of three separate concepts: typological, nondimensional, and multidimensional. Each view has a different implication for the expected variation within species, the extent of distinctions between species, and the methods by which species can be recognized.

2. The speciation process is generally viewed as having several distinct stages: initial subdivision of a gene pool, independent evolution of the separate gene pools, and sometimes later overlap of population ranges.

3. In the study of evolution it is necessary to distinguish between populations ancestral to a number of species, including our own, and populations ancestral to those of modern humans alone.

4. Evidence for reconstruction of human evolution comes from two major sources: living populations and those whose remains are preserved in the fossil record. Both sources have special advantages and limitations, and they complement rather than substitute for each other.

5. The evolutionary end points represented by living primate populations may not show the actual range of potential adaptive patterns; thus, although all living primates can be assigned to one or another evolutionary step, some populations in the fossil record may not fit so neatly.

6. Not all individuals are equally representative of the population from which they were drawn; one or a few unusual primate fossils might represent extreme variants of an already known population rather than a new species.

7. Single traits give an uncertain basis for deciding phylogenetic relationships; the less complete a fossil specimen, the less reliable the conclusions based on it.

8. Not all functional subsystems of an organism evolve at the same rate, and so caution must be used when reconstructing the appearance and way of life of an animal from a partial skeleton.

9. The more complex and far-reaching an evolutionary change is, the less likely it is that it can be reversed to restore the ancestral condition.

10. Resemblances among different populations can be due to parallelism and convergence, but the likelihood that either is the explanation for a resemblance between two fossil specimens must be weighed against the possibility that the specimens are alike because they are from the same lineage.

11. Morphological appearance is not a secure guide to geological age, although morphological criteria may provide a tentative guide to a population's place in an evolutionary sequence.

12. Events within hominid evolution can be dated in a number of ways: relative dating allows different events to be placed in a sequence, and absolute dating ties their time of occurrence to a known chronometric scale.

13. Each dating technique is useful with different raw materials and a particular time range, but often two or more techniques can provide independent estimates of the age of associated hominid remains.

14. In the study of hominid evolution, the limitations of dating methods are important. Even if two specimens have identical dates, the populations from which they were drawn could have lived thousands of generations apart because the ± error factor in dating is so great.

APPENDIX TO CHAPTER 11: DATING METHODS

Relative Dating

The sequence of events in the earth's history has never been uniform everywhere; sediments of a given age may have formed in some areas but not in others. And among those areas where sediments were deposited at the same time, only a few scattered sites may have escaped the leveling effects of later erosion. Geologists have built a view of the past by correlating partial sequences of geological strata in many different parts of the world. From this work, an idealized section of the earth's crust has been constructed in which all the rock layers formed from the beginning of the earth's history until the present are shown in a continuous series. A schematic view of this idealized section, known as the *standard geologic column*, is shown in Table 11-3. Any deposits that are found to contain fossil primates can be compared with it, and thus we can arrange the fossils themselves in a time sequence. When fossil remains, cultural implements, or other items are arranged in a series like this, the process is called *relative dating*. Relative dating allows a reduction in the number of interpretive schemes simply because populations that came later in time cannot be ancestral to those which lived earlier.

One evident feature of the geologic column is that its subdivisions become finer toward the present. This is due largely to the fact that

TABLE 11-3   Carbon-14 dates for historically dated objects

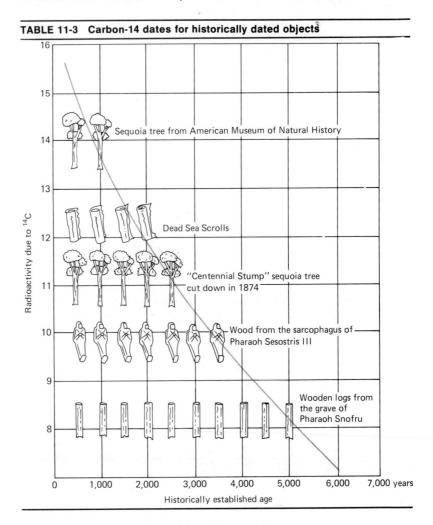

earlier strata are distorted by the weight of more recent deposits that bear down on them and obliterate details, making the record of the distant past increasingly more difficult to read. As a result, relative dating becomes less precise further back in time. The most serious limitation of relative dating, though, is that it provides little or no idea of the amount of time separating two populations which might have been in an ancestor-descendant relationship. Although the time gap separating them is not the only variable to consider, the greater the time available, the more evolutionary change is possible from a given level of selection. For this reason, it would be desirable to measure as precisely as possible the length of these intervening spans of time. In response to this need, the absolute dating techniques described below were developed.

Absolute Dating

Absolute or *chronometric* or *radiometric dating* refers to quantitative measures of time with respect to a known scale. The various chronometric dating techniques are all based on a common assumption: that some natural phenomena, such as the radioactive decay of certain elements, occur at relatively constant, measurable rates. Given knowledge of these rates and the extent to which radioactive breakdown has occurred in the sample under study, it is possible to determine its age in terms of our calendar years. *Radiocarbon dating* was the first of the chronometric techniques to gain acceptance among anthropologists, and it remains the most widely used method for later segments of the fossil record. Since the basic assumptions underlying its use are common to many others developed more recently, they will be presented in some detail.

Radioactive carbon is produced by a chain reaction that begins when cosmic rays bombard the earth's atmosphere, producing uncharged particles (neutrons) of high energy. When these neutrons collide with the nucleus of any atom, they can cause great changes in its structure and properties. Since nitrogen is the most abundant element in the atmosphere (forming nearly 75 percent of the air by weight), atoms of nitrogen are the ones most frequently struck. They are transmuted into carbon-14, or radio (active) carbon. This is a less common form of the element ^{12}C, differing from it in that it contains two additional neutrons in the atomic nucleus. Carbon-12 and carbon-14 are *isotopes,* atoms of a single element that differ in mass.

Because this cosmic ray bombardment has been going on for an extremely long time and because the conversion of ^{14}N to ^{14}C is a reversible reaction, the conditions for an equilibrium exist. Thus, ^{14}C forms a constant, if small, fraction of the total pool of carbon present in the atmosphere, where it exists in the form of CO_2. Much of CO_2 is in living matter because it enters the food chain by means of the photosynthetic activities of plants, which are in turn consumed by animals. When an organism dies, however, carbon intake ceases. Breakdown of ^{14}C then takes place without any compensatory replacement. The rate of this radioactive decay is reckoned in terms of the *half-life* of the element, which is the time it takes, on the average, for half of any given quantity of ^{14}C to be transformed back into ^{14}N. For carbon-14 this is presently believed to be about 5568 years (Grootes, 1978). Thus if a given sample of a carbon-containing substance originally held 2 g (0.07 oz) of ^{14}C, after 5568 years only 1 g (0.035 oz) would be left. A sample of material this old would emit only half as many radiations per minute as a sample taken from a living organism or one which had just died. By monitoring the number of radiations emitted per minute per gram of substance, we can estimate its age in years. Table 11-4 shows several examples of close agreement between dates provided by historical information and the radiocarbon technique.

With very old samples of carbon, the radiation level being measured becomes very small relative to normal background radiation emitted by all objects, including the investigator. This sets a practical limit to the age of specimens which can be dated with any degree of accuracy to about 50,000 years. With the special techniques discussed by Grootes, this can be extended at the present time to about 75,000 years. But even at that outer limit, radiocarbon dating is unfortunately useful only in the very latest stages of the hominid fossil record.

Potassium-argon dating is far more useful for the study of primate evolution because it allows the dating of materials from a much broader span of time. Published dates already include some as old as 4.5 billion years (in all probability prior to the origin of life on earth) and others as recent as 2500 years (within the range of recorded human history). Several factors are responsible for this wide range of usefulness. First, there is a great abundance of potassium; it is found in nearly every mineral in the earth's crust. Second, the radioactive isotope potassium-40 (abbreviated ^{40}K) is very stable, breaking down steadily but only very slowly into calcium-40 (^{40}Ca) and argon-40 (^{40}Ar), an inert gas. The conventionally accepted half-life of ^{40}K is 1.3 ± 0.04 billion years. The development of methods for detecting very minute quantities of argon has made it possible to date even very recently formed minerals.

While chemically stable, argon gas can escape when the rock in which it is contained becomes molten. Measurement of the age of a sample therefore proceeds on the assumption that any argon present must have accumulated after the mineral cooled and hardened. Unfortunately, this assumption is not always valid, and, as a result, not all minerals give equally reliable results. Rocks that lose argon readily by diffusion into the atmosphere would show an artificially recent date. Care must also be taken to ensure that the sample does not include

TABLE 11-4   Chronometric dating techniques and portions of the time scale which they date

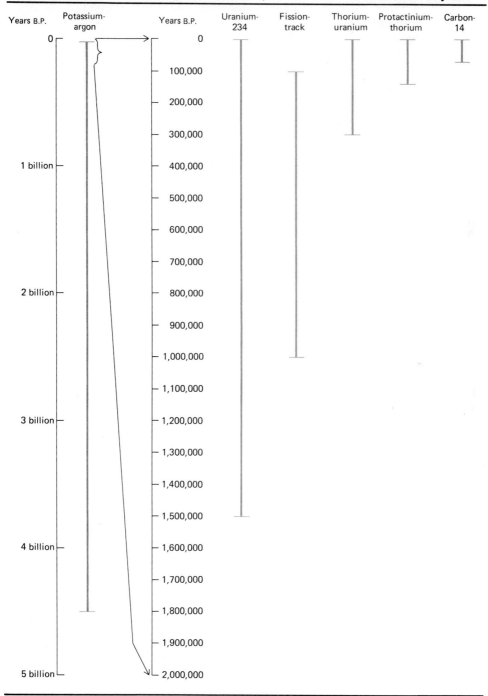

fragments of older minerals which remained unmelted when incorporated into the new mineral. Inclusion of older, unmelted material containing trapped argon would bias the apparent date in the opposite direction, making the sample appear older. Despite these problems, appropriate techniques and procedural safeguards exist to aid in the interpretation of dates arrived at with the potassium-argon method.

Uranium-series dating includes a number of related procedures that all depend on the progressive decomposition of uranium-238 and uranium-234. Both are soluble and typically present at very low concentrations in carbonate solutions. Organisms that build carbonate-containing parts, such as shells, will fix these radioactive elements as well. While the ratio of uranium-234 to uranium-238 varies from one body of fresh water to another, at the present time in ocean waters it is a fairly constant 1.15. With the passage of time, more and more of the ^{238}U is converted to ^{234}U, and so the ratio increases. This makes it possible to estimate the age of carbonate-containing material such as coral. The range of this dating technique is about 2 to 1½ million years.

The *thorium-uranium method* (sometimes also referred to as the *ionium-deficiency method*) is possible because of the further breakdown of uranium-234. While ^{234}U is soluble, one of its breakdown products, thorium-230, is not. Hence it is not included in materials removed from solution as carbonate-containing compounds are formed. Once this process ceases, however, thorium will begin to build up. In any given sample, the amount of thorium present will be a function of the original concentration of ^{234}U and the period of time available for its breakdown to occur. Since thorium (half-life, 75,000 years) itself undergoes radioactive decay at a different rate from that at which ^{234}U (half-life, 248,000 years) undergoes this decay, ultimately an equilibrium is reached. This occurs after about 500,000 years and sets the theoretical outer limit of this technique. In practice the effective limit of material which can be dated is about 200,000 to 300,000 years.

The *protactinium-thorium* (or *protactinium-ionium*) *method* is based on the simultaneous presence of both ^{234}U and ^{238}U in ocean waters. When these decay in sediments formed by precipitation, they give rise to both thorium-230 and protactinium-231 (half-life, 32,500 years). The ratio of these two elements is independent of the original concentration of uranium isotopes. It is therefore a function only of the time elapsed following deposition of the sediments, barring contamination from disruptions or intrusions which could result from later geological events and also barring loss by diffusion after the sediments have been formed. The time range in which samples can be dated by this manner extends back to about 140,000 years.

Fission-track dating shares some points of similarity with the two radiometric techniques. Like potassium-argon dating, this method is usually used to measure the age of newly formed mineral deposits associated with fossil remains. But like that of carbon-14 dating, its usefulness is limited to later phases of the geologic time scale—from about 100,000 to about 1 million years.

Datable with the fission-track method are any of the numerous minerals that contain uranium, even in small quantities. Like other radioactive substances, uranium undergoes a process of decay through time. Usually this occurs gradually as a result of the successive emission of a number of alpha particles, which are fast-moving helium nuclei. Loss of one of these changes the atom which emitted it into another element with an atomic number smaller by 2 and an atomic weight smaller by 4. About 1 in every 2 million atoms, however, breaks down into two nearly equal parts. *Fission tracks* are produced when these relatively massive particles tunnel through the surrounding material. For a given concentration of uranium, the density of these fission tracks increases with the age of the sample. Thus, by counting the number of tracks in a given area and measuring the uranium content, the age of the sample can be calculated. Uranium content is measured by inducing the uranium-235 present to fission by neutron bombardment. Like spontaneous fission, this also produces tracks in the material. Thus the age is calculated from the ratio of spontaneous to induced fission tracks.

The primary value of this technique is that, within a time scale of up to 1 million years, it can provide a measure of the age of some minerals independent of that which can be obtained from the potassium-argon method. Confidence in this independence is enhanced by the observation that the possible sources of error in the two methods are different.

SUGGESTIONS FOR ADDITIONAL READING

Clark, W. E. Le Gros. 1964. *The Fossil Evidence for Human Evolution* (rev. ed.). Chicago: University of Chicago Press. (This is a cogently reasoned introduction to the practices and pitfalls of interpreting fossil remains.)

Mayr, Ernst. 1976. *Evolution and the Diversity of Life.* Cambridge, Mass.: Belknap (Harvard University Press). (This collection of essays is useful in understanding the processes at work in macroevolution. Section V, "Theory of Systematics," is particularly good.)

Rensch, Bernhard. 1959. *Evolution above the Species Level.* New York: Columbia University Press. (This is a thoughtful and somewhat philosophical approach that attempts to outline the major rules by which the process of evolution works to bring about large-scale changes.)

REFERENCES

Clark, W. E. Le Gros. 1964. *The Fossil Evidence for Human Evolution* (rev. ed.). Chicago: University of Chicago Press.

Darymple, G. B., and M. A. Lanphere. 1969. *Potassium-Argon Dating.* San Francisco: Freeman.

Gregory, W. K. 1922. *The Origin and Evolution of the Human Dentition.* Baltimore: William and Wilkins.

Grootes, P. M. 1978. "Carbon-14 Time Scale Extended: Comparison of Chronologies," *Science,* **200**(4337): 11–15.

Mayr, Ernst. 1965. *Animal Species and Evolution.* Cambridge, Mass.: Belknap (Harvard University Press).

Michels, J. 1973. *Dating Methods in Archaeology.* New York: Seminar.

Oakley, K. P. 1966. *Frameworks for Dating Fossil Man* (2d ed.). Chicago: Aldine.

Osborn, H. F. 1910. *The Age of Mammals in Europe, Asia, and North America.* New York: Macmillan.

Tattersall, Ian, and Niles Eldredge. 1977. "Fact, Theory, and Fantasy in Human Paleontology," *American Scientist,* **65**(2):204–211.

Zeuner, F. E. 1958. *Dating the Past.* Darien, Conn.: Hafner.

Chapter 12

Hominid Origins:
The Search for Ancestors

In the study of human evolution it is important to distinguish between two senses in which the term *human ancestors* is used. If an evolutionary view of the universe is correct, humans have ancestors that stretch backward in an unbroken sequence to the very origin of life, over 3 billion years ago. Not all these ancestors, however, can be considered human themselves, unless we are willing to apply the label *human* to certain reptiles, amphibians, and fish. Moreover, in our study of human evolution we focus on the primates, rather than on earlier forms of life. Within this order, we need to differentiate between those primates which are ancestral to a number of later forms, including humans, and those which are members of our own evolutionary line.

Precisely how early these human ancestors existed is still a matter of some debate. Different students of human evolution have suggested the existence of independent lines of human ancestry which began as far back as 40 million years ago or as recently as 5 or 6 million years ago. Debates of this sort have existed for well over a century. As we saw in Chapter 11, problems of this sort are rooted in our reliance on the fossil record for direct evidence on the appearance of earlier primates, the environmental settings which shaped their characteristics, and the time periods during which they existed. Most of the time the fossil remains themselves are quite scarce and fragmentary. Dates, too, are often uncertain; often the best that we can do is to assign a fossil to a geological age that spanned several million years.

Primate fossils have been found in geological deposits of various ages in many different parts of the world. Sometimes they are dug up in the course of an expedition launched for this purpose. This is the ideal situation because then their context can be noted and later used to reconstruct the ecological setting and estimate the age of the deposits. But fossils are also exposed by erosion of the sediments in which they were originally buried. They may be discovered then, or they may be redeposited elsewhere, perhaps to be found at a later time.

Many fossils are found by accident. Among the early discoveries was a lower jaw quite like that of living great apes, unexpectedly found in France over a century ago. In recent years, primate fossil finds have increasingly been made by scientific expeditions, but accident continues to play a significant role. For example, the first lower jaw of a large hominoid primate known as *Gigantopithecus* ("giant ape") was discovered in China's Kwangsi province in 1956 by a peasant digging in a mountain cave for mineral-rich earth to use as fertilizer. And an American expedition organized in the 1930s to search the Gobi desert of Mongolia for fossil primates had as its most notable find several clutches of dinosaur eggs. Serendipity plays a far greater role in science than is generally acknowledged.

In this chapter we will review the fossil evidence that such skill and luck have brought to us so far. Along the way many different taxonomic groups of earlier primates will be discussed and, as far as possible, reconstructed as they appeared in life. Our central concern, however, will be to identify the earliest origin of our own human evolutionary line and the reasons why it began. It is unlikely that general agreement will be reached on these points during the lifetime of anyone reading this book. However, we can narrow the possibilities to a few of the more likely candidates for early humans, and we can learn a good deal about our earlier relatives along the way.

THE WORLD OF THE PAST

How do we evaluate the significance of fossil remains for primate evolution in general and for human evolution in particular? One major step in the process is placing known fossils into a temporal and spatial framework and then referring all new finds to their proper place in this matrix.

Time: Geologic Eras

Our era, the Cenozoic, is the latest of four major geologic eras that have existed since life originated several billion years ago. The Cenozoic (time of recent life) began about 70 million years ago. The preceding Mesozoic era (middle life) lasted 150 million years, from 70 million years ago back to its beginning 220 million years ago. The Paleozoic era (ancient life) was longer still, spanning 180 million years, from 220 to 400 million years ago. The Precambrian era (older than the Cambrian rocks at the base of the Paleozoic) stretches back over a couple of billion years that saw the formation of the earth's crust and the origin of life itself.

It has taken geologists about 200 years of careful study to arrive at this dated system of eras and their finer subdivisions, shown in Table 12-1. This work began with the mapping of outcrops of rock that were

TABLE 12-1   Geological epochs of the Tertiary period

	Millions of years before present (approximate)*
Quaternary Recent	
	0.01
Pleistocene	
	2–3
Tertiary / Cenozoic Pliocene	
	5
Miocene	
	25
Oligocene	
	40
Eocene	
	60
Paleocene	
	75
Cretaceous	
	135

*These temporal boundaries should be considered provisional rather than absolute, since their definitions and locations are still topics of continuing research by geologists. For example, within the last few years the duration of the Pliocene has been substantially reduced. Some geological deposits formerly considered to be lower Pliocene are now instead assigned to the upper Miocene, and some upper Pliocene deposits have been redefined as early Pleistocene. For a fuller discussion of the establishment of the Pliocene-Pleistocene boundary, see the beginning of Chap. 15.

exposed in various places. Study of the maps revealed that the same sequences of layers occurred in different areas, suggesting that the events that produced the layers operated over extensive areas of the earth's surface. From there it was a short—but momentous—step to the realization that the lower layers were deposited earlier than higher ones (just as in a building the bottom rows of brick must be laid before the upper rows). Later study of the fossils contained in the rocks revealed that over broad areas these, too, showed a regular succession. Correlation of strata from all over the world has made it possible to construct the standard geologic column shown in Table 11-3. Now when newly discovered deposits are studied for the first time, their rocks and fossils can be matched with known occurrences elsewhere, and an estimate of their age obtained with the methods discussed in the appendix to Chapter 11. Without this temporal framework, the study of evolution would be haphazard at best, and probably impossible.

Many discussions of primate evolution treat the subject as if this order evolved entirely within the latest era of geological history, the Cenozoic. Instead, recent discoveries indicate that one of the most crucial events— the very origin of the primates—took place sometime during the Cretaceous period of the preceding Mesozoic era. But because we have far fewer specimens of Mesozoic primates than of those from the Cenozoic, we will focus on the latter era here.

Once we know the duration of time represented by each of the respective strata, we can use the relative and absolute dating methods outlined in Chapter 11 to begin to relate primate fossils to one another. We can then set up a temporal sequence of populations and use it to reconstruct lineages. Before we can do this, however, we also need to know how these populations are distributed in space.

Space: Location of Primate Fossils

Primate fossils have been discovered at sites of various ages scattered across Eurasia, Africa, and the Americas. Some of the most important early primates are listed in Table 12-2. The taxonomic status of living and extinct primate families is given in Table 12-3, and their locations are shown in Figure 12-1. In considering lines of evolution, some knowledge is needed of the possibilities for contact and gene flow among the populations that inhabited these various areas. Because of interruptions in the geological record, however, this information is not easy to find. And the gaps may be produced by a variety of causes. An absence of fossils in a given location may mean that primates did not live there when these strata were being formed—perhaps because of a very cool or dry climate. Even if primates once lived in an area, their

TABLE 12-2 Some important pre-Pleistocene fossil primates

Genus	Family	Date	Location	Material discovered	Significance	Animal with comparable body size
Purgatorius	Paromomyidae	Late Cretaceous and early Paleocene	North America (Montana)	Over 50 premolar and molar teeth	Oldest known primate	Shrew or small mouse
Plesiadapis	Plesiadapidae	Middle Paleocene to early Eocene	North America (Colorado) and Europe (France)	Many jaws and teeth, complete skull, nearly all bones of forelimbs and hindlimbs, vertebrae	Best represented Paleocene primate; one of few primate genera to live in both Old World and New World (others are *Phenacolemur*, *Pelycodus*, and *Homo*)	Squirrel or domestic cat
Notharctus	Adapidae	Eocene	North America (Wyoming)	Several relatively complete skulls and nearly entire skeleton (composite)	Prosimian which had attained form like that of modern lemur	Domestic cat
Smilodectes	Adapidae	Eocene	North America (Wyoming)	Skulls and virtually entire skeleton (composite)	Most completely known pre-Pleistocene prosimian; postcranial skeleton identical to *Notharctus* with slightly larger braincase	Domestic cat
Adapis	Adapidae	Eocene	Europe	Numerous skulls and postcranial bones	First primate genus described (Cuvier, 1821)	Domestic cat
Tetonius	Anaptomorphidae	Early Eocene	North America (Wyoming)	Over 100 finds, including one skull, facial fragments, and lower jaws	Provides evidence that enlarged brain found in most living primates appeared very early	Tarsier or mouse lemur
Rooneyia	Anaptomorphidae	Early Oligocene	North America (Texas)	Nearly complete skull	Only Oligocene primate skull found in North America	Squirrel
Necrolemur	Tarsiidae	Middle to late Eocene	Europe (France and Germany)	Approximately a dozen skulls and some postcranial bones	Structurally very similar to living tarsier; hence, marks early appearance in fossil record of a form close to one still living	Tarsier
Branisella	?	Oligocene	South America (Bolivia)	Fragment of upper left jaw containing P^3 through M^2	Earliest South American primate; possible ancestor of platyrrhine monkeys	Squirrel
Oligopithecus	?	Oligocene	Africa (Egypt)	One specimen—left half of lower jaw	Earliest known Old World member of Anthropoidea having catarrhine dental formula	Squirrel

(Continued)

TABLE 12-2 Continued

Genus	Family	Date	Location	Material discovered	Significance	Animal with comparable body size
Apidium	Cercopithecidae	Oligocene	Africa (Egypt)	Two hundred specimens in all, including several dozen jaws, skull fragments, and limb bones	Early Old World member of Anthropoidea with dental formula the same as in platyrrhines	Squirrel
Parapithecus	Cercopithecidae	Oligocene	Africa (Egypt)	About 40 jaws and fragments, of both juveniles and adults	Early Old World member of Anthropoidea, very similar to *Apidium*	Squirrel
Propliopithecus	?	Oligocene	Africa (Egypt)	Three lower jaws and about a dozen teeth	Old World catarrhine primate variously suggested to be ancestral to hylobatids, pongids, and hominids	Squirrel
Aeolopithecus	Hylobatidae	Oligocene	Africa (Egypt)	One lower jaw	Early catarrhine primate with gibbonlike dental adaptations	Squirrel
Aegyptopithecus	?	Oligocene	Africa (Egypt)	One skull, other cranial fragments, several incomplete mandibles, isolated teeth, limb bones, tail vertebrae	Largest and most advanced Oligocene catarrhine primate known	Domestic cat
Pliopithecus (and *Limopithecus*)	Hylobatidae	Miocene and Pliocene	Europe and Africa	Nearly complete skull, cranial fragments, jaws, and postcranial remains	Ancestral to gibbons and possibly other hominoids	Gibbon
Dryopithecus	Pongidae	Miocene and Pliocene	Europe, Asia and Africa	Nearly complete skull, very numerous jaw fragments and isolated teeth, and limb bones	Ancestral to living pongids and hominids	Siamang to large chimpanzee or small gorilla
Oreopithecus	?	Late Miocene	Europe (Italy) and possibly Africa (Kenya)	Over 200 specimens, including one nearly complete skeleton	Large catarrhine primate with adaptations for brachiation	Chimpanzee
Ramapithecus	?	Miocene and Pliocene	Europe, Asia, and Africa	About a dozen jaw fragments and teeth	Forest-living hominoid believed by some to represent earliest hominids	Pygmy chimpanzee
Gigantopithecus	?	Pliocene	Asia (south China and India)	Four jaws and over 1000 teeth plus possible fragment of limb bone	Large terrestrial hominoid primate with similarities to robust early hominids	Gorilla

TABLE 12-3   Classification of primate families

Suborder	Infraorder	Superfamily	Family
Prosimii	Plesiadapiformes	Plesiadapoidea	Paromoyidae
			Picrodontidae
			Carpolestidae
			Plesiadapidae
	Lemuriformes	Adapoidea	Adapidae
		Lemuroidea	Lemuridae
			Indriidae
			Daubentoniidae
			Megaladapidae
	Lorisiformes	Lorisoidea	Lorisidae
	Tarsiiformes	Tarsiioidea	Anaptomorphidae
			Omomyidae
			Tarsiidae
Anthropoidea	Platyrrhini	Ceboidea	Cebidae
			Callithricidae
			Xenothricidae
	Catarrhini	Cercopithecoidea	Cercopithecidae
		Oreopithecoidea	Oreopithecidae
		Hominoidea	Hylobatidae
			Pongidae
			Hominidae

Source: Based on E. L. Simons. 1972. *Primate Evolution*. New York: Macmillian. App. I, pp. 283–289.

remains may not have been preserved. Furthermore, in those areas where sediments contain the bones of primates, only a few scattered sites might have escaped the effects of later erosion.

The difficulty of deciding among these alternatives exists for all times before our own, but it is particularly great for the early part of the Cenozoic era, when widespread uplift resulted in the formation of elevated areas such as the Colorado plateau. In addition to the disruption caused by the raising of some areas, the resulting steep slopes were subject to erosion, and thus few early Paleocene deposits remained intact. The initial differentiation of the primates from an ancestral insectivore stock and the subsequent evolution of members of the primate order therefore took place in a world that was often marked by environmental change rather than constancy.

The map in Figure 12-1 shows the world as it looks today. Most of us are accustomed to thinking of the continents as fixed and unchanging masses. Around the turn of the century, however, the geographer Alfred Wegener proposed that all present land areas had originated from the breakup of a single supercontinent. He called this hypothetical landmass

Equator

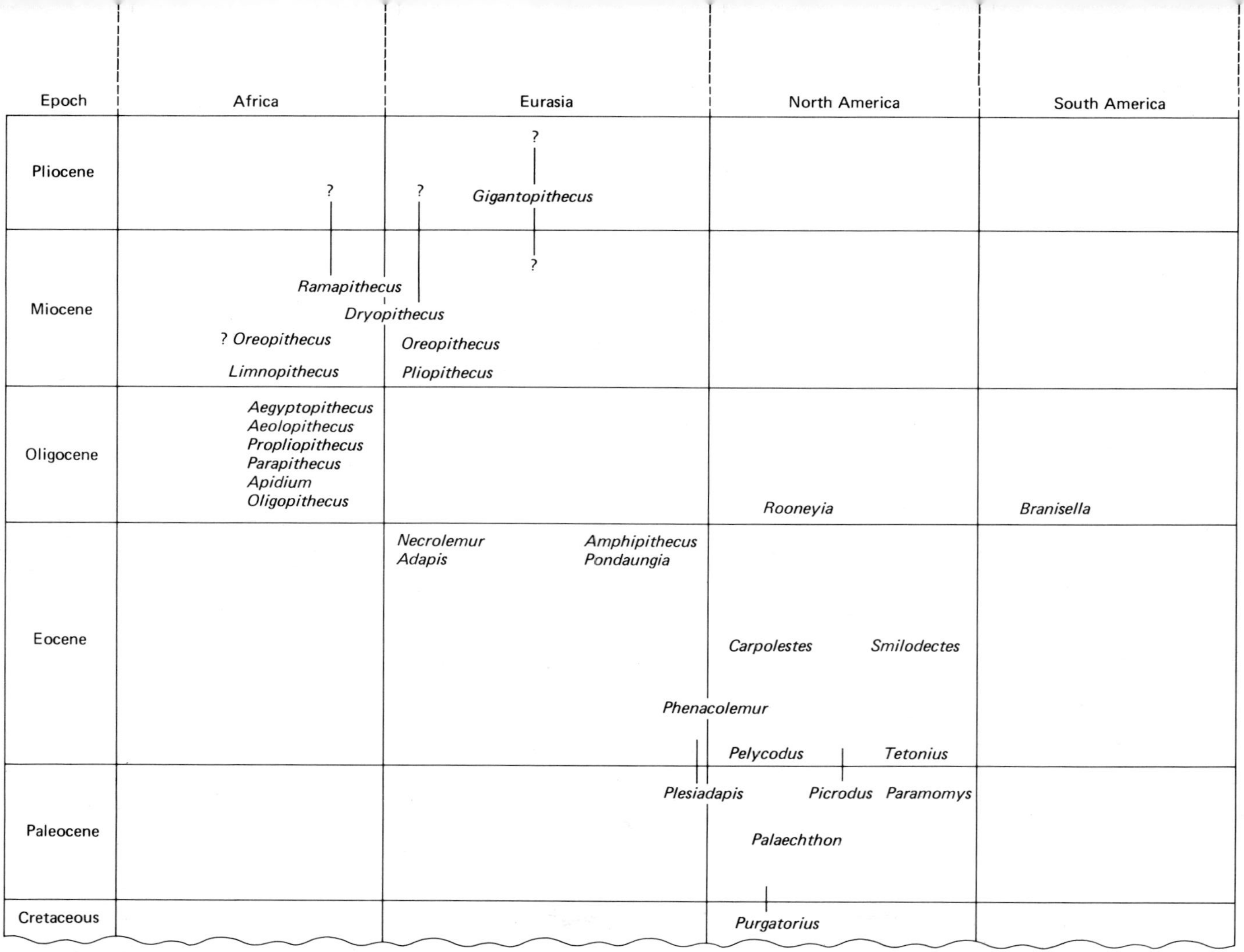

Epoch	Africa	Eurasia	North America	South America
Pliocene	?	? Gigantopithecus ?		
Miocene	Ramapithecus / Dryopithecus / ? Oreopithecus / Limnopithecus	Dryopithecus / Oreopithecus / Pliopithecus		
Oligocene	Aegyptopithecus / Aeolopithecus / Propliopithecus / Parapithecus / Apidium / Oligopithecus			Branisella
Eocene		Necrolemur / Adapis / Amphipithecus / Pondaungia	Rooneyia / Carpolestes / Smilodectes / Phenacolemur / Pelycodus / Tetonius / Picrodus / Paramomys	
Paleocene			Plesiadapis / Palaechthon	
Cretaceous			Purgatorius	

FIGURE 12-1 Spatial and temporal distribution of some important primate genera.

Pangaea ("all lands"). A recent reconstruction of Pangaea's appearance is shown in Figure 12-2. Most geophysicists now support Wegener's belief that the continents were united in the past and then broke apart and moved to their present locations. In fact, the continents are still moving; for example, Europe and North America are drifting apart by about 1 cm (under half an inch) a year—or roughly half the length of a person's body in a lifetime.

Continental drift probably had direct effects on evolution, increasing the extent and tempo of divergence (Hallam, 1972; Kurtén, 1969). During an early phase of the age of reptiles, for example, drift had split Pangaea into only two parts: Laurasia in the Northern Hemisphere and Gondwanaland in the Southern Hemisphere. Kurtén's work shows that the decks of these enormous arks, each with a broadly similar range of adaptive zones, served as stages for independent evolutionary radiations. Comparable niches in both were filled by functionally and sometimes structurally similar reptiles. Crocodiles and lizardlike tuataras were found in Gondwanaland, while Laurasia was the home of the crocodilelike champsoarurus as well as true lizards (Figure 12-3). By the close of the Cretaceous, the period in which the mammals began their great period of expansion, Laurasia and Gondwanaland had split again, giving rise to a continental configuration more like that of today (Figure 12-4). Each of these smaller areas was able to support a largely independent adaptive radiation of its own, and so there was much functional duplication as mammals of different ancestry moved into similar habitats. This led to parallel and convergent evolution on a grand scale—for example, the many similarities between the Australian marsupial and its North American, Eurasian, and African placental counterparts. The increase in mammalian diversity over that shown by the reptiles was enormous. Apparently the mammalian radiations produced half again as many different orders (30) as the reptiles (20) in about one-third the time (approximately 75 million years, as opposed to 200 million).

In addition to shaping the faunal composition of continents into which primates would later spread, continental drift had a more direct impact on the course of primate evolution: it played a large part in channeling the direction and order of primate expansion. By the Cretaceous the continents had already assumed orientations more like those of today. Yet there were still important differences. Australia had not yet separated from Antarctica, and India had not yet collided with the main Eurasian landmass. Of more direct significance for the earliest stages of primate evolution, Greenland was still connected with northern Europe, forming a land bridge to North America, where the earliest known primates have been discovered. Since the climate was warmer then, the land bridge would have provided a corridor for the passage of animals in both

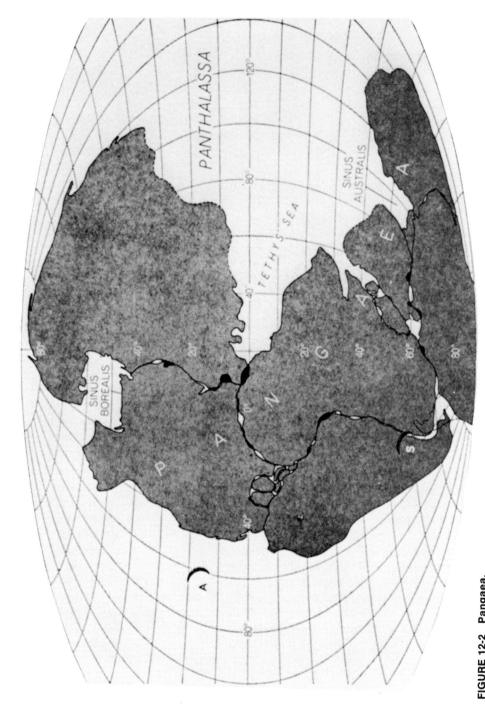

FIGURE 12-2 Pangaea.

Source: R. S. Dietz and J. C. Holden. 1970. "The Breakup of Pangaea," *Scientific American*, **223**(4):30–41. Fig. on p. 34.

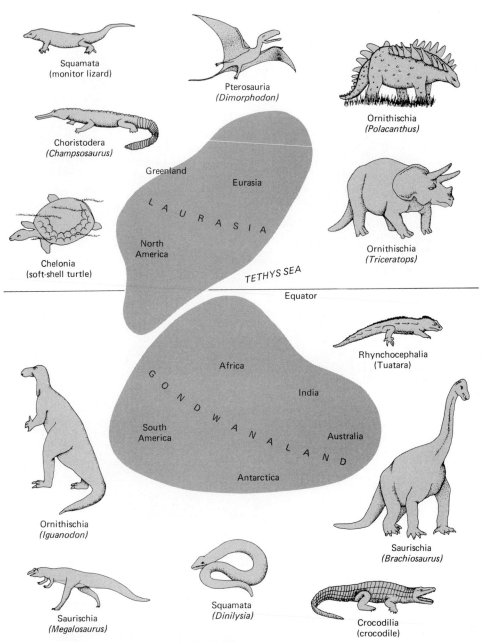

FIGURE 12-3 Continental drift and faunal evolution.
Source: B. Kurtén. 1969. "Continental Drift and Evolution," *Scientific American*, **220**(3):54–63. Fig. on p. 55.

Squamata
(monitor lizard)

Pterosauria
(Dimorphodon)

Ornithischia
(Polacanthus)

Choristodera
(Champsosaurus)

Greenland

Eurasia

L A U R A S I A

North
America

Ornithischia
(Triceratops)

Chelonia
(soft-shell turtle)

TETHYS SEA

Equator

Rhynchocephalia
(Tuatara)

Africa

India

G O N D W A N A L A N D

South
America

Australia

Antarctica

Ornithischia
(Iguanodon)

Saurischia
(Brachiosaurus)

Saurischia
(Megalosaurus)

Squamata
(Dinilysia)

Crocodilia
(crocodile)

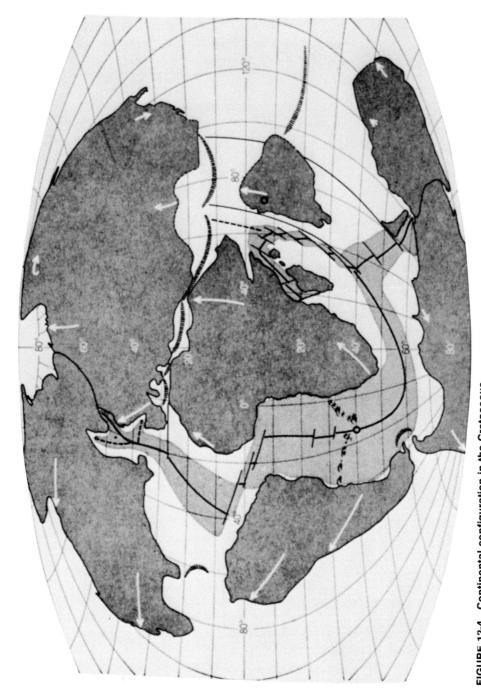

FIGURE 12-4 Continental configuration in the Cretaceous.
Source: J. C. Holden and R. S. Dietz. 1970. "The Breakup of Pangaea," *Scientific American*, **223**(4):30–41. Fig. on p. 37.

directions. This, in turn, should mean a close relationship between North American and Eurasian primates. In fact, during the Eocene some members of the lemurlike prosimian family Adapidae living in Europe and North America were so similar that they can be placed in at least the same genus. The contact between the animals inhabiting these two continental blocks was ended by the rifting that occurred early in the Eocene.

During the very early Tertiary, South America, Africa, Madagascar, and India—all areas in which nonhuman primate fossils have been found—were still separated from the holarctic ("whole northern") faunal region by water gaps of various extents. Primates living in these more southerly areas must therefore have arrived later or crossed the water barriers. Since few primate species are good swimmers, any that arrived before the existence of land connections probably got there on rafts of floating vegetation. This would be a chance affair, with the probability of reaching another piece of land being about inversely proportional to the distance between the two bodies. Because of the unfavorable ratio of winners to losers under such circumstances, this type of colonization of new areas is known as *sweepstakes dispersal*. Madagascar, which separated from Africa before the beginning of the Cenozoic era, may have been populated with prosimians by this means.

During the Oligocene, the continent of Africa began to make contact with Eurasia, a process that continued into the Miocene. India started grinding into the main body of the Eurasian landmass at about the same time, approximately 40 million years ago (Molnar and Tapponnier, 1977). The relationship between North and South America was a fluctuating one. Early in the Tertiary period the two seem to have been connected for a short time and then separated by the Bolívar trench, which divided South America from Central America. Still later, the region of this trench seems to have been raised far enough above sea level to form a mountain range. However, not until the end of the Pliocene, when the Isthmus of Panama was formed, was there a corridor that allowed large-scale faunal interchange between North and South America.

Although continental drift played a central role in shaping the early course of primate evolution, there were other determinants as well: among these were fluctuations in sea level, changes in climate, shifts in types of vegetation used for food and shelter, and the evolution of new predators. For example, during some parts of the early Tertiary, elevated sea levels seem to have converted the African continental block into an archipelago with several large islands and numerous smaller ones; at about the same time, eastern Asia was isolated from direct contact with Europe by the intrusion of a large inland sea. Interchange between these two areas was possible, but perhaps only via the long route through North America.

PRE-PLEISTOCENE PRIMATE EVOLUTION

Primitive humans evolved by at least the beginning of the Pleistocene epoch, which began about 3 million years ago. Anthropologists do not know for certain how much earlier our line originated, but the alternative possibilities will be explored in this chapter. We do know that primates originated from insectivores at some time before the close of the Cretaceous period. We also know that over the course of the Tertiary period (which lasted from the beginning of the Paleocene epoch, about 7 million years ago, to the end of the Pliocene epoch, about 2 million years ago) a great variety of prosimian, monkey, and ape species made their appearance. In the remainder of this chapter we'll look at the evolution of these species, which number our own ancestors among them. The following discussion, which is organized chronologically, gives a summary of the major evolutionary advances by primates in each time period.

CRETACEOUS: *PURGATORIUS*

The insectivore-primate transition apparently took place at some time before the close of the Cretaceous. Late Cretaceous deposits in Montana have been found to contain the tooth of an individual assigned to the prosimian genus *Purgatorius*, which had previously been known from nearby Paleocene deposits (Van Valen and Sloan, 1965). This was a tiny animal with molar tooth crowns roughly the diameter of a pencil lead. The jaws which held them must have been less than 2.5 cm (1 in) long, and the whole animal could not have been larger than a field mouse. Reconstruction much beyond this point would be highly speculative, as there are still only a few teeth to go on. The contemporaries of this earliest known primate included at least six species of dinosaurs. Among the other mammals found with it were marsupials and a variety of primitive placentals, some of which are now extinct—for example, condylarths, which resembled the ungulates (cows, sheep, etc.) and may well have been ancestral to them. In most condylarths the clavicles had already been lost, signaling that flexibility of movement had been sacrificed, probably for a gain in speed and efficiency of running on the ground—a pattern opposite to that of the primates. Some condylarths also had exchanged claws for hoofs, although the feet still retained all five toes; others showed the beginnings of distinctive ungulate dental features. Additional now-extinct forms included multituberculates, small mammals with rodentlike gnawing incisors and elongated molars which bore two or three rows of numerous cusps.

Why is *Purgatorius* thought to be a primate? Most of the reasons have to do with the presence and location of certain cusps and crests on the tooth crowns. Perhaps of most obvious significance is the fact that the cusps on the upper and lower molars are rather bulbous and quite unlike the tall and sharply pointed molar cusps of its insectivorous ancestors (Figure 12-5). Some anthropologists believe that these changes in the teeth of primates resulted from a change in selection pressures that came with a shift in diet. While fine for piercing and slicing insects, needlelike cusps would break easily when used to grind hard, dry seeds or fruits that had tough husks and gritty or fibrous pulp. Those animals in the population born with slightly more rounded cusps would have teeth that were less subject to damage and thus had a chance of lasting longer and enhancing survival and reproductive potentials. The initial dietary modification was a behavioral change that may have been made as some populations exploited the resources of a broad zone rich in fruits and leaves. In the early Tertiary this zone was probably not yet fully utilized, perhaps because the change in vegetation was comparatively recent. Up until the early part of the Cretaceous, there were few flowering plants; ferns, conifers and other gymnosperms predominated. Sometime before the first half of the Cretaceous drew

FIGURE 12-5 Isolated teeth of *Purgatorius*. (a) M^2 of *Purgatorius unio* (Paleocene). (b) M_2 of *Purgatorius ceratops* (Cretaceous).

Source: L. Van Valen and Robert E. Sloan. 1965. "The Earliest Primates," *Science*, **150**:743–745. Copyright 1965 by the American Association for the Advancement of Science.

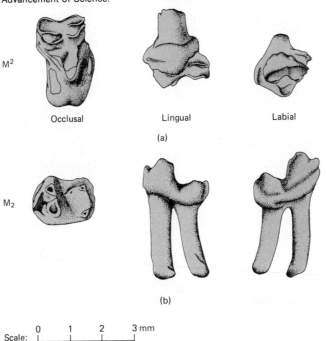

M^2

Occlusal   Lingual   Labial

(a)

M_2

(b)

Scale: 0   1   2   3 mm

to a close, however, many areas showed a great influx of flowering plants, which greatly altered the composition of the vegetation. The distribution of Cretaceous plant remains seems to point to somewhere in what is now the arctic as the place of origin of these flowering plants. During the early part of the upper Cretaceous, the oceans flooded great areas in the Northern Hemisphere. In addition, at the end of the Cretaceous there was a temporary lull in mountain formation and the development of more level areas. It is quite likely that these geological events had effects first on climate and then on plant life. Climate usually differs on the opposite sides of a mountain range, with the windward side often getting more rain and developing more vegetation; absence of mountains often leads to a decrease in the diversity of plants and other features of the habitat. As plants (the primary producers of energy in the organic world) were modified, the animal populations that depended on them were also likely to change.

More level areas may have meant more environmental sameness, and competition for remaining niches might have intensified. What is more, the character of the remaining niches may have changed markedly. For one thing, trees and other woody plants evolved, and many reached enormous size. In the resulting tropical forests, most of the edible vegetation—leaves, fruits, and seeds—was high above the ground. This gave an advantage to animals that were agile enough to climb and yet small and light enough to have a high probability of surviving falls. Known skeletal material indicates that the earliest primates possessed these characteristics. By taking advantage of new ecological opportunities, primate populations subjected themselves to continuing selection for the complex of trends—including increased manipulative skills, visual acuity, and intelligence—that mark later members of the order. The same changes in food supply that created opportunities for the ancestors of the primates were probably responsible for the passing of the dinosaurs. More and more vegetation would have been beyond the reach of the herbivorous dinosaurs, and a decline in their numbers would have drastically reduced the food supply of the carnivorous species. For a fascinatingly detailed discussion of this problem, read the book by Desmond (1976). An alternative model for primate origins has also been proposed. In 1974 Cartmill made the suggestion that *Purgatorius* and its relatives in the superfamily Plesiadapoidea (discussed later in this chapter) might not be primates at all. The molars of plesiadapoids do show detailed resemblances to those of later known primates, and the bony structure of the ear region also shows primate features (Figure 12-6). But it is known that plesiadapoids had already lost certain teeth that were still present in some later primates. Consequently, later primates would have to descend from still unknown earlier forms in which dental reduction had not proceeded so far. Whether these earlier animals were primates depends on how the line

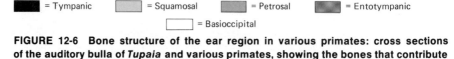

Key: Bones that contribute to structure of the ear region —

■ = Tympanic   ▨ = Squamosal   ▨ = Petrosal   ▨ = Entotympanic
□ = Basioccipital

FIGURE 12-6   Bone structure of the ear region in various primates: cross sections of the auditory bulla of *Tupaia* and various primates, showing the bones that contribute the component parts at various stages of complexity among primates.
Source: E. L. Simons. 1972. *Primate Evolution.* New York: Macmillan. Fig. 20, p. 65.

between primate and preprimate is drawn. Cartmill's suggestion is that the most important selective factor leading to the beginning of primate evolution was preying on insects and small animals in the lower canopy around the forest fringes. This too could have led to the evolution of visual acuity, grasping hands, and a brain to coordinate the functions of these two systems. Cartmill suggests using the characteristics associated with visually directed predation, rather than dental traits, as the basis for drawing a boundary line between the primate order and its immediate predecessors (Figure 12-7). The problem may eventually be resolved by the study of additional Paleocene primate remains and experimental studies of living prosimian behavior.

PALEOCENE: PRIMITIVE PROSIMIANS

Tropical forests were also quite widespread during the Paleocene, despite a slight cooling of the climate. In the forested areas, primate remains are now more abundant. For example, primates make up over 10 percent of the genera present in some upper Paleocene deposits in the Rocky Mountain region. And at one middle Paleocene site in Montana, 20 percent of all identified individuals were primates (Barth, 1950). Not all these early primates, however, were arboreal; some show

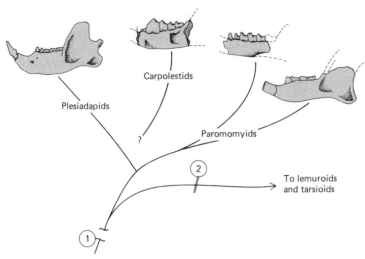

FIGURE 12-7   Alternative boundaries for primate origins. Representatives of the plesiadapoid radiation, *left to right:* **Plesiadapis tricuspidens, Carpodaptes aulacodon, Palaechthon alticuspis, Phenacolemur jepseni. The morphological shift at (1), which established certain dental traits, is usually considered the boundary of the order Primates. But the inferred shift at (2), considered by Cartmill to be a shift toward visually directed predation, could (if monophyletic) serve as the boundary of a more coherent order.**

Source: M. Cartmill. 1974. "Rethinking Primate Origins," *Science,* **184:**436–443. Fig. 4, p. 441. Copyright 1965 by the American Association for the Advancement of Science.

rodent-like characteristics. The adaptive zone of the Paleocene (and Eocene) members of the primate order seems to have been wider than it was at some later points and to have constituted a *primate-rodent* ecological niche. No true rodents are known before the late Paleocene, and no major new group of primates evolved clear rodentlike adaptions after this time. Thus it seems that the record reflects a true absence of rodents, rather than just failure to find or recognize them.

The evolutionary line represented by *Purgatorius* continued, with Paleocene descendants so similar to Cretaceous members of the genus that some who have studied them feel that the later population should not be considered a separate species, though separated in time by several million years. About a dozen different genera (and more species) are presently recognized from the Paleocene. These may all be grouped into four prosimian families: Paromomyidae, Picrodontidae, Carpolestidae, and Plesiadapidae, which have in turn been united into a single superfamily, the Plesiadapoidea. There are no living members of these taxonomic units, but as shown in Table 12-3, they are classified as prosimians. Plesiadapoids have been found in sediments in Europe and

North America, but not yet in other regions, such as Africa and Asia, where they may have existed. This is a handicap in reconstructing the earliest stages of primate evolution, but we simply don't know how great a one. For the present, all we can do is keep our minds open while building on what is known.

Paromomyidae

The Paromomyidae, to which *Purgatorius* is assigned, is one of the better-known families, even though our reconstruction is based largely on fragments of lower jaws and teeth. Only two skulls are known: one, a badly crushed specimen of *Palaechthon* from the middle Paleocene of New Mexico, is the earliest known skull of any primate; the other is an early Eocene skull of *Phenacolemur,* also from New Mexico. Despite the limitations of the evidence, we can make several generalizations about the paromomyids. The most common dental formula in the family is $\frac{2133}{2133}$; you will recall from Chapter 9 that this is a very common prosimian dental formula. In other genera there are still further reductions from the primitive mammalian condition. As in nearly all early primates, the two halves of the mandible remained unfused throughout life, thereby retaining mobility and doing without the greater strength of a fused joint (*symphysis*). Immediately adjacent to the symphysis is the most striking dental feature of paromomyids, a greatly enlarged and projecting pair of central incisors. These were all quite small mammals; the largest was scarcely bigger than a rat, and the smallest was comparable to a shrew.

Picrodontidae

These animals were even smaller than the paromomyids. The family, containing *Picrodus* and another genus, is now believed to have been derived from the Paromomyidae, although some paleontologists think a closer relationship to insectivores or to bats is possible. This uncertainty exists at least in part because although there are several mandibles, only one crushed, incomplete skull has as yet been found. Reduction in the number of teeth has proceeded further in the lower jaw than in the upper, thus giving a dental formula of $\frac{2133}{2123}$. The lower incisors project forward, while those in the upper jaw are planted more vertically. The major deviations from the primitive mammalian condition are seen in the cheek teeth. The premolars are laterally compressed, perhaps for slicing. The cusps of the upper molars are greatly reduced in height; on the lower molars the front part of the crown is enlarged, and the rear part is reduced. The depressed central regions of both upper and lower molars are dotted with numerous tiny enamel projections that greatly increased the area of the contact surfaces. Such low-crowned teeth would have been worn down rapidly if used to grind very abrasive substances. For this reason it has been suggested that the diet of these animals consisted of soft plant material such as pulpy fruit or even nectar.

Carpolestidae

Carpolestes and its relatives were the size of modern mice or rats and appeared after the middle of the Eocene in North America. Because their remains are fragmentary, there is some disagreement about their dental formula. The most recent comprehensive study suggests that it was like that of the middle Paleocene paromomyids, $\frac{2133}{2133}$. The lower central incisors are greatly enlarged, while the following three teeth—which probably represent another incisor, the canine, and a single-rooted premolar—are quite small. The next tooth in line, another premolar, arches high to form a sharp, serrated blade. It thus closely resembles comparable teeth of several other Mesozoic and early Cenozoic mammals, including the extinct multituberculates. A functional interpretation is made possible by comparison with some living Australian marsupial mammals which have similar teeth. Although no detailed studies have been undertaken, it seems that this type of highly crowned blade represents an adaptation to a vegetable diet high in fiber—perhaps roots, bark, or fruits with tough husks.

Plesiadapidae

This family of Paleocene prosimians was squirrel- to cat-sized. Here we are fortunate enough to have not only a complete skull but also skeletal remains. A reconstruction based on these is given in Figure 12-8. Because its dental formula $\frac{10 \,(or\, 1)\, 33}{1033}$ exhibits an extreme degree of reduction in some categories (to the extent that the lower and possibly the upper canines have been lost), this species of *Plesiadapis* is probably not a direct ancestor of later primates. Nonetheless, Figure 12-8 does give us the most complete picture that we have to date of a Paleocene member of our own order.

Plesiadapis embodies some advances over its primitive nonprimate mammalian ancestors. The dental reductions are one specific example of a primate trend. In addition, as shown in Figure 12-6, the inner ear of *Plesiadapis,* like that of higher primates, is surrounded by a capsule (called the *bulla*) derived from one bone (the *petrosal*) that makes up part of the base of the skull; leading into this bulla is a large bony tube

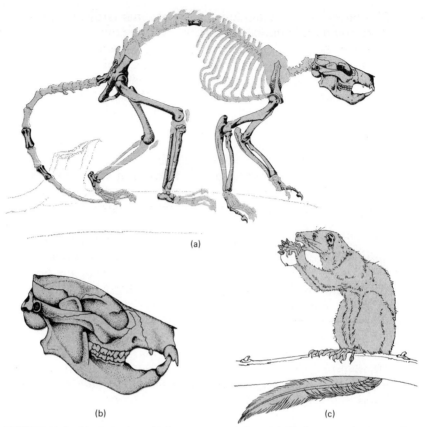

FIGURE 12-8 *Plesiadapis*, a Paleocene prosimian. (a) Skeleton reconstructed on the basis of fossils found in France and North America. (b) Skull. (c) Restoration by analogy with living tree shrews. Species of *Plesiadapis* ranged from squirrel-size to cat-size.

Source: Modified from E. L. Simons. 1964. "The Early Relatives of Man," *Scientific American,* **221**(1):50–62.

(formed from the *ectotympanic* bone). These dental and auditory features are combined with characteristics that seem far less advanced. The snout of *Plesiadapis* is still long, indicating continuing dependence on the sense of smell; the visual apparatus shows little development of stereoscopic sight; and no bar or strip of bone protects the eye. The feet still have claws rather than nails, and the equal length of the forelimbs and hindlimbs indicates a continued reliance on quadrupedal locomotion. This mosaic of advanced and primitive characteristics in *Plesiadapis* should not be disturbing, since "generalized" ancestors exist only in theory. Real, living animals belong to species that are adapted to specific niches in the environment. The morphology of

Plesiadapis shows a particular complex of characteristics that had been developing from long in the past. On balance, these characteristics indicate that *Plesiadapis* had not advanced beyond the evolutionary grade typified by more primitive living prosimians.

Present evidence indicates that other known Palocene primates were comparable to *Plesiadapis* in grade, although even in the Paleocene a wide variety of structurally (and presumably ecologically) differentiated populations existed. Preceding or during the early Tertiary there had been an adaptive radiation, perhaps when the insectivore ancestors of the primates expanded into the broad new adaptive zone characterized by flowering plants, carving it into a variety of well-differentiated sectors.

In assessing the scope of this early prosimian adaptive radiation, we must remember that the number of taxonomic names given to Paleocene primates tends to exaggerate the diversity which may have existed in any one generation. Fossil-bearing beds in Europe and North America have usually been studied by different people, with the result that some closely related forms occurring in both areas have been given quite different names. Still other factors may make the phylogenetic diversity more apparent than real. Not all Paleocene genera and species existed at the same time. For example, several species of the Paromomyidae (*Palaechthon, Paromomys,* and *Phenacolemur*) may represent three successive evolutionary levels within one line. Thus at least half (and possibly more) of the six genera of this family could be ancestors and descendants rather than sympatric or allopatric populations.

EOCENE: MODERN-LOOKING PROSIMIANS

During the early part of the Eocene, the land connection between North America and Eurasia remained intact, and Alaska again had a warm, subtropical climate. The likelihood of continuing faunal exchange is confirmed by the presence of several primate genera (including *Plesia-dapis* and *Phenacolemur*) in both Europe and North America. During this period some Paleocene primates survived, and new forms developed. Among these are the earliest known primates that might be called lemurs and tarsiers. Eocene lemurs are conventionally distinguished from living ones (Lemuridae)) by being placed in a separate family, the Adapidae. The extinct Eocene tarsierlike prosimians are usually all grouped into the family Anaptomorphidae. The family Tarsiidae, which still persists, was also represented in the Eocene. What did these Eocene primates look like? There are so many that not all of them can be described in detail. Let us look at a few to get an idea of the variety of adaptive types that had evolved by this point in time: Adapidae, Anaptomorphidae, and Tarsiidae.

Adapidae

Adapis, the type on which the family Adapidae is based, was the first genus of primates described by Cuvier in 1821 (see Figure 12-9). Another adapid genus, *Pelycodus*, was so common that it serves as an *index fossil* (like an alphabet letter on a file drawer, its presence is a reliable guide to what's contained): the appearance of *Pelycodus* marks the beginning of the Eocene epoch in North America.

FIGURE 12-9 *Adapis,* **an Eocene prosimian.** (a) **Reconstruction.** (b) **Lateral view of skull of** *Adapis parisiensis.* **Note the relatively small braincase, the muscular crests, the unreduced premaxillary region, and the small orbits. Compare this with** (c) **the skull of a modern lemur,** *Lemur catta.*

Sources: (a) W. E. Le Gros Clark, 1959. *The Foundations of Human Evolution.* Eugene, Ore.: Oregon State System of Higher Education (Condon Lectures). P. 61. (b) W. E. Le Gros Clark. 1971. *The Antecedents of Man.* Chicago, Ill.: Quadrangle. Fig. 57, p. 142. (c) Frederic Wood Jones. 1929. *Man's Place among the Mammals.* New York: Longmans, Green. Fig. 18, p. 96.

(a)

(b)

(c)

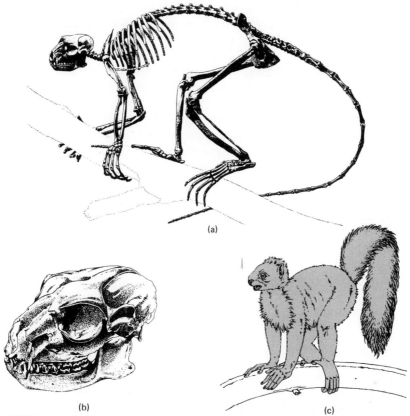

FIGURE 12-10 *Smilodectes,* another Eocene prosimian: (a) skeleton; (b) skull; (c) reconstruction. *Smilodectes* is several million years junior to *Plesiadapis* and far more advanced. Its snout is shorter, the front portion of the brain is enlarged, and its eyes are positioned on the skull so as to allow the visual fields to overlap. Although the notharctine subfamily to which this genus belongs was not ancestral to any living primates, the relatively long hind limbs of *Smilodectes* give it a remarkable resemblance to one modern prosimian, the sifaka, a Malagasy lemur.

Source: E. L. Simons. 1964. "The Early Relatives of Man," Scientific American, **221**(1):50–62.

The most completely known Eocene primate is the genus *Smilodectes* (Figure 12-10). These cat-sized primates show a number of evolutionary advances over their Paleocene ancestors. The most significant of these are concentrated in the skull. Compared with *Plesiadapis* and many other Paleocene prosimians, *Smilodectes* had more teeth; its dental formula is $\frac{2143}{2143}$. Its snout was shorter, suggesting that the sense of smell was reduced in importance, and the eyes were rotated further forward, with the result that the visual fields overlapped to a greater extent than in earlier primates, giving improved depth perception. a postorbital bar provided protection for each eye, further accenting the importance of vision. *Smilodectes* had a large brain, particularly in the frontal region, the area which in living primates controls organized thinking and coordinated activity. The skeleton of *Smilodectes* is that of an agile, lightly built quadrupedal animal, with narrow rib cage and flexible spine extending in a long tail that could have been used for balancing. Its fingers and toes are tipped with flattened nails rather than sharp claws, suggesting increased facility in grasping and manipulation. Hindlimbs longer than forelimbs indicate that it was capable of long leaps. Overall, the impression one gets is that of an animal brighter than its Paleocene predecessors and equipped to make rapid movements and decisions as it moved about in a network of branches high above the ground.

Anaptomorphidae

The members of this family are more similar to tarsiers than to lemurs, but there was less difference between these lineages in the early Eocene than there is now. And in addition, not all anaptomorphid genera resembled tarsiers to the same degree. Anaptomorphids are poorly preserved in the fossil record. The best known is the genus *Tetonius*, which is represented by several hundred specimens, mostly lower jaws and one relatively complete skull. Its dental formula is $\frac{2133}{21(2 \text{ or } 3) 3}$, and overall its dentition resembles that of *Tarsius* (Figure 12-11). It also resembles *Tarsius* in that the orbits of its eyes are very large, suggesting that it was nocturnal. When discovered, the skull of *Tetonius* contained a natural cast of the brain, formed when minerals filled the cranial cavity after death. This brain is strikingly modern in appearance, with the enlarged temporal and occipital lobes and reduced olfactory bulbs typical of all known primates. These advances had already occurred at a time (about 55 million years ago) when the brains of other animals were as primitive as those of living opossums.

The trend in brain development was continued into the Oligocene. A bit of evidence for this is provided by the North American primate genus *Rooneyia*, known from one almost complete skull recently found in Texas. In the reduction of the parts of the brain concerned with smell and in a highly developed visual system, *Rooneyia* approaches the level seen in primitive members of the Anthropoidea.

Tarsiidae

The first known fossil remains of this family were found in deposits of middle Eocene age; thus the lineage has a beginning somewhat more recent than either the Adapidae or the Anaptomorphidae. These early tarsioid genera show such strong resemblances to the living *Tarsius*

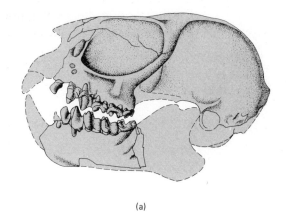

(a)

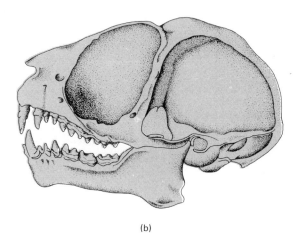

(b)

FIGURE 12-11   (a) **Skull of** *Tetonius homunculus,* **an Eocene prosimian, compared with** (b) **skull of** *Tarsius spectrum.*

Source: W. K. Gregory. 1922. *The Origin and Evolution of the Human Dentition.* Baltimore, Md.: Williams and Wilkins. Figs. 118, 119, pp. 194, 214.

that a vertical classification seems justified. *Necrolemur* is by far the best-known Eocene member of the tarsiid lineage, chiefly through study of about a dozen well-preserved skulls found in France (see Figure 12-12). The orbits of the eyes are greatly enlarged, leaving little doubt that *Necrolemur* was as nocturnal as *Tarsius.* And as in the living species, the postorbital bar has widened so far that it has begun to form a wall behind the eye, thus foreshadowing the bony socket found in higher primates. The muzzle between the orbits has been reduced even more than in earlier primates such as the Adapidae, and so the front part of the dental arch narrows quite sharply.

The dental formula of *Necrolemur* $\left(\frac{2133}{0143}\right)$ is the same in the upper jaw as that of *Tarsius* $\left(\frac{2133}{1133}\right)$, but it shows the loss of some teeth in the

FIGURE 12-12   Reconstruction of *Necrolemur,* a fossil tarsierlike prosimian.

Source: W. E. Le Gros Clark. 1959. *The Foundations of Human Evolution.* Eugene, Ore.: Oregon State System of Higher Education (Condon Lectures). Fig. 4, p. 63.

lower jaw that are still present in the living form. In a strict sense this would mean that *Necrolemur* could not be ancestral to *Tarsius,* unless that lost incisor tooth were regained. Another alternative is that these Eocene tarsiid populations were polymorphic, with a few individuals still having a lower incisor. There are certainly other details of resemblance in the skull and the skeleton. In the living tarsier the two lower leg bones, tibia and fibula, are fused together to form a single structure, the *tibiofibula.* In the same deposits containing the skulls of *Necrolemur* there are fused lower leg bones of this kind, of a size appropriate for this animal. There seems to be little doubt that these ancient primates were capable of the long hops that are such a striking element in the locomotor behavior of living tarsiers. Eocene *Necrolemur* probably pounced on its insect or small vertebrate prey as avidly as its present nocturnal primate counterparts do.

Amphipithecus and *Pondaugnia* are two primate genera founded on the basis of extremely limited remains—one individual each—discovered in 1927 in what are believed to be late Eocene deposits in Burma. Of *Pondaungia,* a fragment of the left maxilla with the first two molars exists, along with parts of the lower jaw containing second and third molars. These fossils show resemblances to the Adapidae, on the one hand; and, on the other, to some of the Oligocene primates from north Africa (discussed in the next section). *Amphipithecus* is represented by only a piece of a left mandible, which preserves parts of three premolars and the first molar. Once again, there are similarities to some of the Oligocene primates of north Africa.

From the preceding account it is evident that a number of major evolutionary changes occurred among Eocene primates. For one thing, the more archaic prosimians of the preceding Paleocene were replaced by prosimians that were more like living lemurs and tarsiers. From the structural characteristics of these more familiar forms we can reconstruct, by analogy, more details of behavior: *Tetonius* surely hopped about in pursuit of its prey, and did so at night. Another conclusion about primate evolution during the Eocene must remain more tentative because it rests on purely negative evidence—the absence of fossils representing advanced anthropoid forms. Some of the more speculative ideas about Eocene primate evolution are explored in "The Tarsier Model for Human Ancestry" on pages 418–419.

OLIGOCENE: ADAPTIVE RADIATION OF THE ANTHROPOIDEA

Primate fossil remains from this epoch are of very limited distribution in time and space. One of the richest sites is in north Africa, 97 km (60 mi) to the southwest of Cairo and 193 km (120 mi) inland from the Mediterranean Sea, in a part of Egypt called the *Fayum.* While fossils have been known from this area since the turn of the century, most of our detailed knowledge comes from the research of Simons and his colleagues in the 1960s. Today the Fayum is a desert environment, but the ancient Egyptians named the area for its lake, which survives in the form of a large body of blackish water. The sands of the present-day desert rest on sediments of mud and sand over 183 m (600 ft) thick, carried there by rivers that flowed to the Mediterranean, which 30 million years ago extended as far inland as the Fayum. During the Oligocene this area was probably tropical forest and delta. Fossil remains show that the waters contained fish, turtles, crocodiles, and dugongs (or sea cows) and that the forest trees were as tall as 31 m (100 ft). There must have been open areas near the rivers as well, for Fayum sediments hold the bones of a variety of large mammals, including early elephants; large, primitive, piglike herbivores; and arsinotheres. These last were huge beasts, now extinct, with two pairs of horns. They must have looked and functioned much like rhinoceroses, although they were not at all closely related to them. Primitive carnivores were present too, chiefly in the form of weasel-like animals.

THE TARSIER MODEL FOR HUMAN ANCESTRY

The possible existence of Eocene hominids was formerly considered seriously by some students of human evolution, and this still serves to show the dangers inherent in extensive theorizing in the absence of substantial direct fossil evidence.

In several books Frederick Wood Jones (1916, 1926), a prominent British anatomist, argued that our direct lineal ancestors were small tarsierlike primates that adapted to life on the ground before any anthropoid apes or Old World monkeys had evolved. Wood Jones ruled great apes out of a position of direct ancestry to humans by invoking the principle of the irreversibility of evolution. For example, he pointed out that, like tarsiers, humans have hindlimbs longer than forelimbs, while in the apes these proportions are reversed. It was evolutionarily unlikely, he believed, that hindlimbs became shorter than forelimbs and then lengthened again. Likewise, Wood Jones stressed that human teeth show relatively little sexual dimorphism. Low amounts of sexual dimorphism are more frequently found in lower primates than in the great apes. On the basis of this coincidence in patterns of dimorphism, Wood Jones reasoned that instead of going through a stage of evolution characterized by greater dental dimorphism than we now have, human ancestors have never shown marked sexual dimorphism in their teeth (an argument revived recently by Kinzey, 1971).

Wood Jones' hypothesis of an especially close relationship between tarsiers and humans was challenged most strongly by William King Gregory (1922). After a detailed review of the dental evidence for primate relationships, Gregory concluded that the resemblances between humans and apes are far more detailed and fundamental than those connecting humans with any of the other primates—down to precise

In the forests lived rodents, bats, and primates. Some of their bones may have been washed into the rivers after the animals died on the ground, since not many of these forest-living animals are preserved. When small mammals die, their carcasses can be completely destroyed quite quickly. In a Midwestern town one hot July afternoon, I watched maggots consume a bat that had died the day before. Within a few hours it was entirely gone: its bones were dissolved by the digestive fluids of the insects, and at the end not even its tiny teeth could be seen in the dust. In tropical regions these processes take place even more rapidly. It is likely that the fossil hunters' luck in the Fayum area comes from the animals' misfortunes. Many of the primates found there are young animals. Less wary than adults, they may have come to the

correspondences of cusp and crest patterns on the molar teeth (as discussed in this chapter). Moreover, even the greatest dental contrasts between apes and humans, as in the size of incisors and canines and in the shape of the dental arch, are reduced in phylogenetic significance by the great amount of variation in these structures within single species of living primates. Again, as shown in Chapter 9, an occasional recent ape may have nonprojecting canines or other hominidlike features.

Gregory's dental evidence in support of a close relationship between hominids and pongids has been buttressed by studies of other systems, most recently by molecular data. In the end, though, Wood Jones' theory was perhaps less crushed by the weight of evidence against it than blown away by a light puff of sarcasm. In a footnote in *Up from the Ape,* Hooton (1931)—then the most prominent physical anthropologist in America—made the following sharp judgment:

For this writer [Wood Jones] the intimate resemblances which exist between man and the giant apes are of no significance except as examples of "convergent adaptation," a term which is little more than scientific jargon for "coincidence." To Wood Jones the anthropoid apes show many pithecoid or catarrhine specializations which make him refuse to recognize them as relatives, although man occasionally, in individual cases, exhibits these same features. The narrow specializations of Tarsius, whom he regards as our nearest primate relative, divergent from human development as they are, disturb him no whit. He strains at the anthropoid gnat and swallows the tarsioid camel (p. 105).

Sources: (1) W. K. Gregory. 1922. *The Origin and Evolution of the Human Dentition.* Baltimore: Williams and Wilkins. (2) E. A. Hooton. 1931. *Up from the Ape.* New York: Macmillan. (3) F. W. Jones. 1916. *Arboreal Man.* London: E. Arnold. (4) F. Wood Jones. 1926. *Man's Place among the Mammals.* London: Longmans. (5) W. G. Kinzey. 1971. "Evolution of the Human Canine Tooth," *American Anthropologist,* **73**(3):680–694.

water's edge for a drink and been dragged to their deaths by waiting crocodiles. Still others may have misjudged a leap and drowned in the waters below.

The approximate age of the Fayum deposits has been measured by the potassium-argon dating technique. Volcanic rocks about 76 m (250 ft) above the top beds appear to be about 25 to 29 million years old; this would be late Oligocene. The lowest fossil-bearing deposits may be as much as 6 million years older, thus reaching back into the early Oligocene. Scientists who have studied the primate remains believe that a number of genera and species are represented, although the number of separate lineages is not known. Given the millions of years of evolution sampled here, some of the taxa may be time-successive populations of a small number of lineages that evolved from Eurasian prosimian ancestors who arrived in Africa during the Eocene.

Oligopithecus

From the lowest reaches of the Fayum beds, *Oligopithecus* is reasonably considered to be the oldest primate genus found there. Only part of the left half of one lower jaw is known as yet, but it holds a continuous series of teeth, including the canine, two premolars, and the first two molar teeth. Although its top has been snapped off, the canine still projects beyond the plane formed by the crowns of the other teeth; the first premolar bears a single major cusp, and its outer front surface is worn from contact with the upper canine. The second premolar is bicuspid, as in living anthropoid primates, and the two molars show a slight degree of cross-cresting. These features are sufficient to place this small animal among the higher primates in the suborder Anthropoidea, but they are not enough—especially because only one specimen exists—to support a more precise placement.

Apidium

This primate comes from geological strata that are higher—hence later—than those in which *Oligopithecus* was found. *Apidium* has the triple distinction of being the first primate discovered in the Fayum (in 1908), the one now represented by the most abundant specimens, and the smallest in overall size. These primates were in the same size range as the smallest living New World monkeys or larger marmosets—about 1 kg (2 lb) in weight. The comparison with South American monkeys is appropriate for another reason: *Apidium's* dental formula is $\frac{2133}{2133}$. A dental formula like that of a New World monkey seems out of place in what is undeniably an Old World primate, but it is also found in many living and extinct prosimians. The incisors of *Apidium* are narrow and not very high, and the canines do not project very far beyond them. Each of the lower premolars bears one major cusp, and each of the upper premolars has two. In each molar series, above and below, the last tooth is smallest. The two sides of the lower jaw are fused at the midline, and a complete cup of bone holds the eye. *Apidium* therefore looks like an animal that has just made the morphological transition from advanced prosimian to primitive monkey, or was not very far in either direction from it. Its external appearance, ecological niche, and behavior were probably quite comparable to those of living marmosets. It can almost be pictured springing about actively in the forest canopy, snatching at insects, pillaging birds' nests, and nibbling at ripening fruits much as living South American primates do.

Unless the stratigraphic information is incorrect or a whole complex of evolutionary reversals has occurred, the existence of *Apidium* long after *Oligopithecus* is excellent evidence that at least two independent lineages were represented among the Fayum primates. It is not impossible for populations with characteristics like those of *Apidium* to have been ancestral to ones like *Oligopithecus*. But if they were, the more primitive group must have lived on long after the divergence. *Parapithecus,* another Fayum primate, comes from the same later geological horizon as *Apidium*. The two taxa show such detailed and pervasive similarities (postorbital closure, fused mandibular symphysis, low-crowned canines, and three premolars) that there is probably a very close genetic relationship between these two early monkeys.

Propliopithecus

Like *Apidium, Propliopithecus* has been known since the beginning of the century, when most of two halves of one lower jaw were discovered in the Fayum by Richard Markgraf, a professional fossil collector. In the last decade several more specimens have been found in strata midway between that of the earlier *Oligopithecus* and that of the later *Apidium*. The few jaws and teeth assigned to *Propliopithecus* show characteristics that can be briefly summarized as follows: dental formula $\frac{????}{2133}$, canine larger than that of *Apidium* but still relatively small when compared with its other teeth, anterior premolar with one larger outer cusp and the faintest trace of a second inner one, and uncrested low-crowned molars. All sorts of phylogenetic interpretations have been based on this evidence. *Propliopithecus* has been seen as a specific lineal ancestor of the modern gibbon, a generalized ancestor of the Miocene dryopithecines (discussed below), and the earliest hominid. The case for its hominid status is considered in "*Propliopithecus* as a Hominid" on page 422.

Aeolopithecus

Another more recently discovered Oligocene primate that has been proposed as a direct ancestor for the gibbon is *Aeolopithecus*. Only one jaw is known from the same later Fayum sediments that yielded *Apidium* and *Parapithecus*. Extrapolating from jaw size to body size would make *Aeolopithecus* comparable to a marmoset. Its jaw shows the typical 2123 catarrhine dental arrangement. The canine tooth is quite robust and projecting, and P_3 exhibits a wear facet from the upper canine on its anterior surface. The mandible is fused where its two halves come together, and the sides of the jaw diverge sharply to the rear, suggesting that the lower part of the face did not project out very

PROPLIOPITHECUS AS A HOMINID

The case in favor of hominid status for *Propliopithecus* has been argued most recently and strongly by Björn Kurtén (1972). Kurtén's evidence is purely morphological, and the argument of his whole book—that hominids are descended from *Propliopithecus,* which he sees as a "generalized" primate—turns on one fine point. Apes, from *Dryopithecus* to the modern anthropoids, typically have a sectorial first lower premolar tooth, one in which the crown bears a diagonal ridge culminating in a single cusp; in living humans this tooth has two cusps. According to Kurtén, our bicuspid tooth cannot possibly be derived from the single-cusped tooth found in apes. *Propliopithecus* had first lower premolars with one large and one very small accessory cusp, and Kurtén (invoking the doctrine of the irreversibility of evolution) argues that no single-cusped ancestor could intervene between this primate and humans.

Such a position ignores modern evolutionary theory, which sees variation introduced into species by infrequent but recurrent random mutations and which sees genes sometimes being retained in populations for long periods without being expressed. Genetic variations can be increased or decreased in frequency by the action of various evolutionary forces. A dental variation such as the presence of an additional premolar cusp could increase fitness under certain environmental conditions. Additional premolar cusps (found in occasional living apes) increase the chewing area of the tooth, an advantage if coarser food must be eaten, which was probably the case when our ancestors became terrestrial.

It is surprising to find Kurtén denying that sporadically recurring dental features can be selected for when this would be advantageous, since in an excellent paper entitled "Return of a Lost Structure in the Evolution of the Felid Dentition" he once described a most dramatic example of just such a change, as noted earlier in "Irreversibility of Evolution: Evidence from Paleontology and Genetics," page 371.

Since it relies on a very narrow application of the doctrine that evolution is irreversible, Kurtén's theory that a separate and distinct hominid line must have begun during the Oligocene, about 30 million years ago, with the marmoset-sized *Propliopithecus* remains a possibility, but not a very compelling one. If Kurtén is correct, all the numerous and detailed anatomical, physiological, biochemical, and behavioral similarities between humans and other hominoids, living and fossil, are to be set aside as due to parallel evolution, overbalanced by the evidence of one small premolar cusp.

Source: B. Kurtén. 1972. *Not from the Apes.* New York: Pantheon.

far. There is no feature that would exclude this animal from being an ancestor of later members of the Anthropoidea, but its precise relationship to both contemporary and succeeding populations awaits the discovery of more evidence. One jaw may not accurately represent a population.

Aegyptopithecus

The largest of the primates found in the upper reaches of the Fayum strata is *Aegyptopithecus,* whose jaws are about one-third longer than those of *Propliopithecus,* placing this genus in the size range of the smaller monkeys (see Figure 12-13). *Aegyptopithecus* is significant in large part because of the abundance and completeness of the remains attributed to it. Although an entire skeleton has not yet been found, enough parts exist to make a reconstruction that leaves less to the imagination than is the case for most Oligocene primates. The skull is monkeylike in general appearance, with a relatively projecting muzzle, fairly small oblong braincase, and partial postorbital closure. The dental formula is anthropid, $\frac{2123}{2123}$. A nearly complete right ulna (one of the two bones from the lower part of the arm), discovered almost a decade ago, is of a size more appropriate to *Aegyptopithecus* than to other primate and nonprimate mammals at the site. In estimated overall length, surface conformation, and a number of other morphological details, this bone looks much like those of the New World howler monkey, *Alouatta.* It seems to have belonged to a comparatively strong arboreal quadruped. Some tail vertebrae found in the same quarry show that as *Aegyptopithecus* climbed about in the trees, its relatively large tail probably helped it maintain its balance.

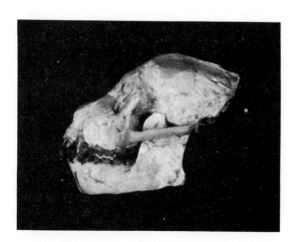

FIGURE 12-13 *Aegyptopithecus,* **an Oligocene member of the Anthropoidea (½ natural size). The generally monkeylike population from which this skull was drawn may have been ancestral to the dryopithecine apes of the Miocene.**
Source: Photograph of cast.

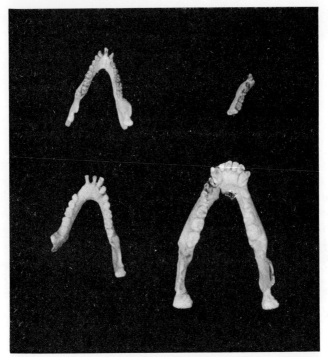

FIGURE 12-14 Comparison of Oligocene primate jaws (½ natural size). *Upper left,* **Apidium;** *lower left,* **Proliopithecus;** *upper right,* **Oligopithecus (fragment);** *lower right,* **Aegyptopithecus.**
Source: Photograph of casts.

Aegyptopithecus has been widely discussed as the earliest dryopithecine, a group named for the genus *Dryopithecus* (discussed later) that lived in the Miocene and Pliocene and was the ancestor of modern pongids and hominids. This view represents classification from a vertical perspective, a focus on what descendants of this population might later have become. If we were to concentrate on the evidence for what the animal itself looked like and how it lived, a horizontal perspective would classify *Aegyptopithecus* as a primitive, monkeylike primate that would make a suitable ancestor for several later lineages. Figure 12-14 gives a view of the lower jaws of several Fayum primates.

Branisella

The Fayum primates of the Oligocene had reached the evolutionary level of present-day New World monkeys or had gone a bit beyond it. For comparative purposes, it is interesting that the only other notable evidence for the existence of Oligocene higher primates comes from

South America. Early Oligocene deposits in Bolivia have recently yielded an upper left maxillary fragment of a primate genus now called *Branisella.* It contains teeth from P³ through M² and shows less resemblance to the skull of *Rooneyia,* an Oligocene anaptomorphid prosimian from North America, than to the African *Apidium.* It would not be impossible to derive *Branisella* and the later New World monkeys from North American prosimians, but we cannot rule out an African origin for South American primates. Africa and South America were closer during the Oligocene than they are now, and primates may have made the crossing on rafts of floating vegetation. This possibility is strengthened by the appearance at the same time in South America of ancestors of the caviomorph rodents (relatives of the guinea pig), for which an African ancestry is also likely (Hoffstetter, 1972), but the issue is still open.

The Oligocene is a particularly tantalizing period of primate evolution; about 15 million years are represented by just a handful of sites. Fortunately, one of these, the Fayum deposits of north Africa, has yielded abundant fossil remains. From these we can see the origins of the Old World monkeys, as well as plausible ancestors for the apes that became abundant during the Miocene.

MIOCENE: SPREAD OF THE HOMINOIDS

There is a gap of several million years in the African fossil record following the early Oligocene deposits in north Africa. Beginning with the early Miocene, fossil primate sites are much more widely distributed, occurring not only in Africa but also in a wide band across Eurasia. We'll consider primates from the more important sites in these areas to get an overview of this significant period, which saw the first appearance of the hominoid stock that gave rise to humans.

Pliopithecus

The oldest Miocene primate fossils from Europe are assigned to the genus *Pliopithecus* (Figure 12-15) and are dated to about 16 million years ago, about the middle of the Miocene. This lineage apparently continued on with relatively little modification for several million years. In *Pliopithecus,* with its dental formula of $\frac{2123}{2123}$ and its moderately projecting canines, many anthropologists see a stage in the ancestry of modern gibbons. Whether anyone would choose to call *Pliopithecus* a gibbon if it were alive today is another matter. The forelimbs of this Miocene primate were not yet as greatly elongated as those of modern gibbons, and so brachiation had not yet become as typical or efficient a means of locomotion. In addition, *Pliopithecus* seems to have had a tail. In animals that lack a tail, the hole in the vertebrae of the lower

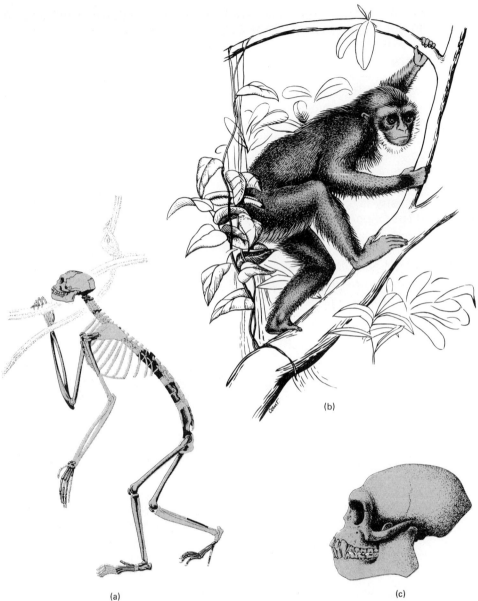

(a)   (c)

FIGURE 12-15 *Pliopithecus,* a Miocene anthropoid. (a) Skeleton, based on a fairly complete specimen discovered in 1957, with known parts represented by dark shading and restored areas by dotted lines. (b) Reconstruction, based on fossil remains of *Limnopithecus,* a close relative. (c) Although *Pliopithecus* may be as much as 20 million years old, its skull is very much like a modern gibbon's.

Sources: (a, c) E. L. Simons. 1964. "The Early Relatives of Man," *Scientific American,* **221**(1):50–62. (b) W. E. Le Gros Clark. 1959. *The Foundations of Human Evolution.* Eugene, Ore.: Oregon State System of Higher Education (Condon Lectures). Fig. 5, p. 65.

back for the spinal cord narrows sharply; in animals with a tail, however, it remains wide because the spinal cord continues through to the tail. *Pliopithecus* had the wide sacral opening typical of tailed primates (Ankel, 1965).

Remains very similar to those of *Pliopithecus* in age and morphology have been found on Rusinga Island in Lake Victoria, Kenya. These are conventionally assigned to another genus, *Limnopithecus*, but although there are some measurable differences from *Pliopithecus*, recent studies question the need for two genera. In either case it would seem that the European and Asian populations of these early Miocene primates occupied much the same ecological niche: that of a small arboreal climbing, swinging, hanging primate much like the present South American spider monkey.

Oreopithecus

Oreopithecus is another Miocene primate with tantalizing implications for human evolution (see "*Oreopithecus* as a Hominid" on pages 428–429); the debate over its phylogenetic position has lasted for over a century. The first lower jaw of *Oreopithecus* was discovered in 1872 in deposits of lignite (a brownish-black fuel midway between peat and bituminous coal in texture) in Tuscany in Italy. Along with *Oreopithecus* fossils are the bones of rodents, dormice, and shrews, as well as those of fish, snakes, lizards, amphibians, crocodiles, turtles, and birds. *Oreopithecus* is the most abundant mammal there, however. There are almost 200 specimens, one of them a nearly complete skeleton discovered in 1958.

From the evidence, the appearance of *Oreopithecus*, its behavior, and the environmental setting in which it existed can be reconstructed better than can be done for most other Tertiary primates. The lignites seem to have been built up from the abundant vegetation that grew in a swamp-forest environment, one in which broad expanses of trees formed a canopy over the still waters beneath. This type of habitat is found today over part of the range of the orangutan in Sarawak, Borneo. Like the orang, *Oreopithecus* had forelimbs much longer than hindlimbs, and there are detailed structural resemblances between these two primates in the femur (the bone from the upper part of the leg) and the humerus (the upper arm bone) as well. It would be odd if structure did not point toward function in this case; until something else is proved, it seems reasonable to assume that *Oreopithecus*, like the orang, was a relatively large, highly arboreal herbivore capable of supporting itself by grasping with several limbs while feeding relatively far out near the ends of branches.

Morphological features can reflect heritage as well as habitus, but

OREOPITHECUS AS A HOMINID

The strongest argument for the hominid status of *Oreopithecus* was made by Johannes Hurzeler, the discoverer of the complete skeleton and of many of the other fossil remains of this animal. Hurzeler stressed that this primate of 10 to 12 million years ago had some strikingly hominidlike features in the postcranial skeleton and skull. *Oreopithecus* had an iliac crest that was longer and more flaring than that of any known pongids, past or present. In its pelvis there was also a well-developed anterior inferior iliac spine, the point of attachment for a ligament important in modern humans because of its function in making upright posture and bipedal locomotion possible. The lower face of *Oreopithecus* projected less than the muzzles of modern pongids and consequently had a more human appearance. Furthermore, both lower premolars (P_3 and P_4) are bicuspid, the canines are relatively small, and there is no diastema in 11 out of 12 specimens that have been studied in detail.

Several other anatomical characteristics are in contrast to the ones listed above. The long arms and short legs of *Oreopithecus* imply the prior operation of selection for a brachiating mode of locomotion. This is entirely consistent with the habitat that can be reconstructed. The volume of the brain of *Oreopithecus,* recently estimated by Szalay (1973), may be as low as 200 cc. If correct and representative, this estimate would place the brain size of this Miocene primate well below that seen in the chimpanzee, an animal of comparable size.

Yet other characteristics of *Oreopithecus* are unlike those of either pongids or hominids. Notable in this respect is the form of the molar teeth, which have thin enamel covering their high and deeply wrinkled crowns. The third molar has an extra central cusp that is not a normal feature of the crowns of hominoid primates; see the illustration opposite. These and other dental features are discussed and illustrated in detail by Butler and Mills (1959).

For these reasons, many anthropologists believe that *Oreopithecus* was neither a hominid nor a pongid, and perhaps not a hominoid at all. Nor does this taxon seem to belong with the Old World monkeys in the Cercopithecoidea. Most anthropologists tend to place it in a superfamily of its own, as shown in Table 12-3. As has been noted in "*Megaladapadis* and *Archaeolemur:* Two Primates of the Past That Do Not Fit Present Patterns," page 360, not all primate evolutionary experiments have

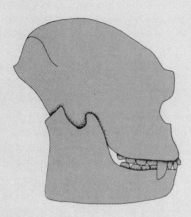

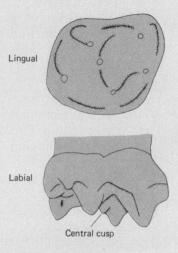

Lingual

Labial

Central cusp

Upper third molar

Rough outline of skull of *Oreo-pithecus*, drawn from a photograph of the skull as it lay in its matrix. This differs from Hürtzler's reconstruction in two respects: the nuchal crest is higher, and the mandible is flared out at the rear corner (as among the leaf-eating langurs), suggesting a specialized diet of soft vegetable matter.
Source: Carleton S. Coon. 1963. *The Origin of Races.* New York: Knopf. Fig. 17, p. 210.

Specialized dentition of *Oreo-pithecus*. The upper right third molar has a small cusp, or conulid, at the center of the crown, in addition to the five cusps characteristic of the dryopithecines, modern apes, and hominids. All its cusps are high and pointed.
Source: Carleton S. Coon. 1963. *The Origin of Races.* New York: Knopf. Fig. 18, p. 211.

survived to the present; *Oreopithecus* too may represent a taxon that left no survivors. This case may be of particular instructive value nevertheless. *Oreopithecus* suggests that during the Miocene there existed some large arboreal primates who, although perhaps not themselves hominids, still possessed features that clearly foreshadowed the pattern of evolution to be seen later in our own direct ancestors.

Sources: (1) P. M. Butler and J. R. E. Mills. 1959. "A Contribution to the Odontology of *Oreopithecus,*" *Bulletin of the British Museum of Natural History,* **4**(1):1–26. (2) C. S. Coon. 1962. *The Origin of Races.* New York: Knopf. Fig. 17, p. 211. (3) F. S. Szalay and A. Berzi. 1973. "Cranial Anatomy of *Oreopithecus,*" *Science,* **180**(4082):183–185.

here *Oreopithecus* is still the source of much debate. Many anthropologists see it as a hominoid, but one unusual enough to merit a distinct family of its own. Others believe it to be a cercopithecoid that evolved a large size and dental morphology and limb proportions comparable to those of the hominoids. A few anthropologists have placed *Oreopithecus* directly among the hominids (see "*Oreopithecus* as a Hominid").

Dryopithecus

There are other hominoid fossil primates in the size range of *Oreopithecus* from the Miocene. The first discovery was made in 1856 near the village of Saint-Gaudens, France, and was described by the French paleontologist Louis Lartet the next year. It comprised several large fragments of the lower jaw, plus a humerus with both ends broken off, leaving only the middle of the shaft. Since the same deposits contained the trunks of oak trees, Lartet gave the primate the generic name *Dryopithecus* ("oak ape") and called the species *fontani,* in honor of its discoverer, the naturalist Fontan. Over the succeeding period of more than a century, fossils similar to those found by Fontan have been discovered at other sites in France; in Spain, Germany, and Austria; and in a broad band eastward across Eurasia through Turkey to China. Rich finds in several countries of east Africa document the expansion of dryopithecine populations down through at least part of that continent. These bones were dug up or found by many naturalists, anthropologists, and paleontologists. In describing a find, each scientist tended to emphasize its distinctiveness from what was previously known. As a result, the bones were attributed to over two dozen different genera and to many more species. Following the extensive review by Simons and Pilbeam (1965), these are now usually all lumped into the original genus, *Dryopithecus*.

No complete dryopithecine skeletons are yet known, and most of the individual fossils are jaws and teeth. One partial cranium has been found, along with limb bones and vertebrae. The composite picture that can be formed from these is one of animals about the size of modern chimpanzees, but varying from somewhat smaller to a bit larger (see Figure 12-16). Probably most dryopithecines were more slightly built than chimpanzees, since their bones are a trifle less heavy (Figure 12-17). Dryopithecines had the capacity for at least as great a range of limb movements as chimpanzees, and perhaps greater. Some studies of the hand, wrist, and limb structure have suggested that monkeylike features were present. These might well have been a heritage from a more quadrupedal ancestor, and they do not by any means bar dryopithecines from a position directly ancestral to the present-day great apes and, indirectly, to humans. Much the same is true for dryopithecine dental traits. These include strongly projecting canines, sectorial anterior lower

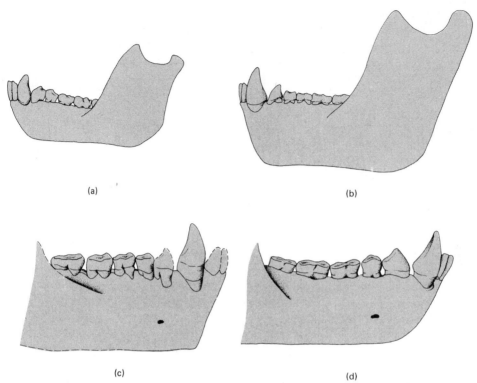

FIGURE 12-16   Range of jaw sizes in African dryopithecines (½ natural size): (a) *Dryopithecus africanus,* (b) *D. nyanzae,* (c) *D. major,* (d) **modern gorilla.**

Source: W. E. Le Gros Clark and L. S. B. Leakey. 1951. "The Miocene Hominoidea of East Africa," in *Fossil Mammals of Africa,* (no. 1). London: British Museum (Natural History). Figs. 10, 17, pp. 46, 58.

premolars, and low-crowned molars with Y-5 cusp and crest patterns. A functional diastema and simian shelf are also commonly present. Less robust dryopithecines lack the expanded muzzle seen particularly in large modern adult male pongids, but this difference, like some of the others, is largely a matter of degree.

Not only were the dryopithecines numerous and widespread, but they also existed for a long period of time. Most sites at which their remains occur have not been precisely or reliably dated, but the few dates that do exist indicate that, following an origin probably early in the Miocene (about 20 million years ago), they lasted on through the epoch and into the Pliocene. Indeed, since the living great apes are their descendants and are not as greatly different from the dryopithecines as we are, in time it should prove possible to reconstruct a continuous phyletic sequence to chimps, gorillas, and orangs spanning the later Pliocene and Pleistocene as well.

Though not hominids themselves, dryopithecines have long been

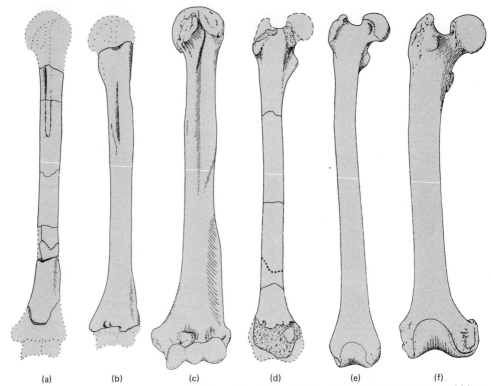

(a)   (b)   (c)   (d)   (e)   (f)

FIGURE 12-17   Limb bones of dryopithecines (⅓ natural size). Left humeri: (a) **African specimen,** (b) **European specimen,** (c) **male chimpanzee. Right femurs:** (d) **African specimen,** (e) **European specimen,** (f) **male chimpanzee.**

Source: W. E. Le Gros Clark and L. S. B. Leakey. 1951. "The Miocene Hominoidea of East Africa," in *Fossil Mammals of Africa* (no. 1). London: British Museum (Natural History). Figs. 23, 24, pp. 93, 96.

considered to be a stage in our ancestry. The evidence for this position was outlined in a classic book by William King Gregory (1922) based on evidence from the preceding half century. The evidence still favors his view that dryopithecines gave rise, through speciation, to at least one distinct hominid lineage. In all the characteristics by which they differ from modern humans—for example, limb proportions, brain size, canine tooth size, and cusp patterns on the molar teeth—dryopithecines are similar to living chimpanzees and differ from modern humans chiefly by degree or frequency (Figure 12-18). What is more, for the bulk of these features dryopithecines differ even less from the Plio-Pleistocene hominids, discussed in Chapter 13, which are nearer to them in time. Given at least several million years of time for the various changes to occur, it is possible to derive Plio-Pleistocene hominids from the dryopithecines. Even though we do not know the modes of inheritance of the specific traits that differ among these Tertiary populations, some relatively complex polygenic traits could be transformed in the time

(a)

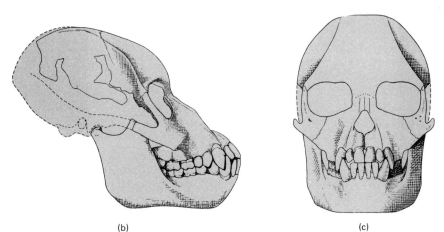

(b)   (c)

FIGURE 12-18 *Dryopithecus,* **a Miocene hominoid. (a) Reconstruction. (b) Skull, side view. (c) Skull, front view.**

Sources: (a) W. E. Le Gros Clark. 1959. *The Foundations of Human Evolution.* Eugene, Ore.: Oregon State System of Higher Education (Condon Lectures). Fig. 6, p. 67. (b, c) W. E. Le Gros Clark and L. S. B. Leakey. 1951. "The Miocene Hominoidea of East Africa," in *Fossil Mammals of Africa* (no. 1). London: British Museum (Natural History). Figs. 2, 3, pp. 20, 21.

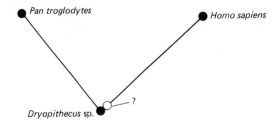

FIGURE 12-19   A cladogram showing probable relationships among some hominoids, and the position occupied by the earliest hominids (indicated by the open circle).

available. Most anthropologists would therefore agree with the placement of the dark circles or nodes (representing specific populations) that represent dryopithecines as ancestral to all recent hominoids in the simple cladogram presented in Figure 12-19.

There is widespread agreement on this general position, but not on any attempt to transform the simple cladogram into a more detailed phylogenetic tree. Disagreement is particularly strong, for instance, over the characteristics of the taxon that should be considered the earliest recognizable hominid (represented by the open circle in Figure 12-19). It is not hard to understand why there is such controversy about this specific detail of hominid phylogeny. Dryopithecines were biologically successful in every conventional sense of the term: they were geographically widespread, apparently abundant in at least some parts of their range, and persistent, and they seem to have left survivors—some (humans) changed considerably by phyletic evolution, and others (the great apes) less so. Each of these factors probably contributed to some extent to the individual differences that can be seen from specimen to specimen among the hundreds of dryopithecine fossils that are known. The existence and extent of the differences can be seen; deciding to what extent these differences represent intraspecific variation or interspecific differences, however, is a matter of interpretation. As was pointed out earlier, there is no general rule concerning how much morphological difference corresponds to genetic isolation. Recognizing the earliest hominids from among their probable near relatives among the dryopithecines is complicated still further by the fragmentary nature of the fossils, which are often parts of the jaws and teeth that do not show the most characteristic features of each group.

Disagreement over the relative qualifications of particular candidates for hominid ancestry has persisted for just about as long as there has been general agreement on the phylogenetic position of the dryopithecine group as a whole. Over this time, majority opinion among anthropologists has favored first one and then another specific taxon as the earliest hominid. In fact, since there is no difficulty in deriving later hominids from some offshoot of the dryopithecine stock, it is not necessary to decide here which particular population from within or near the dryopithecine group is most likely to be an early hominid rather than another variant Tertiary ape. It is perhaps even premature as well as unnecessary to decide just now in favor of one or another candidate for hominid ancestry. When we know more and can solve this problem, however, we will have definite information on the duration and initial heritage of our lineage. The evidence available for several possible Miocene hominids is summarized in "*Sivapithecus* as a Hominid" (on pages 436–437) and "*Ramapithecus* as a Hominid" (on pages 438–440). The cases that can be made for each of these alternatives (and for still another taxon from around the Pliocene, discussed in "*Gigantopithecus* as a Hominid," pages 441–443) are not entirely comparable. The argument that *Sivapithecus* was a hominid, presented here chiefly for historical perspective, would not be considered seriously by most specialists today. *Ramapithecus* is probably the most popular nominee at present.

Beginning from an ancestor perhaps like *Aegyptopithecus* from the preceding Oligocene, dryopithecines increased in size, abundance, and range during the 20 million years of the Miocene. These chimpanzeelike apes had reached the stage from which evolution of a hominid line required just a short step—but one that would have momentous consequences for all life on earth. Pliocene deposits were until recently confused with those of the preceding Miocene or the later Pleistocene. More widespread use of improved dating methods should help considerably here. But in any case, dryopithecines or their phyletic descendants must have continued through this span of time. Those which stayed in the reduced areas of forest survive as the gorilla, chimpanzee, and orangutan of today. In some areas, however, various primate populations were adapting increasingly to life on the ground. The first macaques and baboons were found in the Pliocene of Europe and North Africa, and true baboons or baboonlike monkeys in Africa, India, and China. At the same time, terrestrially adapted apes were apparently evolving as well. A large hominoid primate, *Gigantopithecus,* is known from a lower jaw discovered in Indian deposits that seem to be somewhere between 3 and 7 million years old. This specimen shows unmistakable resemblances to other material (three mandibles and over a thousand teeth) found in China and formerly believed to have been of middle Pleistocene age.

There is little problem in deriving *Gigantopithecus* from dryopithecine ancestors of the Miocene or early Pliocene. Some larger-than-average dryopithecine jaws and teeth are already known from India and Europe. The fate of this large primate, however, is a matter of much greater uncertainty. To account for its strong resemblances to the robust australopithecines (discussed in Chapter 13), many anthropologists suggest that parallel evolution occurred before the lineage that included *Gigantopithecus* became extinct. Others have suggested that *Gigantopithecus* was either itself a hominid or an immediate hominid predecessor that disappeared through phyletic evolution (see "*Gigantopithecus* as a Hominid" on pages 441–443).

SIVAPITHECUS AS A HOMINID

A good example of the tendency of past generations of paleontologists to multiply taxonomic categories, *Sivapithecus indicus* was established by Pilgrim in 1915 after the discovery of a single right third lower molar in the Siwalik Hills of India. *Sivapithecus* ("Siva's ape") was named after one of the supreme gods of Hinduism worshiped in India, the location referred to in the second part of the species name, *indicus*. Several additional specimens found soon afterward were also referred to the same taxon. The largest of these included part of the right half of a lower jaw containing P_4, M_1, and M_2, as well as the roots of the lower canine and P_3. Another relatively large fragment from the front half of the lower jaw contained a canine and parts of the roots of several adjacent teeth. One P_3 was also found and definitely assigned to the same species. These parts are all indicated by dark shading in (b). The rest of the jaw shown in (a) represents Pilgrim's reconstruction, which resembles the shape of the human lower jaw shown earlier in Figure 12-7.

A year later Gregory offered an alternative reconstruction of the same material, shown in (b). Gregory's sketch closely resembles that of a male orangutan, shown for comparison in (c). In support of his drastically different conclusion, Gregory noted that while all the cheek teeth of *Sivapithecus indicus* did resemble to some extent those of modern humans, they resembled even more closely the comparable teeth of *Dryopithecus* and even the orang. The canine tooth, in addition, was far more apelike than human. Gregory also pointed out several unlikely features of Pilgrim's reconstruction. The most notable among these was that the thin midline, or symphysis, seemed inconsistent with the heavy apelike canine and anterior premolar.

In his judgment about the significance of *Sivapithecus,* Gregory did not exclude any relationship with humans. Instead, he said that a mixture of humanlike and apelike characteristics should be expected in a mid-Tertiary ancestor of the hominids. But he also stressed that a few hominidlike dental features were not in themselves a sufficient basis for claiming status as a hominid, and he cautiously counseled waiting for further knowledge about other skeletal characteristics before according hominid status to *Sivapithecus.*

Source: W. K. Gregory. 1916. "Studies on the Evolution of the Primates," *Bulletin of the American Museum of Natural History,* **35**(19):239–355. Figs. 15, 16, pp. 290–291.

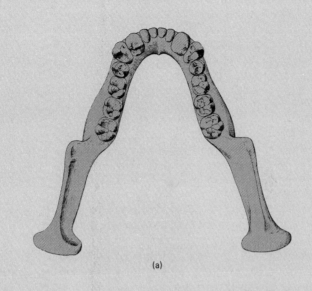

(a)

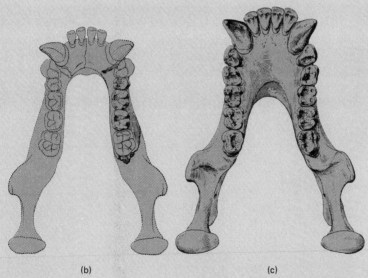

(b)   (c)

(a) **Largest of the specimens from Siwalik, with Pilgrim's reconstruction.** (b) **Gregory's alternative reconstruction.** (c) **Male orangutan.**

Source: W. K. Gregory. 1916. "Studies on the Evolution of the Primates," *Bulletin of the American Museum of Natural History*, **XXXV**:239–355. Figs. 15, 16, pp. 290–291.

RAMAPITHECUS AS A HOMINID

The dispute over the possible hominid status of *Ramapithecus* began nearly as long ago as the debate over *Sivapithecus.* In 1932 G. Edward Lewis, a Yale graduate student, discovered in India seven fragments of fossil jaws and teeth of hominoid primates. In a brief article 2 years later he assigned two of the specimens to existing dryopithecine categories, and then he created four new genera and five new species for the remaining five pieces. One of the new taxa Lewis created was *Ramapithecus brevirostris* ("Rama's ape," after another of the Hindu deities, and "short-snouted" to suggest that the face did not project very much). Though he called it an ape, Lewis stressed the similarity of *Ramapithecus* to modern humans in several features. Among these were a parabolic rather than a U-shaped tooth row in the upper jaw and a relatively small socket for the missing canine tooth. Several years later, in his doctoral dissertation, Lewis argued more strongly for placement of *Ramapithecus* in the Hominidae.

The strongest early disagreement came from Aleš Hrdlička (1935), the prominent and experienced Czech-born founder of physical anthropology in America. Among other things, Hrdlička pointed out that there was not enough of the upper jaw preserved to tell whether it was parabolic or U-shaped, and he noted that the canine was not particularly small once its crushed-in socket (not mentioned by Lewis) was reconstructed. Something of a standoff or compromise followed. In the absence of any new evidence most anthropologists, perhaps following the middle course suggested by W. K. Gregory, apparently accepted the *Ramapithecus* jaw fragment as being among the more hominidlike specimens of the dryopithecine group.

It was not until several decades later that the controversy was again revived by another Yale paleontologist, Elwyn Simons (1961). In the absence of any new fossil remains of *Ramapithecus,* Simons repeated the arguments that Lewis had made. In a related move, Simons argued that some other fossil material earlier assigned by Pilgrim to *Dryopithecus punjabicus* (a species named after the Punjab region of India, where it was found) resembled *Ramapithecus* so closely that it should be assigned to the same genus. In doing this, Simons retained the second part of the name that Lewis had coined, and so today the taxon is known as *Ramapithecus punjabicus.*

Simons's paper did help revive interest in hominid origins. One effect of this was to stimulate the creation of detailed scenarios that suggested how primates with the observed and inferred characteristics of *Ramapithecus* could have evolved. Some of these scenarios depicted *Ramapithecus* as a diminutive terrestrial hunter or scavenger with canines

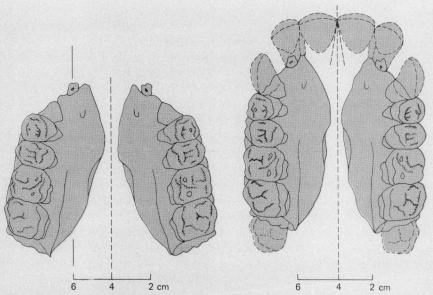

Reconstruction drawings of the upper dental arcade of *Ramapithecus punjabicus* from the maxillary fragment YPM 13799 according to (left) **Simons (1961) and** (right) **Genet-Varcin (1969).**

Source: C. Vogel. 1975. "Remarks on the Reconstruction of the Dental Arcade of *Ramapithecus*," in R. Tuttle (ed.), *Paleoanthropology: Morphology and Paleoecology*. The Hague, Netherlands: Mouton. Pp. 87–98.

reduced in size because they had already been replaced by hand-held weapons. As an alternative, Jolly envisioned *Ramapithecus* as having reduced canines because the animal was a seed eater like the gelada baboons. While all these scenarios were in theory possible, for the most part they preceded the discovery of enough direct evidence to conclusively support one or another alternative, as well as to develop accurate cladograms and phylogenetic trees.

In more recent years the sample of specimens attributed to *Ramapithecus* has expanded to several dozen (Frayer, in press; Simons, 1977). While a few of these specimens represent new discoveries, most had previously been assigned to other taxa now recognized as part of the widespread dryopithecine group. As might therefore be expected, the geographic and temporal distribution of the *Ramapithecus* remains—Europe, India, mainland Asia, and Africa, chiefly during the Miocene—closely parallels that of undoubted dryopithecines. Indeed, many sites that have yielded remains believed to represent *Ramapithecus* also contain unquestioned dryopithecines in the same deposits. These associations make it difficult to reconstruct an ecological niche for *Ramapithecus* that is differentiated in a meaningful way from the niches of these other Tertiary hominoids.

Morphologically, too, *Ramapithecus* remains something of an enigma. Some of the newly assigned specimens exhibit traits—such as a simian shelf in one jaw from Domeli in Pakistan—that were formerly considered to be clearly indicative of pongid status. Of course, if *Ramapithecus* is a taxon that evolved from other Miocene dryopithecines, it is reasonable to expect that it might retain some shared characters indicating this common heritage. However, the fossils referred to *Ramapithecus* come from widely separated points in time and space. The crucial issue to be settled, therefore, is whether the characteristics that these specimens exhibit represent a true total morphological pattern of a unique hominid taxon or a synthetic morphological pattern (in the senses in which these terms were used in Chapter 11) built up from the more hominidlike individual variants in dryopithecine populations. This distinction is particularly difficult to make, since many of the traits used to distinguish early hominids from early pongids—canines small relative to the other teeth, a high degree of wear from friction between adjacent cheek teeth, arching rather than parallel tooth rows, and so on—still appear as occasional variants in pongid populations even today and may have been present in the dryopithecine stock from which hominids speciated.

There is reason to hope that, in the future, questions about the taxonomic status of *Ramapithecus* can be resolved on the basis of models provided by modern population biology (Eckhardt, 1975). By the time populations have reached a level of distinctiveness such that they can be recognized as hominids from a few incomplete fossils, their gene pools should have been separated for a considerable period of time. Convincing evidence for such separation would be provided by the discovery of a number of reasonably complete specimens together in an appropriate ecological setting, unassociated with other dryopithecines.

Sources: (1) R. B. Eckhardt. 1975. "*Gigantopithecus* as a Hominid," in R. Tuttle (ed.), *Paleoanthropology: Morphology and Paleoecology.* The Hague, Netherlands: Mouton. Pp. 105–129. (2) D. W. Frayer. In press. "Is *Ramapithecus* a Hominid Ancestor?" in *Controversies of the Living and the Dead.* (3) E. Genet-Varcin. 1969. *A la recherche du primate ancêtre de l'homme.* Paris: Boubée et Cie. (4) A. Hrdlička. 1935. "The Yale Fossils of Anthropoid Apes," *American Journal of Science,* **229**(166):34–40. (5) G. E. Lewis. 1934. "Preliminary Notice of New Man-like Apes from India," *American Journal of Science,* **227**(159):161–179. (6) E. L. Simons. 1961. "The Phyletic Position of *Ramapithecus*," *Postilla, Yale Peabody Museum,* (57):1–9. (7) E. L. Simons. 1977. "*Ramapithecus*," *Scientific American,* **236**(5): 28–35. (8) C. Vogel. 1975. "Remarks on the Reconstruction of the Dental Arcade of *Ramapithecus*," in R. Tuttle (ed.), *Paleoanthropology: Morphology and Paleoecology.* The Hague, Netherlands: Mouton. Pp. 87–98.

GIGANTOPITHECUS AS A HOMINID

Knowledge about the existence of this large hominoid primate dates back to 1935, when the Dutch paleontologist G. H. R. von Koenigswald found a large third lower molar tooth on the shelf of a Chinese apothecary shop in Hong Kong. In China as well as elsewhere in the Far East, fossil bones and teeth of many animals were ground up and used in medical preparations believed effective against many kinds of diseases. The crown pattern identified the tooth as that of an anthropoid, but its large size suggested to von Koenigswald that he was dealing with a previously unknown population. Accordingly, he named the new taxon *Gigantopithecus blacki,* the generic name referring to the large size of these primates, and the species name in honor of Davidson Black, a paleontologist who had previously done distinguished work in the area.

In 1937 von Koenigswald found another molar tooth resembling the first, and 2 years later he found yet another, these too in apothecary shops—and hence without sufficient geological evidence to gauge their age with any degree of certainty. While their discoverer was interned by the Japanese during World War II, his colleague Franz Weidenreich (1945) described the new specimens. In doing this Weidenreich altered his earlier opinion that the teeth belonged to a giant orangutan, and said instead that their morphological characteristics required hominid status. However, Weidenreich's enthusiastic embrace of *Gigantopithecus* into the hominid family was accompanied by an effective kiss of death. Assuming that body size would be in direct proportion to tooth size (a doubtful proposition from what we know about the allometry of these traits in living primates), Weidenreich estimated that *Gigantopithecus* was twice as large as an adult male gorilla. The incredible image thus conjured up may well have served as an effective block against serious consideration of this primate's other less speculative and less spectacular characteristics.

On balance these other characteristics leave little doubt that *Gigantopithecus* populations had become genetically distinct from their dryopithecine ancestors. To date, over a thousand isolated teeth and three relatively complete lower jaws have been found in China, often in concentrations rather than as isolated specimens. One additional lower jaw, usually referred to as *Gigantopithecus bilaspurensis,* has been recovered from India, and two somewhat similar jaws from Greece (de Bonis, Bouvrain, and Melentis, 1975). However, the controversy that surrounded the discovery of the fossil remains of this hominoid primate over four decades ago continues to the present and still focuses on the same basic issues: the causes of the known and inferred characteristics of the populations, the time during which the populations lived, and the phylogenetic relationships of these populations to undoubted Plio-Pleistocene hominids.

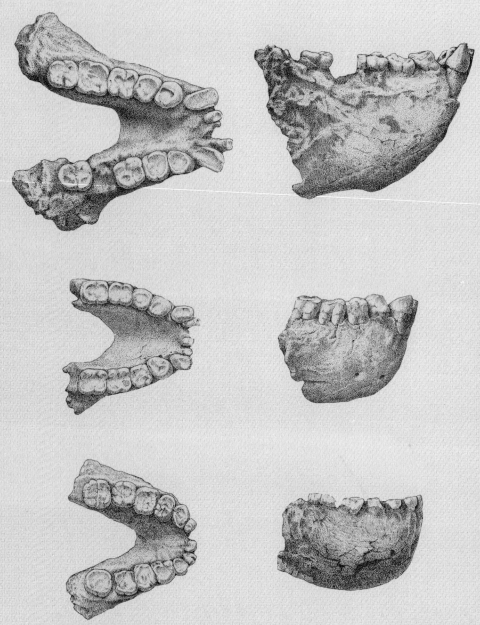

While few *Gigantopithecus* specimens fall within the range of variation of later hominids, some approach it. For example, the moderate-sized jaw of one specimen (perhaps a female) of *Gigantopithecus* from China (*middle*) shows some resemblance to one of the oldest known jaws of a robust early Pleistocene hominid from Ethiopia (*bottom*). The specimen of *Gigantopithecus* at *top*, however, is much larger and less hominidlike in morphology.

Source: R. B. Eckhardt. 1972. "Population Genetics and Human Origins," *Scientific American*, **226**(1):94–103.

The most evident hominidlike features of *Gigantopithecus* include incisors and canines that, while absolutely large, are small relative to the other teeth, particularly in crown height; anterior lower premolars that are bicuspid rather than sectorial; and a dental arcade that in some specimens (though not all) resembles that found in the better-known, more robust Plio-Pleistocene hominids. Postcranial remains that would make it possible to reconstruct posture and locomotion are as yet unavailable. The known features are commonly said to represent parallels to similar features found in more widely accepted hominids. For example, a scenario that gives a functional explanation for these features has been developed by Pilbeam (1970), who sees *Gigantopithecus,* like *Ramapithecus,* as a seed eater. There is good evidence that *Gigantopithecus* lived on open savanna areas, and it is not impossible to imagine this large primate eating some seeds and other small objects. However, the seed-eating hypothesis itself supplies no clue as to why one of these primates became extinct while the other underwent phyletic evolution to give rise to later hominids. In any case, the diets of most primates are quite flexible, and the possibility should be considered that *Gigantopithecus* may have picked up some animal protein in its foraging activities, as baboons and chimps do in some areas today.

The dating of Chinese *Gigantopithecus* populations is quite uncertain; ages from late Pliocene to middle Pleistocene have been suggested by various scientists, but no chronometrically dated materials are known, and the associated faunal remains are inconclusive. The Indian specimen may be middle Pliocene, though there is some uncertainty because this specimen, too, is without proper geological context. The similar Greek lower jaws have tentatively been said to be middle Miocene in age.

Since there are no chronometric dates available for materials associated with any *Gigantopithecus* fossils, there is a fundamental problem in trying to determine whether there would have been sufficient time for any of these large hominoid primate populations to have given rise to Plio-Pleistocene hominid taxa. If all *Gigantopithecus* populations were of middle Pleistocene age, this primate could easily be excluded from our ancestry. Given hundreds of thousands to millions of years of time, however, even quite divergent morphological characteristics—tooth size, body size, etc.—can be transformed. Any more certain conclusion must await the discovery of more complete remains of well-dated populations.

Sources: (1) L. de Bonis, G. Bouvrain, and J. Melentis. 1975. "Nouveaux restes de Primates hominoides dans le Vallésien de Macédoine (Grèce)," *Comptes rendus de l'Académie des Sciences de Paris,* Série D, **281**:379–382. (2) R. B. Eckhardt. 1972. "Population Genetics and Human Origins," *Scientific American,* **226**(1):94–103. (3) D. R. Pilbeam. 1970. "*Gigantopithecus* and the Origins of Hominidae," *Nature,* **225**:516–519. (4) F. Weidenreich. 1945. "Giant Early Man from Java and South China," *Science,* **99**:479–482. (5) F. Weidenreich. 1946. *Apes, Giants, and Man.* Chicago: University of Chicago Press.

It is possible to approach the question of hominid origins through the study of fossil evidence alone. But as was stressed in Chapter 11, other lines of evidence arise from the comparative study of living organisms. Until the fossil record is far more complete than it is today, we must use modern population biology to help bridge the gaps.

SUMMARY

1. Primate fossil remains are distributed widely in time (over at least 75 million years) and space (chiefly North and South America, Eurasia, and Africa), but for many reasons the record is incomplete and uneven.

2. Some dental evidence indicates that the earliest known primate genus is *Purgatorius,* found in Cretaceous deposits in Montana.

3. From the Paleocene of North America and Europe, a number of populations of primitive prosimians are known; these are all grouped into four families: Paromomyidae, Picrodotidae, Carpolestidae, and Plesiadapidae.

4. In the Eocene more modern-looking primates appeared; these included lemurlike forms, some sharing features of both lemurs and tarsiers, and the first members of the family Tarsiidae itself.

5. During the Oligocene the first primates of anthropoid grade evolved. The most abundant remains of these have been found in early and middle Oligocene sediments in north Africa, but their characteristics are foreshadowed by traces found in late Eocene deposits in Burma and shared by primates from the early Oligocene of Bolivia.

6. During the Miocene small gibbonlike apes and the larger pongidlike dryopithecines lived in the forests that covered vast areas of Eurasia and Africa.

7. In the Pliocene these forested areas were reduced in extent; as a result, some primate populations adapted to terrestrial life. Those which did not adapt continued to live in smaller forest habitats or died out.

8. The origin of an independent hominid lineage is still a matter of debate; some evidence can be found in each period of time from the Eocene to the Pliocene.

SUGGESTIONS FOR ADDITIONAL READING

Simons, E. L. 1972. *Primate Evolution.* New York: Macmillan. (This book provides one of the very few comprehensive surveys of the fossil evidence for primate evolution written in recent years. It is well illustrated and has a useful bibliography.)

REFERENCES

Ankel, F. 1965. "Der Canalis sacralis als Indikator fur die Lange der Caudalregion der Primaten," *Folia Primatologia,* **3**:263–276.

Barth, F. 1950. "On the Relationships of Early Primates," *American Journal of Physical Anthropology,* **8**(2):139–149.

Cartmill, Matt. 1974. "Rethinking Primate Origins," *Science,* **184**(4135):436–443.

Coryndon, S. C., and R. J. G. Savage. 1973. "The Origin and Affinities of African Mammal Faunas," in N. F. Hughes (ed.), *Organisms and Continents through Time.* Special Papers in Paleontology, no. 12. London: Paleontological Association.

Desmond, A. 1976. *The Hot-Blooded Dinosaurs.* New York: Dial Press/James Wade. Pp. 121–135.

Fleagle, J. G., E. L. Simons, and G. C. Conroy. 1975. "Ape Limb Bone from the Oligocene of Egypt," *Science,* **189**(4197):135–137.

Gregory, William King. 1922. *The Origin and Evolution of the Human Dentition.* Baltimore: Williams and Wilkins.

Hallam, A. 1972. "Continental Drift and the Fossil Record," *Scientific American,* **227**(5):56–66.

Hoffstetter, Robert. 1972. "Relationships, Origins, and History of the Ceboid Monkeys and Caviomorph Rodents: A Modern Perspective," in T. Dobzhansky, M. K. Hecht, and W. C. Steere (eds.), *Evolutionary Biology.* Vol. 6. New York: Appleton-Century-Crofts. Pp. 323–347.

Kurtén, Bjorn. 1969. "Continental Drift and Evolution," *Scientific American,* **220**(3):54–64.

Molnar, P., and P. Tapponnier. 1977. "The Collision between India and Eurasia," *Scientific American,* **236**(4):30–41.

Radinsky, L. B. 1967. "The Oldest Primate Endocast," *American Journal of Physical Anthropology,* **27**(3):385–388.

Simons, E. L. 1965. "New Fossil Apes from Egypt and the Initial Differentiation of Hominoidea," *Nature,* **205**(4967):135–139.

Simons, E. L. 1967. "The Earliest Apes," *Scientific American,* **217**(6):28–35.

Simons, E. L. 1972. *Primate Evolution.* New York: Macmillan.

Van Valen, L., and R. E. Sloan. 1965. "The Earliest Primates," *Science,* **150**(3697): 743–754.

Chapter 13

Plio-Pleistocene Hominids: Evolution in a Cultural Setting

The hominid fossil record is a gigantic jigsaw puzzle in time and space, with well over 99.9 percent of the pieces missing. Fortunately, because the period in which most hominids lived is closer to the present and because interest in evolution has been intense for over a century, there are at least more specimens of fossil hominids known than of many other groups of mammals. But populations from all parts of the world are not proportionally represented; many more fossils are known from Europe than from parts of the world inhabited far longer, simply because for a long time most scientists who were interested in human evolution lived there and searched for their evidence close to home.

Because hominid remains are so few, their study must be supplemented by other approaches. *Archeology,* the study of stone tools, implements, shelters, and other objects of material culture, is particularly helpful. Each person may have made many times his or her own body weight in tools in a lifetime, and the stone is more enduring than bone. These items of material culture can tell us much about the distribution and ways of life of earlier populations. The known record of material culture that began just under 3 million years ago marks a new phase of evolution that is uniquely human, so much so that our dependence on material culture and technology has become a distinguishing characteristic of our species. Help also comes from *ethnology,* the study of the culture and ways of life of living human groups. Analogies with the ways of life of living hunter-gatherers are useful for the later phases of human evolution. For earlier time ranges, this shades into *ethology,* the study of the behavior of other animals under natural conditions. Genetic and evolutionary theories allow scientists to go beyond the limited sources of material evidence in the reconstruction of human evolution. With the types of population-oriented reasoning characteristic of these fields, it becomes possible to reconstruct from fragments of bone a dynamic picture of the ways of life of all early humans.

TIME FRAMEWORK FOR HUMAN EVOLUTION

It is not yet certain at what point in time hominids had their origin; as discussed in Chapter 12, it may have been as early as the Eocene or as late as the Pliocene. But the existence of hominids first moves out of the realm of conjecture with evidence discovered in the late Pliocene or just before the Pliocene-Pleistocene boundary, depending on where that line is set. In contrast to the 75 million years reviewed in Chapter 12, all the developments discussed from here on took place within a time span of somewhere around 3 million years. The evolutionary processes themselves remain the same, but their results are scrutinized more finely. For this reason it is necessary to begin with a closer focus on the details of the time framework in which evolution operated.

Pleistocene Chronology

The Pleistocene is a period long enough to have included many major alterations of the physical environment, and these in turn have shaped the evolution of many species still living, including our own. But it is recent enough that many more details of the evolutionary record are still preserved reasonably well in some areas.

Since the Pleistocene was first defined over a century ago, its lower and upper boundaries have been established according to a number of very different criteria. The first was based on Charles Lyell's study of mollusk (shellfish) fossils in rocks; by his definition, all strata with 90 to 95 percent of mollusk species still living were considered to belong to the Pleistocene. Vertebrate paleontologists using fossils of land mammals instead of mollusks have commonly considered the Pleistocene to begin with the *Villafranchian*. This is a stage of faunal evolution that starts with the appearance of *Equus, Bos,* and *Elephas,* the three major surviving genera of horses, cattle, and elephants, respectively. Another term introduced into discussions of Pleistocene chronology is *Quaternary.* This division was first based on geological evidence of fluctuations in sea level in marine deposits in France younger than those in the basin of the Seine. Shortly afterward, artifacts of early humans were found in Quaternary deposits, and so the Quaternary is now often taken to be the time of human toolmaking activity. The Quaternary is also equated with yet another set of events: the advance and retreat of glaciers over the face of Europe and wide fluctuations in climate. In the opinion of some experts, these climatic shifts are over, and we are now living in a new epoch, the *Holocene;* many others believe that the present time is just a warm interval between the retreat of the last glaciation and another advance.

From this confusing array of terms that overlap and crosscut one another, some general agreement has begun to emerge in the last few years. One dominant view now is that the base of the Pleistocene is best defined as a biostratigraphic event; that is, it can be recognized on the basis of paleontological evidence for the disappearance of certain marine organisms. Independent chronometric dating techniques indicate that this happened about 2 million years ago and marks the Pliocene-Pleistocene boundary. Acceptance of this date would end a long debate, in which duration of the Pleistocene has been set at as little as 0.6 million years and at as much as 3.5 to 4.0 million years. The shorter scale makes the Pleistocene approximately equivalent to part of the period of glacial advances and retreats; the longer scale would include the entire Villafranchian stage of faunal evolution. As things stand now, with about a 2-million-year duration, the Pleistocene would include only the later part of the Villafranchian; the lower half would be considered late Pliocene.

The Villafranchian stage was first named for a local faunal grouping from shoreline deposits near Villafranca d'Asti in Italy's Po Valley; the name then began to be used for a faunal-based stage over a wide geographic area. This broadening creates problems because it is often difficult to separate faunal differences due to regional geography from changes due to evolution over time. The entire situation is complicated by the strong likelihood that Africa and Europe were at least partially isolated from each other around this time. In Africa the typical Villafranchian animal species suggest a warm, moist climate. Those in Europe may have inhabited subtropical forests too, but a more temperate setting is also possible. There may even have been fluctuations between dense vegetation and more open steppes. There is also evidence that the glacial sequence typical of the Pleistocene had a prelude during the Villafranchian. In the Alps there are traces of a pre-Pleistocene ice sheet, often referred to as the *Donau complex.*

Pleistocene Glacial Sequence

Climatic fluctuations intensified during the Pleistocene; in Europe, ice sheets covered not only the Alpine region but also the Scandinavian landmass and southern England. In North America, glaciers reached as far south as New York City. The cooling may have been triggered by increased volcanic activity. Even one erupting volcano pours out enormous amounts of ash and dust into the atmosphere, reducing the amount of sunlight that reaches the earth's surface and thus cooling the climate. After Krakatoa erupted in 1883, unusually cool weather was noted in many parts of the world for the next few years. Simultaneous eruption of many volcanoes could have pronounced cumulative effects. Volcanic activity on a worldwide scale may also be related to continental drift, discussed in Chapter 12; many active volcanoes are located near contact areas between plates of the crust, and they seem to be activated by periodic plate shifts.

Whatever the causes for cooling of the earth's landmasses, the impact on climate was profound. More of the water evaporating from the oceans fell as snow, adding to the mass of mountain ice caps. These spread down the mountains and across lowland areas, and were added to by increased precipitation captured by the growing ice sheet. Even where glaciers melted, the runoff formed rivers that cooled the oceans into which they flowed. Ice formed in the seas, supplemented by icebergs from glaciers. Eventually the cooling reached a maximum, at which point it reduced the amount of evaporation from the oceans. Moisture-bearing clouds shielded less of the earth's surface from the sun's rays, and warming began. At the same time, precipitation was cut down to the point at which it no longer exceeded or even balanced melting, and glaciers receded. During the Pleistocene, this cycle was repeated four times in Europe. From earliest to latest, the post-Donau glaciations are called *Günz, Mindel, Riss,* and *Würm.* The warmer periods between them are called *interglacials.* These are sometimes given separate names, but they are easiest to remember as hyphenated combinations of the preceding and following glacials. The first Pleistocene interglacial, then, is the Günz-Mindel; the second is the Mindel-Riss; and the third is the Riss-Würm. In addition, the glacial periods themselves were not uniformly cold. Each seems to have contained one or more of the slightly warmer phases known as *interstadials.*

Glacial advances and retreats had great effects on mammalian evolution. As climates changed, some species died out, others became cold-adapted, and yet others shifted location. Migration routes were shaped to a great degree by fluctuations in sea level. At the maximum point of glacial advances, so much water was tied up as ice that ocean levels dropped as much as several hundred feet (about 100 meters or so) (Figure 13-1). Shallow areas along continental shelves became dry land. Among the more important of these were the Bering land bridge, which connected northern Eurasia with North America, and the Sunda shelf, which united many parts of southeast Asia that now exist as scattered islands. Even at the height of glaciation, not all the world's climate became cold; major portions of Africa and Asia remained warm.

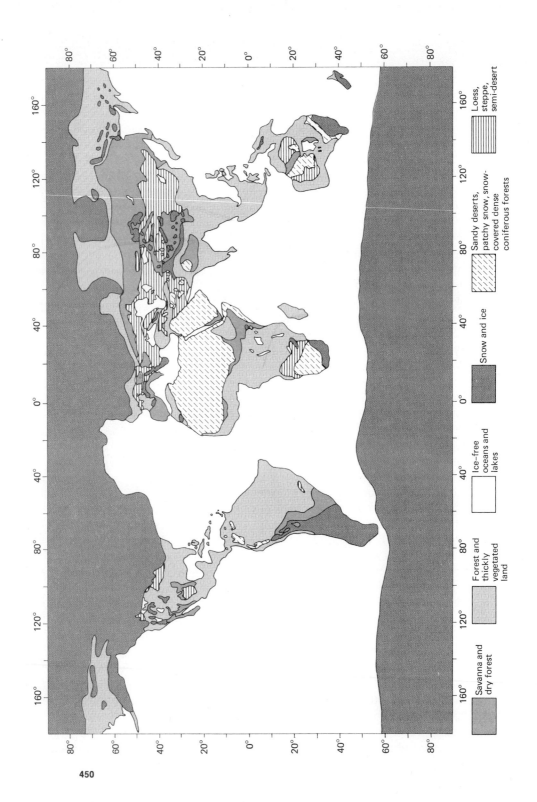

Savanna and
dry forest

Forest and
thickly
vegetated
land

Ice-free
oceans and
lakes

Snow and ice

Sandy deserts,
patchy snow, snow-
covered dense
coniferous forests

Loess,
steppe,
semi-desert

There have been attempts to correlate European glacials and interglacials with African and Asian *pluvials* (wet periods) and *interpluvials* (dry periods). However, major problems still exist here, and a generally acceptable correlation remains to be worked out. For this reason it is important to be able to date the hominid fossils found in each area accurately.

PHYLOGENETIC FRAMEWORK

Reconstruction of the course of hominid evolution involves interpreting known evidence within theoretical frameworks that help us to re-create a more complete picture of early human culture and way of life. Many different phylogenetic reconstructions have been made in the past, and there is as yet no single, universally accepted scheme. One of the simplest approaches is to present the course of hominid evolution as a sequence of four major time-successive grades or stages (see Brace and Montagu, 1977, for an example of this approach). In this framework for ordering the fossil evidence, hominids from the early Pleistocene (and perhaps the late Pliocene as well, depending on where geological and morphological boundaries are drawn) are labeled *australopithecines*, and those from the middle Pleistocene are labeled *pithecanthropines*; these in turn are followed first by neanderthals and then later in the Pleistocene by early representatives of the modern form of humans. Sometimes neanderthals and humans more modern in appearance are grouped together under the formal taxonomic label *Homo sapiens*, pithecanthropines are placed in the taxon *Homo erectus*, and australopithecines are designated as the earliest species in the genus *Homo* or are placed in another genus entirely. There are also more complex interpretations in which, for example, early Pleistocene hominids are split into several contemporaneous, genetically isolated lineages and in which some populations more modern in appearance are believed to precede neanderthals. How many hominid species are there? Over the last quarter century, two general theoretical positions have emerged in answer to this question: the *multiple-species* hypothesis and the *single-species* hypothesis. Each refers to the number of hominid species thought to have existed at any one point in time during the Pleistocene. The contrast can be made clearer still if we refer to them as the *multiple-lineage* and *single-lineage* hypotheses. This avoids confusion between populations of different lineages which existed at the same point in

---

FIGURE 13-1 *Opposite page:* **Surface of the ice-age earth.**
   Source: Modified from CLIMAP Project Members. 1976. "The Surface of the Ice-Age Earth," *Science,* **191**(4232):1131–1137. Fig. 1, p. 1132.

time but did not interbreed and between several time-successive populations in the same lineage which differed as a result of evolutionary changes.

Each of these theoretical positions is associated with a particular view of human evolution. The multiple-lineage hypothesis (which exists in a variety of specific phylogenetic trees) is much the earlier view, as reflected in the variety of names first given to Pleistocene hominids when they were discovered. Adherents of this view point out that most families of mammals consisted of more than one species at some time in their evolutionary history. Why should the Hominidae be an exception? Those who take the single-lineage position reason from the one period we know about with certainty: the present. Only a single polymorphic and polytypic hominid species exists today. It is more widespread than ever before, and thus it is exposed to a greater variety of environments than hominid populations at any point in the past. If there is only one species now, why assume that more than one existed at any time in the past? These contrasting positions may condition the interpretation of new evidence as much as they change in response to what is discovered.

Clearly, not all Plio-Pleistocene specimens are alike. Differences exist among those found at different sites as well as among those found at a single site. Whenever one site has yielded the remains of several individuals, these have varied in appearance. But the step from observation of difference to its correct interpretation is neither small nor automatic—and it is in the step from the observation of real differences to the interpretation of their taxonomic and phylogenetic significance that the single- and multiple-lineage views differ.

Chapter 11 presented evidence to document the point that there is no necessary correlation between the degree of resemblance between two populations and the extent to which they were capable of gene exchange. It is impossible, of course, to perform breeding experiments between Plio-Pleistocene hominids of different appearance. But we do know that most populations (including living humans and nonhuman primates) are polymorphic and polytypic, and the same was probably true for the Plio-Pleistocene hominids. They ranged over many parts of the African continent and probably into the Near East and Far East as well, and thus some degree of geographic variation would be expected.

Living humans also show notable amounts of sexual dimorphism, as nonhuman primates do, particularly the great apes. It is likely that the australopithecines did as well, although the degree is difficult to estimate accurately and no doubt varied from trait to trait. Assumptions about the extent of sexual dimorphism in populations of early Pleistocene hominids have a direct effect on phylogenetic reconstructions. Robinson has maintained that there was little sexual dimorphism in either robust or gracile australopithecines, but other anthropologists have disagreed.

Since living great apes, later fossil hominids, and living humans all show appreciable amounts of sexual dimorphism, it is odd to assume that early Pleistocene hominids would show little. As an alternative, one anthropologist, Loring Brace, has suggested that many of the differences between australopithecine specimens can be accounted for by assuming that the robust ones are generally male, and the gracile ones generally female.

The age of a given fossil hominid at the time of its death would also affect the expression of developmental characteristics, since older individuals have larger and more robust skeletons than younger ones. Age effects might influence interpretations in cases like that of the early Pleistocene "Zinj" and "pre-Zinj" fossils, discussed later in this chapter. The latter are represented chiefly by postcranial bones, and the former by most of a skull. Although there are a few cranial fragments assigned to *Homo habilis,* these are from a juvenile and so may be more gracile for that reason.

In addition to analogies from closely related living primates, theoretically oriented approaches can also be of help in exploring the implications of the single-lineage and multiple-lineage hypotheses. Population genetics might be of some help in exploring the effect of evolutionary change through time (phyletic evolution) on the range of variation in samples of early hominid fossils. The group usually referred to as australopithecines persisted for a very long period of time—from about 3 or 4 million years ago to about 1.5 or 1 million years ago, a span of roughly 2 or 3 million years. This is far longer than any other stage of hominid evolution. Because of the great uncertainty in dating many sites from which Plio-Pleistocene hominid remains have come, it is quite possible that the individual specimens or populations being compared may differ in age by a million years or more. Making allowances for differences due to possible phyletic evolution over such spans of time is difficult because rates of evolutionary change probably were not constant and differed from system to system. But even greater problems may be introduced by failing to take evolutionary change into account. Even if only very slight generation-by-generation changes accumulate over time, appreciable changes in appearance can result. "The Significance of KNM-ER 1470 for Models of Early Pleistocene Hominid Evolution" (on pages 454–456) explores the implications of these changes in more detail, as well as the possible influences of sampling problems and inaccuracies of dating and other procedures.

Although the body of Plio-Pleistocene hominid remains is increasing rapidly and although many new approaches are being used in the reconstruction of hominid evolution, there are still significant gaps in our knowledge. Because of this, it is logically valid and scientifically acceptable to have several alternative explanations for the patterns of variation observed.

THE SIGNIFICANCE OF KNM-ER 1470 FOR MODELS
OF EARLY PLEISTOCENE HOMINID EVOLUTION

In 1973 Richard Leakey described a fragmentary hominid cranium (KNM-ER 1470) that had been discovered the previous year in the area east of Lake Rudolf, Kenya. Two features of this find excited considerable interest: its early date (said to be probably 2.9 million years ago) and its high endocranial volume (reconstructed at 810 cc). In his preliminary assessment of the find, Leakey referred this specimen to an undetermined species of the genus *Homo,* which he considered generically distinct from *Australopithecus.* In support of this placement he said that to include the 1470 cranium found near Lake Rudolf with *Australopithecus* would require an extraordinary range of variation of endocranial volume for that taxon. Endocranial volume in previously known australopithecines (see Table 13-2) ranged from 428 to 530 cc. For a sample of six South African gracile specimens Holloway has estimated a mean of 442 cc. The marked contrast of KNM-ER 1470 with apparently contemporaneous but smaller-brained taxa could be taken as strong evidence for a multiple-lineage hypothesis. Indeed, this is perhaps the most general interpretation of the significance of KNM-ER 1470, and one that may be correct.

An alternative possibility is that KNM-ER 1470 might represent not a more advanced contemporary taxon, but instead a lineal descendant of previously known australopithecines. The potential for this interpretation exists in part because of the uncertainties inherent in assigning dates to early hominid specimens. In this case, the age of the fossil is based on a series of indirect connections. The KNM-ER 1470 skull was not found embedded in a geological layer of known age, but was loose on the surface. Adhering to it was a matrix similar to that on other specimens from sediments buried about 36 m (118 ft) below a geological formation known as the *KBS tuff.* In the region where KNM-ER 1470 was found, KBS tuff yielded samples unsuitable for isotopic dating. The samples of tuff actually dated originated several miles away at another site. For these samples, the date obtained was 2.61 ± 1 standard deviation of about 10 percent. This would include a range of about 520,000 years (from 2.61 + 0.26 to 2.61 − 0.26). What does this range of dates mean? Just this: if all geological inferences were correct, then because of the inherent, irreducible error term, two hominid fossils that each had precisely the same estimated chronometric age could differ by more than half a million years.

If any evolutionary change is taking place, half a million years is a long time. Is it long enough to transform small-brained australopithecines into hominids like KNM-ER 1470? There is no way to answer this question without raising other ones. For example, was KNM-ER 1470 drawn from the mean of its population—or 1, 2, or even 3 standard deviations above or below it? If we assume that the figures given above

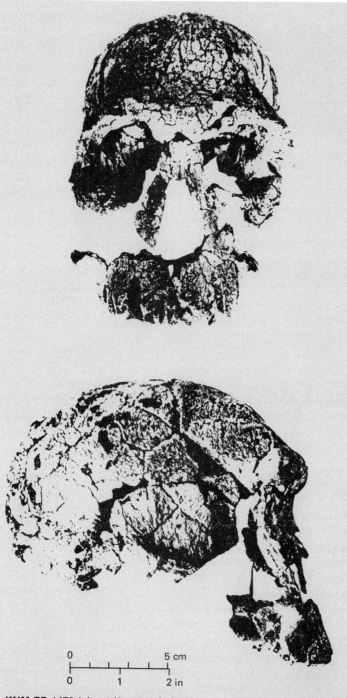

0   5 cm

0   1   2 in

KNM-ER 1470 (about ½ natural size).
    Source: R. E. F. Leakey. 1973. "Evidence for an Advanced Plio-Pleistocene
Hominid from East Rudolf, Kenya," *Nature,* **242:**447–450. Fig. 3B, p. 449.

for gracile australopithecines and for the Lake Rudolf cranium are representative, then average brain size in the two populations differed by about 368 cc. While this sounds like a substantial difference, it is not very great in comparison with the time potentially available for evolutionary change. If an australopithecines generation were about 20 years (probably a conservative overestimate), then half a million years represents a minimum of about 25,000 generations. Another if: if brain size increased by a steady, fixed amount each generation (for other, more complex but probably more accurate means of calculation see Eckhardt, 1976), the amount could be estimated roughly by dividing the difference in endocranial volume by the number of generations; 368 cc ÷ 25,000 generations equals a possible change of 0.015 cc per generation. This is a difference so tiny that it probably could not even be detected in any comparison of two successive generations. More to the point, *higher* rates of change in brain size have been documented in later phases of human evolution. Of course, if KNM-ER 1470 were larger than average for its population, the required rate of evolution would be even less.

Assumptions such as those made here, and the rough calculations based on them, do not in any way prove that KNM-ER 1470's population was descended from, rather than contemporary with, gracile australopithecines. However, data available *at the time* of the Lake Rudolf discovery did make it possible to integrate even this specimen into a single-lineage interpretation of hominid evolution.

It is worth adding that some recent restudies of the KNM-ER 1470 find have raised questions about what were thought to be the basic data. Curtis and his coworkers (1975) present evidence that the KBS tuff may be no older than 1.82± 0.04 million years, and perhaps even younger. If correct, this more recent date would make the large brain size less striking—and less supportive of a multiple-lineage model. Furthermore, the brain may be somewhat smaller than first thought; a recent reconstruction by Holloway places it closer to 770 cc than to 810 cc. On the other hand, the chances that KNM-ER 1470 may be drawn from near rather than above its population mean are increased by the discovery of a similar skull at Lake Turkana, tentatively dated at about 1.5 million years ago (Leakey and Lewin, 1977).

If any conclusion is to be drawn from this exercise in the probing of assumptions, it is that several strikingly different alternative models of hominid evolution are all tenable with the same limited set of data at hand. Additional fossils, more discriminating dating techniques, and a better knowledge of rates of evolution for skeletal traits could all help in narrowing the presently wide-open field of alternatives.

Sources: (1) G. Curtis, T. Drake, and A. Hampel. 1975. "Age of KBS Tuff in Koobi Fora Formation, East Rudolf, Kenya," *Nature*, **258**:395–398. (2) R. B. Eckhardt. 1976. "Observed and Expected Variation in Hominid Evolution," *Journal of Human Evolution*, **5**(4):467–475. (3) Leakey, R. E., and R. Lewin. 1977. *Origins*. New York: Dutton. (4) R. E. Leakey. 1973. "Evidence for an Advanced Hominid from East Rudolf, Kenya," *Nature*, **242**:447–450.

On the basis of the existing evidence, to insist that only one interpretation of patterns of variation is true would be quite simply unscientific. About all that can be said right now is that a number of distinctly different phylogenetic reconstructions are consistent with the existing data. An attempt will be made in this chapter and in Chapters 14 and 15 to summarize those data and then to outline some of the alternative hypotheses that have been advanced to explain the information we now have. We will focus here on reconstructing the appearance, culture, and behavior of the populations that were part of the line of evolution that leads to modern *Homo sapiens.*

At the outset it is important to emphasize the degree to which chance events have shaped our thinking about human evolution. For example, the order in which hominid fossils were discovered is not the sequence in which they evolved. The effect of this accident on the interpretation of human evolution can be appreciated if we use the analogy with a jigsaw puzzle that was mentioned earlier. A jigsaw puzzle is not built systematically from the bottom edge up or from the top edge down. Instead, the first few pieces that fit together become the focus of attention, and efforts are made to build outward from this region of order. Only after much more of the design is built up may it become evident that the first few interlocking pieces are a relatively minor and even unrepresentative part of the whole scene. In the same way, the first hominid fossils found in a given geographic area or time zone were not always representative of the large group that became known from later fossil discoveries.

In any attempt to form an orderly picture out of unconnected bits of information a systematic approach is helpful, of course—but it may not lead straight to the pattern that is guessed at in the beginning. In doing a puzzle it often helps to start by grouping pieces by color. All the blue bits may be sorted together into one area, and all the white ones into another. But when the picture is completed, it may be seen that some of the white and blue pieces do really belong together, as, for example, clouds against an open sky. It could also turn out that the blue pieces must be separated into categories as different as the air of the sky and the water of a pond. The same type of situation could occur in the reconstruction of human evolution. Fossils that look very different from one another may prove to belong in the same species (differing because of the effects of age, sex, etc.), while other specimens that at first appear to be very similar on a few attributes may emerge as members of very different taxa when more complete remains are found.

The other problem is that representatives of all the earlier major hominid grades—australopithecines, pithecanthropines, and neanderthals—had been discovered by the 1920s, a full decade before the modern synthetic theory of evolution had been formulated and nearly four decades before accurate chronometric dating methods were developed. As a result of this differential rate of progress in the separate areas of science, early discussions of hominid evolution took place before the mechanisms of evolutionary change were well understood and before it was realized that the time available for human evolution spanned millions and not tens of thousands of years. Surprisingly, some of the opinions put forth nearly a century ago concerning the course of hominid evolution have persisted, despite dramatic changes in our knowledge.

Here we will introduce each grade of hominid evolution with the initial major discovery representing it and then discuss discoveries of other specimens that lived at about the same time in order to present more fully the distribution and the extent of hominid variation in time and space. We will look first at the physical characteristics of the fossil specimens and the sites where they were found and then at the material culture and ways of life of these populations.

PLIO-PLEISTOCENE HOMINIDS

Fossil remains from this time period are known chiefly from South Africa and east Africa, and possibly from Europe and Asia, though the distribution outside Africa is less certain. No single site has yielded a complete skeleton, but there are enough fossils from different places to allow anthropologists to reconstruct a composite view of how humans looked at a time when they had only the crudest tools, weapons, and forms of shelter. A list of the major Plio-Pleistocene hominid discoveries is given in Table 13-1, pages 460–467.

The Taung Fossils: South Africa

The first evidence for Plio-Pleistocene hominids was recognized over half a century ago, in 1924. For years, mammalian fossil remains had been turning up in limestone quarried at various places near the southern tip of Africa. One of these fossils, the skull of a cercopithecoid monkey from a quarry operated at Taung in Bechuanaland, was shown to Raymond A. Dart, professor of anatomy in the medical school at the University of Witwatersrand in South Africa. Dart recognized its importance, and a geologist colleague, R. B. Young, who was then about to study some lime deposits on a farm adjacent to Taung, was pressed into service. Young brought back a number of new specimens for Dart to study, among them a chunk containing a relatively large endocranial cast, which joined perfectly with another rock fragment in which part of a broken mandible could be seen. Dart suspected that the face might still be preserved within the rock. He proved the hunch corrrect, but only after 73 days of careful chipping to remove the adhering matrix. What he found is shown in Figure 13-2, page 468.

Just 5 weeks later, on February 7, 1925, Dart published his findings in the British journal *Nature*. The article includes a detailed series of observations and the inferences Dart drew from them. For example, the jaws held a full set of deciduous teeth, and the first permanent molars had just erupted. Since this normally occurs at about 6 years of age in humans and between the third and fourth years in chimpanzees, Dart reasoned that the Taung specimen had to be a juvenile. However, the supraorbital tori, or brow ridges, that would have already been evident in an ape of this age were absent, leading Dart to suggest that the face was more similar to that of a human child. The dental features also struck Dart as being humanlike. The canines did not project above the level of the other teeth, nor were they separated from the lower premolars by a gap. Although there was a slight diastema between the lateral incisors and canines of the upper jaw, this was no greater than the diastema seen in some modern humans. The incisors did not project markedly forward, as is common in apes. The teeth formed an even parabolic arch in the upper jaw, and the lower jaw showed no trace of the simian shelf common in living apes. Reinforcing the impression of development beyond the pongid level was the position of the foramen magnum. In the Taung child, this opens much more directly beneath the skull rather than toward the rear, as in present-day apes. As a result, the head would have been supported atop the spinal column with much less muscular effort. Because of these notable differences from any apes then known, Dart held that the group to which the Taung child belonged should be recognized as a new species, which he named *Australopithecus africanus* ("southern ape of Africa"). But because its members were more similar to humans than any pongid yet discovered, he spoke of them as "man-apes" and proposed placing them in a new primate family, the Homo-simiadae, to indicate their intermediate status.

Dart also made a number of educated guesses about the probable behavior of these animals. The better balance of the head implied more erect posture, which in turn suggested that instead of being used for locomotion, the hands would be free for use in offense and defense, the manipulation of objects, and the employment of sticks and stones as tools. This helped explain the absence of large canine teeth, used in other primates such as macaques for fighting and threatening. If their functions were substituted for by the hands or by hand-held tools, projecting fangs would be unnecessary. Increased mastery of the natural environment was inferred from the location of the find. The Taung site was on the fringe of the Kalahari Desert, 3200 km (2000 mi) from the

TABLE 13-1 Plio-Pleistocene hominid fossils

Geographic location	Site	When found	Material discovered
South Africa	Taung	1924	Broken mandible, facial part of a skull, and stone endocast of one juvenile
	Sterkfontein (type site and extension site)	1936 and intermittently to present	Four skulls (one almost complete), several damaged mandibles too robust for the skulls, and over 120 teeth; postcranial remains include parts of a scapula, ribs, pelvis, vertebrae of lower back, upper end of a humerus, and both ends of a femur
	Kromdraai	1938–1941	Much of left half of one fragmentary skull, right half of a mandible, and isolated teeth (four premolars and three molars); postcranial remains include distal end of a humerus, proximal end of an ulna, and finger and toe bones
	Makapansgat	1947–1962; excavation recently reopened	One cranium with face missing, parts of other crania, and mandibles and isolated teeth; among the postcranial remains were pelvic fragments and parts of a humerus, radius, clavicle, and femur
	Swartkrans	1948–1952; currently being excavated again	Robust adult cranial remains include one nearly complete cranium, half of another, and several other partial, crushed crania, as well as three relatively complete mandibles and parts of others (also one adolescent cranium, one of a juvenile, and juvenile mandibles) and about 100 isolated teeth; postcranial remains include the lower end of a humerus, parts of a pelvis, and the upper ends of two femurs; gracile cranial remains include parts of two mandibles and a facial fragment; postcranial parts include the proximal end of the radius and one bone from the hand
East Africa	Kanam (Kenya)	1932	Front part of a heavily damaged mandible containing right P_4 and parts of other teeth
	Garusi (Tanzania)	1939	Maxillary fragment containing P^3, P^4, M^3
	Olduvai Gorge (Tanzania)	1959 to present	Olduvai hominid 5—almost complete cranium and upper dentition from lower Bed I

---

*, †, ‡ Notes appear on pages 466–467.

Number of individuals*	Morphological designation	Material culture, if any	Age of site	Dating method used
1	?	None found	Villafranchian	Faunal correlation
16–111	Gracile	None at type site; at extension site are tools of late Oldowan or early Chelles-Acheulean tradition	Villafranchian	Faunal correlation and comparison of stone tools
3–9	Robust but intermediate in some respects between Sterkfontein and Swartkrans	Five stones that may or may not represent crude tools	Villafranchian	Faunal correlation, but mammalian fossils used for dating came from different locations from those where hominid remains were found
8–13	Gracile	Osteodontokeratic culture	Villafranchian	Faunal correlation
27–218 (probably 75–90)	Robust and gracile†	Several tools of Oldowan type	Villafranchian	Faunal correlation
1	?	None found	Lower Villafranchian, about 4 million years ago	Faunal correlation, based chiefly on fossil elephants
1	Gracile	None found	Villafranchian about 2–4 million years ago	Faunal correlation
1	Robust‡	Bed I and lower Bed II: variety of pebble tools of Oldowan type and a crude semicircular shelter of rocks	Bed I ≃ 2–1.5 million years ago; Bed II ≃ 1.5–1 million years ago	Chiefly potassium-argon and fission-track dating, supplemented by faunal correlation with other areas

(Continued)

TABLE 13-1 Continued

Geographic location	Site	When found	Material discovered
East Africa	Olduvai Gorge (Tanzania)	1959 to present	Olduvai hominid 20—fragment of femur from upper Bed I or lower Bed II Olduvai hominid 15—three permanent upper teeth from middle of Bed II Olduvai hominid 3—two deciduous teeth from upper Bed II Olduvai hominid 10—terminal phalanx of big toe from near top of Bed I Olduvai hominid 7—juvenile mandible with teeth, upper molar, parts of both parietals; some hand bones from Bed I below level of Olduvai hominid 5 Olduvai hominid 8—adult hand bones, most of a foot, and clavicle from same site as Olduvai hominid 7 Olduvai hominid 6—skull and tooth fragments possibly associated with parts of a tibia and fibula from same site as Olduvai hominid 5 Olduvai hominid 13—most of a juvenile skull with upper dentition and mandible with lower dentition from slightly above middle of Bed II Olduvai hominid 14—broken juvenile skull from slightly above middle of Bed II Olduvai hominid 16—badly crushed skull and dentition of a young adult hominid from base of Bed II
	Peninj or Lake Natron (Tanzania)	1964	Nearly complete mandible with all teeth preserved
	Kanapoi (Kenya)	1965	Distal end of a humerus
	Lothagam Hill (adjacent to Lake Rudolf, Kenya)	1967	Mandibular fragment and most of right horizontal ramus, including one molar tooth and roots of two others
	Lake Baringo, Ngorora Formation (Kenya)	1968	Isolated molar tooth, possibly left M^2

*, §, ¶, **, ††, ‡‡ Notes appear on pages 466–467.

Number of individuals*	Morphological designation	Material culture, if any	Age of site	Dating method used
1	Robust	Upper Bed II: Oldowan culture, accompanied by hand axes	Bed I $\simeq$ 2–1.5 million years ago	Fauna of Bed I and lower half of Bed II appears to be early Pleisto-
1	Robust		Bed II $\simeq$ 1.5–1 million years ago	cene, and that from above the
1	Robust			middle of Bed II to be early middle
1	?§ and gracile¶			Pleistocene
1	Gracile			
1	Gracile			
1 or 2	Gracile			
1	Gracile			
1	Gracile			
1	Gracile			
1	Robust	Early Acheulean tools	1.5 million years ago	Potassium-argon; faunal correlation
1	Robust**	None	2.5–3 million years ago	Potassium-argon; faunal correlation
1	Gracile††	None found	3.7 and 8.3 million years ago	Potassium-argon
1	?‡‡	None found	9–12 million years ago	Potassium-argon

(Continued)

TABLE 13-1   Continued

Geographic location	Site	When found	Material discovered
East Africa	Koobi Fora and Ileret, east of Lake Rudolf (Kenya)	1967 to present	Abundant and varied remains ranging from nearly complete crania to isolated teeth; postcranially, most parts of fore-limbs and hindlimbs are represented by fragments or even complete bones
	Chesowanja (Kenya)	1971	Crushed cranium with more than half of upper dentition preserved
Ethiopia	Omo River Valley	1967 to present	Partial juvenile skull, four lower jaws, and about 100 teeth
	Afar region	1974	Over 40 bones of one individual, includ-ing parts of the skull and lower jaw; teeth; part of the spinal column; ribs; portions of a humerus, radius, and ulna; bones of the hand and wrist; a half pelvis with complete left femur; parts of the right tibia and fibula; and ankle and foot bones
North Central Africa	Koro Toro (Chad)	1960	A broken frontal bone and some facial bones
Indonesia	Sangiran (Java)	1939, 1941, 1952	Three mandibular fragments containing premolars and molars

*, §§ Notes appear on pages 466–467.

Number of individuals*	Morphological designation	Material culture, if any	Age of site	Dating method used
Several dozen	Robust and gracile	Pebble tools, flakes, and core tools like those from Bed I at Olduvai	From older than 2.6 million years ago to possibly as recently as 1 million years ago	Stratigraphically lowest hominid fossils are found just above and below a tuff (KBS) dated by potassium-argon technique to 1.82 ± 0.04 million years ago; stratigraphically highest hominids come from deposits that have given dates that scatter widely between 1 and 2 million years ago
1	Robust in appearance but showing some features usually found in gracile crania	None found	About 1.1 million years ago	Potassium-argon dating and faunal correlation
Several dozen	Gracile to robust	Oldowan tools in deposits between about 1.9 and 2.2 million years ago	About 4–1.75 million years ago	Potassium-argon dating
1	Gracile	None found	About 3.5–4 million years ago	Potassium-argon dating of basalt near the site
1	Gracile§§	None found	About 1–2 million years ago	Faunal correlation
3	Robust	None found	Villafranchian or lower Pleistocene	Hominid fossils were found in Djetis faunal beds, which have been correlated with lower Pleistocene on basis of animal remains

(Continued)

TABLE 13-1 Continued

Geographic location	Site	When found	Material discovered
Indonesia	Sangiran (Java)	1939, 1941, 1952	Three mandibular fragments containing premolars and molars
China	Unknown (specimens purchased in drugstores)	Before 1946	Eight teeth, premolars and molars

* At most sites a number of separate fossil fragments have been found. The maximum number of individuals represents the total number of fossils found; the minimum takes into account the possibility that several fossils may have come from the same individual.

† Refers to remains originally designated as *Telanthropus capensis*. These are considered by some to represent a more advanced hominid that either was contemporaneous with the australopithecines or lived later, in which case its bones would be intrusive in the earlier levels.

‡ These are the specimens originally classified as *Zinjanthropus boisei*, of which Olduvai hominid 5 is the type specimen.

§ This specimen has not been assigned to either the gracile or the robust group with certainty.

¶ With the exception of Olduvai hominid 16, these have generally been assigned to the taxon *Homo habilis*, of which Olduvai hominid 7 is the type specimen. Olduvai hominid 16 is also considered by some to represent *habilis*; others consider it an early specimen of *Homo erectus*.

nearest living apes and the luxuriant forest belts that sheltered and fed them. The presence of baboon skulls in the same deposits argued that the climate had been quite as arid when the Taung child lived as it was when the specimen was found. The savanna setting, short on water and vegetation but long on sizable predators, would have provided a challenging environment in which intelligence would have conferred an advantage.

Other authorities, however, came to quite different conclusions. Some of the most eminent scientists of the day declared the Taung fossil to be just an immature ape, while a few preferred to suspend judgment until an adult was found. There were many reasons for this opposition to Dart's view, some objective and others emotional. Probably the most effective barrier to the acceptance of australopithecines as hominids was Piltdown, disovered a decade before. The Piltdown skull supported the then-current theory that the brain had led the way in human evolution. The Taung fossil, with its apelike brain and human dentition, pointed in the opposite direction. As noted in "A Synthetic Morphological Pattern: The Piltdown Hoax," page 366, it was not until 1953 that

Number of individuals*	Morphological designation	Material culture, if any	Age of site	Dating method used
3	Robust	None found	Villafranchian or lower Pleistocene	Tektites from overlying Trinil beds have given potassium-argon dates of about .5 million years ago
1–8	Robust	None found associated	? late Pliocene or early Pleistocene	Rough estimate based on other animal remains also purchased in drugstores

** The Kanapoi humeral fragment is larger than comparable specimens from Kromdraai; however, while this specimen is robust, it is not certain that it is hominid, as the distal end of the humerus is very similar in pongids and hominids.

†† This section of the jaw is not very diagnostic, being extremely similar in pongids and early hominids. Given this and the uncertain date, this specimen cannot be safely taken definitely as a hominid on the basis of existing evidence.

‡‡ Isolated molar teeth of pongids and hominids are extremely similar. Given this and the early date of the specimen, it cannot be considered definitely a hominid.

§§ Originally this was said to be an australopithecine, but some now believe it to represent a population intermediate between australopithecines and pithecanthropines.

Piltdown was conclusively shown up as a piece of trickery, clearing the way for acceptance of the australopithecines as early hominids.

Fortunately, during the intervening 30 years Dart was neither idle nor isolated. He discovered other valuable specimens, thanks in good part to the help of Robert Broom and John Robinson. Broom, trained as a physician, began a second career as curator of the Transvaal Museum in 1936, when he was 69 years old. He located specimens of adult australopithecines at Sterkfontein, a South African lime quarry about 290 km (180 mi) northeast of Taung. Through the late 1930s and 1940s numerous other hominid fossils were found at a variety of similar sites.

The body of evidence from South Africa became quite impressive. In addition to endocranial casts and skull fragments, there were also complete skulls, hundreds of teeth, and most major parts of the skeleton: shoulder blades, pelvises, vertebrae, ribs, limb bones, and bones of the hands and feet (see Table 13-1). Size, shape, and posture could be reconstructed if these pieces were pooled. When Piltdown was struck from the roster of valid early hominids, instead of a void there was a well-documented alternative.

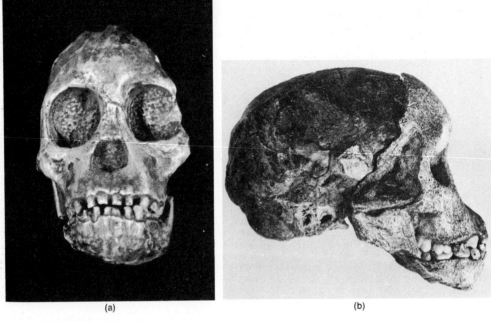

(a)   (b)

FIGURE 13-2   Australopithecine skull: the Taung child, (a) **front** and (b) **side views (½ natural size).**

Source for side view: P. L. Stein and B. M. Rowe. 1978. *Physical Anthropology* (2d ed.). New York: McGraw-Hill. Fig. 14-11.

Physical Characteristics.   By the early 1960s abundant remains of these South African Plio-Pleistocene hominids had been discovered; they were so extensive, in fact, that about as much is known of their skeletal morphology as about that of any fossil hominid group that has lived since.

Skull.   The most important feature of the australopithecine skull had already been noted by Dart. This was the combination, unexpected in a hominid, of a small braincase and large jaws. Both features are compatible with a relatively recent pongid heritage. But in each feature some differences are evident, as can be seen from Figure 13-3. In australopithecines the supraorbital height (BF) is greater, both absolutely and relative to skull length (DE). As a result, in frontal view more of the cranium can be seen rising above the level of the brow-ridges because the brow ridges themselves are less massive and because the skull vault is slightly expanded. The precise degree of expansion is uncertain for a number of reasons. Although several australopithecine skulls have been found, only a few are complete enough to permit an

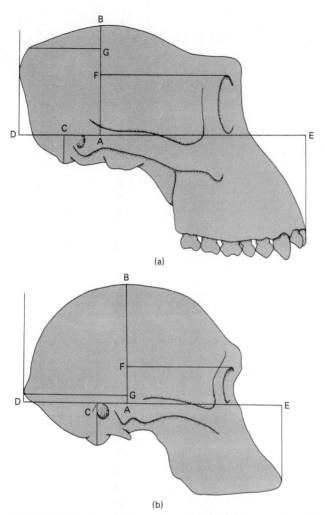

(a)

(b)

FIGURE 13-3 Comparison of australopithecine and gorilla skulls. Outlines of skulls of (a) **a female gorilla and** (b) ***Australopithecus* (Sterkfontein skull V) (⅓ natural size).**

Source: W. E. Le Gros Clark. 1964. *The Fossil Evidence for Human Evolution* (2d ed.). Chicago, Ill.: University of Chicago Press.

estimate of endocranial volume. For eight South African skulls this ranged from 428 to 530 cc, with an average of 460 cc (Table 13-2). As has been shown in Table 9-4, page 299, this is larger than the average for orangutans and chimpanzees and is just below that for gorillas. One adult male gorilla skull has been found to have a cranial capacity of 752 cc, larger than that of any of these eight australopithecines. However, hundreds of gorilla skulls have been studied; the spread for australopithecine brain size given above is the observed range for a very small

TABLE 13-2 Endocranial volumes of South African australopithecines

Specimen	Endocranial volume, in cc*
Taung	440‡
Sterkfontein (STS 60)	428
Sterkfontein (STS 71)	428
Sterkfontein (STS 19/58)	436
Sterkfontein (STS 5)	485
Makapansgat (MLD 37/38)	435
Makapansgat (MLD 1)	500 ± 20
Swartkrans (SK 1585)	530
Range	428–530
Average	460

* These figures should be taken as approximations rather than as precise and accurate measurements; other projections for the adult volume of the Taung skull are 570 cc, 600 cc, and 625 cc. Tobias (1967) estimated the adult size of the Taung skull to be 562 cc, and for a sample that included Taung, four Sterkfontein specimens, and MLD 37/38 his range was 435–562 cc, with a mean of 498 cc.

‡ Projected value for adult; the actual value in the juvenile specimen is 404 cc.

Sources: (1) R. L. Holloway. 1975. *The Role of Human Social Behavior in the Evolution of the Brain.* Forty-third James Arthur Lecture on the Evolution of the Human Brain (1973). New York: American Museum of Natural History. Table 2, p. 24. (2) P. V. Tobias. 1967. *Olduvai Gorge.* Vol. 2. *The Cranium and Maxillary Dentition of Australopithecus (Zinjanthropus) boisei.* London: Cambridge University Press. Table 12, p. 79.

sample. As more Plio-Pleistocene hominid skulls are discovered, some will probably fall outside the range now known, perhaps by an appreciable amount.

The australopithecine skull was balanced fairly easily in an erect position. This is attested to by the central location of the occipital condyles, two smooth facets of bone at the base of the skull which rest on the top of the spinal column. Also, the occipital torus at the back of the skull is quite low. In pongids (particularly adult males) it is much higher and is marked by a prominent shelf of bone, the *nuchal crest*, which provides additional area for the attachment of the powerful neck muscles needed to suspend the large-muzzled skull out ahead of the trunk. Clearly, less muscular effort was required to support the skull in these early hominids.

The facial region, too, shows some structural changes. The upper part forms a flatter, more vertical plane than that seen in either dryopithecines or modern pongids. This difference reflects a smaller dental arch, made possible by the absence of the large canine teeth that mark the far forward corners of the pongid muzzle. Small canine teeth, along with the articulation of the mandible, allowed free movement of the lower jaw, as in later hominids. The flattened crowns of worn teeth reinforce this impression. The attrition on these teeth resulted from heavy use, suggesting a coarse diet and powerful muscles to move the jaws. These muscles were connected to the mandible at one end and were anchored on the skull vault at the other. In some individuals the vault provided an inadequate surface for attachment, and as a functional response it was sometimes expanded by formation of a sagittal crest along the midline of the skull. The mandible was typically robust, but it lacked the internal buttress of the simian shelf found in pongids, as well as the external reinforcement of the chin typical of recent hominids.

Dentition.   The teeth are essentially hominid in character; in adults, the incisors are chisel-shaped, like ours. The canines have a flattened, spatula-shaped outline, and they nearly match the incisors in height to begin with. Matching pairs in the upper and lower jaws do not overlap significantly; with age they wear down from the tip and do not show contact spots along the anterior and posterior surfaces. The anterior lower premolars have taken on the same bicuspid shape as the other premolars that oppose and follow. In outline the molars are roughly rectangular, with rounded corners; their surfaces show the Y-5 dryopithecine pattern or some variant of it, like ours and like those of apes. In size range these teeth overlap those of modern humans, although the average is larger. The larger tooth size may be due to heritage (possibly from good-sized pongid ancestors) as well as habitat. Wear on australopithecine teeth is typically heavy. In one specimen whose unworn third molar had erupted just before death, the enamel of the first molar had already been ground down flat, exposing the dentine underneath.

Skeleton.   Only about 10 percent of Plio-Pleistocene fossil remains are from the postcranial region, although this makes up the bulk of the skeleton. Because of this, interpretations must be made with particular caution. For example, much of the spinal column is present in one major find from Sterkfontein; included are all the lumbar vertebrae from the lower back and the sacrum (Figure 13-4), a few ribs, a nearly complete pelvis, and a good part of one femur. The reconstructed spinal column shows the forward curvature typical for this region in later hominids. Yet in the lower part of the spine there are six lumbar vertebrae, a condition rare in humans and not normally found at all in pongids. The simplest explanation is that the individual australopithecine is atypical in this trait, but until many more specimens are known, we cannot even rule out the possibility that most australopithecines also had six lumbar vertebrae.

The australopithecine pelvis shown in Figure 13-4 is strikingly hominid in appearance. The ilium is short and broad, with its crest extending backward to provide a wide area for the attachment of the muscles that stabilize the trunk in an upright position. Its interior forms a bowl for

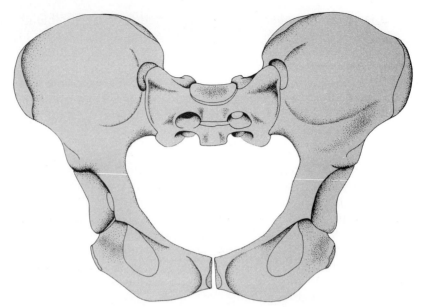

FIGURE 13-4   Restoration of the pelvis of STS 14 (about ½ natural size). Except for the distal portion of the sacrum, this pelvis is essentially complete; restoration involved primarily the correction of some relatively minor distortions.

Source: John T. Robinson. 1972. *Early Hominid Posture and Locomotion.* Chicago, Ill.: University of Chicago Press. Fig. 42, p. 308.

support of the intestines and other internal organs. The point at which the pelvis joins the sacral vertebrae of the spine is lower than in apes. This shifts the center of gravity lower, again lending stability to the upright trunk. It also brings the pelvic joint with the spine nearer to the *acetabulum,* the socket for the upper end of the femur. This makes for greater stability in transmission of weight from the upper body to the legs. The legs themselves are longer than those of apes, and both the knee and the hip joints allow the striding gait needed for efficient bipedal locomotion.

The shoulder and arm bones show an interesting combination of heritage and habitus features. The scapula (shoulder blade) seems to have been set higher on the trunk than in modern humans. In conjunction with the longer upper arm, this strongly suggests that the locomotor behavior of the australopithecines' ancestors included arm-swinging, or brachiation. The hands themselves are capable of a strong grasp and a reasonable degree of manipulation.

Overall body size covers a wide range. Stature estimates for South African specimens range from about 122 cm (4 ft) to a bit over 152 cm (5 ft). Body weights are also distributed over an impressive range, from as little as 18.2 kg (40 lb) to as much as 95.5 kg (200 lb), for a total range of 77.3 kg (160 lb).

At first glance these seem like enormous ranges of variation, suggesting a considerable amount of taxonomic diversity at or above the species level. Some perspective can be provided by an objective comparison with ranges of variation in living hominoid species. If we look back at Table 9-3, page 299, we can see that the range of estimated weights for Plio-Pleistocene hominids is exceeded by the combined weight range for a very large sample of human males and females (91.2 kg, or 200 lb). It is also exceeded by the combined weight range for a much smaller sample of gorillas of both sexes (105 kg, or 230 lb). Therefore, the range of estimated australopithecine body weights cannot be taken as reliable evidence for more than a single lineage.

Interpretations.   How much phylogenetic diversity was represented by the many individual specimens discovered in South Africa? Answers to this question were given each time new specimens were found. As South African fossils were described, the populations they were believed to represent were given formal Latin binomials: *Australopithecus africanus, Australopithecus prometheus, Plesianthropus transvaalensis, Paranthropus crassidens, Paranthropus robustus, Telanthropus capensis,* and so on. There is reason to doubt that the biological implications of these names were realized by many of those who coined them. For example, Broom (1950) remarked: "I think it will be much more convenient to split the different varieties [of South African fossil apemen] into different genera and species than to lump them" (p. 13). But formal taxonomic names should be more than convenient labels. If all those listed here are valid, then at least half a dozen reproductively isolated gene pools must have existed within a radius of about 320 km (200 mi) in South Africa during the early Pleistocene.

Several evolutionary biologists writing at the time these fossils were discovered, including Julian Huxley and Ernst Mayr, objected to this degree of splitting. In 1951 Mayr wrote a paper called "Taxonomic Categories in Fossil Hominids" in which he went to the opposite extreme: he lumped all known hominids into one genus, *Homo*, because all appeared to share one major adaptive characteristic. This was upright posture, occasioned by a shift to living on the ground, which freed the hands for activities that might in turn have stimulated evolution of the brain. Mayr believed that once this new adaptive plateau had been attained, all hominids, from the australopithecines through living humans, belonged to one continuous phyletic line. A few years later, in the mid-1950s, a compromise interpretation of australopithecine taxonomy was put forth by a younger colleague of Robert Broom. John Robinson proposed that on the basis of the evidence available then, all South African australopithecines could be divided into two taxonomic categories: the relatively gracile, or lightly built, *Australopithecus africanus* and a larger form that he called *Paranthropus robustus*. These two taxa were widely believed to contrast strongly in physical characteristics (see Table 13-3). Robinson believed the morphological differences reflected contrasts in behavior and way of life that were tied to different ecological niches. *Paranthropus* was seen as predominantly vegetarian, its large teeth being suited to a diet high in vegetable fiber and low in concentrated nutrients. *Australopithecus* was believed to be more omnivorous, to eat flesh as well as vegetable matter (much the same diet as that of present-day hunter-gatherers). The general inference drawn from Robinson's *dietary hypothesis* was that gracile australopithecines represented a species of an early stage in the main line of human evolution, whereas robust specimens were from a collateral line that became extinct. On the basis of some jaw fragments, Robinson

TABLE 13-3   Comparison of gracile and robust australopithecines

Characteristic	Gracile form	Robust form
Stature	About 122 cm (4 ft)	Over 152 cm (5 ft)
Body weight	88–110 kg (40–50 lb)	A few hundred pounds
Body build	Small and slender	Heavily built and muscular
Cranial capacity	About 500 cc	About 500 cc
Skull	Has low forehead; lacks strongly developed ridges or crests of bone	No trace of a forehead; cheekbones project; sagittal crest is typically present
Mandible teeth*	Robust	Massive
Incisors	Small, chisel-like blades	Similar in size and shape to those of *Australopithecus*
Canines	Bladelike; nonprojecting	Upper canines of same size as those of *Australopithecus*; lower canines smaller
Premolars	Bicuspid	Bicuspid, and similar in size to those of *Australopithecus* (except P^4, which is larger)
Molars	Large	Massive

* It is commonly stated that the anterior teeth of robust australopithecines are smaller than those of gracile australopithecines; however, this is a partial truth. When all the individual teeth of gracile and robust australopithecines are compared, it can be seen that there are differences of morphological significance in only two. Robust australopithecines have lower canines that are smaller, and upper fourth premolars that are larger, than those of gracile australopithecines. For all other individual teeth the differences between gracile and robust australopithecines are less than those which are known to exist within living human populations. If the total crown areas (length X breadth) of all three anterior teeth (I1 through C) are added and divided by the sum of all posterior teeth (P3 through M3), robust australopithecines do have *relatively* smaller anterior teeth. However, this is mainly because the posterior teeth are *absolutely* larger, while the anterior teeth are substantially similar in size in both groups of early hominids. The data on which these statements are based are presented in detail in: Milford H. Wolpoff. 1971. *Metric Trends in Hominid Dental Evolution.* Case Western Reserve Studies in Anthropology, no. 2. Cleveland: Case Western Reserve Press.

also believed that more advanced hominids existed at the same time in South Africa. To the population represented by these he gave the name *Telanthropus capensis* (and later merged it into *Homo erectus*). This general scheme remained the dominant opinion in the field until the mid-1960s.

East African Evidence

At the end of the 1950s the comfortable South African stereotype was thrown into disarray by the discoveries of Louis and Mary Leakey in east Africa. The Leakeys had been working in east Africa for 25 years before they discovered substantial hominid fossil remains. In 1959 their patient, painstaking search paid off with the discovery of a skull (Olduvai hominid 5; Figure 13-5) that was highly significant for the study of human evolution. It was strikingly robust, much more so than australopithecines from South Africa. To emphasize its unique qualities, Leakey called it *Zinjanthropus boisei*. The name is no longer used by most paleoanthropologists, since later studies have shown clearly that "Zinj" is an australopithecine. The skull did extend the range of variation of this grade by an appreciable degree, however, and as a result, some anthropologists concluded either that there were more than two australopithecine species or, alternatively, that one australopithecine lineage was more variable than had been believed.

The Leakeys' discovery disturbed stereotyped interpretations of hominid evolution in two other ways. Minerals associated with Zinj were dated by the potassium-argon technique to about 1.7 million years ago—nearly twice the age previously estimated from geological studies and attempts at faunal correlation. Furthermore, the early robust hominid's remains were found in association with stone tools and remains of other animals. This suggested to some anthropologists that even a robust and primitive-looking early hominid could be a hunter and toolmaker. More gracile-appearing skeletal remains subsequently found at Olduvai were first dubbed "pre-Zinj" because they came from deposits beneath (and presumably earlier than) those in which Olduvai hominid 5 was found. This collection of specimens—a juvenile skull and jaw fragments, plus a hand, a foot, and other limb bones of several adults—were later given the formal label *Homo habilis* by Louis Leakey to indicate that they belonged to a more advanced taxon that that of "Zinj."

Evidence for great diversity or variability among early Pleistocene hominids comes from several other localities in east Africa. For example, the Omo River Valley in Ethiopia has yielded remains, chiefly jaws and teeth, of early hominids that range from rather gracile to quite robust. These fossils apparently span a period from less than 2 million years ago to as much as 4 million years back in time. As shown in *"Gigan-*

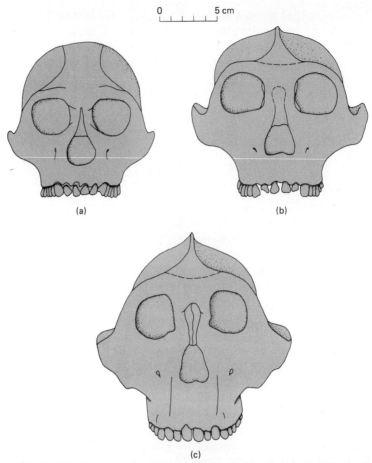

FIGURE 13-5   Facial views of three australopithecine skulls: (a) South African gracile australopithecine (STS 5); (b) South African robust australopithecine (SK 48); (c) Olduvai hominid 5.

Source: Modified from P. V. Tobias. 1967. "The Cranium and Maxillary Dentition of *Australopithecus* (*Zinjanthropus*) *boisei*," in L. S. B. Leakey (ed.), *Olduvai Gorge*. Vol. 2. Cambridge, England: Cambridge University Press.

topithecus as a Hominid," page 441, some of the robust lower jaws from Omo approach those of *Gigantopithecus* in size, though the significance of this similarity is disputed. Ethiopia's Afar region has recently been the site of an important series of discoveries. In 1974 a field crew headed by Donald Johanson recovered over 40 bones of a single gracile specimen (Figure 13-6), which makes it one of the most complete as well as one of the earliest (3.0 million years) hominids yet discovered. Johanson estimated this individual to have been quite small, about 91 to 107 cm (3 to 3.5 ft) in stature, though other anthropologists think it may have been as much as 31 cm (1 ft) taller. Details of pelvic

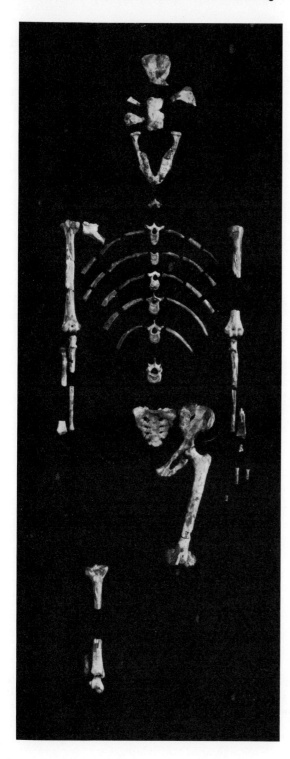

FIGURE 13-6   Skeletal remains of Lucy (AL 288-1), a Plio-Pleistocene hominid from the Afar region of Ethiopia.

Source: D. C. Johanson and M. Taieb. 1976. "Plio-Pleistocene Hominid Discoveries in Hadar, Ethiopia," *Nature,* **260**(5549):293–297. Fig. 5, p. 296. Photograph, Cleveland Museum of Natural History.

morphology suggest that this early hominid, nicknamed "Lucy" by its discoverer, was a female; males of the same population would have averaged larger.

Enough of Lucy's skeleton is preserved to provide the basis for a meaningful discussion of the total morphological pattern of a Plio-Pleistocene hominid, as long as we keep in mind that a single specimen is not necessarily typical of its population. Lucy's expanded ilium provides the best evidence for upright posture and, by this criterion, hominid status. As discussed recently by Christie (1977), other bones from the leg can be used to reconstruct the way in which hominids at this stage of evolution stood and walked. From the shape of the joint surfaces on the distal ends of the tibia and fibula and the upper surface of the talus (which supports the tibia), Christie inferred that, as in modern humans, Lucy's leg and trunk turned inward just before she pushed off for each step. However, in Lucy the angle of rotation was two to three times greater than that in modern humans, and so her locomotion might have been characterized by a more rolling gait than ours. In standing as well as walking, her feet were probably planted more widely apart than in modern adults, while her knees were closer together. This combination would have produced a knock-kneed posture more like that seen in today's children. The functional explanation in both cases is probably the same: the resulting wide stance produces a broader base for support, increasing balance and stability. Lucy's arm bones were relatively long in relation to her overall body size and leg length. These proportions could suggest the existence of a prehominid stage in which the forelimbs played a greater role in locomotion than they did after the assumption of upright posture on the ground.

Less is preserved of Lucy's skull than we could wish for, but a few observations and inferences can be made. The presence of an elongated premolar with one major cusp behind the canine in the lower jaw in itself points backward toward the condition seen in dryopithecines at an earlier stage of evolution. Like modern apes, dryopithecines had a single cusped premolar tooth that is sometimes called a *sectorial* premolar because it shears against the upper canine. If Lucy's missing canines were consistent with this, we would feel safe in assuming that they projected beyond the level of other teeth more than those of modern humans do, though how far we would not know. Some independent, though not solid, support for this assumption may already exist. Pictures of a palate discovered by Johanson and his colleagues show canine teeth that do project a bit more than those of later hominids (Figure 13-7). If the position of these teeth is not due to postmortem distortion of the specimen, it would be consistent with the general impression built up from Lucy's remains. To a population biologist, however, perhaps the most exciting of the Afar discoveries still awaits

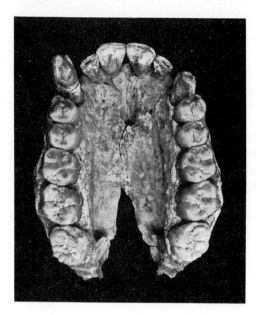

FIGURE 13-7   Maxilla of a Plio-Pleistocene hominid from the Afar region of Ethiopia (about ½ natural size). The greater projection of the right canine appears to be due to loosening from the socket.

Source: M. Taieb, D. C. Johanson, and Y. Coppens. 1975. "Expédition internationale de l'Afar, Éthiopie (3° Campagne 1974); découverte d'hominides Plio-Pléistocènes à Hadar," *Comptes rendus des séances de l'Academie des Sciences,* 3 novembre. Plate II, no. 1. Photograph, Cleveland Museum of Natural History.

full analysis. In 1975 Johanson found over 150 bones of two children and three to five adolescents and adults. Their association in a stream bed suggested to him that they may have been killed by a single flash flood. The sample they make up is extremely important because it holds the possibility for measuring, on a limited scale, within-population variation of a single contemporaneous group.

Although east African early hominids retain a few traits that provide clues to their ancestry, at least a few show traits that foreshadow future advances. Perhaps the most striking example of the second sort is a cranium, KNM-ER 1470, discovered east of Lake Rudolf, Kenya, by Richard Leakey. (The implications of this specimen for the study of human evolution are discussed in "The Significance of KNM-ER 1470 for Models of Early Pleistocene Hominid Evolution," page 454.) These east African fossil hominids discovered by Johanson and Richard Leakey in deposits from the early Pleistocene show diversity in traits that reflect the workings of numerous functional systems: the brain, the face and jaws, the manipulative forelimbs, and the handlimbs, which are responsible for posture and locomotion. Measurement of the extent of this diversity, and interpretation of its significance, will not be complete for quite some time, especially because the pace of discovery is now so rapid. Among the few safe generalizations we can make is that these new specimens have extended the range of variation in Plio-Pleistocene hominids well beyond that known from South Africa.

PLIO-PLEISTOCENE MATERIAL CULTURE
AND WAY OF LIFE

Cultural evolution, although discussed separately here for purposes of simplicity, does not take place apart from biological evolution. Human culture evolved in primates with certain biological characteristics, including reasonably large body size, which provided at least relative safety on the ground; a hand with sufficient manipulative ability to allow rudimentary tool use; a brain capable of remembering the outcome of past interactions with members of a reasonably large social group; a liking for the taste of fresh meat; and so on. These traits are still present in living great apes such as chimpanzees, and they were probably also found in the dryopithecines that were the common ancestors of all living hominoids. This is the heritage that Plio-Pleistocene hominids had to build on. Let us see how scientists go about trying to reconstruct the behavior of humans they have never seen.

Pinpointing the Origin of Hominid Culture

If the evolutionary record were complete, it would probably be impossible to draw a clear dividing line between nonhominid primates and hominids because of the underlying genetic continuity from one generation to the next. The same may prove to be the case for material culture as well, since studies of captive and free-living primates over the last few decades have documented toolmaking and tool use among many species other than our own. But tool use by nonhuman primates also makes it easier to understand the roots of hominid culture. Occasional modification of natural objects is part of the flexible adaptive strategy of some anthropoid primates; the behavior of the African apes in particular indicates that they are preadapted to a hominid way of life. It is reasonable to assume that the present level of ability found in these nonhuman primates was matched by our primate ancestors. What remains to be explained is how these behavioral tendencies could have been amplified as part of a shift in way of life.

The extent and direction of this adaptive shift can be seen by looking at the differences between human hunter-gatherers and living apes. These differences, which emphasize the central importance of the food quest as a focus for hominid behavior, have been summarized by the paleolithic archeologist Glynn Isaac (1978). Far and away the most striking contrast is the ability of humans to communicate with a spoken language, which makes it possible to discuss the results of strategies used in the past, to plan for the future, and to regulate social relationships among group members and with members of other groups in the present. Although apes can communicate, they lack a language that would give them the ability to make cooperative arrangements of any great complexity. Central to the plans that hunter-gatherers make is a home base, a camp or other location to which group members can return; no continuing spatial focus of this sort governs the relationships of apes. One of the primary activities that take place at the home base of human hunter-gatherers is the sharing of food between adults and the young. The closest that apes come to food sharing is what Isaac calls "tolerated scrounging" by one chimp from another.

Humans are able to bring food to their home base for regular sharing because their hands do not have to be used for locomotion; they are free to carry not only food but also tools and other objects that would seriously hinder apes in long-distance travel. Although chimpanzees do occasionally hunt and kill other animals for food, humans devote much more time to the quest for meat, fish, and other high-protein foods. There is also a dividing line in the size of prey taken; only humans habitually feed on prey over 15 kg (33 lb) in weight. The ability of humans to kill, cut up, and carry the meat from these larger animals is due in large part to the great variety of weapons, tools, and containers they have. Other items of material culture make it possible for humans to gather a whole range of smaller items of protein food, including eggs, young birds, lizards, and turtles, as well as roots, tubers, and other plant foods that are all but inaccessible to chimps because their tools are too unsophisticated. After the foodstuffs have been brought to the camp, humans often cut, pound, grind, and cook them; only the most rudimentary preparations, such as stripping off husks or shells, are done by apes.

There is a substantial body of evidence indicating that Plio-Pleistocene hominids were the first primates to be significantly dependent on material culture for survival rather than on physiological and anatomical adaptations alone. Early hominids therefore show the beginnings of an important hominid trend: increasing control over the physical environment, even to the extent of creating a new ecological niche in the process. This cultural influence over the natural world must have had rather modest beginnings that are now difficult or impossible to detect. Most tools used by nonhuman primates are made of sticks or leaves. Similar implements might have been made by Plio-Pleistocene hominids for thousands or millions of years without leaving a trace in the archeological record, for only materials such as stone or bone endure over long spans of time. For another thing, implements made from decay-resistant materials might be so little modified from natural state or so simply made that they would not be recognized as tools. In defining the earliest stages of cultural evolution, therefore, context and distribution are probably as important as substance, form, and design.

Dart's Osteodontokeratic Culture

When Raymond Dart first suggested that the australopithecines were hominids, he based his case entirely on anatomical clues. But Dart was also the first to provide what may possibly be evidence that members of these Plio-Pleistocene populations made and used tools. His theories about australopithecine cultural activity originated from later analyses of bone-containing deposits (*breccias*) associated with the australopithecine fossils. The quantities of these deposits are staggering. At Makapan alone, they amounted to well over 10 metric tons (10 tons) and occupied thousands of cubic meters at the site. The debris consisted largely of the smashed bones of turtles, birds, insectivores, rodents, baboons, and small antelope. The agent or agents responsible for the accumulation of this material remain a matter of dispute, as has been the case for the last four decades. Dart believed that the deposit was built up by the australopithecines themselves over long periods. He said the faunal remains represented "the careful and thorough picking of an animal, which did not live to kill large animals but killed small animals to live" (p. 2). Others have said that the bones were dragged into caves by other animals, including hyenas, porcupines, and leopards. Some have even suggested that the deposits did not form in caves at all, but in limestone sinkholes that were open at the top and thus formed natural garbage cans for whatever might have been dropped or washed into them (see Figure 13-8). However, at least a large proportion of the bones that accumulated must have been covered with flesh, since mixed with the mammalian skeletal fragments were fossilized blowfly larvae that multiply in meat.

Whether the Plio-Pleistocene South African breccias accumulated in caves, sinkholes, or both is an issue that has yet to be resolved. Recent studies by anthropologists on what happens to animal bones at living sites abandoned by modern human hunter-gatherers may soon help us make sense of past bone accumulations; at the moment we cannot be sure of any single answer. It is possible—even likely—that some Plio-Pleistocene hominids fell victim to carnivores such as leopards and that the remains of the predators' meals might have found their way

FIGURE 13-8 *Opposite page:* (a) **Reconstruction of Swartkrans cave as it may have existed at the time when bones accumulated in it. Of 88 australopithecine fossils in the cave, some show marks of carnivore teeth.** (b) **One skull has two puncture marks 33 mm (1.3 in) apart, a distance which matches the spacing between a leopard's canines.** (c) **C. K. Brain has theorized that leopards may have dragged their hominid prey into trees, out of reach of hyenas and other scavengers. The bones might then have fallen into the cave. The presence of stone tools in the cave also suggests at least periodic occupation by hominids.**

Source: R. J. Trotter. 1975. "From Endangered to Dangerous Species," *Science News*, **109**:74–76.

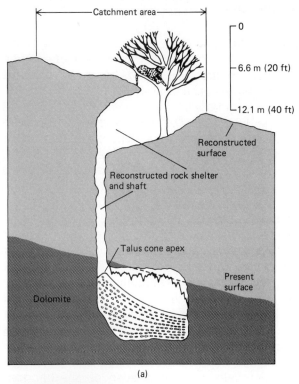

Catchment area

0

6.6 m (20 ft)

12.1 m (40 ft)

Reconstructed surface

Reconstructed rock shelter and shaft

Talus cone apex

Present surface

Dolomite

(a)

Puncture holes

(b)

(c)

into openings in the limestone. But none of these possibilities settles decisively the question of whether Plio-Pleistocene hominids did or did not modify natural objects for use as tools. According to Dart, some other observations do.

Ad Hoc Tool Use. In 1934 Dart published the results of a study on baboon skulls found in the same deposits as the australopithecines. There were 58 baboon skulls in all; 21 came from Taung, 22 came from Sterkfontein, and 15 came from Makapansgat. About 80 percent of these showed evidence of having sustained a crushing blow prior to fossilization. In many cases the shape of whatever struck the skull can be reconstructed, since it left its outline in the form of a depressed fracture (Figure 13-9). Apparently a double-headed object with a groove between the two heads caused the damage. Dart combed the breccia for objects of a size and shape to match the indentations. His searches turned up a number of long ungulate bones; of these, the humerus almost always showed ends that had been battered and chipped before becoming mineralized. Although the damage could have been caused by rock falls, a pathologist experienced in doing postmortems on miners injured in accidents has argued strongly against this. Pressure from earth or rock in mine cave-ins produces general crushing and distortion, not the sharply localized damage seen on the baboon skulls. Some creature with manipulative ability, reasoned Dart, had bashed the baboons with bone clubs.

The pattern seems to have differed from site to site. At Sterkfontein attackers preyed mainly on infant baboons and indiscriminately battered

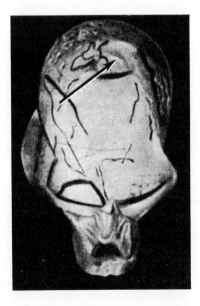

FIGURE 13-9 Skull of fossil baboon showing effects of a crushing blow that left traces shaped like the distal end of an ungulate humerus.

Source: R. A. Dart. 1949. "The Predatory Implemental Technique of *Australopithecus*," *American Journal of Physical Anthropology*, n.s. **7**(1):1–38. Plate 1, p. 16.

the heads of a few adults in several places. Neither of these feats can be considered beyond the level of ability of australopithecines, given the many recent observations of the capture, killing, and consumption of infant and young baboons by chimpanzees. At Taung the majority of victims, 12 out of 21 baboons, were adults; five were juveniles, one was an infant, and three were of unknown age. The injuries on most of them were due to a single stroke delivered at a vital spot. At Makapansgat, the same pattern holds: of the 15 baboon skulls showing injuries, 11 were of adults, 2 were of juveniles, and 2 were of infants.

Dart has also reconstructed additional details of the baboon kills. Three specimens each from Taung and Makapansgat bear wounds on the back of the skull, indicating attack from the rear and perhaps implying stealth. Most were injured in the face, though, suggesting frontal assault, and by far the majority of blows were delivered to the left side. This angle of delivery suggests that the club was wielded by a right-handed attacker. Is it reasonable to accept the idea that the hunters were australopithecines? The answer to this question involves two important factors: (1) the relationship between brain size and behavioral capabilities and (2) beliefs about hominid taxonomy.

At about the time the australopithecine fossils were discovered, the eminent British anatomist Arthur Keith had set 750 cc as the minimum brain size that could be accepted as human. Anything below this was by definition outside the hominid range—automatically excluding the South African Plio-Pleistocene fossils. The few specimens complete enough to allow reconstruction of brain size all fell well below the specified threshold value and were in the same general range as the great apes. At that time, reliable information on the behavior of wild pongids was virtually nonexistent, and few anthropologists would have considered them capable of making tools; most would not even have speculated on the possibility. Against this background, Dart's belief that the australopithecines used the bones, teeth, and horns of other animals as tools (making up what he referred to as an "osteodontokeratic culture") seemed fantastic (see Figure 13-10). Although crude stone tools (Figure 13-11, pages 488–489) were known from the same areas and geological horizons as those associated with the australopithecines, manufacture of these simple artifacts was also thought to be beyond the abilities of such small-brained creatures. If stone tools existed and australopithecines could not have made them, then the obvious answer was that members of some more advanced hominid population must have been responsible.

Systematic Selection of Raw Materials.   Dart saw toolmaking as just one step in the evolution of cultural behavior. To him it was likely that the savanna-living ancestors of the australopithecines had been occasional users of whatever natural objects came to hand. The australopithecines

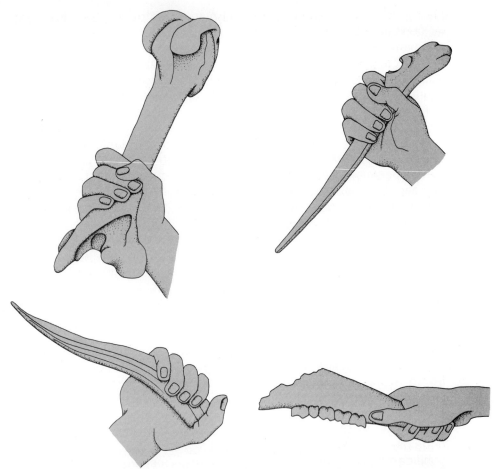

FIGURE 13-10   Reconstruction based on Dart's view of the uses of animal remains as tools and weapons by early hominids.

Sources: (1) E. Genet-Varcin. 1969. *A la recherche du primate ancêtre de l'homme.* Paris, France: Editions N. Boubee. (2) After R. A. Dart. 1957. *The Osteodontokeratic Culture of* Australopithecus prometheus. Pretoria, South Africa: Transvaal Museum Memoirs (no. 10).

themselves went beyond this to purposeful selection and modification of familiar raw materials. This was no casual speculation on his part, but a conclusion reached after a monumental task of analysis. At Makapansgat, Dart and his colleagues began with some 5000 tons of breccias discarded by limestone quarriers. From this they selected 20 tons that seemed to have the greatest concentration of animal remains. Chipping these out took 6 years and yielded 7159 fossils. Many were too small to be identified, but 4560 proved to be recognizable parts that could be allocated to specific genera.

The completed inventory provided several points of support for Dart's theory that the australopithecines used bones as tools: (1) Of all the bones present, 11 percent were the humeri of antelope; these bones of the front legs were five times more frequent than those of the hind legs, even though the latter would have had more meat on them. (2) Of four size categories of antelope represented, ranging from the tiny duikers to the huge kudu, 30 percent of the bones and 60 percent of the humerus fragments belonged to the medium category, which would have made the most convenient size of club. (3) Of these medium-sized antelope humeri, the distal ends—which fit the indentations in the baboon skulls—were more than 30 times as common as the proximal ends. Some other bones also showed strikingly nonrandom distributions. Although few of their limb bones were present, the small antelope were represented by parts of the skull and by the horizontal sections of the lower jaws. Dart's view was that horns attached to the skulls would have made good stabbing tools, while the teeth in the mandibles could have served as serrated cutting tools.

From Bone to Stone.   Over time, Dart believed, australopithecines replaced some of their osteodontokeratic implements with others made of stone. In support of this he described a block of breccia from Makapansgat that held a variety of broken bones, some stalactite fragments that looked like hand axes, and a shaped piece of chert—the last apparently brought in from elsewhere. He thought the Makapansgat hominids were substituting stalactites for teeth and horns as penetrating tools and for bones as pounding utensils.

Oldowan Culture

Whether Dart's theories are correct or not, stone tools have been found at other sites. The oldest known stone artifacts were found by Mary Leakey at the Koobi Fora site on the east side of Lake Turkana, Kenya; these are dated to approximately 2.6 million years ago. Other early sites containing stone tools are the upper beds at Hadar, Ethiopia, and the Shungura beds (E and F) in the Omo River Valley of Ethiopia. All are from about 2 to 2.5 million years old. By far the most complete early cultural sequence known at present comes from Olduvai Gorge in east Africa. There, a variety of chipped stone tools have been found in sites dating from the base of the Pleistocene. Some of the tools were fashioned from rounded or egg-shaped water-worn pebbles. A few *flakes* were usually whacked off one end to make a cutting edge, leaving a larger *core*. Others were made from jagged chunks of stone.

The Oldowan Industry.   The Oldowan industry consists of a set of shaped stones that were made by similar techniques and are found together at various sites. The stone objects that belong to this industry have been classified by Mary Leakey (1971) into four major categories: tools, utilized material, debitage, and manuports. (The first two of these categories are illustrated in Figure 13-11.)

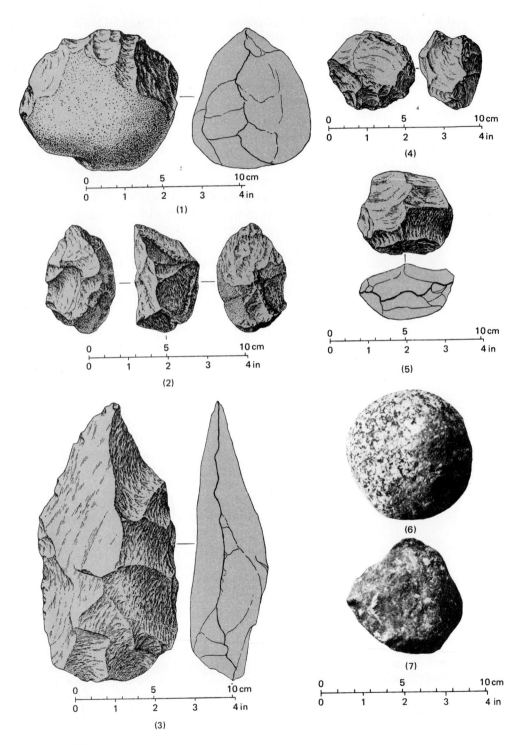

(1)

0   5   10 cm
0   1   2   3   4 in

(2)

0   5   10 cm
0   1   2   3   4 in

(3)

0   5   10 cm
0   1   2   3   4 in

(4)

0   5   10 cm
0   1   2   3   4 in

(5)

0   5   10 cm
0   1   2   3   4 in

(6)

(7)

0   5   10 cm
0   1   2   3   4 in

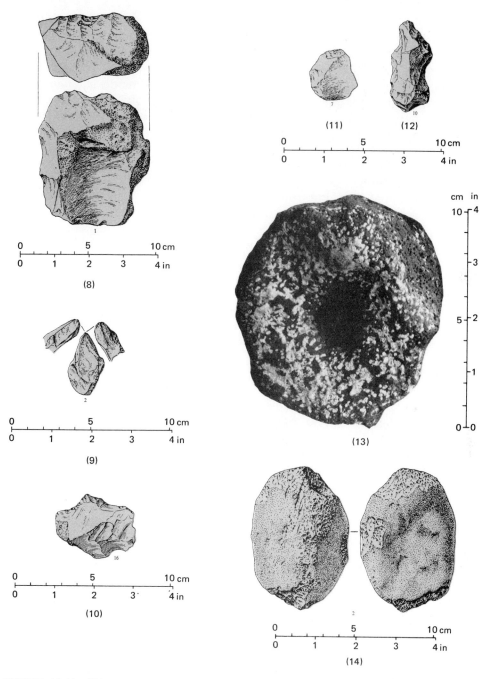

FIGURE 13-11 Oldowan stone tools. *Tools:* (1) **choppers,** (2) **protobifaces,** (3) **bifaces,** (4) **polyhedrons,** (5) **discoids,** (6) **spheroids,** (7) **subspheroids,** (8) **scrapers,** (9) **burins,** (10) **awl,** (11) **flaked tool,** (12) **laterally trimmed flake.** *Utilized material:* (13) **anvil,** (14) **hammerstones.**

Source: M. D. Leakey. 1971. *Olduvai Gorge.* Vol. 3. Cambridge, England: Cambridge University Press.

Definitions of the four categories of Oldowan objects are as follows:

Tools.   Through much of the sequence—for 2 million years, in fact—the chopper remains the basic tool. This was made in at least five different basic forms, with many individual variations. The most common of these is shown in Figure 13-11. But in even the earliest days at Olduvai this tool is just one element in a kit whose components increase in variety through time. Other tools include scrapers for working hides, chisel-like burins for shaping wood, rough stone balls (subspheroids) that may have served as hammers or missiles, and larger rounded stones that—judging from the degree of battering—were perhaps used as anvils on which other stones were placed while being trimmed to shape. Some tools—such as discoids—have uses that are harder to reconstruct with much certainty.

Utilized material.   These are stones which show marks indicating use but which seem not to have been end products themselves. In all probability they were used as aids in making other tools. Included here are hammerstones, light-duty flakes and other fragments, anvils, and cobblestones, nodules, and blocks.

Debitage.   This includes unmodified flakes and other fragments. Although irregular, these are probably not waste materials from the manufacture of other tools, since at certain sites they seem to have been used as cutting tools.

Manuports.   These show no sign of modification. They consist of lava cobblestones and nodules, as well as blocks of quartz or quartzite that appear to have been transported to the sites by hominids.

So much for the kinds of tools made by these early Pleistocene hominids. What were their makers doing with them? How did the people live? Here the Olduvai story provides some clear answers, a few of them astounding in what they reveal about the cultural capabilities of these hominids.

Who Were the Toolmakers?   One site (FLK I) from the middle of Bed I at Olduvai seems to have been an open-air camp. A central area covering about 28 m² (roughly 300 ft²) has a high concentration of small, sharp tools—chiefly flakes. With these were bones that had been smashed, perhaps to get at the marrow. Around the outside of this area were scattered heavier chopping-type tools, along with jaws and other bones that would have contained little marrow. The site is evidently one at which a large amount of meat was processed, with activities differentiated by areas: rough butchering and dismembering was done around the outer fringes, while the bulk of fine cutting and marrow extraction went on in the center. Unlike the situations described in preceding sections, in this case an early Pleistocene site had yielded undisputable cultural remains. But who made and used the stone tools is still in question. Louis Leakey's initial view that a robust hominid (Olduvai hominid 5) made the stone tools found at the site has been questioned by Tobias and others. These doubts were reinforced by the discovery of additional hominid remains near the base of Bed I at Olduvai Gorge, below those of Olduvai hominid 5. Some anthropologists believe that these lower remains represent a more advanced hominid (which they have designated *Homo habilis*) and that this, not a robust australopithecine, was the toolmaker. Here we have one of the many points at which the interpretation of cultural evolution is influenced by ideas about the course of biological evolution.

Evidence from Olduvai Gorge does point to one undisputable conclusion: some early Pleistocene hominids made simple stone tools according to a *tradition*, a recognizable pattern that persisted for many generations. Beyond this point, interpretations must become more tentative. It is difficult to flatly rule out the possibility that even the smallest-brained australopithecines yet known made tools. All of them had brains far larger than those of chimpanzees, a species which modifies a variety of natural materials to make tools of traditional patterns and which hunts and kills other animals for food. However, if a good case can be made for the existence of two or more contemporaneous lineages of hominids in the early Pleistocene, then the belief that members of only the more advanced one made and used tools could prove correct.

Tool Use: Food and Shelter.   Regardless of which hominids made the stone tools, someone put them to good use. In the lower levels of Bed I a stone circle 3.7 to 4.3 m (12 to 14 ft) in diameter was discovered. The structure was made of rocks ranging in size from less than 10 cm (4 in) to over 25 cm (10 in) in diameter. In a few places these were still piled into heaps nearly 31 cm (1 ft) in height. The surviving traces of this crude arrangement are strikingly like temporary shelters still made by some present-day nomadic peoples in Africa (see Figure 13-12). Today the stones support upright branches and anchor skins or grass used to cover the resulting framework.

Some evidence on the food sources of early hominids is provided at one site in upper Bed I, where excavators uncovered minute bone fragments of birds, reptiles, and small mammals such as rodents and insectivores. These concentrations very likely represent fecal pellets— but whose? Wild dogs, cats, and hyenas can be ruled out because the strong stomach acids of these true carnivores would have dissolved the

(a)

FIGURE 13-12 (a) **Stone circle from Bed I, Olduvai Gorge.** (b) **Its modern counterpart, a rough shelter of branches and grass with stones supporting the bases of the branches, made by the Okombambi people of southwest Africa.**

Source: M. D. Leakey. 1971. *Olduvai Gorge.* Vol. 3. Cambridge, England: Cambridge University Press. Plates 2, 3.

bones. A large primate seems to be the other logical choice (chimps, you recall, chew up and swallow the bones of their prey), and among the larger primates hominids seem most likely because the patches were scattered among stone artifacts and other occupational debris.

The other end of the faunal scale is also represented among the food sources of these early hominids. At one site in upper Bed I excavators uncovered a nearly complete skeleton of an ancient type of elephant (*Elephas recki*) in close association with a number of artifacts (Figure 13-13). Though not quite mature (the epiphyses, or end tips, of its limb bones had not yet fused to the shafts, indicating that the animal was

(b)

still growing), the beast was already fully as large as a modern adult elephant. Its bones had been scattered about, producing a pattern similar to that seen when scavengers have tugged at a carcass.

We can tell how the animal fell—most of the bones on its left side were beneath those on its right—but not why. An early hominid hunting pack might have driven it into a swamp, where, bogged down, it could be slaughtered easily. Most of the other bones were whole or nearly so, but the skull was broken to tiny bits, and the mandible was crushed— perhaps by the battered cobblestones found at the site. Alternatively, the elephant may have already been dead when found, and its skull smashed open to get at the brain as part of the butchering operation. Associated with the skeleton were over a hundred tools. They ranged from heavy-duty choppers down to small, sharp flakes—everything needed to reduce the pile of protein into portable portions. But this was

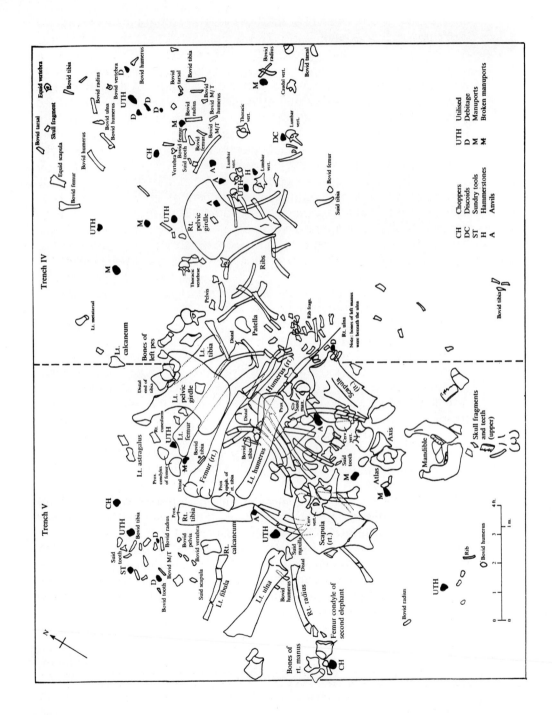

not a regular occupation site, for there was no cultural material beyond that needed for the job at hand. This large-mammal butchering site was not the only one found. In lower Bed II the skeleton of a *Deinotherium* (a member of a family closely related to elephants and about as large) was found. Here the evidence is especially strong that the animal had died while mired in a swampy spot, since its foot bones were buried well below the rest of the skeleton, as if the animal had sunk in an erect position. Hominid butchers reached it, bog or not; the head was severed and dragged aside, and one chopper was left within the pelvic girdle.

It is unlikely that a lone nuclear family could have killed an elephant or used any significant fraction of its meat before it spoiled or was carried off by hyenas or vultures. The exploitation of large mammals as food suggests that early hominids operated—at least at times—as large groups. Cooperative hunting makes economic sense. The field ecologist George Schaller, for example, found from his studies of spotted hyenas that a pair working together ended their hunt successfully with the capture of prey three times more frequently than one hyena alone. Schaller also noted that social carnivores such as wild dogs practice effective division of labor, with one adult staying behind to guard the pups while the others hunt together. After the prey has been captured and devoured, pack members return to home base and regurgitate food for the young and the guard. Perhaps as early humans became part-time carnivores, they too might have modified previous primate behavior patterns toward greater food sharing and other forms of cooperation.

In fact, Isaac (1978) emphasizes food sharing as the central organizing theme for early hominid social behavior. It integrates several separate contributing factors, such as the ability to make tools, to hunt, to gather vegetable foods, and to carry objects. Isaac assumes that early hominids divided subsistence labor by sex, with males and females each tapping a different range of food resources, which were then carried back to the camp, prepared, and shared. The sharing behavior itself could have played an important part in the development of the systems of reciprocal social obligations that characterize all present human societies. For systems of exchange to function, humans must be able to keep track of complex economic and social obligations that link past and future behavior. The need to calculate and record contingencies and obligations of this sort could have provided an important stimulus for the evolution of the human brain.

FIGURE 13-13 *Opposite page:* **Elephant-butchering site at Bed I, Olduvai Gorge.**
Source: M. D. Leakey. 1971. *Olduvai Gorge.* Vol. 3. Cambridge, England: Cambridge University Press. Fig. 32, p. 65.

SUMMARY

1. Though hominids originated earlier, most of the major genetic and morphological changes in hominid evolution have taken place since the beginning of the Villafranchian faunal period, about 3 million years ago.

2. The opening of the Pleistocene epoch is now set at about 2 million years ago; this is the period when some regions of the world, including parts of Europe and North America, were covered periodically by glaciers.

3. The European glacial advances included (from earliest to latest) the Günz, Mindel, Riss, and Würm; another glaciation, the Donau, may have preceded the Günz. Successive glacial periods were separated by the warmer interglacials; temporarily warmer phases within glacial phases are called *interstadials.*

4. During glacial maximums, sea levels dropped as ocean water became tied up in the form of ice; shallow areas were exposed along continental shelves, sometimes forming land bridges between areas that had been isolated by water barriers at other times.

5. Though abundant in comparison with the remains of many other groups of mammals, hominid fossils are still relatively rare, and the human evolutionary record is somewhat incomplete; consequently, the study of the fossils themselves must be supplemented by the use of evidence and reasoning from other related fields such as archeology, ethnology, and ethology.

6. At present, several alternative phylogenetic reconstructions fit the data from hominid fossil remains; these alternatives include single-lineage and multiple-lineage models.

7. The first fossil hominid from the time range of the Plio-Pleistocene boundary was discovered at Taung in South Africa in 1924. Hominids like this one exhibit the structural characteristics expected in a very early stage of hominid evolution: a somewhat reduced face and smaller canine teeth than in later dryopithecine apes, coupled with a small brain.

8. Additional fossil discoveries from east Africa have extended the ranges of variation of Plio-Pleistocene hominids even further than those previously known from South Africa.

9. Though the earliest stages of cultural evolution are hard to detect, it is reasonable to assume that early hominids made tools of more perishable materials before stone was used systematically; while not the only mammals to make tools, early hominids may have been the first primates to have been significantly dependent on material culture for survival. The evidence from Olduvai Gorge in east Africa demonstrates that by early Pleistocene some hominids were already making simple stone tools to a traditional pattern.

10. The very real and appreciable morphological differences among early Pleistocene hominids are a matter of observation, but the phylogenetic reconstructions based on this variation are matters of interpretation. Among the factors that must have contributed to the known variation, evolutionary change through time is likely to have been particularly important, though it may have been supplemented by differences due to age, sex, and environment.

SUGGESTIONS FOR ADDITIONAL READING

Clark, W. E. Le Gros, 1967. *Man-Apes or Ape-Men?* New York: Holt, Rinehart and Winston. (This is a brief (150 pages) and pleasantly readable account of the history of the discovery of Plio-Pleistocene hominid fossils in South Africa and east Africa. As suggested by its title, the book discusses the controversy over the hominid status of australopithecines.)

Coppens, Y., F. C. Howell, G. L. Isaac, and R. E. F. Leakey (eds.). 1976. *Earliest Man and Environments in the Lake Rudolf Basin.* Chicago: University of Chicago Press. (This is an edited collection of papers that resulted from a conference held in 1973 at the Kenya National Museum. Although written at a technical level, the papers provide a good overview of recent developments in stratigraphy, geochronology, the morphology and evolutionary biology of nonhominid and hominid taxa, and hominid behavior and ecology. Together they provide serious students with a challenging introduction to the professional literature.)

Isaac, G. L., and E. R. McCown (eds.). 1976. *Human Origins. Louis Leakey and the East African Evidence.* Menlo Park, California: Staples Press of W. A. Benjamin. (This book, the third volume in a series titled "Perspectives on Human Evolution," was prepared in honor of Louis Leakey, the founding father of the study of human evolution in east Africa. Particularly valuable are some of the lists and descriptions of the newer discoveries of hominid material from this section of the African continent).

Mann, A. E. 1975. *Paleodemographic Aspects of the South African Australopithecines.* Philadelphia: University of Pennsylvania Publications in Anthropology, no. 1. (This technical monograph provides a careful analysis of the number of hominids at each of the major South African sites, and it draws from this record some interesting conclusions about developmental patterns in early humans.)

Tobias, P. V. 1967. *Olduvai Gorge.* Vol. 2. *The Cranium and Maxillary Dentition of Australopithecus (Zinjanthropus) boisei.* London: Cambridge University Press. (This monumental monograph devoted to a fossil find of major importance is an excellent example of detailed description and careful analysis of data bearing on problems of hominid evolution.)

Wolpoff, M. H. 1971. *Metric Trends in Hominid Dental Evolution.* Case Western Reserve Studies in Anthropology, no. 2. Cleveland: Case Western Reserve Press. (This is an extensive compilation of data on tooth size in early hominids. Since teeth are the most abundant fossil remains from all earlier periods, Wolpoff's summary is a useful reference for those interested in comparison of norms and variations in several earlier groups.)

REFERENCES

Brace, C. L., and M. F. A. Montagu. 1977. *Human Evolution* (2d ed.). New York: Macmillan.

Broom, R. 1950. "The Genera and Species of the South African Fossil Ape-Men," *American Journal of Physical Anthropology,* n.s. **8**(1):1–13.

Butzer, K. W. 1971. *Environment and Archeology* (2d ed.). Chicago: Aldine.

Christie, P. W. 1977. "Form and Function of the Afar Ankle," *Abstracts of the Forty-Sixth Annual Meeting of the American Association of Physical Anthropologists,* p. 5.

Clark, W. E. Le Gros. 1964. *The Fossil Evidence for Human Evolution* (2d ed.). Chicago: University of Chicago Press.

Clark, W. E. Le Gros. 1967. *Man-Apes or Ape-Men?* New York: Holt, Rinehart and Winston.

Curtis, G., T. Drake, and A. Hampel. 1975. "Age of KBS Tuff in Kobi Fora Formation, East Rudolf, Kenya," *Nature,* **258:**395–398.

Dart, R. A. 1925. *"Australopithecus africanus:* The Man-Ape of South Africa," *Nature,* **115**(2884):195–199.

Dart, R. A. 1949. "The Predatory Implemental Technique of Australopithecus," *American Journal of Physical Anthropology,* n.s. **7**(1):1–38.

Dart, R. A. 1962. "Substitution of Stone Tools for Bone Tools at Makapansgat," *Nature,* **196**(4852):314–316.

Isaac, G. L. 1978. "The Food-Sharing Behavior of Protohuman Hominids," *Scientific American,* **238**(4):90–108.

Leakey, M. D. 1971. *Olduvai Gorge.* Vol. 3. Cambridge, England: Cambridge University Press.

Mayr, E. 1951. "Taxonomic Categories in Fossil Hominids," *Cold Spring Harbor Symposium on Quantitative Biology,* **15:**109–117.

Schaller, G. B., and G. Lowther. 1969. "The Relevance of Carnivore Behavior to the Study of Early Hominids," *Southwestern Journal of Anthropology,* **25**(4):307–341.

Chapter 14

Middle Pleistocene Hominids: Cultural Elaboration and Human Variation

The current controversies over the place of Plio-Pleistocene hominids in our own ancestry are in many ways replays of similar debates at the end of the nineteenth century touched off by the discoveries of Eugene Dubois, a Dutch army doctor who in the 1890s took a government post in Indonesia for the specific purpose of finding human fossils. Dubois appears to have been better at discovering fossils than at classifying them, though hindsight gives us an unfair advantage in making such judgments. In 1891, in a cave along the banks of the Solo River near Trinil on Java, he found an upper right third molar that he said belonged to a chimpanzee. He came to the same conclusion about a skullcap found 3 months later at a spot 3 m (10 ft) away from where the tooth was discovered. But a fossil femur found the next year in the same stratum 4.6 m (15 yd) away was so modern in form that it was attributed to a taxon more closely related to our own. On his return to Europe, Dubois published a paper on his discoveries, in the process altering his viewpont in several ways. All the bones mentioned above, plus another molar, were assigned to a single taxon, *Pithecanthropus erectus* ("erect ape-man").

Both views attracted supporters and critics, and a fierce scientific debate continued for years. Dubois added to the confusion by changing his mind once again, in later years saying that the fossil bones were those of a giant gibbon. Actually the features shown by the Trinil find are more those of a primitive hominid, although one not equally archaic in all respects. Even the first few fossils found are enough to show this mosaic pattern. The skullcap, thick and robust, is constricted behind the heavy supraorbital torus; its virtually nonexistent forehead slopes sharply backward into a low-crowned vault. But in size and shape the femur can be matched by many found in modern human populations (the match is so close that some anthropologists still feel it really is modern and should not be included with the middle Pleistocene remains).

The formal taxonomic name coined by Dubois, which exaggerates the differences between his find and modern humans, has now been abandoned, but we can still use the convenient colloquial label *pithecanthropine*. The Latin binomial *Homo erectus* is also widely used to refer to members of pithecanthropine populations. However, since at least some populations of these middle Pleistocene hominids are our own direct lineal ancestors, from whom we evolved through phyletic evolution, the formal Latin binomial probably artificially divides what was a genetic continuum. These terms apply not only to populations represented by the Javan specimens but also to remains found in Europe, Asia, and Africa (see Table 14-1, pages 502–505) that show the same combination of a primitive-looking skull with a functionally modern skeleton.

These fossil remains demonstrate that pithecanthropines were distributed widely in time and space. Members of these populations that were discovered later extend the range of biological and cultural characteristics from this period in human evolution far beyond that originally known from the first Javan specimens. By the middle Pleistocene, wide areas of Eurasia and Africa were inhabited by erect, robustly built hominids with brains of a size that sometimes overlaps those of present-day humans, though their average cranial capacity was lower than ours by several hundred cubic centimeters.

Let us begin our study of these middle Pleistocene hominids with the Javan specimens, and then look at variations on their theme that have been found in China, Kenya, Algeria, and Germany. Then we will go on to study the cultural advances—including new varieties of stone tools, the elaboration of woodworking, and the control of fire—made by these ancestors of ours and to see how this cultural evolution had an impact on some of their biological characteristics.

PITHECANTHROPINE FOSSIL RECORD

Javan Pithecanthropines

Between 1889 and 1941, over 30 fossil hominids were found in Java, a sample large enough to allow anthropologists to reconstruct the appearance of these middle Pleistocene hominids.

Physical Characteristics

Skull.   The skull shows striking differences from the skulls of modern human populations. When it is seen in side view, a perpendicular line touching the rear of the orbit divides the long, low cranial vault from what was a massive, projecting face (Figure 14-1, page 506). The pithecanthropine skull had a forward-flaring, or *prognathous*, upper jaw and a large nasal opening, flanked by the thick frontal processes of the *zygomatic arch* and topped by a robust supraorbital torus that continues the thick nasal root. The forehead is low, and the cranial vault does not rise far above the brows (Figure 14-2, page 506). The skull is widest near the base, at about the ear level (Figure 14-3, page 506). From there, it slopes inward and upward in two major planes, like the roof of an old-fashioned barn. In some skulls the frontal bone has a slight keel, or thickening, along the midline, though there is no actual sagittal crest (that central ridge of bone seen in the skull of most gorillas and some australopithecines); the temporal lines do not reach the midline, but instead are strongly marked on the parietal bones. Seen from the top, the Javan pithecanthropine skull has a sharp constriction of the brain-case behind the brow ridges. In profile from above, the skull looks like an hourglass with a small upper chamber and a larger lower one.

These features seem to make up a long and rather formidable list. They become easier to remember and understand when we look at them as particular expressions of two central functional features: (1) well-developed muscles that were used in chewing and (2) a brain that was still relatively small, with an average endocranial volume of 931 cc (see Table 14-2, page 507). Javan pithecanthropines had brains intermediate in size between those of australopithecines and modern humans.

Dentition.   The teeth also show some interesting features. On the average the incisors and canines are larger than those of australopith-ecines. Some Javan pithecanthropines have canines that even project a bit beyond the level of the adjacent premolars and a diastema, or gap, of 5 to 6 mm (about 0.2 in) between the upper lateral incisor and canine. The cheek teeth (premolars and molars), arrayed in what is almost a straight line along the sides of the jaw, are smaller on the average than those of australopithecines, though still larger than those of most modern humans.

Skeleton.   From the neck down, these middle Pleistocene hominids do not differ very much from people now living. Their stature, reconstructed from limb bones, has been found to be about 160 to 168 cm (5 ft 3 in to 5 ft 6 in), which would place them in the low end of the size range of modern human populations. Their body weight would have been about equal to, or a bit heavier than, that of modern humans of the same stature, since their bones are more robust.

TABLE 14-1   Middle Pleistocene hominids

Geographic location	Site	When found	Material discovered	Number of individuals
East Africa	Olduvai Gorge	1960	Olduvai hominid 9—fragmentary thick skullcap lacking face and base	1
North Africa	Ternifine	1954–1955	Right parietal bone of immature individual and three mandibles	3–4
	Littorina cave, in quarry of Sidi Ab der-Rahman (Casablanca, Morocco)	1953	Two mandibular fragments with P_3 and $M_1–M_3$	1
	Smuggler's cave (Temara, Morocco)	1958	Body of mandible	1
	Rabat (Morocco)	1933	Fragments of skull (with maxilla) and mandible	1
Israel	Tell Ubeidiya (Jordan Valley, near outlet of Lake Tiberias)	1969	Five, possibly six, fragments: two of parietal bone and one of temporal bone; teeth; and possibly fragment of occipital bone	1
Europe	Vértesszöllös (Hungary)	1965	Occipital bone (of adult male)	1
	Mauer sands near Heidelberg (Germany)	1907	One mandible with all categories of teeth present (left $P_3–M_2$ broken)	1
	Přezletice, near Prague (Czechoslovakia)	1969	One fragment of a molar tooth	1
China	Choukoutien	1921	Two molars	1–2
	Choukoutien	1927	Molar	1

Morphological designation	Material culture, if any	Age of site	Dating method used
Pithecanthropine	At same geological level about 90 m (100 yd) away, numerous hand axes and bones of large animals broken possibly for extraction of marrow	Middle Pleistocene; about 5.6 m (15–20 ft) below top of Bed II	Faunal correlation and potassium-argon dating
Pithecanthropine	Early Acheulean industry with many hand axes, choppers, chopping tools, and flakes	Early middle Pleistocene	Faunal correlation
Pithecanthropine	Evolved Acheulean	May correspond to Riss glaciation in Europe	Sea-level fluctuations correlated with glacial periods
Pithecanthropine	Final Acheulean or transition to Aterian flake culture	Probably contemporaneous with Littorina cave	Correlation of cultural evidence
Pithecanthropine	None found	Probably equivalent to end of last interglacial in Europe	Sea-level fluctuations correlated with European glacial chronology
Indeterminate	Primitive stone tools made from water-worn pebbles, including roughly trimmed stone balls, trihedral flint points, and chopping tools	Villafranchian or middle Pleistocene	Faunal correlation
Pithecanthropine or possibly modern human	Chopper—chopping tool complex, evidence of use of fire	Middle or later part of Mindel glaciation	Faunal correlation
Pithecanthropine	No tools found	Post-Villafranchian, before second (Mindel-Riss) interglacial (probably first interglacial or interstadial within Mindel)	Correlation of fauna, which suggests a warm, temperate climate
Pithecanthropine	Around 50 crudely chipped stone tools	Günz-Mindel interglacial	Faunal correlation
Pithecanthropine	No tools found	Middle Pleistocene	Various methods, including faunal correlation and potassium-argon dating
Pithecanthropine	No tools found	—	—

(Continued)

TABLE 14-1 Continued

Geographic location	Site	When found	Material discovered	Number of individuals
	Choukoutien	1928–1937	Fourteen skulls or skull fragments, eleven mandibles, sixty-four isolated teeth, and seven femoral fragments	Over 30
	Choukoutien	1949	Mandible	1
	Choukoutien	1959	Mandible	1
	Choukoutien	1960	Five teeth and two fragments of a humerus and tibia	1 to several
	Lantian, Shensi province	1964	Cranium (bones of skullcap plus face)—frontal, large part of upper jaw with right M^2 + M^3, left upper molar	1
	Lantian, Shensi province	1959–1963	Mandible	1
Java	Trinil	1891	Skullcap and two teeth; left femur (*Pithecanthropus* I)	1–4
	Trinil	1898	Premolar	1
	Trinil	1900	One complete femur plus four fragments of femurs	Up to 4
	Sangiran	1936	Half a lower jaw containing four teeth (mandible B)	1
	Sangiran	1937	Skullcap (*Pithecanthropus* II)	1
	Sangiran	1938	Part of cranium (*Pithecanthropus* III)	1
	Sangiran	1939	Maxilla and rear portion of skull (*Pithecanthropus* IV)	1
	Sangiran	?1938–present	Parts of four skulls (*Pithecanthropus* V, VI, VII, VIII)	4
	Kedung Brubus (40 km, or 25 mi, from Trinil)	1890	Fragment of lower jaws near symphyseal region (with socket for canine, root of first premolar, and part of socket of second premolar)	1
	Modjokerto (near Surabaya)	1936	Braincase without face, with part of base missing	1

504

Morphological designation	Material culture, if any	Age of site	Dating method used
Pithecanthropine	Chopper-tool industry; tools (made from coarse-grained quartz and greenstone) consisted of a few cores and numerous flakes; ash heaps and pieces of charcoal suggest use of fire	Middle Pleistocene, possibly 300,000 to 400,000 years ago	Various methods, including faunal correlation and potassium-argon dating
Pithecanthropine Pithecanthropine Pithecanthropine			
Primitive (or early) pithecanthropine	Several cores and flakes from soil levels just above skullcap	700,000 years ago	Faunal and pollen analysis
Pithecanthropine	Large chisel-pointed pebble tool found 1000 m (300 ft) away (earliest known tools in east Asia)	300,000 years ago	Faunal and pollen analysis
Pithecanthropine	None found	Middle Pleistocene, perhaps 550,000 years ago	Faunal analysis; potassium-argon dating of tektites and basalt in deposits believed to correlate with those at Trinil
Pithecanthropine Pithecanthropine	None found No tools found		
Pithecanthropine	No tools found	Djetis faunal beds; age estimates vary widely, from lower Pleistocene to early middle Pleistocene	Faunal correlation
Pithecanthropine	No tools found	Early middle Pleistocene	Stratigraphic and faunal correlation
Pithecanthropine	No tools found	Early middle Pleistocene	Stratigraphic and faunal correlation
Pithecanthropine	No tools found	Djetis	Stratigraphic and faunal correlation
Pithecanthropine	None found		
Pithecanthropine	None found	Possibly middle Pleistocene	Faunal correlation
Possibly Pithecanthropine	—	Djetis beds	Faunal correlation

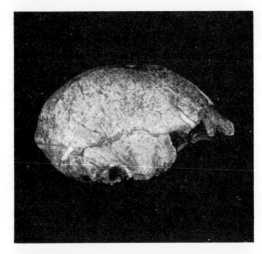

FIGURE 14-1 Skull of a Javan pithecanthropine found in 1937 by G. H. R. von Koenigswald; side view.
Source for Figures 14-1, 14-2, and 14-3: Photographs of cast.

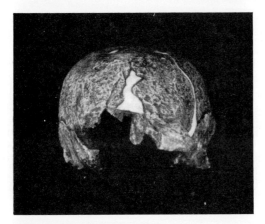

FIGURE 14-2 Skull of Javan pithecanthropine; front view.

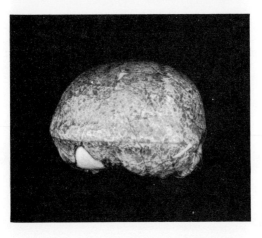

FIGURE 14-3 Skull of Javan pithecanthropine; rear view.

TABLE 14-2 Endocranial volumes of Javan pithecanthropines

Specimen	Endocranial volume, cc
Pithecanthropus I	953
Pithecanthropus II	815
Pithecanthropus IV	900
Pithecanthropus VI	855
Pithecanthropus VII	1059
Pithecanthropus VIII	1004
Javan sample range	815–1059
Javan sample average	931

Even though the skeletal remains from Java are not very extensive, they do include one interesting case of abnormal variation, discussed in "Abnormal Variation in a Middle Pleistocene Hominid: The Trinil Femur" on pages 508–509.

Variations: Djetis and Trinil. The Javan pithecanthropines were not all drawn from a narrow time range, as shown in Table 14-3. Chronometric ages are rarely known with any accuracy, because none of the specimens came from stratified living sites. Many fossils had been carried by rivers to the beds where they were discovered. The best we can do is place them as members of two major faunal assemblages (collections of bones of different animals that are found together in the same stratum), Djetis and Trinil, both of which lasted for very long periods of time.

TABLE 14-3 Distribution of Javan pithecanthropines in time and space

Chronology	Geological divisions	Faunas	Site
Upper Pleistocene	Notopuro	Ngandong	Ngandong
Middle Pleistocene	Kabuh	Trinil	Trinil skull I Sangiran skulls II, III, VI, VII, VIII
Lower Pleistocene	Putjangan	Djetis	Sangiran skull IV, mandible B Modjokerto skull V
	Kalibeng	Kali Glagah	
		Tjidjulang	

Sources: (1) W. W. Howells. 1973. *Evolution of the Genus* Homo. Reading, Mass.: Addison-Wesley. (2) S. Sartono. 1975. "Implications Arising from *Pithecanthropus* VIII," in R. Tuttle (ed.), *Paleoanthropology: Morphology and Paleoecology*. The Hague, Netherlands: Mouton. Pp. 327–360.

ABNORMAL VARIATION IN A MIDDLE PLEISTOCENE HOMINID: THE TRINIL FEMUR

The femur found by Eugene Dubois in his 1891–1892 field season was of primary importance in establishing the existence of hominids in Java several hundred thousand years ago. From a phylogenetic viewpoint, the extensive bony growth near the upper end of the femoral shaft is of some importance, since it shows, once again, that not all fossils represent population norms. Beyond this, however, the abnormal form of the Trinil femur is useful in tracing the history of human disease patterns.

The bony growth on the Trinil femur has long been recognized as pathological. Rudolf Virchow, the famous nineteenth-century German pathologist, diagnosed its cause as syphilis, but later workers have cast doubt on this interpretation, as well as on the possibility that the projecting mass of bone represents a badly healed break.

Recently M. Soriano of the University of Barcelona, Spain, discovered a strikingly similar bony growth on the femur, as well as on other bones, of a former alcoholic man undergoing autopsy following his death from cirrhosis of the liver. The bone disease in this man was not an isolated occurrence. Dr. Soriano and his associates had observed similar growths in several patients, and eventually they discovered what all the afflicted shared in common. All had been habitual drinkers of a table wine to which the manufacturer had fraudulently added sodium fluoride (which is absolutely tasteless and colorless) to control fermentation. Chemical analysis of the patients' bones confirmed the hunch; for example, the femurs of the autopsied alcoholics contained over 8 parts of sodium fluoride per million, twice the level usually considered diagnostic of bone fluorosis.

The similarity of the lesion on the Trinil femur also suggested fluoride poisoning, but how could a middle Pleistocene pithecanthropine have ingested enough of the element to trigger the abnormal growth? In searching the scientific literature for possible mechanisms, Soriano discovered that sheep grazing in a volcanic region of Iceland also develop bone fluorosis. During eruptions, large quantities of gaseous fluoride compounds are discharged into the atmosphere and later settle onto plants and grasses eaten by the animals. Immediately after volcanic eruptions the concentrations are great enough to cause death in a few

Potassium-argon dating of minerals from the earlier Djetis beds shows that these deposits extended back as far as about 2 million years. The morphological characteristics of the robust hominid mandibles found near Sangiran are consistent with this age. Originally called *Meganthropus palaeojavanicus* ("large man from ancient Java") by their

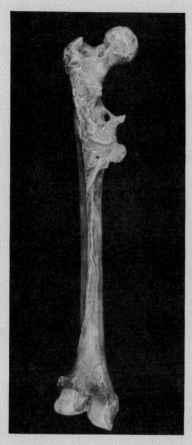

Trinil femur, showing pathological growth.
Source: Photograph of cast, University of Pennsylvania.

days to a few weeks. After a time they become diluted enough to allow survival, but with deformation of the bones.

In closing the link of his paleopathological diagnosis, Soriano pointed out that the Javan hominid remains were found associated with thick beds of volcanic ash. The fruits and other vegetation consumed by these early humans may have been contaminated by an ancient form of natural air pollution.

Source: M. Soriano. 1970. "The Fluoric Origin of the Bone Lesion in the *Pithecanthropus erectus* Femur," *American Journal of Physical Anthropology*, **32**(1):49–58.

discoverer, the Dutch paleontologist G. H. R. von Koenigswald, these specimens quite possibly belonged to a population of early Pleistocene hominids less advanced than the better-known pithecanthropines. The upper part of a small skull from the Modjokerto site (also found by von Koenigswald, in 1936) is more difficult to interpret. The extremely thin walls of this braincase suggest that it belonged to a small child, although the remains of australopithecine and pithecanthropine children are far rarer than those of adults. Whether the skull belongs to either of these populations or to an intermediate group will remain uncertain until more comparative material becomes available. At least one undoubted pithecanthropine specimen comes from the lower Pleistocene Djetis beds. This is skull IV from Sangiran. Predictably enough, this specimen is not only geologically older than the bulk of Javan pithecanthropines but also more primitive in appearance, and its endocranial volume of 900 cc falls below the average for the Javan group.

The upper parts of the Trinil beds contain tektites and other minerals datable by the potassium-argon method to a range from about one-half to three-quarters of a million years ago. Trinil pithecanthropines show a wide range of morphological variation, as Figure 14-4 illustrates. This impression is reinforced by the data on endocranial volumes (see Table 14-2); they range from 815 to 1059 cc. Of course, these are estimates based on just six incomplete skulls; the actual population range was almost certainly much greater than this.

Chinese Pithecanthropines

Choukoutien Sample.   Three decades after the first of the Javan fossils came to light, more pithecanthropine remains were discovered at Choukoutien, near Peking. The Peking hominids were not carbon copies of their Javan predecessors, and the differences in appearance led at first to their being placed in a different taxon, *Sinanthropus pekinensis*. However, members of the Choukoutien population were at least similar enough to confirm the widespread existence of a stage in hominid evolution more primitive than that of either living humans or neanderthals (later humans with heavy faces but larger brains, discussed in Chapter 15). Furthermore, the Peking pithecanthropines came from stratified living sites, and the elements of material culture (stone tools, evidence of control of fire, etc.) found with them showed that they were not too primitive to be considered humans. Pithecanthropines could no longer be dismissed as giant gibbons.

Figures 14-5 to 14-7 show the reconstructed skull of a Choukoutien pithecanthropine. Comparison of these with the illustrations of Javan skulls shows pervasive similarities and a few contrasts. Perhaps the most important difference is in brain size, as reflected in endocranial volumes and in the slightly higher dome of the Chinese skull. A comparison of the endocranial volumes of the two samples (Tables 14-4 and 14-2) shows that those from Choukoutien averaged about 100 cc larger, a difference of about 10 percent. Moreover, the upper end of the range of endocranial volumes of the Chinese pithecanthropines overlaps the lower end of the range of modern populations; surprising as it may seem, some of these fossil hominids had brains larger than those of some living humans.

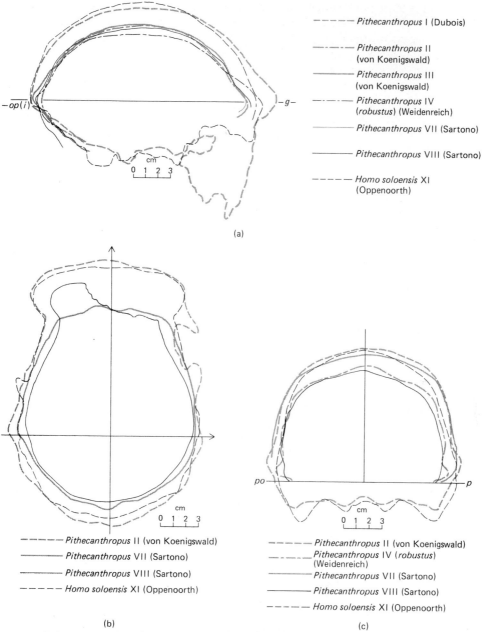

(a)

(b)   (c)

FIGURE 14-4   Cranial variation in Javan pithecanthropines. (a) **Side view: comparison of cross sections from front to rear along midline of skull.** (b) **Top view: comparison of outlines of skulls in horizontal plane.** (c) **Front or rear view: comparison of cross section at point near midline of skull between front and rear.**

Source: Modified from S. Sartono. 1975. "Implications Arising from *Pithecanthropus* VIII," in R. Tuttle (ed.), *Paleoanthropology: Morphology and Paleoecology.* The Hague, Netherlands: Mouton. Figs. 1, 2, 3, pp. 332, 333, 334.

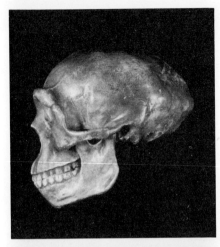

FIGURE 14-5 Chinese pithecanthropine skull, side view.
    Source for Figures 14-5, 14-6, 14-7: Cast, University of Pennsylvania Museum, representing a reconstruction based on the original specimen.

FIGURE 14-6 Front view of Chinese pithecanthropine skull.

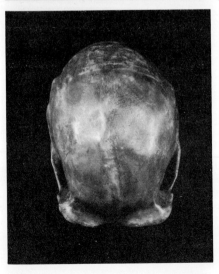

FIGURE 14-7 Top view of Chinese pithecanthropine skull.

TABLE 14-4   Endocranial volumes of Chinese pithecanthropines

Specimen	Endocranial volume, cc
Choukoutien II	1030
Choukoutien III	915
Choukoutien X	1225
Choukoutien XI	1015
Choukoutien XXI	1030
Choukoutien sample range	915–1030
Choukoutien sample average	1043
Lantian	780

Source: P. V. Tobias. 1971. *The Brain in Hominid Evolution.* New York: Columbia University Press. Table 13, p. 90.

Below the brain, the upper jaw juts forward from the base of the nose, while the chinless lower jaw sweeps upward to meet it. As a result, the maximum forward projection of the skull is marked by a point where the upper and lower incisors meet in an edge-to-edge bite. Not evident from the illustrations here is the presence in the Chinese pithecanthropines of a well-developed *mandibular torus,* a lingual (tongue-side) thickening of the horizontal part of the lower jaw in the region between the canine and the first molar (Figure 14-8). This torus was absent or rare in preceding fossil hominid populations, but it is present not only in all the lower jaws of Choukoutien pithecanthropines but also in many modern populations in northern Asia and northern Europe. Frequencies in these living groups range from 15 percent in Chinese to 97 percent

FIGURE 14-8   Mandibular tori. (a) **"Natural" cross section through the right side of the mandibular torus of a Choukoutien pithecanthropine (natural size).** (b) **Cross section through a particularly large mandibular torus of a recent prehistoric Chinese from Kansu Province (natural size).**
Source: Modified from F. Weidenreich. 1936. "The Mandibles of *Sinanthropus pekinensis,*" *Palaeontologia Sinica,* **VII** (ser. D, fascicle 3): 52, 55, Figs. 39, 41.

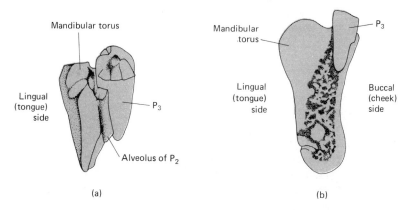

(a)   (b)

in Eskimos. This thickening may be an adaptation to stresses such as those imposed by grinding up coarse food.

The teeth of the Choukoutien pithecanthropines also show some features not seen in the Javan group (see Figure 14-9). The upper incisors are frequently shovel-shaped and have a basal tubercle, or lump, of enamel. In some individuals, the canines project beyond the other teeth and have quite long roots that extend down into the jaw. In addition, the roots of most of the permanent premolars and molars are fused, thus enlarging the pulp cavities of these teeth. This *taurodontism* (as well as shovel-shaped incisors) also appears in some later populations, including north African pithecanthropines and European neanderthals, as well as in modern Eskimos, American Indians, and South African Bushmen. The slightly projecting canine could be interpreted as a characteristic of pongid heritage, though this explanation would be suspect because there is no trace of the diastema, or gap, that is part of the same functional complex in apes. Furthermore, it is more efficient to look for a single common explanation that might underlie all these apparently separate traits. One simple functional interpretation comes to mind: all these features—shovel-shaped incisors, slightly elongated canines, and taurodont cheek teeth—increase the mass of the tooth crowns. For this reason they would lengthen the functional life of the teeth without increasing the area they occupied in the dental arch. In fact, most of the teeth of the Choukoutien pithecanthropines are smaller in surface area than those of their Javan counterparts and so would not require large jaws to hold them. Nearly all the Choukoutien teeth fall within the range of tooth size for modern populations.

The skeletons of the Choukoutien population are much like those of modern humans. Males are estimated to have been 156.2 cm (5 ft 1½ in) tall, and females to have been 143.5 cm (4 ft 8½ in) tall. This is within the low end of the range of stature for living humans. Their bones differ little from those of present human populations except that they are more robust. In particular, the long bones have thicker outer walls of compact bone and more pronounced roughened areas for muscle attachment. The overall picture that we can reconstruct ties together biology and behavior. We have here a population whose members were not large, but were sturdily built for their size, probably because they led an active hunting-and-gathering life.

Earlier Chinese Pithecanthropines.   Some earlier Chinese pithecanthropines are known from the vicinity of Lantian, in Shensi province (Aigner and Laughlin, 1973). The fact that only two individuals were found makes this an extremely small sample to work from; moreover, the two are drawn from populations that were separated in time by roughly 400,000 years. The earlier find is a partial cranium of what is probably a female over 30 years of age. The low forehead, pronounced postorbital constriction, and thick cranial walls mark her as a member of a primitive

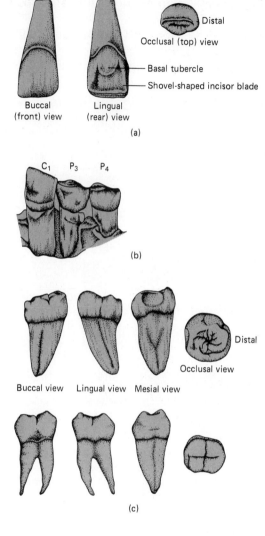

FIGURE 14-9 **Dental characteristics of Chinese middle Pleistocene hominids.** (a) **Left upper central incisor of** *Sinanthropus* **1, showing shoveling and basal tubercule.** (b) *Sinanthropus* **G 1, left lower jaw, showing height of worn crown of canine.** (c) *Above,* **right lower molar of** *Sinanthropus* **45, showing fusion, of the roots (taurodontism);** *below,* **right lower molar of recent human. (All natural size.)**

Source: Modified from F. Weidenreich. 1937. "The Dentition of *Sinanthropus pekinensis:* A Comparative Odontography of the Hominids," *Palaeontologia Sinica*, n.s. **D**(1) (whole series no. 101). Plate 1, fig. 1; plate VI, fig. 54; plate XIX, figs. 161, 162.

pithecanthropine population. The skull has a cranial capacity of about 780 cc (Table 14-4), which is lower than that of the Choukoutien population and entirely consistent with earlier chronological age. The lower jaw from Lantian, found in deposits dated to about 300,000 years ago, is closer in time to Choukoutien. It belonged to an adult of advanced age who lacked third molars. X-rays of the fossil reveal no tooth germs for these teeth, suggesting congenital absence. One explanation for this absence is provided by the appearance of the jaw itself; there is simply not enough space behind the second molars for the third molars to have fit. Congenitally missing third molars occur as uncommon variants in living human populations, and the trait reaches its highest frequency in some Asian populations with flattened, shallow faces and relatively short mandibles. There is thus a suggestion of continuity between the Lantian pithecanthropines and modern Mongoloid populations.

African Pithecanthropines

These, like their counterparts in Java and China, existed over a very long period of time. From present evidence, in fact, pithecanthropines have a greater time span in Africa than anywhere else. Among the earliest and most primitive of the African pithecanthropines is a population represented by a fragmentary skullcap, lacking both face and base, from Olduvai Gorge (see Figure 14-10). This specimen, Olduvai hominid 9, came from about 5 to 6 m (15 to 20 ft) below the top of Bed II. Faunal correlation with other sites dated by the potassium-argon method suggests an age that may be as great as 1 to 1½ million years ago, not very much later than the more recent date calculated for KNM-ER 1470 (see "The Significance of KNM-ER 1470 for Models of Early Pleistocene Hominid Evolution," page 454).

Somewhat later in time come three or four individuals from a population that lived at Ternifine in Algeria during the early middle Pleistocene. The scattering of bones—a right parietal of an immature individual and three lower jaws—documents the existence in this region of humans whose cranial and dental traits overlap those of pithecanthropine populations elsewhere. Three other sites have been found in

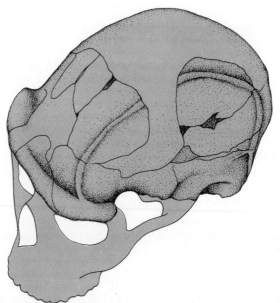

FIGURE 14-10 Olduvai hominid 9, an African pithecanthropine (about ⅓ natural size).

Source: Redrawn from reconstruction made by Wenner-Gren Foundation.

Morocco. The first two, Littorina cave and Smuggler's cave, contain only jaw fragments from populations believed to be comparable in age to the Riss glaciation in Europe. The third site, Rabat, is probably later—perhaps equivalent to the Riss-Würm interglacial. It has yielded a lower jaw and associated pieces of a cranium and upper jaw that were blasted out of a quarry. Probably part of a skull intact until that moment, they are now too fragmentary to be reassembled into a braincase. The teeth, however, show shoveling of the incisors and other features that link this north African population to other pithecanthropine groups.

European Pithecanthropines

Known remains of European pithecanthropines are fewer and are restricted to a narrower time range than the African and Asian specimens. It is difficult to estimate their age with any degree of accuracy; about all that can be managed at present is to place them within the European glacial chronology. The thin line of European hominid evolution begins with the fragment of a molar found at Přezletice near Prague, Czechoslovakia, in sediments that probably come from the Günz-Mindel interglacial (Fejfar, 1969). An occipital bone of an adult male from Vértesszöllös, Hungary, continues the trace. The Vértesszöllös specimen seems to represent a population that lived a bit later, around the middle or latter part of the Mindel glacial period. Reconstructions of the endocranial volume of this specimen range from about 1400 to about 1600 cc, which would be extremely large for a pithecanthropine. Contemporaneous with this or a bit earlier is the better-known fossil from the Mauer sands near Heidelberg (Figure 14-11). This is a complete

FIGURE 14-11   Heidelberg mandible, representing European pithecanthropines: (a) top view; (b) side view.

Source: Photographs of casts, University of Pennsylvania.

(a)

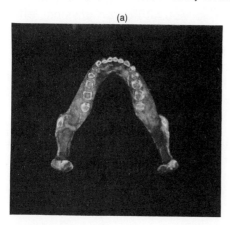

(b)

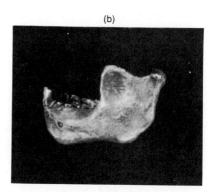

and quite robust jaw. The horizontal parts (referred to as *rami*, or branches) that hold the teeth are thick, and the ascending rami that hinge with the base of the skull are quite wide and low. Contrasting with the heavy bone structure is the relatively small size of the teeth, which are within the upper end of the range of living humans and match rather closely those of Australian Aborigines. Only the molars are noticeably larger, falling within the size range of the Choukoutien pithecanthropines and sharing with them the moderately taurodont pulp cavities. The incisors show just a trace of robustness, being swollen a bit at the base of the inner margin.

The most significant recent discovery of what may be a European pithecanthropine was made in 1959 by several amateurs at Petralona in Greece. The find is a well-preserved skull, minus its lower jaw. At present there are widely varying estimates of its cranial capacity (from 1220 to 1440 cc; see Howells, 1973) and age (from pre-Mindel to early Würm). Until definitive studies are published, this is yet another fossil that can be fitted into several quite different interpretations of the relative rates of human evolution in different areas.

MATERIAL CULTURE AND WAY OF LIFE

New Tools, New Techniques

We can never see our ancestors of hundreds of thousands of years ago in their everyday activities of hunting, eating, mating, caring for children, and playing. However, the broad outlines—and occasionally tantalizing details—of these categories of behavior can be reconstructed from the tools and bone fragments left behind. From these it seems that the middle Pleistocene was a time of increasing technological diversity, with new tools being made and used side by side with some that had served well over long periods in the past. At least as important as the new tools themselves are the new and more efficient techniques that were used to make them. Crafts began to substitute for physical effort, and behavioral diversity helped to open new areas of occupation, with new challenges to be met by the adaptive processes.

There is continuity from the early to the middle Pleistocene in traditions of stoneworking. Oldowan-type tools, first associated with australopithecines, are also found with the remains of pithecanthropines. These associations can be documented from Olduvai Gorge in east Africa, which provides the world's longest continuous record of hominid cultural and biological evolution. Pebble-based choppers, scrapers, and other tools of the Oldowan tradition are known from the earliest (lower Bed I) levels at this site, from which their name is derived. As shown in Table 14-5, several other basic tool forms continue with little or no

TABLE 14-5 Tools of the Oldowan tradition

Tools	Bed I			Bed II		
	Lower	Middle	Upper	Lower	Middle	Upper
	Oldowan			Developed Oldowan A	Early Acheulean / Developed Oldowan B	
Choppers (T)						
Discoids (T)						
Scrapers (T)						
Hammerstones (U)						
Light-duty flakes and other fragments (U)						
Debitage						
Cobblestones, nodules, and blocks (U)						
Subspheroids (T)						
Polyhedrons (T)						
Anvils (U)						
Burins (T)						
Manuports						
Spheroids (T)						
Protobifaces (T)						
Awls						
Modified battered nodules and blocks (T)						
Bifaces (T)						
Heavy-duty flakes (U)						
Outils écaillés (flaked tools) (T)						
Laterally trimmed flakes (T)						

Note: T = tool; U= utilized material.

Source: Based on M. D. Leakey. 1971. *Olduvai Gorge*. Vol. 3. Cambridge, England: Cambridge University Press. Table 1, p. 3, and information in the text.

change for over a million years. But the Oldowan tool kit was not static; as early as middle Bed I a new class of tools, polyhedrons, was added. These are angular stones with three or more working edges, easier to describe than to guess a specific function for. In upper Bed I two new items swell the cultural inventory: spheroids and protobifaces. Spheroids are stone balls that may have served as hammers or missiles. They are much more smoothly rounded than the subspheroids from earlier levels. Protobifaces are made from cobbles, as can be seen from their thick, unworked butts; however, the remainder of the stone has been flaked from both sides to produce two rough cutting edges that taper to a sharp tip. These cutting tools seem to be intermediate between choppers and the true bifaces described below; they could represent crude attempts to make a hand-held ax. Protobifaces are significant because they mark the beginning of a technological trend that later becomes dominant. Awls also show up in lower Bed II. These stone tools are made from light-duty flakes deeply notched at two places along one edge to leave a short, pointed projection. The points are often blunted by use, and some have been snapped off. Awls may mark the further elaboration of leatherworking. Hides could be stripped from a carcass with a chopper and cleaned with a scraper. The awls could be used to punch holes in the prepared skins, making it possible to join them together for use as shelter or clothing.

By the middle of lower Bed II, the original Oldowan tool kit had become elaborated to a point where, according to Mary Leakey, it becomes possible to recognize a new tradition, *Developed Oldowan A.* This tradition is marked by the addition of modified battered nodules and blocks, which are angular fragments of irregular form. There is also a great increase in the number of smoother subspheroids and spheroids and a variety of light-duty tools. Overall, the diversity of the Developed Oldowan A tool kit is greater than that of the earlier Oldowan from which it continued. *Developed Oldowan B* first appears in the upper part of middle Bed II and continues into upper Bed II; this assemblage, or set of tools usually found together, is for the most part the same as the Developed Oldowan A tool kit but includes a few true bifacial tools, those with two cutting edges. Although these first bifaces are crudely made and variable in form, they are important because bifacial tools later became very much more common and regular in shape. From the upper part of middle Bed II also come the first tools that are classed as *Acheulean;* this term comes from the site of Saint-Acheul in the Somme Valley of France, where tool kits containing bifaces were first found. To be classed as Acheulean, at least 40 percent of the tools must be bifaces (Figure 14-12). Like those of Developed Oldowan B, these show considerable variation in design. The basic biface pattern is that of a symmetrical stone with two regularly flaked cutting edges that taper to form a sharp point at one end; at the other they are battered down to form a blunt base so that they can be held in the hand and used safely—

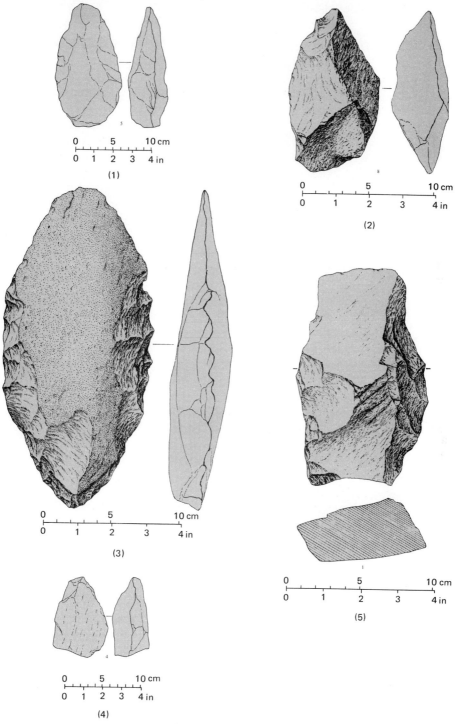

FIGURE 14-12 Acheulean bifaces: (1) **irregular ovate,** (2) **trihedral,** (3) **double-pointed,** (4) **flat or square-butted,** (5) **cleaver.** (Continued on page 522.)

Source: M. D. Leakey. 1971. *Olduvai Gorge.* Vol. 3. Cambridge, England: Cambridge University Press. Figs. 61-5, 100-8, 63, 61-4, 66-1, 67, 101.

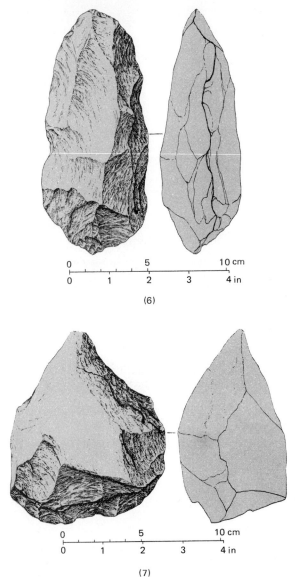

FIGURE 14-12 (continued)   **Acheulean bifaces: (6) oblong pick, (7) heavy-duty pick.**

hence the older term *hand axe*. Like choppers and scrapers, Acheulean hand axes were made in about half a dozen common forms, from cleavers suitable for hacking through sinewy meat and bone to picks used to grub roots or insects from the ground.

Assemblages of the Developed Oldowan B tradition occur at Olduvai in sites later than those containing Acheulean tool kits. Since the

FIGURE 14-13   Distribution of Acheulean and Oldowan assemblages in Africa and Asia.
Source: Modified from H. L. Movius. 1948. "The Lower Paleolithic Cultures of Southern and Eastern Asia," *Transactions of the American Philosophical Society*, n.s. **38**, part 4. Map on p. 409.

Acheulean tradition became a dominant one in other parts of the world (Figure 14-13), some anthropologists have suggested that the late Oldowan B sites were technological backwaters. This is a plausible explanation, but not the only one, as indicated by an analogy from our own time. During World War II United States forces in Italy used asses and mules to haul field artillery pieces and supply wagons. This primitive means of transportation was used not because trucks were unknown but because these more advanced vehicles could not negotiate the narrow roads. The old ways—under certain conditions—were superior to the new. The Acheulean and Oldowan B tool kits may also have been used by the same groups at different times or in different places (Acheulean sites are concentrated in the eastern part of Olduvai Gorge); the components of the tool kits might have varied with the resources being exploited. There is at least good independent evidence, for example, that different prey animals were taken at different seasons (Speth and Davis, 1976) and that different hominid bands at the same level of biological evolution may have had little contact with one another. It is even possible that the different tool kits may have belonged to distinct species, though there is no compelling evidence for this interpretation.

Tools of the various Oldowan traditions may have been preferable for some jobs, but those of the Acheulean had some real technological advantages. Archeologist Karl Butzer (1971) has shown that Acheulean tools have far sharper edges than Oldowan artifacts. Furthermore, Acheulean artisans stretched 20 cm (7 in) of cutting edge from 0.5 kg (1 lb) of stone, about four times as much as Oldowan toolmakers. They did this by preparing a stone block carefully so that each flake knocked off it could be used as a tool with little or no further working—and consequently very little waste. Part of the greater efficiency was made possible by the selection of stones that would flake more predictably. Another part may have been due to the greater creativity, learning ability, and manual dexterity of the workers themselves. One indication of the potential long-term interplay of cultural and biological evolution is that brain size approximately doubled between the australopithecine and the pithecanthropine stages, though toolmaking would have provided only part of the impetus.

Olduvai has also yielded modified animal bones; at least 105 specimens from Bed II show modification. There are parts of long bones, some of the larger ones split lengthwise and flaked and worn on the broken ends; shoulder blades with chipped and battered edges; massive kneecaps and foot bones, usually of elephants, with battered and pitted surfaces, fragments from long bone shafts with pointed or spatula-shaped ends; canines and incisors from hippos and pigs showing chipping at the tips; bifacially flaked fragments; and crania and limb bones showing depressed fractures (see Figure 14-14). These flaked and battered bones are important for two reasons: (1) as evidence that technological diversity at this stage of evolution extended to the use of different materials as well as to the manufacture of a wide variety of products and (2) as firm evidence at this later time for what remains debatable at the earlier South African sites—boneworking by early hominids.

The general level of cultural evolution seen at middle Pleistocene sites of Olduvai was duplicated by pithecanthropines elsewhere. Torralba, in Spain, is a classical Acheulean kill site at least 300,000 years

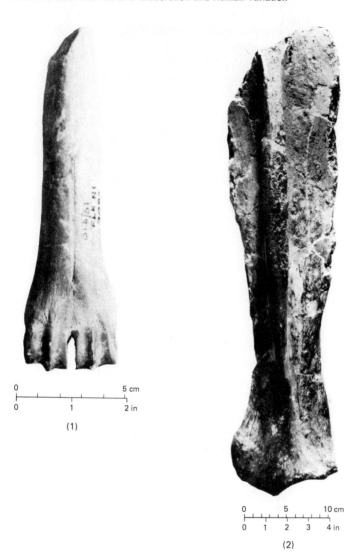

FIGURE 14-14 **Utilized bones from Olduvai Gorge. (1) Long bone (antelope metacarpal) worn on broken ends. (2) Shoulder blade (giraffe) damaged by use.** (Continued on pages 526–527.)

old. It was used only intermittently, possibly when large mammals made their seasonal migrations through the river valley. At Torralba, deer, horses, rhinos, and elephants were hunted, killed, and cut up. Characteristic Acheulean hand axes were left behind with their bones. Also left was a material that must have been used widely but seldom survives—wood, used here as stakes and perhaps as handles for weapons.

0   5 cm
0   1   2 in

(3)

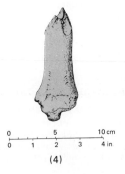

0   5   10 cm
0 1 2 3 4 in

(4)

cm in
10   4

     3

5   2

     1

0   0

(5)

0   5   10 cm
0   1   2   3   4 in

(6)

0   5   10 cm
0   1   2   3   4 in

(7)

FIGURE 14-14 (continued) **Utilized bones from Olduvai Gorge. (3) Massive bone (elephant axis), battered, possibly from being used as an anvil. (4) Fragment of long bone shaft (bovid tibia) flaked to a point. (5) Hippopotamus incisor damaged at tip. (6) Bifacially flaked bone fragment (hippopotamus tibia). (7) Cranium of antelope with depressed fracture.**

Source: M. D. Leakey. 1971. *Olduvai Gorge.* Vol. 3. Cambridge, England: University of Cambridge Press. Plates 35, 37, 38, 40, 41; fig. 109, p. 237; fig. 110-1, p. 238.

Control of Fire

Wood was also used as a fuel. At Torralba charcoal is not concentrated in a few places, as would be expected for hearths, but is scattered widely. Perhaps these early humans fired the brush and grass to drive big game animals to places where they could be killed with considerably less effort and danger than when they were in organized herds.

Middle Pleistocene dwellings have been found by de Lumley (1969) at Terra Amata, near Nice, on the Mediterranean coast of France (Figure 14-15). Though this is a balmy resort area today, it was cold 300,000 years ago, when pithecanthropines lived there. They took shelter in huts, each about as big as a good-sized modern living room, from 4 by 6 cm (12 by 20 ft) to over twice as large. As in the structures at Olduvai, walls were built on poles driven into the ground, with the bases buttressed by piles of stones. The larger Terra Amata structures show a new architectural feature: a row of heavier posts up the middle, probably to support a roof. This roof may have had a small hole somewhere along its length to let smoke escape, since there were hearths inside for heating and cooking. The floors near the fire are clear of debris and were probably sleeping areas. Skins covered parts of the dirt floor and left outlines that can still be seen.

Hearths also appear in other areas, such as northern China. Choukoutien was at one of the northern fringes of the pithecanthropine range, which had widened far beyond the tropical and warm temperate regions to which australopithecines had been limited. At Choukoutien, fire, besides providing warmth, also gave protection against large predatory animals such as bears. Chinese pithecanthropines also used fire for cooking: the thick layers of ash left behind contain great quantities of burned animal bones. Seventy percent of these were of fallow deer (a species now limited to warmer and wetter climates than that of present-day Peking); the remainder run the faunal range from fish and frogs to camels, buffalo, and other humans. Choukoutien pithecanthropines were cannibals. Whether this was for nutritional or ritual reasons is not known, nor is it known whether those eaten were group members or captives from other bands. Bones of individuals of both sexes and all ages have been found in the remains of cooking fires. The survival of hackberry seeds and other plant materials among the debris shows that the pithecanthropine diet was balanced as well as exotic.

Regardless of what animal or vegetable foods went into the fire, all came out softer, since heat softens fibers and proteins. Cooked foods would have required less chewing, decreasing wear on the teeth and allowing a reduction in the size of the jaw muscles. Individuals with smaller-than-average teeth and jaws would therefore have had a slight

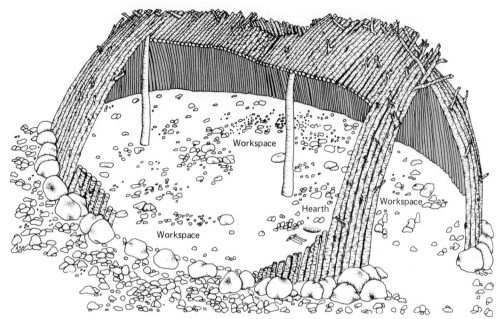

FIGURE 14-15   Structure at Terra Amata. Oval huts, ranging from 8 to 15 m (26 to 49 ft) in length and from 4 to 6 m (13 to 20 ft) in width, were built at Terra Amata by visiting hunters. A reconstruction shows that the walls were made of stakes, about 8 cm (3 in) in diameter, set as a palisade in the sand and braced on the outside by a ring of stones. Some larger posts were set up along the long axes of the huts, but how these and the walls were joined is unknown; the form shown is conjectural. Hearths were protected from drafts by a small pebble windscreen.

Source: Henry de Lumley. 1969. "A Paleolithic Camp at Nice," *Scientific American*, **220**(5):42–50.

advantage, especially when food was in short supply. The same expansion into colder regions that made the control of fire advantageous and led to a reduction in the size of cheek teeth may also have set in motion selective forces that brought about increases in the size of anterior teeth. In contemporary northern hunting-and-gathering groups such as Eskimos, women prepare scraped hides by chewing them. This crude tanning process makes the skins soft, but it wears down the incisors rapidly. Any dental variation that increased the mass of the anterior teeth could thus have conferred a selective advantage on its possessor.

The technological capabilities of the middle Pleistocene pithecanthropines were considerably beyond those of australopithecines, and the tempo of cultural evolution had accelerated. Human populations expanded into new areas with new ecological challenges, and in responding to them they encountered new sets of selective pressures. The result was increased interaction between cultural and biological evolution—a situation that has continued to our own time.

RECONSTRUCTING APPEARANCE:
INTERACTION OF CULTURE AND BIOLOGY

Like pithecanthropines, human populations today differ in some skeletal characteristics. In living humans, however, skeletal variations are dwarfed by the many variations in soft tissues that leave little or no trace in the fossil record: differences in blood groups and serum proteins, skin color, hair color and texture, eye color, and so on. It is reasonable to assume that there was also some variation in these characteristics at earlier stages in human evolution, but reconstructing it is another matter. Some hold that any reconstruction would be pure speculation and therefore of no value. Others are not so sure. If we operate on the assumption that changes in morphological characteristics are approximately proportional to time, then characteristics of populations between the presumed beginning stage in an evolutionary sequence and the present condition can be interpolated. For example, humans now have little visible body hair, while modern pongids (less-changed descendants of our dryopithecine predecessors of the Miocene) have a lot. Should it be assumed that hominids intermediate in time had half as much hair as modern apes and twice as much as modern humans? Should the pithecanthropines of about a million years ago have had, in proportion to time elapsed, half as much hair as their early Pleistocene ancestors? Possibly, but not necessarily; some uncertainty must remain because the assumption that evolutionary change is proportional to time does not always hold true. We know that for skeletal characteristics, different character complexes evolve at different rates, producing mosaic evolution. Even greater problems exist for soft tissues, since beginning points are often not clear. What of characteristics such as skin color, for which modern humans show a considerable range, from light to dark? Was the ancestral condition some intermediate shade, or was it one that was beyond either of the present extremes?

Another approach to reconstructing characteristics that leave no direct trace in the fossil record is through genetic theory. With knowledge of a trait's mode of inheritance and how it is acted on by the forces of evolution, we can estimate its rate of evolution. Combining this estimate with evidence from the study of living primates and the record of hominid cultural evolution can give us a less speculative view of our ancestors' appearance. An example of this type of approach was provided several years ago in Livingstone's (1969) work on skin color. Skin color is a polygenic characteristic, and the bulk of observed variation in present human populations can be explained in terms of genetic combinations at four unlinked loci. Assuming no dominance at any of these loci, just a 6 percent difference in fitness value between the optimum genotype and those selected against most strongly can produce the entire range of differences in human skin color in just 800 generations. With dominance, the rate of evolution would be slower, but even so, 1500 generations would be enough to bring about the present diversity. Given a human generation length of about 20 years, these models suggest that skin color could be changed from dark to light (or the reverse) in from 16,000 years (with no dominance) to 30,000 years (with dominance). Livingstone found that different rates of mutation from one allele to the other had no detectable effect on the attainment of equilibrium, which suggests that mutation has been of little importance in the evolution of differences in skin color. And since the calculated rates of evolution took place in the face of a theoretical 5 percent gene flow between adjacent populations, isolation was evidently not a prerequisite for substantial genetic or phenotypic differences.

Spans of 16,000 to 30,000 years are quite brief in relation to the known hominid fossil record. But in order to reconstruct the appearance of our ancestors at different times and places, it is not enough to know that characteristics can change rapidly; we also need to know what environmental factors have determined the direction and intensity of selection. Weiner (1973) has summarized the major types of evidence for a tropical origin for the human lineage: (1) the concentration of dryopithecine and Plio-Pleistocene fossil remains in tropical and semi-tropical areas, (2) the archeological evidence for the earliest tools in Africa, (3) the discovery that among all living primates humans are morphologically and genetically most similar to the African apes, and (4) the retention in the human species of a number of physiological characteristics that evidently evolved in response to life in hot climates. These characteristics include the ability to dissipate large amounts of heat, up to 400 or 500 kcal per hour, largely through the evaporation of sweat—as much as 1 to 2 L (about 1.8 to 3.6 pt) per hour. The sweat is produced by about 2 million eccrine glands distributed over the body surface of humans. Other primates have many fewer eccrine glands, and these are confined largely to the palms, soles, and other pads, where they function mainly to provide a better grip. Even the great apes have few eccrine glands scattered over the skin surface. Humans also have special adaptive mechanisms that dilate the blood vessels in the skin, ensuring that a large proportion of the blood pumped by the heart passes beneath the body's surface. In addition to its other functions, then, the blood acts as a coolant, carrying heat generated by metabolic activities inside the body to the skin surface, where it is lost as sweat evaporates. With physiological acclimatization to work in hot regions, the volume of blood increases because of retention of both water and salt. These physiological characteristics, which were discussed in Chapter 8 as adaptive mechanisms to life in the tropics, have persisted in all human populations.

One other important biological adaptation to life in equatorial regions is a dark-colored body surface, which results from the presence of a high density of melanin granules. Human populations long native to equatorial regions have dark skins; populations in more temperate regions have lighter skins. Weiner reasoned that since most human physiological characteristics favor a tropical origin, the lighter skin of populations in higher latitudes is a secondary modification of a darker-skinned ancestral condition. But when did some human populations become light-skinned, and how long were their ancestors dark? In answering the second part of the question, anthropologists can use hominoid primates now living in equatorial areas as a starting point. Gorillas have dark skin and hair; chimpanzees, though polymorphic for skin color, also have dark hair. The melanin in these coverings protects the individual against the harmful effects of ultraviolet radiation from whatever sunlight filters through the forest vegetation. With a shift to life in more open habitats, the heat load from ultraviolet radiation would be increased, and the covering of hair would be an advantage. Yet in our ancestors the insulating coat of hair common to all other hominoid primates was replaced by a set of mechanisms that ensure a high rate of heat loss. Why? The traditional explanation has been that body hair was lost in connection with hunting. Although most large predators have special sensory abilities that enable them to hunt at night, humans must track and run down their prey, sometimes for hours, in the heat of the sun. Loss of body hair, some believe, would increase the rate of evaporation of sweat under such conditions. Cultural evidence suggests that humans were hunting large mammals on a regular basis by at least the early middle Pleistocene, and quite possibly much earlier. Selection for reduction in the hair coat could have taken place relatively rapidly under such circumstances. But Newman (1970) has stressed that a hair coat does not interfere with the evaporation of sweat and has suggested that our ancestors' hair covering may have been lost even before they began hunting.

If Livingstone's calculations concerning the rate of change in skin color in response to selection are not totally wrong, then the human populations that first evolved in tropical regions could well have been dark-skinned within a few tens of thousands of years after losing the bulk of their outer covering of hair. As hominids ranged beyond tropical regions, the fitness values of different genotypes would have been reversed in some populations. This would have occurred because too little ultraviolet radiation is as harmful as too much, though for a different reason. Some ultraviolet radiation is needed by our bodies to manufacture vitamin D, which is necessary if calcium is to be absorbed from our intestines and deposited in the bones of growing skeletons. Insufficient vitamin D produces bowed legs, inward curving of the knees (knock-knee), and twisted spines—the symptoms of the disease known as *rickets*.

Loomis (1967) estimated that hominid populations expanding out of the tropics would not have had difficulty obtaining enough vitamin D until they extended their range north of the Mediterranean Sea and latitude 40°N, where most of the needed ultraviolet radiation would have been removed from the sun's rays as it was filtered through the atmosphere. On the basis of this argument, which is buttressed by much observational and experimental evidence, Loomis felt it was probable that early hominids inhabiting western Europe had lost much of the pigmentation in their hair and skin by half a million years ago.

The expansion into more northern areas of Asia and Europe would very likely not have been possible without cultural innovations such as fire and the manufacture of clothing from skins with a variety of stone tools. Here, then, is a case in which cultural evolution allowed an expansion of human populations into areas where entirely new patterns of selection were encountered—patterns that led to long-term evolutionary changes in the human gene pool. Even this single example shows that, by combining our extensive knowledge of modern human biology with the tools of population genetics, it is possible to reconstruct a more complete picture of our ancestors' appearance and way of life in the distant past.

SUMMARY

1. The first middle Pleistocene hominid remains, discovered during the 1890s on the island of Java, show a mosaic of features combining a postcranial skeleton like that of modern humans with a heavy face and a thick, low-crowned skull that held a brain intermediate in size between those of australopithecines and modern humans.

2. Pithecanthropines inhabited a geographic area wider than that known to have been occupied by early Pleistocene hominids; they ranged over cooler regions such as northern China and Europe, as well as over Africa and the warmer parts of Asia. Geographic influences probably caused some of the skeletal variation seen at this stage of hominid evolution, although increased cultural capabilities are likely to have produced some regional differences as well; for example, some reduction in tooth size may reflect reduced selection pressures made possible by the use of stone tools and fire in food preparation.

3. The material culture of pithecanthropines was more advanced and complex than that of early Pleistocene hominids; a greater diversity of stone tools, the working of wood and bone, and control of fire all stand out as significant technological achievements.

4. The technological accomplishments that allowed middle Pleistocene hominids to inhabit a wider area also exposed them to new selective factors that probably changed the frequencies of genes for phenotypic characteristics, such as skin color, that leave no direct fossil record.

SUGGESTIONS FOR ADDITIONAL READING

Coon, C. S. 1962. *The Origin of Races.* New York: Knopf. (Chapter 9, "Pithe-canthropus and the Australoids," and Chapter 10, "Sinanthropus and the Mongoloids," contain useful surveys of the material evidence for middle Pleistocene hominid evolution.)

Shapiro, H. L. 1974. *Peking Man.* New York: Simon and Schuster. (This is a discussion of the discovery and loss of the important middle Pleistocene fossil remains from Peking, well told by an experienced physical anthropologist.)

Tuttle, R. H. (eds). 1975. *Paleoanthropology: Morphology and Paleoecology.* The Hague, Netherlands: Mouton. (Section 5 of this volume contains five up-to-date articles on Asian middle Pleistocene hominids.)

REFERENCES

Aigner, J. S., and W. S. Laughlin. 1973. "The Dating of Lantian Man and His Significance for Analyzing Trends in Human Evolution," *American Journal of Physical Anthropology,* **39**(1):97–109.

Boule, M., and H. V. Vallois. 1957. *Fossil Men.* New York: Dryden.

Brodrick, A. H. 1964. *Man and His Ancestry.* New York: Fawcett.

Butzer, K. W. 1971. *Environment and Archeology* (2d ed.). Chicago: Aldine.

Clark, W. E. Le Gros. 1964. *The Fossil Evidence for Human Evolution* (2d ed.). Chicago: University of Chicago Press.

Fejfar, O. 1969. "Human Remains from the Early Pleistocene in Czechoslovakia," *Current Anthropology,* **10**(2–3):170–173.

Howells, W. W. 1973. *Evolution of the Genus* Homo. Reading, Mass.: Addison-Wesley.

Leakey, M. D. 1971. *Olduvai Gorge.* Vol. 3. Cambridge, England: Cambridge University Press.

Livingstone, F. B. 1969. "Polygenic Models for the Evolution of Human Skin Color Differences," *Human Biology,* **41**(4):480–493.

Loomis, W. F. 1967. "Skin-Pigment Regulation of Vitamin-D Biosynthesis in Man," *Science,* **157**(3788):501–506.

de Lumley, H. 1969. "A Paleolithic Camp at Nice," *Scientific American,* **220**(5): 42–50.

Newman, R. W. 1970. "Why Man Is Such a Sweaty and Thirsty Naked Animal: A Speculative Review," *Human Biology,* **42**(1):12–27.

Speth, J. D., and D. D. Davis. 1976. "Seasonal Variability in Early Hominid Predation," *Science,* **192**(4238):441–445.

Weiner, J. D. 1973. *The Tropical Origins of Man.* Addison-Wesley Module in Anthropology, no. 44. Reading, Mass.: Addison-Wesley.

Chapter 15

Origin of Modern Humans

The pithecanthropine stage of human evolution is represented by relatively abundant remains of fossils from all over the world and from a long span of time. These various populations of pithecanthropines were not identical everywhere—neither, for that matter, are all populations of modern humans. But everywhere the general appearance of middle Pleistocene hominids was substantially the same, and over the whole area they occupied the same ecological niche, that of a large-brained, bipedal carnivore which hunted in groups with stone weapons and fire. This niche represents a refinement of the basic way of life of the Plio-Pleistocene hominids, at least some of whom also lived in groups and hunted with stone tools.

Beginning with the Mindel-Riss (M-R, or second) interglacial, hominid fossils become relatively more scarce, a situation that persists through the Riss glacial period. With the coming of the third interglacial (between Riss and Würm), fossil remains are again found in much greater abundance; and the more abundant remains, at least in Europe, are those of neanderthals. Neanderthals seem to be particularly debatable hominids. Anthropologists argue about their toolmaking abilities, modes of thought, and ability to talk, and even about the spelling of their name (-*thal,* from German *Thal*, "valley," is now often spelled -*tal* as a result of changes in German orthography). From the standpoint of human evolution, however, the most important disagreement is over the role of neanderthals in human evolution: were they direct ancestors that lived all over the Old World for a time, local variants that interbred to a limited extent with more modern-looking contemporaries, or an extinct side branch that contributed no alleles to our gene pools? Because of the controversies surrounding the place of the neanderthals in hominid evolution, the small sample of their Mindel-Riss and Riss predecessors takes on an importance that may be out of all proportion to its actual size.

TRANSITIONS: SWANSCOMBE, STEINHEIM, AND TAUTAVEL

Fossil remains of hominids from the Mindel-Riss interglacial are limited to two major finds. Both were discovered decades ago, and both have been debated and discussed ever since.

Swanscombe is the less complete of the two—at present. That qualification is needed because the story of this discovery is one of astoundingly good luck that perhaps has not yet run out. On June 29, 1935, the English prehistorian A. T. Marston discovered the occipital and left parietal bones (the back and one side of the skull) of a hominid in the Barnfield gravel pit, 732 m (800 yd) north of the Swanscombe church in Kent. With the hominid bones were those of a variety of mammals. Those of the wolf, lion, horse, fallow and red deer, and hare have modern counterparts. Others, including a straight-tusked elephant, a Merck's rhinoceros, a giant deer, and a giant ox, do not. Acheulean hand axes and flake tools were found with the fossils, and the whole complex strongly suggests a Mindel-Riss date.

In 1955, a group visiting the site found another part of a skull vault 23 m (75 ft) away. Almost unbelievably, this was a right parietal that fit perfectly with the two bones found 20 years earlier. Enough of the skull vault is present to permit a reconstruction of the endocranial volume, which is estimated to be about 1300 cc, a value that could fit into any grade from pithecanthropine to modern human. This is not surprising, since these grades are believed by many to be time-successive stages of a single continuous lineage. Anthropologists would like to know just where in our lineage the Swanscombe population belongs, but as you know by now, that is not an easy thing to determine from just one specimen. A value of 1300 cc is low, but not uncommon among modern humans. The same value is at the high end of the pithecanthropine range, but not impossible to accommodate within it. The sex of the one individual known is important in making a phylogenetic judgment, since female human skulls are on the average smaller in volume than those of males. Because the nuchal area of the Swanscombe occipital shows no excessively heavy ridges, most anthropologists who have studied the skull say it was that of a female.

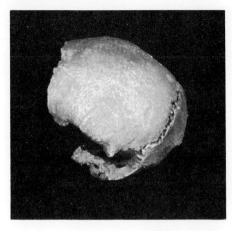

FIGURE 15-1 Swanscombe skull fragments, side view.
Source: Cast, University of Pennsylvania Museum.

Some other features tell us that the population was not completely modern. The parietals are thick, a primitive feature found in earlier pithecanthropine skulls. Further, the skull has its greatest breadth low and to the rear (see Figure 15-1), a feature common to skulls of some later European populations, including neanderthals. Unfortunately, the parts of the Swanscombe specimen that are missing are at least as important as those we have. Because there is no face or forehead, Swanscombe can be reconstructed to fit almost any conceivable form, from archaic to advanced. Few such possibilites have been ignored, but new excavations in the area could still turn up the debated parts and settle the issue.

Steinheim, near Stuttgart, Germany, is the location of a Mindel-Riss hominid skull more complete than Swanscombe, though unfortunately unaccompanied by any tools. The skull has been crushed sideways and is distorted and damaged as a result. Part of the left side of the face is missing, but enough is present on the right to reconstruct the whole with a high degree of confidence. The face of the Steinheim specimen projects ahead of the skull vault less than that of most pithecanthropines. It has a wide nasal opening and a small *canine fossa*. The canine fossa is a groove on the front part of the upper jaw, beneath the eye socket; it is a polymorphic feature usually present in modern humans and more rarely found in earlier populations. The third molars are also quite small, but these teeth are so variable in size and occurrence that it is difficult to base any reliable conclusions on them. Above the face the brow ridges are very thick over each eye and are only slightly thinner over the bridge of the nose. Behind this shelf of bone, the forehead slopes back sharply to a low vault. At the rear of the skull, the occipital region is rounded and is without a heavy torus. If the latter feature suggests that the Steinheim skull is a female, this is flatly contradicted by the presence of heavy brow ridges, which tip the scales in a male direction (Figure

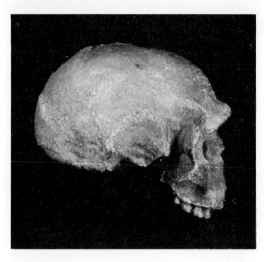

FIGURE 15-2   Steinheim skull.
Source: Cast, University of Pennsylvania Museum.

15-2). Estimates of Steinheim's endocranial volume vary, but most fall in the range of 1200 to 1300 cc, close to that of Swanscombe. Like that British member of a second interglacial human population, Steinheim has been variously classed as everything from a morphological intermediate between pithecanthropines and neanderthals to one of the earliest members of an anatomically modern group.

Until just a few years ago, the list of transitional middle to upper Pleistocene hominids would have ended with Swanscombe and Steinheim. Recently, however, it has been extended substantially with the discovery of a number of Rissian hominids (Table 15-1). Of these, by far the most informative sample comes from Caune de l'Arago at Tautavel, in the Pyrenees of France. Here Henry and Marie Antoinette de Lumley discovered finger bones, loose teeth, parts of two lower jaws, parietal fragments, and the entire face of a hominid from the Riss glacial period (Figure 15-3). In time, these specimens fall between pithecanthropine populations and the upper Pleistocene populations of neanderthals and anatomically more modern humans. The question is: Toward which of these latter groups do they point?

The Tautavel face shows a pronounced supraorbital torus, but this is depressed a bit more at the midline than that of Steinheim, and the forehead does not dip as much behind the brow ridges. The endocranial volume of the Tautavel specimen is said to be larger than Steinheim's, though no numerical values are yet available. All known features mark Tautavel as evolutionarily more advanced than the earlier pithecanthropines, but none of them point decisively toward either neanderthals or their successors. A few other traits, however, do. The eye sockets of the Tautavel face are large and are set even lower than those of neanderthals. Tautavel, like neanderthals, had a deep palate and a robust, projecting

TABLE 15-1 Transitional middle to upper Pleistocene hominids

Geographic location	Site	Material discovered	Number of individuals	Morphological designation	Material culture, if any	Age of site	Dating method used
Europe	England Swanscombe	Occipital and both parietal bones	1	Possibly preneanderthal	Clactonian variety of Acheulean	Mindel-Riss	Faunal correlation
	France Caune de l'Arago (near village of Tautavel)	Cranium, cranial fragments, two mandibles, teeth, and phalanges	About 20	Intermediate between pithecanthropines and neanderthals	Tayacian variety of Levallois technique	Riss	Faunal correlation
	La Chaise, Abri Suard	Cranial and postcranial fragments	16	Intermediate between pithecanthropines and neanderthals	Denticulate variety of Mousterian	Late Riss	Correlation of fauna, material culture, and sediments
	Le Lazaret	Parietal and teeth	≥ 1	?		Late Riss	
	Montmaurin (La Niche cave shaft)	Mandible, four teeth, and one vertebra	≥ 1	Possibly intermediate between pithecanthropines and neanderthals		Mindel-Riss (?)	Faunal and cilmate correlation
	Montmaurin (Coupe-Gorge)	Maxilla	1	?	Pre-Mousterian stone tools	Early Riss	Faunal correlation
	Orgnac-l'Aven	Teeth	1	?		Late Riss	Faunal correlation
	La Rafette	Cranial fragments	?	?	?	Riss (?)	Faunal correlation
	Germany Steinheim	Cranium	1	Intermediate between pithecanthropines and later hominids	None found	Riss I or II (?)	Correlation of fauna found in gravel deposits
	Spain Cova Negra		1	Similar to Swanscombe, Fontéchevade II, and neanderthals of early Würm	Tayacian variety of Levallois technique	Riss III	Faunal correlation

Source: Based on data supplied by E. Trinkaus.

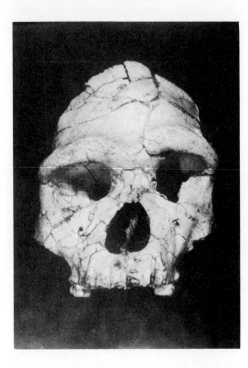

FIGURE 15-3   Tautavel skull fragment.
Source: M. de Lumley. 1975. "Ante-Nean-
derthals in Western Europe," in R. Tuttle
(ed.), *Paleoanthropology: Morphology and
Paleoecology.* The Hague, Netherlands: Mouton. Plate 1.

upper jaw without a canine fossa. On the whole, not only is Tautavel
intermediate in time between pithecanthropines and neanderthals, but
it seems morphologically intermediate as well.

LATER PLEISTOCENE HOMINIDS:
NEANDERTHALS AND THEIR SUCCESSORS

In mentioning the later, more abundant fossil hominid remains of the
third (Riss-Würm) interglacial and the Würm glacial, we also touched
on the neanderthals and the controversy about their place in human
evolution. Who were these hominids? Neanderthals evolved from an-
tecedent pithecanthropine populations: they had skulls visibly different
on the average from those of living humans, and they lived during the
Würm glacial period (and possibly the preceding interglacial) at least
in Europe, with evidently similar and genetically related populations in
the Near East and with other relatives elsewhere.

Although much is known about the appearance and way of life of
neanderthals, many disagreements remain. Were neanderthals preceded
by populations that were more modern in appearance and continued
to coexist with them? What is the phylogenetic significance of the
differences between neanderthals and their successors (including us)?
Was there a wider distribution in time and space of populations like the

European neanderthals? The scientific controversy over neanderthals has actually been shaped by an accident of history. Two men frequented the same valley, millennia apart; one left behind his bones, and the other his name. By this coincidence the label *neanderthal* came to be applied to many later Pleistocene hominids, with a notable impact on ideas about human evolution during this time period.

The Neanderthal Discovery

Until the middle of the nineteenth century, *Neanderthal* was simply the name of a particular valley near Düsseldorf that was a favorite place of the locally revered seventeenth-century composer Joachim Neander. In 1856 workers clearing out a cave in the valley came on a human skeleton. When first discovered, it was probably complete or nearly so, but the workers discarded all but a few of the larger bones and the skullcap. A local schoolmaster named Fuhlrott gathered these together and forwarded them to Professor Schaafhausen in Bonn, who published a monograph describing them in 1858.

These unique remains attracted little attention until 3 years later, when George Busk translated Schaafhausen's monograph into English. Shortly afterward (in 1863) Thomas Huxley, the splendidly successful statesman of British science, examined a cast of the skull and delivered a carefully hedged opinion: the neanderthal specimen was a primitive variety of humans different from sapiens "but by no means so isolated as it appears to be at first" (p. 149). This served as a middle ground from which views diverged in two directions. One sizable body of opinion held that the neanderthal skeleton was that of a recent human who had suffered from some abnormality; thus one Dr. Gibb suggested that the subject had been afflicted with "hypertrophic deformation," a bit of jargon that refers to no medical condition known before or since. Even Rudolph Virchow, the most eminent German pathologist of his day, published a detailed list of the specimen's supposedly pathological characteristics. The other extreme in interpretation, that this skeleton was the first known member of a distinct taxon, was put forth in 1864 by William King, a professor at Queens College, Galway. King suggested that the German fossil represented a new species of our genus: *Homo neanderthalensis.* Not all anthropologists today would accept the degree of genetic distinctiveness suggested by King's taxonomic designation, but most would agree that to King must go the credit for first realizing, or at least having the courage to say, that neanderthal was a normal individual representative of a population that differed on the average from those now living. Confirmation of King's view was not long in coming. In reasonably short order, fossil remains similar to those from the Neanderthal valley turned up elsewhere: in Gibraltar, France, Belgium, and several other places (see Table 15-2 for a list of major neanderthal sites).

TABLE 15-2   Middle Paleolithic hominids

Geographic location	Site	Material discovered	Number of individuals	Age of site
Europe	Belgium			
	Engis	Crania, teeth, and postcranial bones	3	Early Würm
	Fond-de-Forêt	Upper molar tooth and left femur	1	Early Würm
	La Naulette	Mandible, ulna, and right metacarpal	1	Early Würm
	Spy	Two partial skeletons and other postcranial remains	4–5	Early Würm I
	British Isles			
	Saint Brelade	Occipital and teeth	1	Early Würm
	Czechoslovakia			
	Gánovce	Endocranial cast with cranial fragments and postcranial bones	1	Late Riss-Würm
	Kůlna	Cranial fragments of immature individual	1	Lower Würm
	Ochoz	Mandible, parietal, and temporal fragments	1	Early Würm
	Šala	Frontal bone	1	Würm
	Šipka	Mandible	1	Early Würm I
	France			
	Angles sur l'Anglin	Left upper central incisor	1	Würm
	Arcy-sur-Cure:			
	Grotte du Loup	Teeth and parietal fragments	3	Early Würm
	Grotte de l'Hyene	Mandible, maxilla, and teeth	1	Early Würm I
	Grotte du Renne	Teeth	1	Early Würm I
	Castel Merle Abri des Merveilles	Molar tooth	1	Early Würm
	Caminero	Fragmentary skeleton	1	Early Würm
	La Cave	Postcranial skeletal fragments	1	Early Würm
	La Chaise Abri Bourgeois-Delaunay	Cranial fragments, teeth, and postcranial fragments	1	Riss-Würm
	La Chapelle-aux-Saints	Partial skeleton	1	Lower Würm

TABLE 15-2 Continued

Geographic location	Site	Material discovered	Number of individuals	Age of site
Europe	Châteauneuf-sur-Charente	Cranial fragments, mandible, and incisor	3	Early Würm
	Combe-Grenal	Cranial fragments, mandible, and postcranial bones	1	Lower Würm
	La Crouzade	Humerus and phalanges of an immature individual	1	Lower Würm
	La Croze del Dua	Teeth	1	Lower Würm
	La Ferrassie	Skeletons	7 (2 adults and 5 children)	Lower Würm
	Fontéche-vade	Metatarsal	1	Würm (?)
	Fontéche-vade	Frontal fragments and calotte	1	Riss-Würm
	Genay	Cranial fragments and teeth	1	Early Würm I
	Hortus	Teeth and fragmentary bones, including two immature mandibles	38	Lower Würm
	Malarnaud	Mandible and vertebra	1	Riss-Wrm
	Marillac	Mandible, parietal, and occipital fragments	1	Lower Würm
	La Masque	Cranial and postcranial fragments of an immature individual	1	Early Würm I (?)
	Monsempron	Individual skull fragments	7	Early Würm I
	Le Moustier	Partial immature skeletons	1	Lower Würm
	Pech de l'Azé	Immature skull	1	Early Würm
	Le Petit-Puymoyen	Skull and postcranial fragments, many immature	> 6	Lower Würm
	Placard	Teeth	1	Lower Würm
	Portel	Teeth	?	Lower Würm
	Grotte Putride	Phalanx	1	Early Würm (?)
	La Quina	Cranial and postcranial material	> 22	Early Würm
	Régourdou	Mandible and postcranial fragments	1	Early Würm I
	René Simard	Teeth and postcranial fragments of immature individuals	3	Early Würm

(Continued)

TABLE 15-2   Continued

Geographic location	Site	Material discovered	Number of individuals	Age of site
Europe	Rigabe	Deciduous upper central incisor tooth	1	Early Würm I
	Rivaux	Molar tooth	1	Lower Würm
	Roc du Marsal	Immature skeletal fragments	1	Early Würm
	Soulabe-Les-Maretas	Teeth	1	Riss-Würm or early Würm I
	Vergisson	Teeth	1	Lower Würm
Germany				
	Neanderthal	Calotte and postcranial bones	1	Undated
	Neuessing	Right deciduous central incisor	1	Early Würm
	Salzgitter-Lebenstedt	Occipital and parietal	1	Early Würm
	Taubach	Teeth	1	Riss-Würm
	Weimar Ehrings-dorf	Calotte, three mandibles, parietal fragments, and femur fragments	3	Riss-Würm
	Wildscheuer	Parietal fragments	1	Early Würm
Gibraltar				
	Devil's Tower	Child's skull fragments	1	Early Würm
	Forbes' quarry	Cranium	1	Undated
Greece				
	Petralona	Cranium	1	Middle Pleistocene: Riss-Würm (?)
Hungary				
	Subalyuk	Mandible, cranium, and postcranial fragments	1	Early Würm
Italy				
	Archi	Immature mandible	1	Early Würm
	Bisceglie	Right femur fragments	?	Early Würm
	Camerota	Left upper molar tooth	1	Early Würm I
	Ca'verde	Frontal fragments	1	Early Würm
	Monte Circeo	Skull and two mandibles	1	Early Würm I
	Leuca	Left upper second molar	1	Early Würm
	Pofi	Ulna and tibia fragments	1	Riss-Würm (?)
	Quinzano	Occipital	1	Riss-Würm (?)

TABLE 15-2   Continued

Geographic location	Site	Material discovered	Number of individuals	Age of site
Europe	Saccopastore	Cranium and face	2	Riss-Würm
	Sedia del Diavolo	Femur fragments and metatarsal	1	Riss-Würm
	Uluzzo	Left deciduous molar tooth	1	Late Riss-Würm
Spain				
	Bañolas	Mandible	1	Undated
	Carigüela	Two parietal fragments and immature frontal fragments	1	Early Würm
	Lezetxiki	Humerus	1	Early Würm (?)
Soviet Union				
	Akshtyr'	Molar tooth and metatarsals	1	Early Würm (?)
	Dzhruchula	Molar tooth	1	Early Würm (?)
	Kiik-Koba	Hand, foot, tibia, and fibula of an adult; postcranial fragments of a child	1	Early Würm (?)
	Rozhok	Molar tooth	1	Early Würm interstadial
	Starosel'e	Skeletal fragments—infant mandible, radius, and humerus	1	Early Würm
	Teshik-Tash	Skull and postcranial fragments of immature individual	1	Early Würm
	Zaskalnaya	Immature mandible and postcranial remains	1	Early Würm (?)
Yugoslavia				
	Krapina	Teeth and cranial and postcranial fragments	> 30	Riss-Würm to early Würm
Southwest Asia	Afghanistan			
	Darri-i-Kur	Right temporal	1	30,000 ±1900 years ago
Iran				
	Bisitun cave	Radius fragments	1	Upper Pleistocene
	Tamtama	Femur fragments	?	Upper Pleistocene
Iraq				
	Shanidar	Six adult skeletons and one infant skeleton	7	About 60,000–45,000 years ago

(Continued)

TABLE 15-2   Continued

Geographic location	Site	Material discovered	Number of individuals	Age of site
Southwest Asia	Israel			
	Amud	One skeleton and cranial fragments	5	Correlated with middle Würm interstadial
	Jebel Qafzeh	Adult and immature skeletons	16 (10 adults and 6 infants)	Early upper Pleistocene
	Mugharet es-Skh'ul	Skulls and postcranial material	10 (7 adults and 3 immature individuals)	Upper Pleistocene, about 30,000 years ago
	Mugharet et Tab'un	Skulls and postcranial bones	2	39,500±800 years ago 40,900±1000 years ago
	Mugharet el Zuttiyeh	Frontal fragments and sphenoid	1	Upper Pleistocene
	Skukbah	Temporal and frontal fragments, molar, distal femur, and talus	1	Upper Pleistocene
	Lebanon Grotte d' Antelias	Fragments of a fetus	1	Upper Pleistocene
	Ksâr'Akil	Maxilla and fragments of a child's skeleton	2	Upper Pleistocene
	Turkey Karain Adala	Deciduous molar	1	Upper Pleistocene
East Asia	China Chang-yang	Left maxilla and Premolar	1	Middle Pleistocene
	Mapa	Cranium	1	Late middle Pleistocene
	Sjara-Osso-Gol	Incisor tooth, right parietal, and distal femur		Upper Pleistocene (?)
	Tingtsun	Teeth		Upper Pleistocene (?)
	Java Solo	Eleven crania and two tibiae	11–13	Upper Pleistocene (Ngandong beds)
Africa	Ethiopia Bodo	Cranium	1	Pleistocene
	Dire-Dawa	Mandible	1	Upper Pleistocene

TABLE 15-2   Continued

Geographic location	Site	Material discovered	Number of individuals	Age of site
Africa	Omo-Kibish	Crania and postcranial remains	3	Upper Pleistocene
Libya	Haua Fteah	Mandibles	2	40,000–43,000 years ago
Morocco	Jebel Irhoud	Crania and an immature mandible	2–3	Upper Pleistocene, over 32,000 years ago
	Temara	Mandible	1	Upper Pleistocene
Tanzania	Eyasi	Skull fragments	2	Upper Pleistocene
South Africa	Florisbad	Cranium	1	About 35,000 years ago
	Makapansgat Cave of Hearths	Mandible and radius	1–2	Upper Pleistocene
	Saldanha	Cranium and mandibular fragments	1	Middle to upper Pleistocene
Zambia	Broken Hill (Kabwe)	Cranium, maxilla, parietal, and postcranial bones	1 or more	Middle to upper Pleistocene

Source: Derived chiefly from data supplied by E. Trinkaus, revised from A. Mann and E. Trinkaus. 1973. "Neandertal and Neandertal-like Fossils from the Upper Pleistocene," *Yearbook of Physical Anthropology*, **17**:169–193.

Specimens from the sites listed in Table 5-2 were all quite similar to one another and came in time to typify a population: the *classic neanderthals,* the label given to those from the Würm of western Europe. What did they look like?

Physical Appearance of the Neanderthals

Skull. The most distinctive neanderthal physical characteristics are concentrated in the skull (Figure 15-4). The vault is low, but it is long enough so that the total volume enclosed is high, ranging from about 1200 to as much as 1750 cc; the average of known specimens exceeds

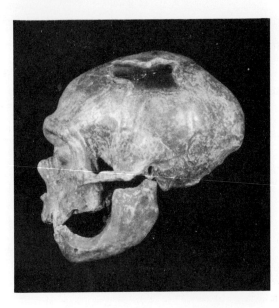

FIGURE 15-4   Neanderthal skull from La Chapelle-aux-Saints, France.
Source: Cast, University of Pennsylvania Museum.

that of modern humans. It may be that neanderthals really did have larger brains than we do, or the difference may be due to sampling error, with the more robust male skulls (which coincidentally have a higher volume) enjoying a better chance of preservation. The rear of the neanderthal skull projects outward, forming a pronounced occipital bun. In front the supraorbital torus is very strongly developed. Beneath it are orbits that are more rounded, as well as larger and set lower on the face, than those of modern humans. Under the eye sockets there is a more rounded angle where the cheekbones blend into the upper jaw. The nasal opening is large, and the entire upper face looks as if someone had grasped the nose and tugged on it while the bones behind were still soft. It projects very far forward at the midline and sweeps back toward the sides. The palate is very deep.

All categories of teeth closely approximate those of the pithecanthropines in size and form; the incisors are a bit larger, and the taurodont molars are, if anything, somewhat more frequent. The lower jaw typically lacks a chin, but it is robust enough to do without this added buttressing. One last feature concerns the position of the teeth in the lower jaw. As a rule, in neanderthals the third molar is rooted well forward of the base of the ascending ramus, and its crown is fully exposed in side view. In jaws of more recent humans, the third molar, in contrast, is usually at least half hidden behind the ascending ramus, another indication of the retraction of the dental apparatus and face beneath the forward part of the braincase in modern populations.

Skeleton.   Neanderthals are the archetypes of the comic-strip cavemen. But though they did indeed have the low-vaulted skulls, sloping foreheads, beetling brows, and craggy faces described in the preceding section, the rest of the skeleton was remarkably free of primitive, apelike features. Cartoonists have perpetuated a myth that has its source in the work of a French anthropologist, Marcelin Boule. In 1911, Boule described an extremely complete neanderthal skeleton found at La Chapelle-aux-Saints. He showed La Chapelle stooped over in a semierect posture, with head slung far forward on a short, thick neck at the top of a spine that lacked the typical hominid S curve (Figure 15-5). Furthermore, Boule showed this neanderthal walking with bent knees (because the joints could not be straightened) on the outer edges of his feet like an orangutan; widely divergent great toes completed the apelike image.

Later research on neanderthal anatomy flatly contradicts Boule's reconstruction on point after point. Some of his errors can be attributed to La Chapelle's advanced age, which brought arthritic deformations to parts of the skeleton. But most of the simian features seem to be traceable to Boule's opinion that any hominid with a skull so brutish-looking must have had a body to match and could have had no place in our ancestry. Boule's work did at least serve to demonstrate that neanderthals were distinctly different from living humans. Perhaps it would be fairest for us to consider his exaggerations to have been the price necessary, in a scientific age different from our own, to shake the belief that fossil humans did not exist. Now that the fossil record is more complete, it is easier to see that mosaic evolution had produced a skeleton substantially like that of modern humans all over the world by the middle Pleistocene in populations ancestral to neanderthals. At the same time, the skull retained many features that dated back to those of earlier populations.

Stereotypes aside, what did neanderthals really look like? The abundance of skeletal evidence makes it possible to do a remarkably complete job of reconstruction. La Chapelle was (5 ft 4½ in) 164 cm tall, just about 1 cm (½ in) taller than the modern Frenchmen who lived in the region when his bones were discovered. Other males from Neanderthal, Spy, and La Ferrassie ranged from 163 cm (5 ft 4 in) to 166 cm (5 ft 6 in). The skeleton of a female from La Ferrassie was 148 cm (4 ft 10 in) tall, with a smaller skull to match. Neanderthal males probably averaged about 73 kg (160 lb). Some of the ribs were ribbonlike, with flat cross sections like those of modern humans, but La Chapelle and Krapina had ribs which were round to triangular in cross section and which were also robust, to match the deep chest. The femurs of neanderthals are of medium length (44 cm, or 17.3 in, in a sample of four presumed males) for modern European populations, and they are

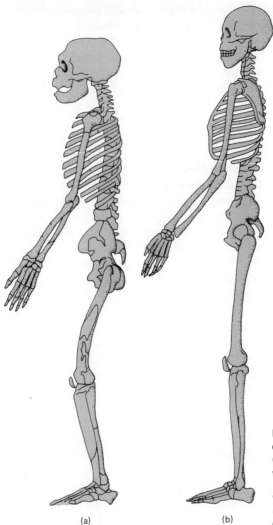

FIGURE 15-5 (a) Boule's reconstruction of the man of La Chapelle-aux-Saints, compared with (b) a modern human.
    Source: Based on M. Boule. 1913. "L'homme fossile de la Chapelle-aux-Saints," *Annales de Paleontologie*, **8**:1–67.

(a)   (b)

like them in having femoral heads of large diameter, consistent with a moderately heavy body build. Three main features distinguish the upper leg bones of neanderthals. First, neanderthals have a lower angle, 118 to 123°, between the neck of the femur and its shaft than modern Europeans, whose range is from 121 to 133°; the functional significance of this difference remains open to investigation. Neanderthals also show a weakly developed *pilaster,* or ridge along the back of the femoral shaft, and the femur has a strongly bowed shaft like that seen in modern peoples who squat on their heels. The tibias, the large bones of the lower leg, share this curvature.

The neanderthal arm bones present a picture comparable to that of the legs. The elbow joint at the distal end of the strongly developed humerus has a deep pit (the *olecranon fossa*). This receives the *olecranon process,* the projecting peak of the ulna's "funny bone." This process is longer in neanderthals than in modern European populations, giving the attached triceps muscle a great mechanical advantage in extending the forearm. The ulna's shaft is strongly curved, as is that of the radius; between them is ample space for the heavily developed forearm muscles. The hand itself is fully human, with a tendency to be short and stubby, and would have been capable of a very powerful grasp with little, if any, sacrifice of fine manipulative ability.

Many of these features were probably in part developmental responses to a way of life that required considerable strength, at least on the part of able-bodied males. The small adult skeletons are less robust and even have slightly different angles at some joints. These anatomical differences suggest that a division of labor allocated the heavier tasks to males. The injured and infirm were also freed from the heavier tasks. For example, one neanderthal's left humerus had been injured early in life and remained underdeveloped. A male from La Ferrassie had such severe arthritic erosion of the jaw that it is doubtful he could have chewed his own food. La Chapelle could not have hunted and had only two teeth left. All this, and evidence gathered from other cases too numerous to list here, documents a high level of mutual aid and cooperation.

Variation among Humans of the Middle Paleolithic

Not all later Pleistocene hominids fit the morphological pattern of classic neanderthals very closely. Those who did not have been given different labels. Some European finds from earlier geological levels (Swanscombe and Fontéchevade, for example) were thought to depart from pithecanthropines, but in the direction of present-day humans rather than classic neanderthals; these were usually tagged with the term *presapiens*, reflecting the belief that they were ancestors of anatomically modern *sapiens* who had diverged from the neanderthal lineage before the Mindel-Riss interglacial. The tag *generalized* (or *progressive*) *neanderthal* is sometimes also used for the same presapiens specimens plus others from the third interglacial (Saccopastore, Ehringsdorf) and the Würm glacial period (Krapina) that do not completely fit the classic type. Finally, just to complicate things a bit more, the term *neanderthaloid* is sometimes used to refer to specimens that look much like neanderthals but come from outside Europe (Mapa, Solo, Broken Hill, Saldanha). The examples given for each category do not make a complete list, because, among other reasons, various physical anthropologists differ in their schemes of classification; one person's presapiens may be another's generalized neanderthal, and still another's generalized neanderthal may be yet another's classic. Despite the inconsistencies in their use, these terms are important because they have shaped discussions of the extent and causes of variation among middle Paleolithic hominids. Some authorities who hold that variation within the sample of classic neanderthals is very low have offered various explanations for this. However, an alternative explanation for why classic neanderthals look much the same could be that any fossils which did not fit the stereotype were placed in other categories.

Underlying the interpretation that classic neanderthals represent an isolated hominid lineage is the belief that western Europe was cut off by glaciers during the Würm. Because of the supposed isolation under glacial conditions, western European neanderthal populations are said to have been subjected to extreme directional selection (which can in theory reduce genetic variation) for certain characteristics such as facial form and to have been prime candidates for the operation of genetic drift. It is now possible to judge these explanations against independent evidence on the world's climate and land areas (see Figure 13-1). Even after the ice-covered areas are deducted from the land available for hominid habitation in Europe, as much as half a million square kilometers (or approximately several hundred thousand square miles) is left. Most of this territory was covered by savanna and dry grasslands, and population densities of hunter-gatherers in comparable regions today range from about 2 or 3 to 20 or 25 people per square kilometer (1 to 10 per square mile). On the basis of these figures, we can estimate, very roughly, that the population of Europe during the middle Paleolithic was on the order of tens of thousands. As you will recall from calculations in Chapter 6, this is not a small number from the standpoint of potential for genetic drift, as long as there was gene exchange among the subdivisions of the entire population. Shared cultural elements suggest that social contacts did exist, and this is a picture completely consistent with today's hunting-and-gathering peoples.

It is sometimes argued that Europe was much less densely inhabited during the Würm, with a total population numbering in the thousands and with subpopulations widely dispersed and rarely in contact. This remains a possibility, though a less likely one, and the full implications of this model for hominid evolution should be examined. There would be greater scope for drift to occur within each of the small subdivisions, which would number in the hundreds or below. But to the extent that drift could occur, it should lead to differences among the various subpopulations within Europe rather than the supposed high degree of uniformity. Both reconstructions are unfavorable to the idea that the morphological characteristics of western European neanderthals can be accounted for in terms of genetic drift. So is one further observation: the map in Figure 13-1 shows an ice-free corridor stretching across western Europe, and so this subcontinent was not isolated from the rest of the world when neanderthals lived there. The idea that European neanderthals differed cranially and facially from their contemporaries elsewhere because of a unique complex of selective factors also seems less likely now than it did formerly. Populations elsewhere—for example, in northern Asia—must have been exposed to much the same environmental conditions, and the potential for gene flow between both regions would have made it likely that genes advantageous in both areas could have diffused from one place to the other, regardless of point of origin. The discovery of the neanderthal-like skullcap from Mapa, China, could be taken as evidence for such shared genes between two environmentally similar areas.

But what about populations in other regions? Neanderthal-like fossils have also been found in Greece (Petralona), the Near East (Tabun, Skhul, Shanidar), and Africa (Rhodesia, Saldanha). The climates in these areas were surely not like that of northern Eurasia. Assuming that neanderthal-like characteristics evolved in response to a complex of selective forces, these forces must have been distributed rather widely— as widely as the middle Paleolithic cultural elements that these populations shared.

Whether all the middle Paleolithic populations should be called *neanderthals* is another question. Some authorities have objected on the ground that doing this applies a name originally given to a local population to more distantly related groups distributed over several continents (Howells, 1974). This is, however, a recurring problem in the study of human evolution. *Australopithecine* and *pithecanthropine* are both terms applied more widely now than they were originally. A more substantive objection to the wider use of the term *neanderthal* is that members of middle Paleolithic populations all over the world did not look just like the European neanderthals. But for that matter, neither do all allopatric populations of the human species today look alike. Australian Aborigines have lower-domed, larger-faced skulls than North American Indians, which are in turn somewhat more robust cranially than most Europeans and Africans.

Some of the variation seen among middle Paleolithic populations may also be allochronic. However, it is very difficult to estimate what this fraction is because precise dates are often lacking. There are several reasons for this shortcoming of the fossil record. Many of these sites were excavated before modern dating techniques were invented, and some, including that of Neanderthal, have been destroyed. In his treatment of the neanderthal problem, Loring Brace stresses that no well-preserved western European neanderthal can be given a chronometric date with any greater accuracy than ±20,000 years. This means that two specimens with the "same" date could have lived as much as 40,000 years apart. In such a period of time, even very moderate levels of selection could have brought substantial evolutionary changes.

Outside Europe, the dating is also far from precise as a rule, although in some regions at least general trends can be made out. One such area

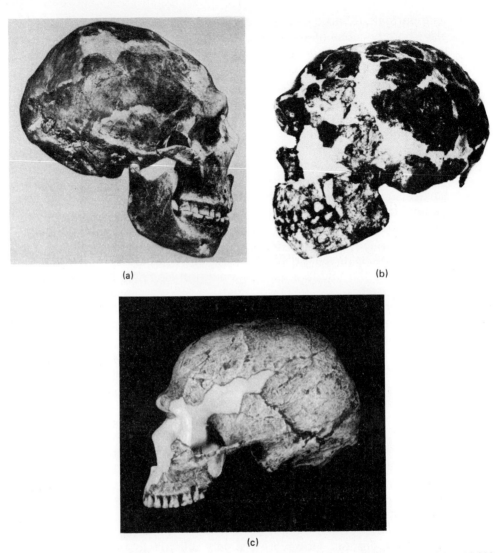

(a)

(b)

(c)

FIGURE 15-6   Near Eastern neanderthals through time. (a) Shanidar: about 60,000 to 45,000 years before present (B.P.). (b) Tabun: about 40,000 years B.P. (c) Skhūl: about 30,000 years B.P.

Sources: (a) R. Solecki. 1971. *Shanidar, the First Flower People*. New York: Knopf. Plate xxii. (b) B. Campbell. 1976. *Humankind Emerging*. Boston, Mass.: Little, Brown. Pp. 370, 371. (c) Photograph of cast, Wenner-Gren Foundation.

is the Near East, where three major sites can be roughly rank-ordered by age: Shanidar (Iraq), with skeletons ranging in age from about 60,000 to 45,000 years ago; Tabun (Palestine), with skeletons from about 40,000 years ago; and Skhūl (Palestine), with skeletons from approximately 30,000 years ago. As shown in Figure 15-6, these populations shared some morphological features with neanderthals, but there is something of a trend toward mixture, with aspects of more modern appearance in the more recent specimens. Two alternative explanations have been offered for the blend of characteristics seen here: (1) hybridization between native neanderthals and more modern populations coming in from elsewhere and (2) gradual evolution by local neanderthal populations into those more modern in appearance. Often overlooked, though, is a third possibility that is not excluded by either of the other two hypotheses. Any hybridization or long-term temporal cline in morphology that could have led to between-site differences should have been imposed on a normal range of expected population variation at each site. The within-site variations would have been due to differences in sex, age at death, and even some limited degree of allochronic variation (the extent of the last factor depending on how long a period of occupation is sampled by the known fossil remains). About all that can be said as yet is that morphological transformation of earlier populations in the direction of those which appeared later would be possible without any extensive gene flow and would not necessarily require very high rates of genetic change. But without knowing the mode of inheritance of the particular skeletal characteristics, we cannot be any more specific.

Material Culture and Way of Life

Half a century ago Aleš Hrdlička, the Czech-born founder of physical anthropology in America, noted that "the only workable definition of Neanderthal man and period seems to be, for the time being, the man and period of the Mousterian culture" (1927, p. 251). The *Mousterian culture* appeared in Africa and Europe about 100,000 years ago and lasted until about 30,000 or 40,000 years ago. It appears to have developed at least in part from the Acheulean tradition, and some Mousterian deposits retain a small percentage of hand axes. The major development that marks the Mousterian stage is the *Levallois technique* of stoneworking, in which a nodule of flint was trimmed to shape in five separate stages (Figure 15-7). The very indirectness of this procedure is a significant commentary on the working of the human mind at this stage in its evolution. Unlike the much earlier Oldowan techniques, in which handy rocks were used as is or with slight modification, the Levallois technique produced tools that resemble the raw nodule of flint as little as a statue resembles the block of marble from which it was carved. The same forethought marked selection of raw materials; the best flints were brought from sources over 1.6 km (more than 1 mi) from the living sites where they were worked, while inferior materials closer to home were passed up. Like the step-by-step flaking technique itself, this discrimination in the choice of flints argues that neanderthals, like modern humans, imposed their own symbolic order on the natural world.

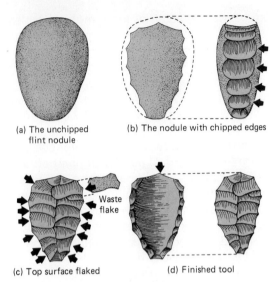

(a) The unchipped flint nodule

(b) The nodule with chipped edges

Waste flake

(c) Top surface flaked

(d) Finished tool

FIGURE 15-7 Steps in making a Levallois flake tool. The Levallois flake has a distinctive, predetermined shape. (a) **The toolmaker begins with a nodule.** (b) **This is prepared by trimming the sides.** (c) **The core is further refined by flaking small chips from both surfaces.** (d) **A final brisk blow at one end removes the finished flake, already sharp and in no further need of retouching.**

Source: B. Campbell. 1976. *Humankind Emerging.* Boston, Mass.: Little, Brown. Fig. 15-1, p. 321.

The value attached to the good-quality flint shows up in the efficient use made of it. Karl Butzer's studies show that Mousterian artisans using the Levallois techniques were able to produce tools with over 100 cm (40 in) of sharp cutting edge from about 0.5 kg (1 lb) of stone. This was five times more efficient than the Acheulean techniques, which in turn had been four times better than the Oldowan.

The middle Pleistocene trend toward more cultural diversity, reflected in a greater variety of tools and tool kits, progressed still further in the later Pleistocene Mousterian culture. Statistical analysis of thousands of Mousterian implements has identified at least 60 different tool types and five distinctive tool kits (Table 15-3). Finally, in the Mousterian cultures there is also considerable regional diversity. North African Mousterian sites have assemblages that closely resemble those of Europe; from there, tools made with the Levallois technique spread gradually east and south, reaching Cape Province over 100,000 years later. In South Africa the Fauresmith tradition eventually developed as Mousterian cultural elements were adapted to fit a dry grassland environment. In the more heavily wooded western and central parts of Africa, the Sangoan tradition predominated, with hand axes being replaced by heavier picks and supplemented by other tools evidently suited to woodworking.

The cultural record of neanderthals goes far beyond stone tools; in fact, anthropologists know far more about neanderthals than they do about any other past group of hominids. What is more, we know more because of the cultural attainments of the neanderthals themselves. Neanderthals lived in caves, and so deep layers of cultural debris were built up in rather protected places. A few neanderthals died in the caves

TABLE 15-3   Five types of Mousterian tool kits

Type	Components	Suggested activity	Type of activity	Analogy to Bordes' types
I	Typical borer Atypical borer Bec Atypical burin Typical end scraper Truncated flake Notch Miscellaneous tools Simple concave scraper Ventrally retouched piece Naturally backed knife	Manufacture of tools from nonflint materials	Maintenance tasks	Typical Mousterian
II	Levallois point Retouched Levallois point Mousterian point Convergent scraper Double scraper Simple convex scraper Simple straight scraper Bifacially retouched piece Typical Levallois flake Unretouched blade	Killing and butchering	Extractive tasks	La Ferrassie
III	Typical backed knife Atypical backed knife Naturally backed knife End-notched piece Typical Levallois flake Atypical Levallois flake Unretouched flake	Cutting and incising (food processing)	Maintenance tasks	Mousterian of Acheulean tradition
IV	Utilized flake Scraper with abrupt retouch Raclette Denticulate	Shredding and cutting (of plant materials?)	Extractive tasks	Denticulate Mousterian
V	Elongated Mousterian point Simple straight scraper Unretouched blade Scraper with retouch on ventral surface Typical burin Disk	Killing and butchering	Extractive tasks	La Ferrassie

Source: Modified from L. R. Binford and S. Binford. 1966. "A Preliminary Analysis of Functional Variability in the Mousterian of Levallois Facies," *American Anthropologist*, **68**(2):238–295. Table VII, p. 259.

too, killed by chunks of cave roof that buried them at the same time. But the main reason we know so much about neanderthals is that most of their burials were purposeful, not accidental. Neanderthals buried their dead, often in a flexed position and accompanied by the things of life. Among these were tools and bones of other animals, the latter perhaps all that is now left of food for the dead.

Evidence of complex beliefs and rituals is very abundant. Some neanderthal customs were inextricably bound up with death. When a cave that had been sealed for some 60,000 years at Monte Circeo in Italy was accidentally opened, archeologists found a neanderthal skull lying on its left side, nestled in a shallow basin that had been scooped in the earth floor. Around the skull was an oval ring of fire-blackened stones, and beneath it were two foot bones, one of an ox and the other of a deer. One or more heavy blows to the skull had been the cause of death, but not the end of the mutilation. The foramen magnum had been hacked open into a hole about 7.6 cm (3 in) across, almost certainly for removal of the brain. Presumably this was eaten, as brains are eaten by humans in some regions today. Most present-day cannibalism takes place in a ritual context; part of an enemy, often the brain, is eaten so that one may partake of the powers of the deceased. Monte Circeo was not the only neanderthal skull to have its base broken open. Similarly damaged skulls, often unaccompanied by any other parts of the skeleton, have been discovered at many sites from this time period. Not all the evidence for violence is limited to single cases or ritual settings, however. At Krapina in Yugoslavia, remains of 20 individuals—men, women, and children—were found broken to bits and charred. These may be the remains of a slaughter and cannibal feast that followed a hostile encounter between two groups, one of which was beaten decisively. The behavior has its counterpart in that of recent peoples who still live in small-scale societies.

There is also evidence of conflict with symbolic overtones, this between neanderthals and other animals such as the cave bear, a now-extinct species that outweighed present-day grizzlies. Neanderthals hunted cave bears and preserved their skulls. At a living site found at Drachenloch in the Swiss Alps, there was an astonishing trove: a stone box 0.9 m³ (nearly a cubic meter, or 30 ft³) covered with a massive stone slab (Figure 15-8). Inside were the skulls of seven cave bears, all facing the mouth of the cave, and six more bear skulls were set in niches in the walls. Similar finds come from many caves in western Europe. At the Mousterian site of Drachenhöhle in Austria, thighbones of 54 bears were all aligned in the same direction. The bones of 20 bears were found at Regourdou in France in a rectangular pit, again covered by a single massive slab of stone. That these are not random collections of bones is beyond doubt, but their purpose can only be conjectured. Cave bears were by far the most formidable animals that neanderthals must

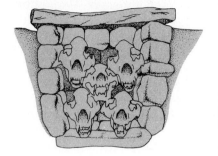

FIGURE 15-8 Skulls of cave bears stacked in a stone chest in Drachenloch, Switzerland.

Source: Redrawn; based on F. C. Howell. 1965. *Early Man*. New York: Time-Life. P. 126.

have encountered, and perhaps killing one was a test of manhood. A similar tradition in Africa pitted a prospective Masai warrior against a lion. Killing a bear may even have become a substitute for killing a human, another criterion of adulthood in some tribes. Whatever their purpose, skulls and bones of cave bears were treated as objects of significance. And this treatment is another piece of evidence that neanderthals had evolved to the point where they could fill the natural world with creatures of the mind.

Phylogenetic Position

All the uncertainties about neanderthals are aspects of one central question: What is their place in human evolution? It might be good to set problems of terminology aside while examining the various possibilities. There are really just three major ones: (1) that no neanderthal populations contributed to the gene pool of living humans (neanderthals were a lineage that became extinct without issue), (2) that neanderthal populations distributed widely over the Old World constituted a general stage in human evolution that gave rise to modern humans, and (3) that anatomically modern humans are descended mainly from a lineage whose members never developed the anatomical features of classic neanderthals but who occasionally interbred with populations that did. Discovering which of these comes closest to decribing what actually happened in the past will occupy anthropologists for years to come. It is a less important issue now than it once was because of a general consensus that has come about over the last decade or so. Neanderthals and neanderthaloids are now generally accepted by most anthropologists as a subspecies of modern humans—that is, as genetically distinct but not genetically isolated. It seems quite likely that present human populations everywhere had ancestors with general neanderthaloid characteristics. There are several reasons for this growing agreement. For one thing, neanderthals now seem less different from us than they once did. The growing body of cultural evidence shows that neanderthal patterns of culture very closely match those of living hunting-and-gathering peoples; neanderthals had minds that worked similarly enough to our own that their behavior seems almost familiar.

The increasingly abundant fossil remains from throughout the Pleistocene have been important too. If australopithecines and pithecanthropines can be accepted as stages in our evolution, why not neanderthals, who had brain sizes that matched our own? This question cannot be answered in a time-free framework, of course. Neanderthals were closer in time to modern humans than either australopithecines or pithecanthropines when they disappeared from the fossil record and were replaced by populations more like us. Whether this replacement was evolutionary or revolutionary depends on the length of the time gap between neanderthals and anatomically modern humans. If it is long enough, gradual evolutionary change from neanderthals is possible; if it is very short, displacement of neanderthal populations in whole or in part would be a more acceptable explanation. Many neanderthal morphological characteristics do seem to fit somewhere between those of pithecanthropines and modern humans. If neanderthal fossils hadn't been found, populations with their characteristics would have had to be reconstructed as intermediates in any case. But such populations have been found over much of the Old World during the third interglacial and the Würm. Aside from a few fragmentary and much-debated pieces of evidence, we have no other hominids from this time period. It would be peculiar indeed if so far we had found extensive evidence for an extinct side branch and little or none for populations in the mainstream of hominid evolution. Finally, there are numerous biological and cultural resemblances between neanderthals and the upper Paleolithic populations that succeeded them. The possibility that these resemblances represent genetic continuities is explored below.

END OF THE PLEISTOCENE: HUMANS OF THE UPPER PALEOLITHIC

When did human populations reach an anatomical form just like that of living humans? This question cannot be answered until we establish what "just like" means. There are substantial differences among living human populations in many traits, genetically simple and genetically complex, some of which affect skeletal anatomy and many of which do not. One feature that does *not* clearly differentiate middle from upper Paleolithic populations is brain size. On the average, neanderthals had endocranial volumes as large as, or larger than, those of modern humans. Though it is admittedly a crude criterion, attainment of modern average endocranial volume does go along with the development of a way of life that is regionally diverse and culturally complex. These anatomical and behavioral criteria were met by neanderthals. How, then, did their successors differ?

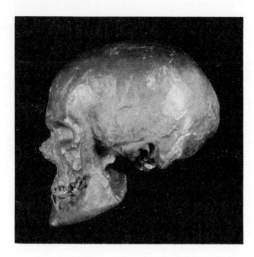

FIGURE 15-9   Cro-Magnon skull.
Source: Cast, University of Pennsylvania Museum.

Cro-Magnon: A Familiar Face

We differ from neanderthals chiefly in having skulls that are shorter and more highly domed, a face that is greatly reduced in size and tucked more beneath the forward part of the braincase, and a lower jaw buttressed externally by a chin.

Human populations that share these cranial details with us appeared in the evolutionary record at the end of the Würm. Of the fossil hominid populations that can be categorized as modern in facial form as well as brain size, perhaps the best known was that found by Louis Lartet in 1868 at the Cro-Magnon rock shelter in France (Figure 15-9). The discovery consisted of five skeletons of adults, males and females, plus that of a premature infant. All had been intentionally buried, some elaborately, and their bones painted with red ochre. They were accompanied by a complex of items, including worked bone and large quantities of perforated shells, that differed from those found with neanderthal skeletons. Just as the find in the Neander Valley came to typify a population, so did that at Cro-Magnon. As other upper Paleolithic sites were found in Europe—in France (Abri-Pataud, Chancelade, Combe-Capelle), Germany (Obercassel), England (Paviland), Italy (Grimaldi), Czechoslovakia (Brno, Predmost), and elsewhere—they were held to belong to a "Cro-Magnon race" of evolutionarily advanced humans. The label is again somewhat misleading, since as anthropologists expanded their searches into other areas of the world, they found upper Paleolithic populations with similar traits outside Europe (Table 15-4).

Physical Appearance

Skull.   The specimen most frequently described is that of the so-called "old man" of Cro-Magnon, who was probably just under 50 years of age at the time of his death. When seen from above, the skull is roughly

TABLE 15-4 Upper Paleolithic hominids

Geographic location	Site	Material discovered	Number of individuals	Material culture, if any	Age of site	Dating method used
Europe	France					
	Cro-Magnon (near Les Eyzies, Dordogne)	Adult skeleton plus fragmentary fetal and infant bones	5 adults plus several children	Aurignacian variety of upper Paleolithic flint tools and pierced seashells	Würm III	Correlation of fauna and material culture
	Chancelade	Skull and partial skeleton	1	Magdalenian variety of upper Paleolithic; body covered with powdered red Hematite iron ore	Würm II–Würm III	Correlation of material culture
	Combe-Capelle (Dordogne)	Nearly complete skeleton	1	Aurignacian	Würm II–Würm III	Correlation of material culture
	Italy					
	Grimaldi Grotte des Infants	Skulls and parts of postcranial skeleton	2 (1 adolescent boy and 1 woman)	Chatelperronian variety of upper Paleolithic; also shells and painted pebbles	Late Würm	Correlation of material culture
	Grotte du Cavillon	Skeleton	1 (adult male)	Burial with ornaments of shells and teeth	Late Würm	Correlation of material culture and hominid morphology
	Barmu Grande cave	Skeletons	4	Aurignacian burials in pits lined with stones and ochre, with shells and an ivory implement	Late Würm	Correlation of associated remains
	Czechoslovakia					
	Brno (Moravia)	Skeletons	3	Burials, two with red ochre	Würm	Correlation of associated remains

TABLE 15-4 Continued

Geographic location	Site	Material discovered	Number of individuals	Material culture, if any	Age of site	Dating method used
Europe	Predmost (Moravia)	Skeletons	29	Intentional burials with horn and bone implements and clay figurines	About 26,000 years ago but perhaps spanning a long period	Stratigraphic correlation
Africa	Algeria Afalou-Bou-Rhummel rock shelter	Skeletons in upper level	34	Mouillian variety of upper Paleolithic	About 10,000 years ago	Correlation with Taforalt cave in Morocco
	Morocco Taforalt cave	Skeletons	182 (96 babies, 6 adolescents, and 80 adults)	Mouillian variety of upper Paleolithic	11,900 ± 240 years ago	Carbon-14 dating
	Sudan Singa	Skull	1	Levallois variety of upper Paleolithic	About 23,000 years ago	Faunal correlation
	Transvaal, South Africa Boskop	Skull fragmentary jaw, and parts of limb bones	1	Stone tool similar to those of late middle Paleolithic	?	Not datable
	Tanzania Olduvai hominid 1	Complete skeleton		Aurignacian artifacts	Upper Pleistocene about 10,000 years ago	Stratigraphic correlation; Bed V burial intrusive into Bed II
Asia	China Choukoutien upper cave	Skulls (three relatively complete: one male and two female)	7	Upper Paleolithic burials with implements including carved bone needles, painted stones, and pierced shells	Late Pleistocene	Correlation of cultural material and associated fauna

(Continued)

TABLE 15-4 Continued

Geographic location	Site	Material discovered	Number of individuals	Material culture, if any	Age of site	Dating method used
Asia	Borneo Niah cave	Skull	1	Chopping tools and coarse flakes	39,600 ± 1,000 years ago	Carbon-14 dating
	Java Wadjak	Skulls	2	None	Uncertain; possibly late Pleistocene or more recent	Faunal correlation; fauna said to be modern, but site destroyed long ago
Australia	New South Wales Lake Mungo	Cremated skeleton	1	Hearth with burnt bones of birds, fish, and mammals; shells and stone tools	Age of skeleton is between 24,500 years ago; other sites in the area were occupied by 32,000 years ago	Carbon-14 dating
	Victoria Keilor	Skull	1	None	8,000–15,000 years ago (probably toward mid to early end of ranges)	Carbon-14 dating (mixed sample)
	Queensland Talgai	Skull	1	None found	About 10,000 years ago	Carbon-14 dating

Source: Based on K. P. Oakley. 1964. *Frameworks for Dating Fossil Man.* Chicago: Aldine. (Some data from other sources.)

pentagnonal in outline, with its long axis running from front to back. The enclosed endocranial volume is nearly 1600 cc, close to the average for all male skulls from this time period in Europe. Skulls of females average closer to 1400 cc. The face, while broad, is short from top to bottom; its rectangular orbits are edged at the top by a ridge of bone weaker than the heavy supraorbital torus of neanderthals (but stronger than that of many living humans) and are topped by a steep forehead. The lower jaw has a parabolic outline and is buttressed at the front with a sharply projecting chin. In this particular specimen the teeth are missing, a fact that underscores the appearance of advanced age.

Skeleton.   Those who have studied this specimen agree that Cro-Magnon was taller than preceding populations, but they differ on the extent of the increase in stature. Estimates of the height of the "old man" range from 168 cm (5 ft 6 in) to over 183 cm (6 ft). Both reconstructions fall within the range of humans of the past several generations. And as in our own time, women from the upper Paleolithic were shorter than men. The general tendency in the upper Paleolithic skeletons is toward a leaner body build than that of European neanderthals. Footprints left on cave floors and hand outlines (made by blowing pigments through a bone tube around the edges of the fingers and palm) on walls look like those of slightly built modern humans.

Variation among Upper Paleolithic Populations

Populations of relatively modern form made their appearance about 40,000 to 50,000 years ago, not only in Europe, but all over the Old World: Africa (Boskop, Afalou), Borneo (Niah cave), Java (Wadjak), China (Peking upper cave), and so on. The upper Paleolithic populations in these areas often showed affinities with preceding populations and with present inhabitants. A good example of this is the sample of human skeletal material from the upper cave at Choukoutien, near Peking. In all, skeletal parts of at least seven people were found there in deposits about 12,000 years old. Three of the upper cave skulls have been described (Figure 15-10); the skull of an adult male was characterized as Ainu-like, and those of two women as like a Melanesian and an Eskimo, respectively. The male's lower jaw, like the jaws of earlier and later Asian populations, had a mandibular torus and taurdont molars. Much of the variation is actually due to cranial deformation before death and to side-to-side crushing afterward. Although the upper cave remains are sometimes said to present yet another case of hybridization among representatives of geographically diverse populations, it is more likely that they are a good example of intrapopulation variation, perhaps inflated somewhat by alterations before and after death. Whenever reasonably large samples turn up from the same site, much the same pattern of variation is seen.

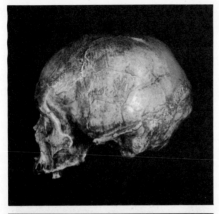
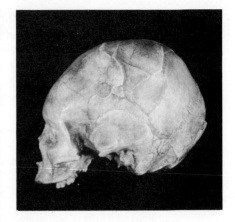
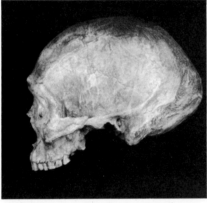

FIGURE 15-10 Variation within an upper Paleolithic site: three skulls from the upper cave, Choukoutien.
Source: Casts, University of Pennsylvania Museum.

Material Culture and Ways of Life

Upper Paleolithic cultures used blades, stone flakes at least twice as long as they are wide. Since blades are thinner than other types of flakes, more blades can be made from a given amount of raw material. Furthermore, their thinness would have resulted in a concentration of force per unit of area, increasing penetrating power. The frequency of burins also increased. These burins were in all probability used in working wood, bone, and antlers—materials used to a much greater extent by upper Paleolithic hunters than by their neanderthal predecessors. It has long been said that upper Paleolithic tools (Figure 15-11), like their makers, originated somewhere in the East, outside western Europe. This remains a possibility for tools of what is called the *Aurignacian tradition*, the name given to a set of tools that includes blades and bone implements and recurs from place to place over western Europe. Predecessors of these have been found in eastern Europe, suggesting immigration or influential cultural contacts with populations of that region. The Perigordian tradition is a more localized

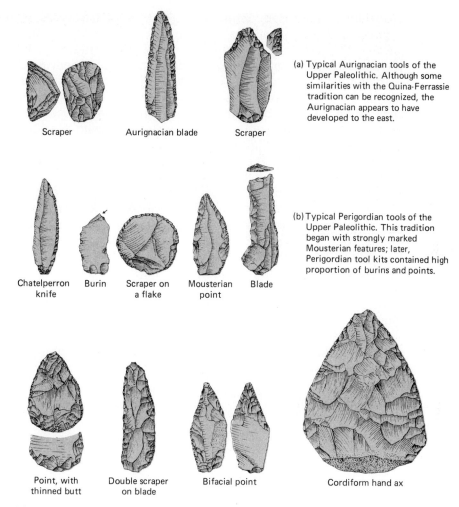

(a) Typical Aurignacian tools of the Upper Paleolithic. Although some similarities with the Quina-Ferrassie tradition can be recognized, the Aurignacian appears to have developed to the east.

Scraper   Aurignacian blade   Scraper

(b) Typical Perigordian tools of the Upper Paleolithic. This tradition began with strongly marked Mousterian features; later, Perigordian tool kits contained high proportion of burins and points.

Chatelperron knife   Burin   Scraper on a flake   Mousterian point   Blade

Point, with thinned butt   Double scraper on blade   Bifacial point   Cordiform hand ax

(c) Typical tools of the Mousterian of Acheulean tradition, which appeared during the Middle Paleolithic. The presence of up to 40 percent hand axes characterizes this variety of Mousterian industry.

FIGURE 15-11 Upper Paleolithic stone tools: (a) **Aurignacian;** (b) **Perigordian;** (c) **Mousterian of Acheulean tradition.**

Source: B. Campbell. 1976. *Humankind Emerging.* Boston, Mass.: Little, Brown. Figs. 17-2, 17-3, 17-4, pp. 367, 368, 369.

variant from southwest France. It is in all likelihood a local development from Mousterian cultures of the Acheulean tradition; at least it is often found immediately above Mousterian assemblages. Blade tools, it turns out, are not the all-or-none cultural element they were once believed to be. Small percentages of them are found in some Mousterian sites, and the frequency of blades appears to increase through time.

There are some other cultural continuities between middle and upper

Paleolithic sites. Red ochre, found in pencil-like sticks with some neanderthal skeletons, was used to color bones of Cro-Magnon and some other later burials. Interest in cave bears persisted as well. In the Pyrenees, Montespan cave contains a headless model of an animal large enough to be a bear, with a bear's skull on the ground between the forepaws. Whether this was a target for practice, an object of veneration, or a device to frighten young initiates is not known, but it is an indication of continuing preoccupation with this animal. The interest in animals spills over into a much more widespread phenomenon: cave painting. Pictures of large mammals between 28,000 and 10,000 years old cover the walls of many caves in France and Spain (Figure 15-12). Though accurate enough that the species can be recognized, these representations are not static. The animals are shown in lively poses, which a slight exaggeration of certain parts and a reduction of others that make them seem almost more real than the living animals themselves. The paintings that have survived are buried deep in subterranean caverns, where they have been protected from weathering. More exposed rock shelters may have been decorated as well, as they are today by some hunter-gatherers. The deeper caves, however, were not used as dwellings; they would have been as damp and inaccessible then as they are today, and furthermore they are free of occupational debris. Probably such places had a ritual function of some sort, in all likelihood related to hunting. The secrecy associated with rituals that still exists today evidently has its roots tens of thousands of years in the past.

FIGURE 15-12   Prehistoric cave painting from Altamira, Spain.
Source: American Museum of Natural History, New York.

Continuity and Change

Upper Paleolithic people differed physically and culturally from their middle Paleolithic predecessors. The extent of the difference can be measured, though there is sometimes disagreement about what scales are proper to use and whether the changes are quantitative or qualitative.

The most enduring controversy concerns the reasons for the similarities and differences. Do these reflect the accumulation of gradual systematic changes in a largely continuous sequence of populations or abrupt replacement by groups that then converged? Both biological and cultural evidence must be used to resolve this point.

Remains of upper Paleolithic populations are often referred to as *australoids*. Like living Australian Aborigines, they have larger faces, somewhat lower skull vaults, and larger teeth than other present human populations. The presence of these features, often referred to as *neanderthaloid elements* in otherwise *sapiens* skulls, are explained in two quite different ways: as survivals of neanderthal heritage and as the result of hybridization between neanderthals and early *sapiens* that evolved elsewhere (usually in an unspecified area) and moved in to replace them in most areas. Both models have counterparts in what we know of the history of modern human populations. In the New World alone, over just a hundred years Europeans have nearly completely replaced American Indians in North America, while hybridization between the two groups has been extensive in South America, with measurable impact on traits in the present population. Technological superiority, then, sometimes allows one population to prevail over another.

The examples from the New World suggest that the sizes of the populations coming into contact are important, as is the degree of technological superiority of one group over another. However, the European take-over in North America represented the incursion of an expanding agricultural population surplus into the territory of horticulturalists and hunter-gatherers. It is unlikely that middle or upper Paleolithic groups could ever have generated a comparable population surplus capable of inundating other groups with a similar subsistence pattern. Furthermore, the technological difference between Mousterian and Upper Paleolithic toolmaking traditions was nowhere near as great as the contrast in material culture between the Europeans and American Indians. Of course, it cannot be said that replacement of one Paleolithic population by another never occurred. However, the expansion of the European population from an agricultural base and equipped with a burgeoning industrial revolution does not provide a very suitable analogy for judging the likelihood of displacement of neanderthals by other hunter-gatherers.

INTERPRETING THE PAST: PROBLEMS AND QUESTIONS

The study of human evolution has been marked by a recurring pattern: the first discoveries of each stage have served as the focal points of discussion, usually with little regard for the play of chance. The features of each major find are taken to represent a type, the unique aspects of which are stressed, and even exaggerated. Later specimens that are similar serve to reinforce the stereotypes, while those which differ to any extent have often been viewed as new types. This pattern is now being replaced by a population approach that attempts to estimate for past groups the effect of influences (pathology, age, sex, environment) that are known to cause variation in living populations. For past groups it will also be necessary to estimate the influence on variation of evolutionary change through time.

Each successive stage of human evolution that is recognized at present has had a shorter duration than the one preceding—often markedly so. This does not necessarily mean that the tempo of human evolution is increasing; it may mean that the stages themselves have been drawn somewhat arbitrarily on the basis of the few fossil finds that first came to light over half a century ago. It is likely that in the future there will be a reassessment of the fossil material and a different evaluation of new discoveries. When this is done, three questions will be paramount: (1) How representative is the specimen of the population from which it was drawn? (2) How representative is this local population of the overall state of human evolution over the world at the time? (3) How different is the local population from the last local group known from the same area, and can the differences be accounted for by known forces of evolution operating at reasonable rates? These are not the only points to consider in the study of our past, but they are questions of fundamental importance, for unless we make these comparisons with care and precision, we will generate only new debates, not bases for new knowledge.

SUMMARY

1. Following the pithecanthropine stage of human evolution, the Mindel-Riss period has yielded few hominid remains. As a result, the transition from pithecanthropines to modern humans is not clear and is a source of controversy.

2. The few Riss hominids known make plausible morphological successors to pithecanthropines.

3. Most neanderthal fossils came from the period corresponding to the Würm glaciation in Europe, although there are a few from the preceding Riss-Würm interglacial.

4. Neanderthals had robust postcranial skeletons, rugged skulls and facial features, and an average cranial capacity higher than that of modern humans.

5. The characteristic material culture associated with neanderthal remains is that of the Mousterian, which includes about 60 different tool types divided among five different tool kits; this suggests considerable technological specialization and regional cultural diversity.

6. Anatomically modern humans succeeded neanderthals all over the world approximately 40,000 to 50,000 years ago; the "australoid" physical features (large faces and teeth, lower skull vaults) present in many of these populations are generally believed to suggest some genetic continuity with neanderthals, though the extent and causes of this are a matter of continuing debate.

SUGGESTIONS FOR ADDITIONAL READING

Coon, C. S. 1963. *The Origin of Races*. New York: Knopf. (Chapters 8 through 12 of this book provide a reasonably detailed review of the fossil evidence for later hominid evolution.)

Howells, W. W. 1975. "Neanderthal Man: Facts and Figures," pages 389–407 in R. Tuttle (ed.), *Paleoanthropology: Morphology and Paleoecology*. The Hague, Netherlands: Mouton. (This is a recent review of the history of the neanderthal problem and the various assessments of the possible role played by these hominids in our ancestry.)

Solecki, R. 1971. *Shanidar: The First Flower People*. New York: Knopf. (This book gives a highly personal account of the circumstances surrounding the discovery of the neanderthal remains at Shanidar cave in Iraq and the 9 years of labor on excavations in the area. The work of Solecki's group provided some of the most detailed evidence for the rich cultural life of neanderthals.)

REFERENCES

Binford, L., and S. Binford. 1966. "A Preliminary Analysis of Functional Variability in the Mousterian of Levallois Facies," *American Anthropologist*, **68**(2): 238–295.

Boule, M., and H. Vallois. 1957. *Fossil Man*. New York: Dryden.

Brace, C. L. 1962. "Refocusing on the Neanderthal Problem," *American Anthropologist*, **64**(1):729–741.

Brace, C. L. 1964. "The Fate of the 'Classic' Neanderthals: A Consideration of Hominid Catastrophism," *Current Anthropology*, **5**(1):3–43.

Broderick, A. H. 1964. *Man and His Ancestry*. New York: Fawcett.

Brose, D., and M. Wolpoff. 1971. "Early Upper Paleolithic Man and Late Paleolithic Tools," *American Anthropologist*, **73**(5):1156–1194.

Butzer, K. 1964. *Environment and Archaeology*. Chicago: Aldine.

Campbell, B. 1976. *Humankind Emerging*. Boston: Little, Brown.

Coon, C. S. 1962. *The Origin of Races*. New York: Knopf.

Day, M. 1965. *Guide to Fossil Man*. Cleveland: World Publishing.

Howell, F. C. 1951. "The Place of Neanderthal Man in Human Evolution," *American Journal of Physical Anthropology*, **9**(4):379–416.

Howell, F. C. 1957. "Pleistocene Glacial Ecology and the Evolution of 'Classical Neanderthal' Man," *Quarterly Review of Biology*, **32**(4):330–347.

Howell, F. C. 1960. "European and Northwest African Middle Pleistocene Hominids," *Current Anthropology*, **1**(2):195–232.

Howell, F. C. 1965. *Early Man*. New York: Time-Life.

Howells, W. W. 1974. "Neanderthals: Names, Hypotheses, and Scientific Method," *American Anthropologist*, **76**(1):24–38.

Hrdlicka, A. 1927. "The Neanderthal Phase of Man," *Journal of the Royal Anthropological Institute*, **57**:249–274.

Keith, A. 1925. *The Antiquity of Man*. London: Williams and Norgate.

Keith, A., and T. McCown. 1939. *The Stone Age of Mount Carmel*. Vol. 2. *The Fossil Human Remains from the Levallois-Mousterian*. Oxford: Clarendon.

de Lumley, H., and M. de Lumley. 1973. "Pre-Neanderthal Human Remains from Arago Cave in Southeastern France," *Yearbook of Physical Anthropology*, **17**:162–169.

McCown, T., and A. Keith. 1939. *The Stone Age of Mount Carmel*. Vol. 1. *The Fossil Human Remains from the Levallois-Mousterian*. Oxford: Clarendon.

Mann, A., and E. Trinkhaus. 1973. "Neanderthal and Neanderthal-like Fossils from the Upper Pleistocene," *Yearbook of Physical Anthropology*, **17**:169–193.

Maringer, J. 1960. *The Gods of Prehistoric Man*. London: George Weidenfeld and Nicholson.

Marshack, A. 1972. *The Roots of Civilization*. New York: McGraw-Hill.

Morant, G. 1938. "The Form of Swanscombe Skull," *Journal of the Royal Anthropological Institute*, **68**:67–97.

Solecki, R. 1971. *Shanidar: The First Flower People*. New York: Knopf.

Straus, W., and A. Cave. 1957. "Pathology and the Posture of Neanderthal Man," *Quarterly Review of Biology*, **32**(4):348–363.

Vallois, H. 1954. "Neanderthals and Presapiens," *Journal of the Royal Anthropological Institute*, **84**:111–130.

Weiner, J., and B. Campbell. 1964. "The Taxonomic Status of the Swanscombe Skull," in C. Ovey (ed.), *The Swanscombe Skull*. London: Royal Anthropological Society of Great Britain. Pp. 175–209.

Chapter 16

Knowledge of the Past, Control over the Future

By the end of the Pleistocene, humans had attained the major features of modern physical form, as far as we can tell from the relatively abundant skeletal evidence. It is also likely that most of today's geographic diversity in physiological and biochemical characteristics—eye color, hair form, and similar soft-tissue features—already existed as well. Can we conclude, then, that human biological evolution has ceased? As is the case with most questions about human evolution, there are different answers. At one extreme are those who point out that human culture has become so elaborated, at least in some areas of the world, that it has effectively replaced biological evolution. When some new stress arises in the future, they say, humans will use their technological capabilities to modify the challenge so that no differential fertility or mortality could possibly result, and thus no genetic change will take place. Statements in the opposite direction are also sometimes made with about the same degree of assurance. Partisans of this viewpoint assume that evolution continues and that its future direction can be predicted on the basis of past trends. For example, since the human brain has roughly trebled in size over the last 3 or 4 million years, it can be expected to increase in size in the future. The mechanism by which the proposed future enlargement could take place has been outlined. Some British studies have shown that there is a positive correlation between body size and intelligence (Lerner, 1968) and that brain size also correlates positively with body size. The geneticist Joshua Lederberg has suggested that brain size in humans has not increased since the time of Cro-Magnon because a large skull at birth is incompatible with the limited size of the mother's pelvic opening, through which the child must pass. The psychologist Louis Terman found that while under 2 percent of all babies born in California weigh over 4.5 kg (10 lb), some 17 percent of those having IQs in the genius range had birth weights in excess of this amount. A uniform policy of delivering all babies by cesarean section would permit a higher proportion of large-brained infants to survive without any injury or impairment to the mental function of the child through birth trauma (the effect on the mothers was not discussed, though there would seem to be some basis for expecting that they might harbor opinions on the matter).

Stature is another physical characteristic for which future trends are sometimes projected. Average statures in many populations in Europe, the United States, and Japan have risen steadily for at least the last century. Improved diet is usually cited as the major reason for these gains, although better medical care probably has helped as well, since modern antibiotics greatly reduce the impact of infectious diseases that can retard growth. Hulse (1958) has provided some evidence that the breakdown of genetic isolates due to improvements in transportation may have reduced inbreeding levels, and so some fraction of the jump in stature may be due to hybrid vigor. Whatever the causes, some of the gains in size have been impressive, with differences between children and their immigrant parents of the same sex sometimes reaching as much as 10 to 15 cm (4 to 6 in). The biological statistician Karl Pearson reconstructed the statures of a large series of seventeenth-century skeletons excavated at Moorfields and Whitechapel in London and calculated an average stature of 165 cm (5 ft 5 in) from them. Sir Francis Galton found that, two centuries later, English males had reached an average height of 173 cm (5 ft 8 in), and the trend continues. Just about exactly the same magnitude of increase took place in the United States, with skeletons of sailors of the Revolutionary War era averaging a reconstructed stature of 165 cm (5 ft 5 in), and soldiers a stature of 168 cm (5 ft 6 in). By the time of the Civil War, Union soldiers averaged 174 cm (5 ft 8½ in). These and other data suggest that over the period studied the average increase in stature was on the order of 2.5 to 3.8 cm (1 to 1½ in) per generation. Should this trend maintain its present rate into the future, humans will be very much larger, and food requirements will be very much higher per person than they already are in some countries.

Not all visions of future characteristics of humans sound as enticing as bigger brains and taller bodies. Brace and Montagu (1977) have recently pointed out that many modern humans have small teeth, congenitally absent lateral incisors or third molars, and enamel that decays rapidly on a sugar-rich diet. However, since dentists have proved so adept at saving the afflicted persons from any real harm (other than financial), in the opinions of those authors future generations can look forward to fewer and fewer teeth, and perhaps eventually none at all. This prediction lies in the realm of possibility, if those with diminishing dentitions have any advantage in fitness values.

Surveying one last trend might help us put all the others into some more useful perspective. In the United States today young women experience the *menarche* (first menstruation) about 1½ years earlier than their mothers did. Should this trend be projected at the same rate over a few generations into the future, it would not be very long before female infants would be menstruating shortly after birth. This is absurd, of course, and realizing that it is should lead us to reconsider the reliability of some of the other trends mentioned above. There is an alternative to uncritical extension of past trends into the foreseeable future, an alternative that can be founded on our study of past primate and human evolution. The lesson from the past is that, over long periods of time, major developments in evolution are shaped to a very considerable extent by the constraints and opportunities offered by the environment. It is a reasonably safe assumption that future evolution will be influenced in the same way.

There is one difference between the past and either the present or the future, however. In addition to the evolution of physical characteristics, human populations have also evolved mentally, emotionally, and culturally. As a result, the world that we experience is a very different place from that which shaped the actions of our ancestors of 5 million or even 500,000 years ago. Wells and Huxley noted in their classic work, *The Science of Life* (1929), that the world of many primitive animals is spaceless and timeless, consisting only of stimuli. Probably no animal below the apes can call up images of past events in the way we can, in order and sequence. Our human world is one of space and time, of things and events joined together in webs of cause and effect that make sense of our experiences. Because of our cultural evolution, we are now capable of exerting control, and at times awesome control, over some of the processes at work in the world around us. It is now our obligation to distinguish between categories of events that are inevitable (such as the eventual flaming out of our sun), probable (the continued increase in numbers of our species for a few generations), and merely possible (the creation of new humans from old ones by cloning rather than sexual reproduction). Since we are time-binding animals, able to perceive the play of cause and effect, we should become increasingly able to predict what the consequences of our actions will be.

Part of the strategy for shaping the future of *Homo sapiens* (and that of the rest of the life on this planet) will require that we learn from past mistakes, which often seem to result from the failure to foresee all the consequences of any particular action. It was easy to predict that large-scale use of DDT would greatly reduce the numbers of mosquitoes that spread malaria and would hence bring about a sharp decline in the number of people suffering and dying from that disease. In contrast, not many people considered the possibility that new strains of DDT-resistant insects would evolve, requiring the development of new types of chemicals to keep their populations in check. What is perhaps even more astonishing is that few experts in science or government looked ahead and saw that a greatly reduced death rate from malarial infections would mean that many more children would live long enough to face starvation. When this prospect began to become reality, some countries in Africa and Asia plunged into vast dam-building and irrigation projects to step up food production. The projects have lessened the threat of immediate starvation, but they have had still other side effects. The newly flooded areas have greatly increased the breeding grounds of the liver fluke *Schistosoma,* which is transmitted by aquatic snails. Humans wading in infested waters to plant crops pick up the parasites even through unbroken skin. Now schistosomiasis has become one of the top three health problems in the tropics, gradually replacing malaria. As the economist Milton Friedman has remarked, "There's no free lunch." Every action has its consequences, and every benefit has a cost that must be paid.

By now we know, however, that such biological consequences of cultural developments are nothing new in our evolution. To this extent the study of human evolution in the past can supply a guide for the future. Our strategy for the future must be rooted in a knowledge of the complexity of our ties with the rest of the natural world.

SUGGESTIONS FOR ADDITIONAL READING

The study of human evolution, like the rest of biological anthropology, is in a period of rapid scientific growth. Those who are interested in learning about new developments will find that many articles of interest appear first in the scientific journals of the field, particularly *Human Biology* (published by Wayne State University Press), *The American Journal of Physical Anthropology* (published by Wistar Institute Press), and *The Journal of Human Evolution* (published by Academic Press). Important papers on human biology and evolution are also printed, though less frequently, in *Science* and its British counterpart, *Nature*. *Scientific American* and *Natural History* often have less technical articles on the same subjects. All the above-mentioned journals publish reviews of recently published books in the field, enabling readers to select more comprehensive works of further interest to them.

REFERENCES

Brace, C. L., and M. F. A. Montagu. 1977. *Human Evolution.* New York: Macmillan.
Hulse, F. S. 1958. "Exogamie et heterosis," *Archives Suisses d'Anthropologie Generale,* **22**(1):103–125.
Lerner, I. M. 1968. *Heredity, Evolution, and Society.* San Francisco: Freeman.
Shapiro, H. L. 1974. *Peking Man.* New York: Simon and Schuster.
Wells, H. G., and J. S. Huxley. 1929. *The Science of Life.* New York: Literary Guild.

Glossary

Acetabulum   The socket in the hipbone (pelvis) that receives the head of the femur.

Acheulean   A toolmaking tradition containing bifacially flaked stone hand-axes; named after the site of Saint-Acheul in the Somme Valley of France.

Achondroplasia   An inherited defect in bone growth that reduces the length of the arms and legs.

Adaptability   The capacity of a species to adjust to and overcome environmental challenges.

Adaptive radiation   The evolution of several species with very different ways of life from one ancestral species in a short period of geologic time.

Agglutination   The formation of a visible clump of red blood cells when surface antigens are linked together by antibodies.

Alkalosis   A reduction in CO_2 in the lungs produced by deeper and more rapid breathing, leading to a shift in the acid-base balance of the blood and symptoms such as dizziness, nausea, and vomiting.

Alleles   Slightly different forms of one gene that can occupy a particular locus.

Allen's rule   The generalization that, in a widely distributed species of warm-blooded animals, populations in colder regions have shorter extremities than those in warmer regions.

Allochronic populations   Populations that live at different times.

Allopatric populations   Populations that live in different areas.

Analogous   Having a similar function but differing in evolutionary and developmental bases.

Anthropoidea   The suborder of primates that includes New World and Old World monkeys as well as apes and humans.

Antibodies   Proteins produced in the body in response to the presence of unfamiliar antigens.

Antigens   Complex chemical molecules (some located on the surface of red blood cells) capable of stimulating the production of antibodies.

Apocrine gland   A type of sweat gland that produces an oily secretion.

Arboreal   Living habitually in trees.

Archeology   The study of the material culture of past populations.

Assortative mating   Choice of a mate based on a preference for a particular characteristic.

Autosomes   Pairs of chromosomes with two members that are identical in length, types of loci, and centromere location.

Balanced polymorphism   The maintenance of two or more alleles in a population as a result of selection favoring heterozygotes.

Band   A unit of organization in some societies consisting of several families (often about 20 to 60 people).

Base   One of the three molecular building blocks of DNA or RNA (the other two being phosphate and sugar molecules). In DNA the four common bases are adenine, guanine, cytosine, and thymine; in RNA, uracil replaces thymine.

Base analog   A synthetic chemical that can substitute for a naturally occurring base, often leading to mutation.

Bergmann's rule   The generalization that, in a widely distributed species of warm-blooded animals, body size is larger in cooler areas and smaller in warmer areas.

Binomial   The two-part Latin name given to a species.

Biota   All the plants and animals that lived together at the same time and in the same place.

Blade   A stone flake that is at least twice as long as it is wide.

Blood group   A class of red blood cell antigens characterized by the same reactions to a set of antibodies.

Brachiation   Locomotion by swinging hand over hand while hanging beneath horizontal branches.

Breeding population (or **Mendelian population** or **deme**)   The pool of potential mates, any of whom may be chosen with equal probability.

Bulla   A hollow capsule of bone in the ear region at the base of the skull.

Canine   A single-cusped tooth located behind the incisors.

Catastrophism (catastrophe theory)   The belief that differences between fossils from one stratum to the next were caused by events that destroyed all life in the area, followed by repopulation by other organisms.

Cenozoic era   A period of geologic time that spans approximately the last 75 million years; often referred to as the *age of mammals*.

Centromere   The point of attachment between the members of a pair of chromosomes during cell division.

Cephalic index   A measure of head shape: (*head breadth* × *100*)/head length.

Cerebral cortex   The outer layer of the brain, composed of gray matter that is highly folded in humans. It is responsible for most complex mental functions.

Character displacement   A phenotypic divergence of two closely related populations in an area where they overlap.

Character state   The phenotypic expression believed to be the usual one in a population.

Chromosome mutation   A structural change in the genome that can be seen under a microscope.

Chromosomes   Paired linear sequences of nucleic acids and proteins within the nucleus of a cell that are capable of self-duplication and transmission of hereditary characteristics.

Chronometric dating   The use of techniques that allow estimating the intervals of time that separated specimens and assigning approximate dates to steps in a sequence.

Cladogram   A diagram that shows the distribution of characteristics within a group of related organisms.

Clavicle   A small, S-shaped bone that connects the sternum to the shoulder joint.

Cline   A systematic change in frequency of a gene or physical feature from population to population across a geographic area.

Codon   See *triplet.*

Community   The set of animal and plant species in a given habitat.

Convergence   The evolution of greater similarity between members of separate phyletic lines whose ancestors were less alike.

Core   A piece of stone made to a desired shape by the removal of one or more flakes.

Correlation coefficient   A statistical measure of the frequency with which two features are found in association with each other.

Correlation, principle of   The belief that the parts of an organism are interdependent to the extent that a change in one will cause changes in others.

Cranial capacity (or endocranial volume)   The internal volume of the skull; an indirect estimate of the size of the brain.

Cultural anthropology   The study of the culture and ways of life of living human groups.

Culture   The knowledge, beliefs, habits, customs, and material objects acquired by humans through learning from other members of their society.

Cusp   A raised bump or point of enamel on the surface of a tooth.

Cytoplasm   The material outside a cell's nucleus containing many different structures and chemical molecules.

Deciduous teeth   The first set of teeth in mammals, usually containing fewer than the adult set.

Deletion   A loss of genetic material resulting from an unrepaired break in a chromosome.

Dental formula   An abbreviation of the numbers and kinds of teeth on one side of the face.

Derived character   A character that is found in one or more species that are descended from a common ancestor, but not in the ancestor itself.

Developmental plasticity   The ability of a genotype to respond to environmental influences during the growth period.

Diastema   A gap between two teeth, most often found between the upper lateral incisor and the canine in higher nonhuman primates.

Diploid (2N) cells   Cells that contain two haploid sets of chromosomes, one from the mother and one from the father.

Display behavior   Activity meant to convey a social signal.

DNA   Deoxyribonucleic acid, the genetic material.

Dollo's law   The generalization that the direction of evolution cannot be reversed.

Dominant gene   A gene that masks the expression of its allele in heterozygotes.

Dominance hierarchy   A set of ranked relationships among the members of a social group.

Down's syndrome (or trisomy-21 or mongolism)   A genetic abnormality resulting from the presence of an extra chromosome on the twenty-first pair.

Duplication   An increase in the length of a chromosome due to repetition of a section through unequal crossing-over.

Eccrine gland   A type of sweat gland that produces a watery secretion.

Ecological niche   The interaction of a species with its habitat, food sources, predators, diseases, and other factors.

Effective population size (N_e)   A measure of the number of individuals of reproductive age in a population.

Electrophoresis   The separation of proteins by size and electric charge.

Endogamous mating system   A mating system in which most partners are chosen from one's own social group.

Enzyme   A biological catalyst that controls the rate of a chemical reaction in a cell.

Eocene   The second epoch of the Cenozoic era, lasting from about 60 to 40 million years ago.

Equilibrium   A stable, unchanging state.

Ethnocentrism   The belief in the basic superiority of one's own culture.

Ethnography   The systematic description of the characteristics of a particular culture.

Ethnology   The comparative study of the culture and ways of life of living human groups.

Ethology   The study of animal behavior under naturalistic conditions.

Evolution, chemical   The building of complex organic molecules from simpler ones prior to the origin of life on earth.

Evolution, cosmic   Change in the structure and organization of the universe.

Evolution, organic   Change in the inherited characteristics of organisms from generation to generation.

Exogamous mating system   A mating system in which the majority of partners are chosen from outside one's own social group.

Extinction   The failure of a phyletic line to leave any surviving descendants.

Fauna   All the animals that lived together at the same time and in the same place.

Femur   The single large bone in the upper part of the leg.

Fetalized (pedomorphic)   Resembling the immature form of a related taxonomic group.

Fibula   The smaller of the two bones in the lower half of the leg.

Fitness value   The relative reproductive ability of any genotype measured against others in the population.

Flake   A chip of stone knocked off a central core; may be discarded as waste or used as a tool.

Flora   All the plants that lived together at the same time and in the same place.

Foramen magnum   The large hole at the center of the base or back of the skull where the spinal cord enters to connect with the base of the brain.

Fossils   Remains of organisms that lived in the past, usually mineralized shells or bones or impressions these have left in rock formations.

Founder effect   Random genetic drift arising through sampling error when a small group buds off from a larger parent population.

Frame shift   A mutation caused by the addition or deletion of one or more bases in a DNA molecule.

Gamete   A reproductive cell from the male (sperm) or female (egg) that contains a haploid set of chromosomes.

Gene   A unit of heredity consisting of a sequence of DNA that produces a polypeptide or protein.

Gene flow   Transfer of alleles from one gene pool to another through interbreeding between members of two populations.

Gene frequency (or allele frequency)   In a population with two or more alleles, the relative proportion of one allele, usually expressed as a percent or decimal.

Gene pool   All the genes at one locus in a population during a particular generation.

Genetic drift   Random fluctuatons in gene frequency from one generation to the next due to chance events in a small population.

Genetic isolate   A relatively small breeding population which has little or no gene exchange with other populations.

Genome   All the DNA contained in the chromosome set of a cell.

Genotype   An individual's genetic constitution, established at the time of fertilization.

Genus   A taxonomic category that includes one or more related species characterized by a similar adaptive level.

Gestation period   The period during which the embryo develops inside the mother's body.

Grade   A level or stage of evolution.

Habitat   The setting provided by the physical environment.

Haploid (1N) cells   Cells that contain one member of each pair of chromosomes.

Hardy-Weinberg law   The generalization that when there are two alleles (A and a) in a population, three diploid genotypes will exist in a constant ratio of p^2AA : $2pq$Aa : q^2aa as long as certain ideal conditions are met (random mating and no change in gene frequency).

Hemizygous   Having unpaired genes on the single X chromosome in males.

Hemolytic disease of the newborn   A disease produced by destruction of fetal red blood cells by antibodies manufactured in the mother's body.

Heritability   The genetic fraction of the total phenotypic variation in a characteristic, specific for a particular population.

Heterosis (or hybrid vigor)   Increased vigor or fitness produced by heterozygosity at some loci.

Heterozygous cell   A cell in which the two members of a pair of alleles are different.

Histogram   A graph in which the height of bars represents the frequency of each class.

Holocene (or recent)   The most recent epoch of the Cenozoic era, spanning the period from about 10,000 years ago to the present.

Homeostasis   The maintenance of a relatively constant set of physical and chemical states within the body.

Hominid   Any member of the Hominidae.

Hominidae   A taxonomic family in the order Primates that includes living humans, earlier members of our lineage, and any other adaptively similar related lineages that may have existed in the past.

Homologous   Having a common evolutionary origin and developmental basis; homologous structures or characteristics may be similar or different in function.

Homologous chromosomes   Members of the same pair of chromosomes that are of the same length and control the same biological processes.

Homo sapiens   Populations with the characteristics of our present human species.

Homozygous cell   A cell in which both members of a pair of alleles are identical.

Horizontal classification   A taxonomic grouping of populations that share a common level or grade of evolution.

Humerus   The single bone of the upper arm.

Hybridization   Intermarriage on a large scale that blends two previously distinct populations into a single group.

Hypothesis   A testable idea about the explanation for a set of data.

Hypoxia   An oxygen deficit that makes breathing difficult; a characteristic stress due to high altitude.

Inbreeding   Mating between individuals who share a common ancestor (usually in the recent past).

Inbreeding coefficient (F)   The probability that two alleles in a homozygous individual are both copies of the same gene in a common ancestor.

Incest   Mating with close biological relatives.

Incisor   A flattened or peglike tooth near the midline of the jaw.

Independent assortment   The process by which genes located on one pair of chromosomes pass into gametes without influencing the distribution of genes on other pairs of chromosomes.

Interglacial   A warmer period between two successive advances of glaciers.

Interstadial   A slightly warmer phase of brief duration during a period of glacial expansion.

Isolating mechanisms   Factors that decrease the chances of mating between members of different species.

Klinefelter's syndrome   Sex chromosome trisomy (XXY).

Knuckle-walking   Quadrupedal walking in which the weight transmitted through the arms is supported on the middle segments of the fingers.

Lamarckiansim   The incorrect belief that characteristics developed by an organism during its lifetime can be transmitted to future generations (see *pangenesis*).

Law   A scientific theory that has become established by repeated successful testing.

Lethal mutation   A mutation that causes death.

Levallois technique   A system of stoneworking in which a nodule of flint is trimmed to the shape desired in five separate stages.

Linguistics   The study of the structure and function of language.

Linkage   The connections between genes on the same chromosome.

Living fossil   A species that has remained virtually unchanged over a long period of geologic time.

Locus (plural, loci)   The position that a particular gene can occupy on either member of a pair of homologous chromosomes.

Lumpers   A colloquial term applied by some taxonomists to others who tend to assign specimens to a relatively small number of categories.

Lysis   The rupture of cells.

Macroevolution   Large-scale transformations in the characteristics of organisms that take place usually over thousands or millions of years.

Mammals   Animals that maintain a constant body temperature and give milk to their young.

Mandible   The lower jaw.

Mandibular torus   A thick ridge of bone on the inner (tongue) side of the horizontal part of the lower jaw between the canine and the first molar.

Marsupials   Mammals such as the opossum and kangaroo that do not nourish their young through a placenta.

Maternal chromosomes   The chromosomes contributed to a fertilized egg cell by the mother.

Maxilla   The upper jaw.

Meiosis   The division of a cell's nucleus to produce four cells, each with a haploid set of chromosomes.

Menarche   The first menstruation.

Mental foramen   A small hole in the lower jaw near the chin through which one of the facial nerves passes.

Mesolithic (or **middle Stone Age**)   A brief transitional period between the Paleolithic and the Neolithic.

Microevolution   Small-scale inherited changes that take place over a few generations.

Migration   A mass movement of people from one area to another.

Miocene   The fourth epoch of the Cenozoic era, lasting from about 25 to 5 million years ago.

Mitosis   The division of a cell's nucleus to produce two daughter cells that each have an identical set of chromosomes.

Model   A simplified arrangement of the important components of some more complex feature of the world.

Molar   A broad-surfaced tooth used for grinding food.

Mosaic evolution   The pattern produced when different subsystems of organisms evolve at divergent rates.

Mountain sickness   A temporary or chronic disease resulting from a combination of hypoxia and alkalosis.

Mousterian culture   A flake-tool industry derived at least in part from the Acheulean tradition.

Multiple allelic series   Three or more alleles that can occupy the same locus, with the result that one individual can have any two of the series.

Mutagens   Agents (x-rays and certain chemicals) that can increase the frequency of mutations.

Mutation   A permanent change in the structure or amount of genetic material in a cell.

Naturalistic explanation   An account of events that does not require the influence of superhuman powers.

Natural selection   The process by which, in every generation, members of each species survive and reproduce with differing degrees of success, with the result that those with the most advantageous hereditary characteristics leave behind more of their kind.

Neolithic (or **new Stone Age**)   The time when tools made of ground or polished stone replaced those made of chipped stone, prior to the introduction of metalworking; also marked by the presence of agriculture, animal husbandry, and potterymaking.

Neutral mutation   In theory, a heritable genetic change that has no influence on probability of survival or reproduction.

Normal curve   A symmetrical, bell-shaped distribution seen for continuous characteristics in many populations.

Norm of reaction   The range of phenotypic expression that a particular genotype may have under a range of environmental conditions.

Nuchal crest   A shelf or ridge of bone at the back of the skull for attachment of muscles that help support the head in some nonhuman primates.

Nuclear family   A population unit typically consisting of a man, his wife, and their dependent children.

Nucleotide A chemical molecule consisting of one molecule each of phosphate, sugar, and nucleic acid base.

Nucleus The region of a cell that contains the chromosomes.

Oldowan culture The earliest known types of stone tools, many crudely made from pebbles; the name derives from the discovery of such tools at the site of Olduvai Gorge in east Africa.

Olecranon fossa A pit or depression at the lower (distal) end of the humerus; receives the olecranon process of the ulna.

Oligocene The third epoch of the Cenozoic era, lasting from about 40 to 25 million years ago.

Osteodontokeratic culture A set of tools that Raymond Dart first suggested were made by australopithecines from bone (*osteo-*), teeth (*-donto-*), and horns (*-keratic,* from *keratin,* the main protein ingredient of horns).

Paleoanthropology The study of the earliest origins of humans.

Paleocene The first epoch of the Cenozoic era, lasting from about 75 to 60 million years ago.

Paleolithic (or old Stone Age) The period in which stone tools were made by chipping or flaking.

Paleontology The study of ancient forms of life by analysis of their fossilized remains.

Pangenesis The incorrect belief that tiny particles from each structure in a parent's body are transmitted to the offspring.

Parallel evolution Maintenance of about the same degree of resemblance by two lineages that both went through a similar series of evolutionary changes at approximately the same rates.

Paternal chromosomes The chromosomes contributed to a fertilized egg cell by the father.

Patrilineal Traced through the male line of descent.

Patrilocal Located in or near the territory of the husband's father.

Pelvis The part of the skeleton that transmits the weight of the body from the spinal column to the hindlimbs.

Pentadactyl Having five digits.

Peptide Two or more amino acids linked together by a chemical bond produced by loss of one molecule of H_2O.

Permanent teeth The second set of teeth found in adult mammals.

Phenocopy An environmentally caused phenotype that mimics an inherited condition.

Phenotype An individual's appearance, produced by the combined influence of genotype and environment.

Phyletic evolution Change through time in successive descendants.

Phyletic line An unbroken ancestral-descendant sequence of populations that exists for a long period of time.

Phylogenetic tree A branching diagram that is meant to express the nature of the evolutionary relationships between different populations.

Physical anthropology The study of the biological characteristics of past and present humans and of their relatives among the primates.

Placenta   A membrane that grows from a developing mammalian embryo and attaches to the wall of the mother's uterus; it allows the offspring to obtain food and oxygen and eliminate waste products.

Plasma   The yellowish fluid part of blood that remains after the cells have been removed.

Pleiotropism   The production of multiple phenotypic effects by a single gene.

Pleistocene   The sixth epoch of the Cenozoic era, lasting from about 2 or 3 million years ago to about 10,000 years ago; it was a time of generally cooler climates and glaciation over some parts of the world.

Pliocene   The fifth epoch of the Cenozoic era, lasting from about 5 to 2 or 3 million years ago; this estimate is very approximate because of continuing revision of the Miocene-Pliocene and Pliocene-Pleistocene boundaries.

Polymorphic populations   Populations that contain several sharply differing forms, each at a frequency above that which could be maintained by recurrent mutations.

Polypeptides   Peptides made up of more than 10 amino acids.

Polytypic populations   Populations that differ substantially from others of the same species.

Population density   The number of individuals living on a given unit of land area.

Population structure   The way in which breeding groups are subdivided by age and sex, as well as how they are distributed over the regions they inhabit; both factors influence their systems of mating and genetic composition.

Postorbital bar   A flat strip of bone behind the orbit of the eye of all living and most fossil primates.

Premolar   A molarlike cheek tooth located behind the canine.

Primates   The order of mammals that includes humans, apes, monkeys, and prosimians.

Primitive character   A character that is found in the common ancestor of a group of related species.

Prognathous   Having the jaws projecting or flaring forward.

Prosimians   A suborder of primates that includes all species that lived prior to the Oligocene and some that survive today; most living prosimians are small-brained, nocturnal, and relatively solitary.

Protein   A peptide consisting of over 100 amino acids.

Quadrupedal   Four-footed.

Quantitative characteristics   Phenotypic features (such as stature) that show continuous variation and are under the influence of several genetic loci and environmental factors.

Quaternary   A geologic period that began at the opening of the Pleistocene; often considered to roughly parallel the span of human toolmaking activity.

Race   A concept sometimes applied to populations of various size below the species level; races are defined in a variety of ways, including similar gene frequencies, physical resemblance, and common language or culture.

Radius   The smaller of the two bones of the forelimb, located on the thumb side between elbow and wrist.

Random mating (or panmixis)   The choice of mates without regard to genotype or phenotype, with the result that genotypes of resulting offspring are formed in proportion to the frequencies of alleles in the gene pool.

Recessive gene   A gene whose expression is masked by another allele in heterozygotes.

Recombination   An exchange of segments during cell division by the two members of a pair of homologous chromosomes.

Relative dating   The placement of objects, specimens, or populations into a sequence without knowledge of how much time separated each step.

Rhinaruim   An area of granular skin at the tip of the snout of some mammals.

RNA   Ribonucleic acid, a chemical similar to DNA that exists in several forms (messenger RNA and transfer RNA).

Sagittal crest   A ridge of bone located on the top of the skull in some pongids and early hominids, providing additional area for muscles used in chewing.

Sampling error   The chance that a small group will not be representative of the larger one from which it was drawn.

Savanna   Dry tropical grassland with few trees.

Scapula   A flattened bone shaped like a right triangle that lies along the back of the rib cage.

Scenario   A verbal explanation of why certain evolutionary events may have taken place.

Sectorial premolar   A tooth typically with a single sloping cusp that shears against the upper canine.

Selective coefficient   A measure of the extent to which the reproductive rate of one genotype falls below that of the genotype with the highest fitness value in a population.

Serum   The straw-colored part of blood plasma that is left after formation of a clot.

Sex chromosomes   The pair of dissimilar chromosomes (X and Y) that influence development of a male or female phenotype.

Sexual dimorphism   Difference in appearance between males and females; in many primate species males have larger bodies and canines than females do.

Sibling species   Populations which are phenotypically similar or undistinguishable but which are reproductively isolated.

Silent mutation   A mutation that is not phenotypically detectable.

Simian shelf   A buttress of bone that grows on the inner surface of the lower jaw at the midline.

Situational dominance   Status temporarily higher than the usual rank in the dominance hierarchy.

Society   A group of interbreeding organisms that can last longer than the life of any of its individual members, whose behaviors are structured by the roles they fill.

Somatic cell   A cell in the body containing a diploid set of chromosomes.

Specializations   Biological features that have evolved to suit a particular way of life.

Speciation   The process by which one gene pool is subdivided into two or more productively isolated gene pools.

Species   A group of plants or animals characterized by shared biological distinctions that are maintained by reproductive isolation.

Splitters   A colloquial term applied by some taxonomists to others who tend to assign specimens to a relatively large number of categories.

Spontaneous generation   The evolution of simpler forms of life from nonliving chemical systems through an increase in complexity.

Standard geologic column   A composite sequence of all the successive layers that have been deposited in the earth's crust.

Steatopygia   The development of large fat deposits localized over the buttocks; a condition characteristic of certain populations.

Stereoscopic vision   The ability to judge depth through overlap of the visual fields of the two eyes.

Sternum   The flat bone that forms the middle of the chest.

Strata   Separate layers of rock in the earth's crust.

Sublethal mutation   A mutation that reduces the ability to survive and reproduce.

Subordinate   An animal that ranks lower than one or more others in a dominance hierarchy.

Supraorbital torus   A ridge of bone above the eye sockets.

Sweepstakes dispersal   The accidental crossing of hazardous barriers by animals or plants; few survive the journey, but those who do often multiply rapidly.

Sympatric populations   Populations that live in the same area.

Synchronic populations   Populations that live at the same time.

Synthetic morphological pattern   The nonexistent composite that can be reconstructed when parts of different organisms are mistakenly assigned to the same taxon.

Taxon   A name applied to a category of real organisms at any level in a system of classification.

Taxonomy   The systematic arrangement of organisms to express their degrees of relationship.

T-complex   A set of dental features including reduced canines and expanded surfaces of the cheek teeth evolved by the gelada baboon (*Theropithecus*) for grinding abrasive foods.

Terrestrial   Living habitually on the ground.

Territory   An area that is defended against intrusion by members of the same species.

Theory   One of several possible explanations for a set of observations.

Tibia   The larger of the two bones in the lower half of the leg.

Tibiofibula   The unit formed by fusion of the tibia and fibula in the lower half of the leg; found in some prosimians.

Tooth comb   A structure, used in grooming the fur, formed by the front teeth in some prosimians.

Total morphological pattern A complex of structures that are functionally related.

Tradition A recognizable pattern of toolmaking that persists for many generations.

Transient polymorphism A genetic variation that exists while one gene is replacing another.

Translocation A shift of a block of genes from one chromosome to a member of a nonhomologous pair.

Tribe A social group composed of several bands that share a common language and occupy adjacent geographic areas (often about 500 people).

Triplet A linear sequence of three DNA nucleotides that specify the presence of a particular amino acid in a protein.

Turner's syndrome A sex chromosome abnormality (XO).

Ulna The larger of the two bones in the forelimb.

Uniformitarianism The belief that the forces that shaped the earth and life on it in the past were the same as the natural processes we see going on today.

Vertical classification Taxonomic grouping of populations that are related as ancestors and descendants, regardless of their grades or levels of evolution.

Vertical clinging and leaping A type of locomotion for crossing gaps between one upright tree trunk and another.

Vibrissae Large whiskers often found on the snout and sometimes at the wrists of some mammals; they are in contact with many nerves and give a highly developed sense of touch.

Villafranchian A stage of faunal evolution that has been defined in several ways; now usually believed to be marked by the first appearance of the surviving genera of *Equus* (horses), *Bos* (cattle), and *Elephas* (elephants).

Y-5 molar pattern (*Dryopithecus* pattern) The arrangement of the major cusps and grooves often found on the lower molars in pongids and hominids.

Zygomatic arch A bridge or strip of bone that bridges the side of the face from the cheekbone (malar) to just ahead of the opening for the ear.

Zygote A cell formed by fusion of a haploid egg and sperm.

Name Index

Subject Index